AF595066

High Performance Concrete

High Performance Concrete

Editor

Sukhoon Pyo

MDPI • Basel • Beijing • Wuhan • Barcelona • Belgrade • Manchester • Tokyo • Cluj • Tianjin

Editor
Sukhoon Pyo
Ulsan National Institute of
Science and Technology (UNIST)
Korea

Editorial Office
MDPI
St. Alban-Anlage 66
4052 Basel, Switzerland

This is a reprint of articles from the Special Issue published online in the open access journal *Materials* (ISSN 1996-1944) (available at: http://www.mdpi.com).

For citation purposes, cite each article independently as indicated on the article page online and as indicated below:

LastName, A.A.; LastName, B.B.; LastName, C.C. Article Title. *Journal Name* **Year**, *Volume Number*, Page Range.

ISBN 978-3-0365-3090-1 (Hbk)
ISBN 978-3-0365-3091-8 (PDF)

Cover image courtesy of Sukhoon Pyo

Contents

About the Editor

Sukhoon Pyo received a PhD in Civil Engineering at the University of Michigan, Ann Arbor, in 2014; was a senior researcher at the Korea Railroad Research Institute from 2014–2018; and has been working as an assistant professor at the Ulsan National Institute of Science and Technology (UNIST) since 2019. Sukhoon Pyo works on ultra-high-performance concrete, sound absorbing concrete, carbon-neutral high-performance concrete, and their structural applications. He is the author of more than 50 publications in scientific journal.

Preface to "High Performance Concrete"

The innovations in construction materials that have been made due to the development of different varieties of concrete have led to innovations in structural applications and design. This Special Issue mainly focuses on state-of-the-art research progress in high-performance concrete, including the effect and characteristics of fibers on the properties of high-performance concrete, the CO_2 curing efficiency of high-performance cement composites, and the effect of nano materials when used in ultra-high-performance concrete. This Special Issue also contains two comprehensive review articles covering the following topics: the role of supplementary cementitious materials in ultra-high-performance concrete and recent progress in nanomaterials in cement-based materials. Readers working towards conducting research on innovative construction materials will be exposed to findings related to this topic in this Special Issue.

Sukhoon Pyo
Editor

Article

Behavior of Colloidal Nanosilica in an Ultrahigh Performance Concrete Environment Using Dynamic Light Scattering

Douglas Hendrix [1], Jessica McKeon [2] and Kay Wille [3,*]

[1] Department of Materials Science and Engineering, University of Connecticut, Storrs, CT 06269, USA; douglas.hendrix@uconn.edu
[2] Department of Chemistry, University of Connecticut, Storrs, CT 06269, USA; jessica.mckeon@uconn.edu
[3] Department of Civil and Environmental Engineering, University of Connecticut, Storrs, CT 06269, USA
* Correspondence: kay.wille@uconn.edu; Tel.: +1-860-486-2074

Received: 23 May 2019; Accepted: 13 June 2019; Published: 19 June 2019

Abstract: The dispersion quality of nanosilica (NS) is an essential parameter to influence and control the material characteristics of nanosilica-enhanced concrete. In this research, the dispersion quality of colloidal nanosilica in simulated concrete environments was investigated using dynamic light scattering. A concrete environment was simulated by creating a synthetic pore solution that mimicked the ionic concentration and pH value of ultrahigh-performance concrete in the fluid state. Four colloidal nanosilica samples were used, ranging in particle sizes from 5 to 75 nm, with differing solid contents and stabilizing ions. It was found that the sodium stabilized 20 nm NS sol remains dispersed at a solid concentration of 2 wt % through a variety of pH values with the inclusion of potassium ions. Calcium ions are a major contributor to the agglomeration of NS sols and only small concentrations of calcium ions can drastically affect the dispersion quality.

Keywords: ultrahigh-performance concrete; nanosilica; dynamic light scattering; zeta potential; pore solution

---

1. Introduction

The development of concrete research has accelerated in the areas of mechanical and durability properties due to potential benefits of using nanoparticles in concrete mixture design, including nanosilica (NS), nanosized TiO_2, carbon nanotubes, and graphene oxide [1–6]. NS has been widely used due to its small particle size, spherical particle shape, and pozzolanic reactivity which are promising properties to further densify the microstructure of concrete [7–10]. In previous research, parameters of interest included the type of NS, the size of the particles or the surface area, the concentration of NS, addition or replacement of cement, and water to cement ratios [11–17]. One of the key challenges in densifying and, therefore, strengthening the microstructure lies in the dispersion quality of the nanoparticles during the mixing process. Figure 1 demonstrates this concept through three systems. System I is a well-dispersed cementitious system with cement and microsized silica, representative of ultrahigh-performance concrete, exhibiting a high particle-packing density and thus enhanced mechanical and durability properties in comparison to conventional concrete [18]. The addition of nanoparticles is a logical step to further increase the particle-packing density. System II highlights what commonly occurs with the inclusion of nanoparticles. These nanoparticles rapidly and irreversibly agglomerate, resulting in an undesired reduction of particle-packing density and thus leading to an inconsistency of mechanical properties. System III is the motivation for this research: Increasing particle-packing density through uniform dispersion of nanoparticles. The control of the dispersion quality will control the concrete's microstructure and thus will have a direct effect on the material's mechanical and durability performance.

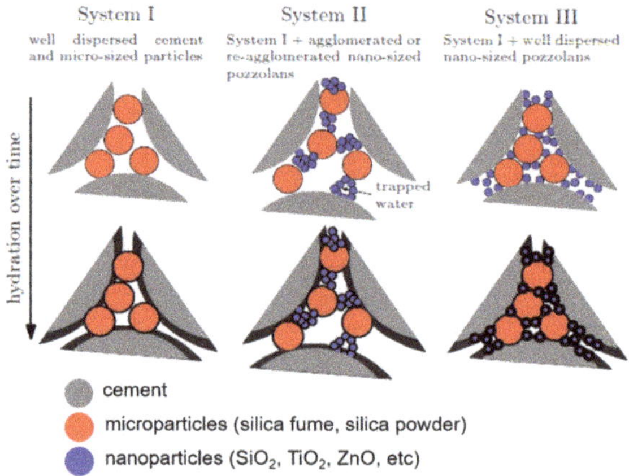

Figure 1. Illustrating the effects of poorly dispersed and well dispersed nanosilica (NS) in a cement system.

The use of NS in concrete has been proven to be a controversial topic due to significant variations of experimental results. Different researchers have reported that nanosilica increases, decreases, and has no effect on the mechanical properties of conventional concrete [19–21] and ultrahigh-performance concrete (UHPC) [22–24]. UHPC is a cementitious composite characterized by a compressive strength in excess of 150 MPa and enhanced ductility and durability properties as compared to conventional concrete. UHPC contains several constituents of varying sizes including sand, cement, silica powder, and silica fume, decreasing in size respectively. These different size gradations are essential to obtain dense particle-packing and favorable mechanical properties. [25–27]. With UHPC containing nanoparticles, the reason for discrepancies in mechanical properties is hypothesized to be a result of the variation in the dispersion quality of NS throughout the matrix. While high dispersion quality is expected to lead to enhanced material properties, poor dispersion quality as illustrated in Figure 1 (System II) compromises the improvement of mechanical properties [28,29].

NS particles are susceptible to form large agglomerates due to their very high specific surface area (50–750 m^2/g). Colloidal NS, as opposed to a dry powder, provides NS particles in a well-dispersed state, free of significant agglomeration. These discrete particles have hydroxylated surfaces and achieve stability by possessing a net negative charge on the surface, strong enough to repulse the van der Waals attractive forces. If different ions are introduced to the sol, this surface charge can be altered and can lead to agglomeration and gelation of the NS [30–34]. Therefore, adding well-dispersed NS to the concrete mixture can lead to agglomeration and gelation resulting from the release of ions during cement hydration.

Understanding the dispersion mechanisms of NS is essential in the engineering and design of novel nanomaterial enhanced cementitious composites. Besides enhanced packing density, improved pore structure, and thus enhanced mechanical and durability properties, these new composites can exhibit specific functionality by inheriting properties of the added nanoparticles, such as air depolluting, elasticity control, and acceleration of hydration kinetics [35–38].

The goal of this research is to understand the mechanisms that cause NS destabilization and agglomeration in UHPC. It is hypothesized that even well-dispersed colloidal NS could result in agglomerates in the concrete environment of UHPC. The challenge is to define and investigate the dispersion quality of colloidal NS sols under concrete mixing conditions. In this paper, particle size distribution (PSD) through dynamic light scattering (DLS) and zeta potential (ZP) were measured and analyzed to evaluate the dispersion quality. NS sols were diluted, pH was altered, and added to a synthetic pore solution to further understand NS destabilization.

1.1. Experimental Theory

1.1.1. Colloidal NS

Most commercially-available NS sols are stabilized at a pH range of 7–10. Silica particles are negatively charged as a result of the silanol groups (Si–O–H) on the surface. This provides a repulsive force between adjacent particles. Since the silica must be electrically neutral, counter ions such as potassium, sodium, and ammonia are inserted into the solution. These positively charged ions balance the charge by being attracted to the negatively charged surface of the silica particles [39]. NS sols can become destabilized for various reasons and once the van der Waals attractive forces are greater than the repulsive forces, the particles irreversibly agglomerate [40]. Silica can also be stabilized at low pH by removing all ions, resulting in slightly negatively charged particles, as this is the case with NS-20a used in this study.

The Derjaguin-Landau-Verwey-Overbeek (DLVO) theory predicts the stability of charged particles in a liquid medium by factoring in van der Waals attractive forces and electrostatic repulsion. DLVO theory predicts that if the electrostatic forces are greater than the van der Waals forces, the colloid will not agglomerate. DLVO provides a fundamental theory of the stability of particles in different pH and different ionic concentrations. However, in experimental practice with silica, DLVO breaks down at low pH. DLVO predicts at the isoelectric point at a pH of 2, the system should lack stability. However, at or near this point, colloidal silica has demonstrated stability [40]. This indicates that there are forces not considered by DLVO that significantly affect the stability, such as the steric hindrance effect. These properties of silica have made it difficult to predict stability, especially with the use of multivalent ions, varying particle sizes, and polymers. Therefore, experimental studies are necessary.

The pH of NS sols can affect the stability of the sol. At high pH (pH > 11), silica starts to dissolve, forming silicate ions, leading to destabilization [41]. At low pH, the repulsive forces are suppressed, resulting in a lower stability. Another potential source for NS destabilization is the ionic composition and concentration in the medium surrounding NS particles [42]. Monovalent and divalent ions, such as potassium (K^+) and calcium (Ca^{2+}) used in this study, alter the ionic charge of sols. This results in two mechanisms of agglomeration. First, the adsorption of potassium and calcium ions can lead to a reduction in the electrical double layer (further discussed in Section 1.1.3.). This reduces the electrostatic repulsive forces and attractive forces dominate. In addition to compressing the electrical double layer, these ions can form bridges between NS particles. A positively charged ion can neutralize a site on the silica surface. If two of these uncharged surfaces collide, the ion can coordinate with the oxygens of the silanol groups and the surface bonded water, forming a physical bridge between two silica particles [41].

For example, at a pH between 6 and 9, alkali metals at a high enough concentration can result in agglomeration. However, at a higher pH (pH > 10), it has been observed that potassium, rubidium, and cesium do not cause agglomeration at high concentrations. It is hypothesized that these ions form a complete double layer around the particle effectively shielding the silica surfaces from coming into contact [41]. Why this happens at a higher pH and not a lower pH is unknown, again supporting the observation that DLVO theory is not perfect.

These two effects of agglomeration are amplified with divalent ions such as calcium. A divalent ion neutralizes one negative surface charge and thus releases a hydrogen ion, while the originally divalent ion retains a positive charge. Since the ion still retains a positive charge, it can neutralize an additional site on an adjacent silica particle, forming a physical bridge between particles. The adsorption of calcium results in a mosaic of positive and negative charges on the surface which facilitates the agglomeration of particles as there are many sites for opposite charges to attract each other, as proposed by Goodman [43] and Iler [44]. When there are two different ions adsorbed onto the surface, such as potassium and calcium, coagulation can occur more rapidly. The addition of potassium reduces the overall repulsive forces by neutralizing some of the surface charge. Calcium then causes bridging and coagulation at lower concentrations compared to sols without salt [45,46].

1.1.2. Dynamic Light Scattering

DLS was used to determine the PSD of colloidal NS particles or their agglomerates. DLS measures time-dependent fluctuations in the intensity of scattered light of particles in a suspension which move in a random Brownian motion [47,48]. These fluctuations are processed into an autocorrelation function and applied to fitting algorithms [49]. In this research, the cumulant method [50] has been used. It is the most common method to obtain a particle size distribution from the autocorrelation function.

Two measurements from DLS are primarily used in this research: Size (z-average) and polydispersity index (PDI). The z-average is the intensity-weighted mean hydrodynamic size of the particles, which is derived from the cumulants analysis. The PDI is a dimensionless number calculated from a fit to the cumulant data. It is a measure of how narrow or broad the PSD is. Smaller values indicate a narrow size distribution. Values greater than 0.5 indicate a very broad distribution and results become more qualitative than quantitative.

In this research, a lower z-average, closer to the manufacturers supplied value, indicates a better dispersion quality. A higher z-average indicates a lower dispersion quality, as agglomeration has occurred. In addition, lower PDI values are preferred, indicating a small range of particle sizes.

1.1.3. Zeta Potential

Obtaining the zeta potential (ZP), or the electrokinetic potential, is a technique to quantify potential stability of suspended particles. By measuring the electrophoretic mobility of a particle, the ZP is calculated. A higher magnitude of ZP corresponds to a higher degree of electrostatic repulsion between the charged particles and, thus, a higher dispersion stability, or stronger agglomeration resistance. As a general rule of thumb, magnitudes greater than 30 mV indicate stability and magnitudes less than 30 mV indicate higher potential for agglomeration and coagulation [47,51].

The distribution of ions in a solution is affected by charged particles. These charged particles result in a higher concentration of ions close to the surface. In this region, known as the Stern layer, ions such as Na^+ and K^+ are strongly bound. An outer layer, known as the diffuse layer, contains ions that are less strongly bound. These two layers form an electrical double layer around each particle. These two layers extend a finite distance from a particle surface and are influenced by several factors including pH, ionic concentration and composition. The potential at the edge of electrical double layer with the surrounding environment is the ZP [40,47,52]. The thickness of the diffuse layer is the Debye length. The Debye length is a measure of how far the electrostatic forces persist from the particle surface. A larger Debye length keeps particles further apart which increases sol stability. In aqueous solutions of monovalent ions, the length is reciprocally proportional to the square root of the ionic concentration. In a 0.1 M solution (the maximum concentration of KOH in this work) the Debye length is 0.96 nm. In a 0.4 M solution of divalent ions, the Debye length is 0.28 nm [51]. This is a very short distance, which is one reason why NS can be difficult to stabilize in some environments.

2. Materials and Methods

2.1. Experimental Parameters

2.1.1. Colloidal NS

Four commercially available colloidal NS sols were used in this study. The compositions and average physical properties provided by the manufacturer of the NS sols are shown in Table 1. Figure 2 shows their spherical shape.

Table 1. Properties of NS sols.

Property	NS-5	NS-20a	NS-20b	NS-75
Particle Size (nm)	5	20	20	75
Surface Area (m^2/g)	600	150	150	40
% SiO_2	15	34	50	40
pH	9.0	2.8	9.0	8.4
Specific Gravity	1.09	1.23	1.39	1.29
Viscosity (cP)	<10	<10	55	10
Stabilizing Ion	Ammonium	—	Sodium	Sodium
% Na_2O	0.02	0.04	0.40	0.30

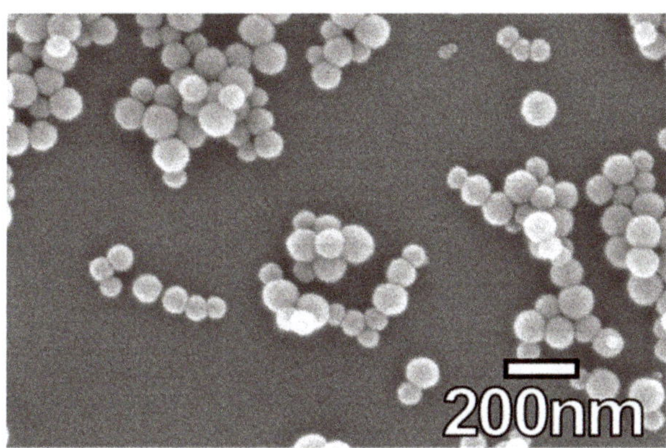

Figure 2. Scanning electron microscopy (SEM) image of NS-75.

2.1.2. Dynamic Light Scattering

The PSD of NS was obtained by DLS using a Zetasizer Nano ZS (Malvern Instruments Ltd., Malvern, UK). Measurements were performed using disposable polystyrene cuvettes at room temperature (25.0 °C). The dispersant (water) properties were set to default values of viscosity: 0.8872 cP and refractive index: 1.330. The material absorption coefficient was set to 0.001 and the refractive index to 1.50 [53]. These values were kept constant throughout all measurements. Samples were thermally equilibrated for 30 s and the angle of detection was set to 173° (Non-Invasive Backscatter (NIBS)). Three measurements were performed for each sample at 11 runs per measurement, 10 seconds per run. The measurements reported in the results section are the mean of the corresponding three measurements. The general purpose algorithm and the method of cumulants were used for analysis. All measurements were taken within one hour of sample preparation. Raw materials were stored at room temperature.

2.1.3. Zeta Potential

ZP was measured using the Zetasizer Nano ZS with the dip cell accessory. The same material parameters were used as in the particle size measurements. Four measurements were performed for each sample at 30 runs per measurement. The voltage selection was set to automatic, with values between 1 V and 20 V. ZP measurements were taken from the same sample as the corresponding DLS measurement.

2.1.4. Sample Preparation

The parameters for testing the four NS sols from Table 1 included varying solid concentration by adding deionized (DI) water, varying pH value by adding 0.1 M aqueous potassium hydroxide (KOH) or 0.1 M hydrochloric acid (HCl), varying amounts of calcium nitrate ($Ca(NO_3)_2 \cdot 4H_2O$), and adding

UHPC-like synthetic pore solution (PS) as outlined by Schrofl et al. (9.72 g $Ca(NO_3)_2{\cdot}4H_2O$ dissolved in 148.5 g of 0.1 mol/L aqueous KOH solution) [54]. An overview of the test matrix is shown in Figure 3 highlighting the four test parameters. The concentration of NS in Figure 3 is by weight percent. It was observed that PS causes irreversible agglomeration. The two ingredients of the PS were isolated to further understand the system.

NS sols at 2% solid content were chosen to investigate the effect of pH value, concentration of $Ca(NO_3)_2$, and ionic concentration of the PS. The alterations and concentrations are summarized in Table 2. The synthetic PS contained 0.1 M of KOH and 0.4 M of $Ca(NO_3)_2{\cdot}4H_2O$ in accordance to Reference [54]. KOH6, Ca8, and PS10 correspond to the molar concentration of the as received synthetic PS. Each successive concentration step down was half of the prior concentration. The pH value of the samples was measured using pH paper in increments of 0.5. DI water, aqueous KOH, aqueous $Ca(NO_3)_2$, and pore solution were filtered to 0.1 μm using Anotop (GE Whatman, Marlborough, MA, USA) syringe filters before mixing with NS.

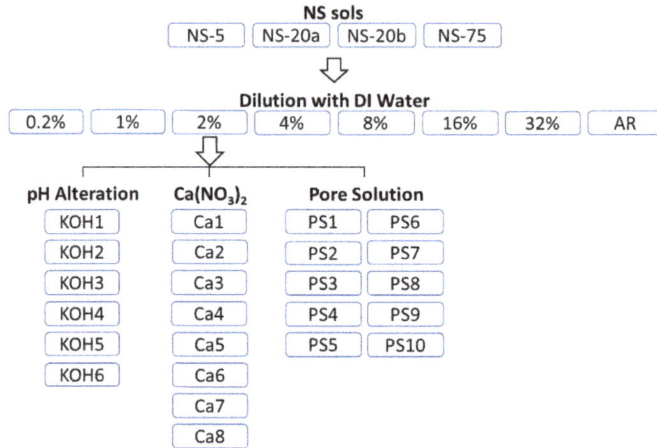

Figure 3. Experimental design DI: Deionized. AR: As received.

Table 2. Test specifications for NS alteration.

2% Solid NS					
pH Alteration	**Molarity (mol/L)**	**$Ca(NO_3)_2$**	**Molarity (mol/L)**	**Pore Solution**	**Molarity of KOH (mol/L)**
KOH0	0	Ca0	0	PS0	0
				PS1	9.77×10^{-5}
				PS2	1.95×10^{-4}
		Ca1	3.13×10^{-3}	PS3	7.81×10^{-4}
		Ca2	6.25×10^{-3}	PS4	1.56×10^{-3}
KOH1	3.13×10^{-3}	Ca3	1.25×10^{-2}	PS5	3.13×10^{-3}
KOH2	6.25×10^{-3}	Ca4	2.50×10^{-2}	PS6	6.25×10^{-3}
KOH3	1.25×10^{-3}	Ca5	5.00×10^{-2}	PS7	1.25×10^{-2}
KOH4	2.50×10^{-3}	Ca6	1.00×10^{-1}	PS8	2.50×10^{-2}
KOH5	5.00×10^{-3}	Ca7	2.00×10^{-1}	PS9	5.00×10^{-2}
KOH6	1.00×10^{-3}	Ca8	4.00×10^{-1}	PS10	1.00×10^{-1}

2.1.5. CryoSEM

Cryogenic scanning electron microscopy (cryoSEM) was used to visualize the NS particle dispersion after mixing with cement paste to be able to determine the technique's efficacy for samples of this type. Cement paste was dropped on a 2 mm rivet holder and shock frozen by immersing into a liquid nitrogen slush. Afterwards, samples were fractured, etched, and sputter-coated with Au/Pt in a Leica EM MED020 with QSG10 (Leica Microsystems Inc., Buffalo Grove, IL, USA). Samples were then transferred using the Leica EM VCT100 (Leica, Wetzlar, Germany) and imaged with an FEI Nova NanoSEM 450 (FEI, Hillsboro, OR, USA). The accelerating voltage was 10 kV and the working distance was 5.5 mm.

A cement paste sample was mixed using white Portland cement (w/c ratio 0.3) (Type I, ASTM C150, Lehigh White Cement Co., Waco, TX, USA) and NS-75 (1.5% bwoc) [26]. The cement paste was mixed in a LabRAM Mixer (Resodyn Acoustic Mixers, Butte, MT, USA) at an intensity of 50% for a total mixing time of 300 s. The time between initial hydration and freezing was 1 h.

3. Results

3.1. Diluting NS

All four NS sols were altered and measured for particle size distribution (PSD) and zeta potential (ZP). The effect of the concentration of NS particles was first studied to isolate the potential influence of the number of particles in a sol. All sols except NS-75 in their as-received state were found to be polydisperse, with a PDI greater than 0.5. Interestingly, the PSD for NS-20b changed as it was diluted. It exhibited a bimodal distribution at higher concentrations and at concentrations below 8 wt % solid, it exhibited a monomodal distribution, as shown in Figure 4, indicating a better dispersion quality. At 2 wt %, the z-average was 24.7 nm (standard deviation (SD) = 0.4), which is close to the provided value of 20 nm by the manufacturer. This suggests that at higher concentrations, the particles tend to weakly agglomerate. By reducing the concentration, there is a greater time and distance between particle collisions, resulting in a better dispersion quality.

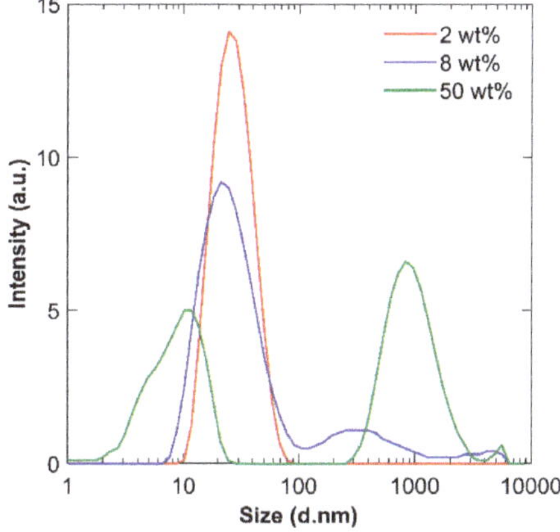

Figure 4. Selected particle size distribution (PSD) of dilution of NS-20b demonstrating the transition from bimodal to monomodal distributions.

NS-20a was too polydisperse for DLS until it was diluted to a concentration of 0.2 wt %. The z-average was 68.3 nm (SD = 1.5), about three times the supplied value. NS-5 was too polydisperse for quantitative results throughout all concentrations. It is hypothesized that these two sols were polydisperse because of the increased specific surface area of the smaller particle diameters. The increased surface energy was not strongly influenced by the reduction in concentration. The complete results are reported in the Supplementary Materials.

The ZP as a function of solid concentration is shown in Figure 5. In the cases of the two sodium-stabilized NS sols, NS-20b and NS-75, the size and ZP trends were directly correlated. As the size decreased, indicating a better dispersion quality, the ZP increased, indicating a higher stability. On this plot, a lower, more negative value indicates a greater ZP. At a concentration between 2% and 4%, the sols experienced the largest ZP, indicating the highest stability.

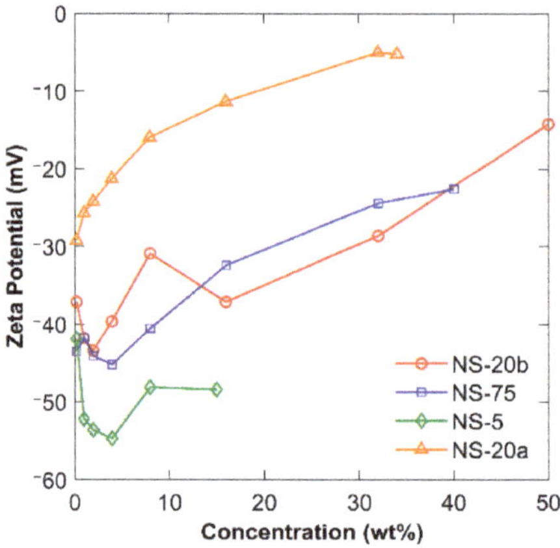

Figure 5. Zeta potential of NS sols through a range of dilution.

Based on the experimental results, there exists an ideal solid concentration for the greatest stability in DI water. Based on the PSD and ZP measurements, this can be concluded to be about 2 wt % solid NS. This conclusion can only be made for NS-20b and NS-75, as the two other sols were not suitable for DLS due to their high polydispersity and produced inconclusive results.

3.2. *pH Adjustment*

The pH of the NS sols was adjusted from about 2 to 12 at 2 wt % solid using HCl and KOH. Throughout pH adjustment, the average size of the sols was remarkably consistent. For the two sodium-stabilized sols, the difference between the maximum and minimum particle size was 7.5 nm (NS-20b) and 7.4 nm (NS-75). For the ammonia-stabilized NS-5, the difference was 5.4 nm. The non-stabilized NS-20a experienced a difference in size of 8.9 nm.

NS-20b had its minimum size at a pH of 8: 23.0 nm (SD = 0.4). The maximum magnitude of ZP was at a pH of 9: −43.3 mV (SD = 3.3). It is not surprising this was observed near its supplied pH value of 9 since the sol is optimized by the manufacturer to experience the greatest stability at this pH. From Figure 6 it is apparent that the average size reaches its greatest value at the ends of the pH scale. The other three sols also generally experienced the largest sizes at the ends of the pH spectrum.

However, the ZP trends were not as apparent. Qualitatively, the ZP was lower at lower pH and higher at higher pH, as is loosely observed in Figure 6. One reason is because the HCl used to lower the pH compressed the electrical double layer. Compared to KOH, the HCl had a negative effect on the ZP. For NS-20b, a reduction in ZP was measured but agglomeration did not occur as would be expected from DLVO theory, exhibited in Figure 6. Similar trends were seen for NS-5 and NS-75. The complete data is presented in the Supplementary Materials.

In a similar experiment, KOH was also added in concentrations relative to the PS to isolate its effect. At the two lowest concentration, KOH1 and KOH2, NS-20b exhibited some agglomeration on the order of magnitude of 3000 nm, while at higher concentrations it experienced a better dispersion quality. NS-20a had a polydisperse size distribution at 2% solid, however, with the introduction of KOH, the dispersion changed to monodisperse distribution. It reached an agglomeration state of about 40 nm, or about twice the manufacturers supplied value, which remained fairly consistent during significant pH adjustment. The addition of KOH altered the electrical double layer and resulted in a metastable state, where slight agglomeration occurred to yield the average size of 40 nm. ZP measurements of NS-20a experienced a similar trend, remaining at about −32 mV. NS-5 exhibited a similar trend of reaching a metastable agglomeration state, although it took an increased amount of KOH to reach a monodisperse distribution. The initial agglomeration state was about twice the supplied manufacturer's value and approached three times that value at the highest concentration of KOH.

Above pH 11, silica was expected to start dissolving. Effects of dissolution were not apparent. Several samples were observed with DLS over a period of 96 h at a pH above 12. The average particle size remained consistent, indicating that the particles were not dissolving. The remarkable stability of the NS sols through a wide range of pH adjustment and over periods of time at high pH suggests that the manufacturer's proprietary stabilization method is fairly robust, which should be useful for achieving uniform dispersion in a concrete mix.

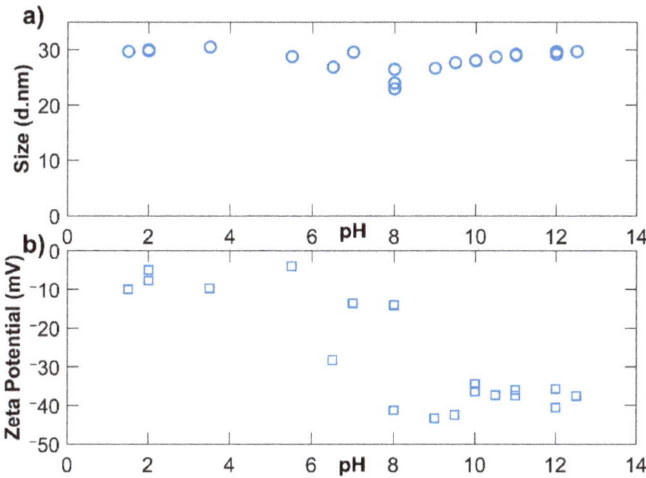

Figure 6. pH adjustment of NS-20b with 0.1 M HCl and 0.1 M KOH; (**a**) displays z-average while (**b**) reports zeta potential.

3.3. Calcium Nitrate

Calcium nitrate was added in several concentrations to isolate its effect in the PS at 2 wt % solid. For NS-20b and NS-5, the sols were able to tolerate the addition of calcium nitrate up to Ca4. Above Ca4, there was significant agglomeration and DLS could not provide meaningful data. The other two sols, NS-75 and NS-20a, had a fairly constant size throughout the range of Ca^{2+} concentrations. From the average size alone, it is apparent that $Ca(NO_3)_2$ has a greater influence on agglomeration than KOH. After 24 h, all NS-$Ca(NO_3)_2$ samples were observed to sediment to the bottom of the cuvette, indicating coagulation had occurred. It was hypothesized that the particle size and stabilizing ion would have an influence on the agglomeration with the addition of Ca^{2+}. However, no conclusions could be drawn based on the PSD.

ZP measurements provide greater insight into the stability of these samples. All four sols experienced a reduction in the magnitude of the ZP and switched from an initially negative ZP to a positive ZP. NS-75 at Ca8 had a ZP of 29.2 mV (SD = 12.3), a difference of almost 78 mV without $Ca(NO_3)_2$. Therefore, it can be concluded that Ca^{2+} significantly changes the surface charge of the silica particles. This happens more rapidly than the K^+ ions due to the higher valency of Ca^{2+}. These ions offset the surface charge and reduce the electrostatic repulsive forces. The bridging of particles with calcium is hypothesized to be the main cause of agglomeration. Small concentrations (<Ca2) of

$Ca(NO_3)_2$ resulted in large changes to the ZP. These results were consistent to those of Iler [41] and the hypothesized mechanisms of agglomeration can be reported with confidence. These findings illustrate the importance of DLS and ZP which provide different information on the same sample.

3.4. Synthetic Pore Solution

Synthetic PS was added in several concentrations to study the effects of the two previous parameters combined. The concentration of the PS was reduced by dilution with DI water. Again, this was added to the 2% solid.

NS-20b and NS-75 both experienced agglomeration beginning at PS4, with significant agglomeration above that concentration. When KOH and $Ca(NO_3)_2$ were combined to create the synthetic PS, its propensity to cause agglomeration was more significant than either component alone, as predicted by Plank et al. [55]. The effects of increasing pH and the introduction of two cations, K^+ and Ca^{2+}, accelerated the agglomeration. The magnitude of the ZP was greatly reduced and the reduction occurred quicker with PS compared to either component alone. It is interesting to note that in no case was the ZP positive, as was observed with the addition of Ca^{2+} only. Before ZP could approach 0, significant agglomeration occurred. These results were expected from the hypotheses of Iler [41] and Tadros [46] and can report these mechanisms with confidence.

PS10 is the pore solution concentration that would be equivalent in pH and ionic concentration to an actual UHPC environment. It is apparent from Figure 7d and Table S13 that NS stability in an actual concrete environment is far from being achieved. Since this pore solution only contains water, KOH, and $Ca(NO_3)_2$ it can be inferred that the $Ca(NO_3)_2$ is the main contributor to NS agglomeration, due to its bridging effect between silica particles.

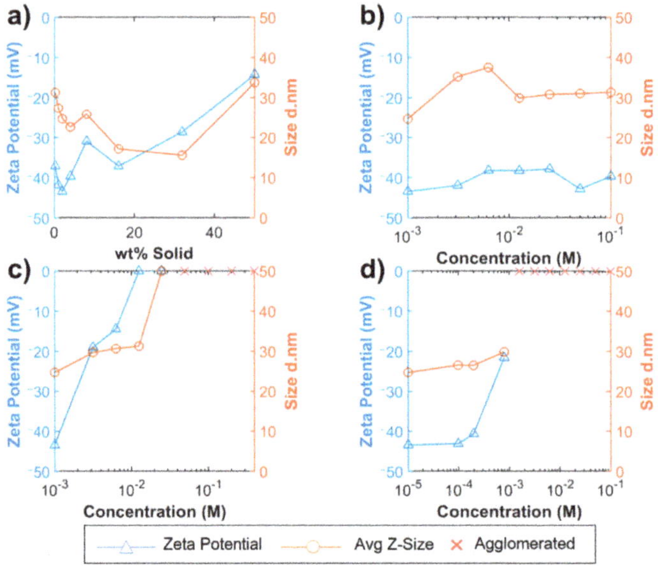

Figure 7. ZP and size measurements for NS-20b: (**a**) solid concentration varied with DI water; (**b**) potassium hydroxide concentration varied with DI water; (**c**) calcium nitrate concentration varied with DI water; (**d**) pore solution concentration varied with DI water.

3.5. NS Visualization

An important aspect of understanding dispersion and agglomeration of NS in a true cementitious environment is to directly observe what happens in a concrete sample. This can often be challenging as NS can lead to an increased amount of C-S-H gel and therefore making individual NS particles indistinguishable. Observation of NS was explored by using cryoSEM, as shown in Figure 8. Since the hydration was stopped after 1 hour, some NS was unreacted and visible. No NS particles were observed to be completely separate from another NS particle and the NS appears to be spread out over about half the cement grain. This data was not used quantitatively but proved that this will be a viable method to explore the effects of cementitious pore solution on the dispersion of NS. Future data analysis could include calculating the number of particles per unit area to quantify the dispersion quality. Based on direct observation, the effects of superplasticizers on the dispersion of cement, silica fume, and NS can be investigated. Ongoing research will utilize cryoSEM to visualize the effects of superplasticizers tailored specifically for improved dispersion of NS.

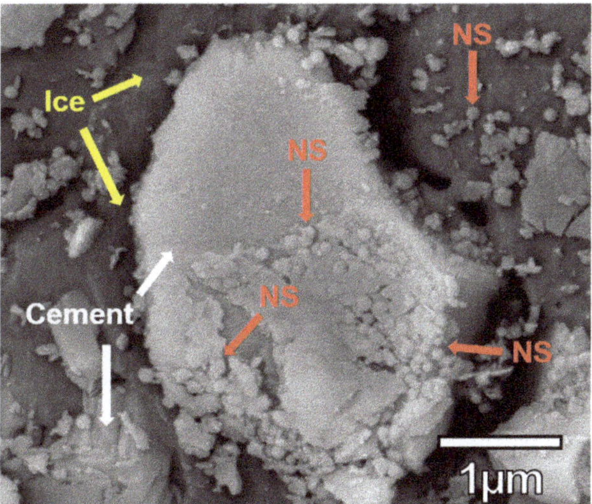

Figure 8. CryoSEM image showing unreacted NS on a cement grain.

4. Conclusions

In this research, the stability of NS in a concrete environment has been studied. The optimal concentration for the best dispersion quality of colloidal NS in a cementitious environment is 2 wt % solid. This concentration allows for sufficient screening of short-range van der Waals attractive forces. The supplied NS sols are fairly pH-robust since they tolerate a wide range of pH changes without agglomerating. This is especially the case with the addition of KOH. The magnitude of the zeta potential remained high and this helped to improve the dispersion quality. On the contrary, Ca^{2+} ions are the main contributor to the agglomeration of NS in pore solution. Its higher valency further compresses the electrical double layer, causing the ZP to shift and thus results in a positive surface charge. This neutralization of negative sites results in a mosaic of positive and negative surface charges, leading to bridging between silica particles. By analyzing the size and ZP measurements, NS-20b experienced the best dispersion quality compared to the other three sols. This is due to its intermediate surface area, compared to higher (NS-5) and lower (NS-75) surface areas. The smaller particles were more susceptible to agglomeration due to the higher surface energy. Achieving stability in full strength PS, representing UHPC, is far from being accomplished. NS sols of similar stabilization methods and size will be of greater focus for future studies to obtain stability in PS. Methods to neutralize the effects of Ca^{2+} will be specifically targeted to improve the dispersion quality in the presence of PS. Finally, cryoSEM proved to be a valuable tool for cementitious material characterization in the fluid state. The ability to stop hydration and image the microstructure in the fluid state will be critical to understanding the dynamics of NS interaction, and thus their dispersion quality in a cementitious environment.

Supplementary Materials: The following material is available online at http://www.mdpi.com/1996-1944/12/12/1976/s1.

Author Contributions: Conceptualization, D.H. and K.W.; investigation, D.H. and J.M.; formal analysis, D.H.; writing—original draft preparation, D.H.; writing—review and editing, D.H. and K.W.; supervision, K.W.; funding acquisition, K.W.

Funding: This research was supported by the National Science Foundation CMMI Grant 1454574, CAREER: Understanding Behavior and Properties of Nano-Sized Particles in Cement-Based Materials.

Acknowledgments: This work was performed in part at the Bioscience Electron Microscopy Facility of the University of Connecticut with Xuanhao Sun with the support of NSF Grant #1126100. The authors would also like to acknowledge the support in terms of material used for experiments from the following companies: Nalco, Lehigh White Cement Company, LafargeHolcim, and Chryso.

Conflicts of Interest: The authors declare no conflict of interest.

References

1. Sobolev, K.; Flores, I.; Hermosillo, R.; Torres-Martinez, L. Nanomaterials and Nanotechnology for High-Performance Cement Composites. *Proc. ACI Sess. Nanotechnol. Concr. Recent Dev. Futur. Prospect.* **2006**, *254*, 91–118. [CrossRef]
2. Silvestre, J.; Silvestre, N.; de Brito, J. Review on Concrete Nanotechnology. *Eur. J. Environ. Civ. Eng.* **2016**, *20*, 455–485. [CrossRef]
3. Aggarwal, P.; Singh, R.P.; Aggarwal, Y. Use of Nano-Silica in Cement Based Materials—A Review. *Cogent Eng.* **2015**, *2*, 1078018. [CrossRef]
4. Chen, J.; Kou, S.; Poon, C. Hydration and Properties of Nano-TiO_2 Blended Cement Composites. *Cem. Concr. Compos.* **2012**, *34*, 642–649. [CrossRef]
5. Metaxa, Z.S.; Konsta-Gdoutos, M.S.; Shah, S.P. Carbon Nanotubes Reinforced Concrete. In Proceedings of the Nanotechnology of Concrete: The Next Big Thing is Small, Session at the ACI Fall 2009 Convention, New Orleans, LA, USA, 8–12 November 2009; pp. 11–20. [CrossRef]
6. Chuah, S.; Pan, Z.; Sanjayan, J.G.; Wang, C.M.; Duan, W.H. Nano Reinforced Cement and Concrete Composites and New Perspective from Graphene Oxide. *Constr. Build. Mater.* **2014**, *73*, 113–124. [CrossRef]
7. Janković, K.; Stankovic, S.; Dragan, B.; Stojanovic, M.; Antić, L. The Influence of Nano-Silica and Barite Aggregate on Properties of Ultra High Performance Concrete. *Constr. Build. Mater.* **2016**, *126*, 147–156. [CrossRef]
8. Kontoleontos, F.; Tsakiridis, P.E.; Marinos, A.; Kaloidas, V.; Katsioti, M. Influence of Colloidal Nanosilica on Ultrafine Cement Hydration: Physicochemical and Microstructural Characterization. *Constr. Build. Mater.* **2012**, *35*, 347–360. [CrossRef]
9. Land, G.; Stephan, D. The Influence of Nano-Silica on the Hydration of Ordinary Portland Cement. *J. Mater. Sci.* **2012**, *47*, 1011–1017. [CrossRef]
10. Lim, S.; Mondal, P. Effects of Incorporating Nanosilica on Carbonation of Cement Paste. *J. Mater. Sci.* **2015**, *50*, 3531–3540. [CrossRef]
11. Hou, P.; Cheng, X.; Qian, J.; Zhang, R.; Cao, W.; Shah, S.P. Characteristics of Surface-Treatment of Nano-SiO_2 on the Transport Properties of Hardened Cement Pastes with Different Water-to-Cement Ratios. *Cem. Concr. Compos.* **2015**, *55*, 26–33. [CrossRef]
12. Torabian Isfahani, F.; Redaelli, E.; Lollini, F.; Li, W.; Bertolini, L. Effects of Nanosilica on Compressive Strength and Durability Properties of Concrete with Different Water to Binder Ratios. *Adv. Mater. Sci. Eng.* **2016**, *2016*. [CrossRef]
13. Ji, T. Preliminary Study on the Water Permeability and Microstructure of Concrete Incorporating Nano-SiO_2. *Cem. Concr. Res.* **2005**, *35*, 1943–1947. [CrossRef]
14. Jo, B.-W.; Kim, C.-H.; Tae, G.; Park, J.-B. Characteristics of Cement Mortar with Nano-SiO_2 Particles. *Constr. Build. Mater.* **2007**, *21*, 1351–1355. [CrossRef]

15. Berra, M.; Carassiti, F.; Mangialardi, T.; Paolini, A.E.; Sebastiani, M. Effects of Nanosilica Addition on Workability and Compressive Strength of Portland Cement Pastes. *Constr. Build. Mater.* **2012**, *35*, 666–675. [CrossRef]
16. Khaloo, A.; Mobini, M.H.; Hosseini, P. Influence of Different Types of Nano-SiO_2 Particles on Properties of High-Performance Concrete. *Constr. Build. Mater.* **2016**, *113*, 188–201. [CrossRef]
17. Madani, H.; Bagheri, A.; Parhizkar, T. The Pozzolanic Reactivity of Monodispersed Nanosilica Hydrosols and Their Influence on the Hydration Characteristics of Portland Cement. *Cem. Concr. Res.* **2012**, *42*, 1563–1570. [CrossRef]
18. Richard, P.; Cheyrezy, M. Composition of Reactive Powder Concretes. *Cem. Concr. Res.* **1995**, *25*, 1501–1511. [CrossRef]
19. Sonebi, M.; García-Taengua, E.; Hossain, K.M.A.; Khatib, J.; Lachemi, M. Effect of Nanosilica Addition on the Fresh Properties and Shrinkage of Mortars with Fly Ash and Superplasticizer. *Constr. Build. Mater.* **2015**, *84*, 269–276. [CrossRef]
20. Yu, R.; Spiesz, P.; Brouwers, H.J.H. Effect of Nano-Silica on the Hydration and Microstructure Development of Ultra-High Performance Concrete (UHPC) with a Low Binder Amount. *Constr. Build. Mater.* **2014**, *65*, 140–150. [CrossRef]
21. Hou, P.; Kawashima, S.; Wang, K.; Corr, D.J.; Qian, J.; Shah, S.P. Effects of Colloidal Nanosilica on Rheological and Mechanical Properties of Fly Ash–cement Mortar. *Cem. Concr. Compos.* **2013**, *35*, 12–22. [CrossRef]
22. Bi, J.; Pane, I.; Hariandja, B.; Imran, I. The Use of Nanosilica for Improving of Concrete Compressive Strength and Durability. *Appl. Mech. Mater.* **2012**, *204*, 4059–4062. [CrossRef]
23. Qing, Y.; Zenan, Z.; Deyu, K.; Rongshen, C. Influence of Nano-SiO_2 Addition on Properties of Hardened Cement Paste as Compared with Silica Fume. *Constr. Build. Mater.* **2007**, *21*, 539–545. [CrossRef]
24. Rong, Z.; Sun, W.; Xiao, H.; Jiang, G. Effects of Nano-SiO_2 Particles on the Mechanical and Microstructural Properties of Ultra-High Performance Cementitious Composites. *Cem. Concr. Compos.* **2015**, *56*, 25–31. [CrossRef]
25. Sanjuán, M.Á.; Argiz, C.; Gálvez, J.C.; Moragues, A. Effect of Silica Fume Fineness on the Improvement of Portland Cement Strength Performance. *Constr. Build. Mater.* **2015**, *96*, 55–64. [CrossRef]
26. Wille, K.; Naaman, A.E.; Parra-Montesinos, G.J. Ultra-High Performance Concrete with Compressive Strength Exceeding 150 MPa (22 Ksi): A Simpler Way. *ACI Mater. J.* **2011**, *108*, 34–46.
27. De Larrard, F.; Sedran, T. Optimization of Ultra-High-Performance Concrete by the Use of a Packing Model. *Cem. Concr. Res.* **1994**, *24*, 997–1009. [CrossRef]
28. Bagheri, A.; Parhizkar, T.; Madani, H.; Raisghasemi, A.M. The Influence of Different Preparation Methods on the Aggregation Status of Pyrogenic Nanosilicas Used in Concrete. *Mater. Struct.* **2013**, *46*, 135–143. [CrossRef]
29. Zabihi, N.; Ozkul, M.H. The Effect of Colloidal Nano-Silica as a Cementitious Material, on Durability and Mechanical Properties of Mortar. In Proceedings of the 11th International Congress on Advances in Civil Engineering (ACE 2014), Istanbul, Turkey, 21–25 October 2014.
30. Böschel, D.; Janich, M.; Roggendorf, H. Size Distribution of Colloidal Silica in Sodium Silicate Solutions Investigated by Dynamic Light Scattering and Viscosity Measurements. *J. Coll. Interface Sci.* **2003**, *267*, 360–368. [CrossRef]
31. Hayrapetyan, S.S.; Khachatryan, H.G. Control of the Growth Processes of the Silica Sols Colloidal Particles. *Microporous Mater.* **2005**, *78*, 151–157. [CrossRef]
32. Quercia, G.; Lazaro, A.; Geus, J.W.; Brouwers, H.J.H. Characterization of Morphology and Texture of Several Amorphous Nano-Silica Particles Used in Concrete. *Cem. Concr. Compos.* **2013**, *44*, 77–92. [CrossRef]
33. Oertel, T.; Hutter, F.; Helbig, U.; Sextl, G. Amorphous Silica in Ultra-High Performance Concrete: First Hour of Hydration. *Cem. Concr. Res.* **2014**, *58*, 131–142. [CrossRef]
34. Quercia, G.; Hüsken, G.; Brouwers, H.J.H. Water Demand of Amorphous Nano Silica and Its Impact on the Workability of Cement Paste. *Cem. Concr. Res.* **2012**, *42*, 344–357. [CrossRef]
35. Sanchez, F.; Sobolev, K. Nanotechnology in Concrete–a Review. *Constr. Build. Mater.* **2010**, *24*, 2060–2071. [CrossRef]
36. Sobolev, K.; Gutiérrez, M.F. How Nanotechnology Can Change the Concrete World. *Am. Ceram. Soc. Bull.* **2005**, *84*, 14.

37. Rashad, A.M. A Comprehensive Overview about the Effect of Nano-SiO_2 on Some Properties of Traditional Cementitious Materials and Alkali-Activated Fly Ash. *Constr. Build. Mater.* **2014**, *52*, 437–464. [CrossRef]
38. Rashad, A.M. A Synopsis about the Effect of Nano-Al_2O_3, Nano-Fe_2O_3, Nano-Fe_3O_4 and Nano-Clay on Some Properties of Cementitious Materials–A Short Guide for Civil Engineer. *Mater. Des.* **2013**, *52*, 143–157. [CrossRef]
39. Iler, R.K. *The Colloid Chemistry of Silica and Silicates*; Cornell University Press: New York, NY, USA, 1955.
40. Bergna, H.E.; Roberts, W.O. *Colloidal Silica: Fundamentals and Applications*; CRC Press: Boca Raton, FL, USA, 2005; Volume 131.
41. Iler, K.R. *The Chemistry of Silica: Solubility, Polymerization, Colloid and Surface Properties and Biochemistry of Silica*; John Wiley & Sons, Inc.: Hoboken, NJ, USA, 1979.
42. Reches, Y. Nanoparticles as Concrete Additives: Review and Perspectives. *Constr. Build. Mater.* **2018**, *175*, 483–495. [CrossRef]
43. Vrij, A.; Sonntag, H.; Tezak, B.; Kitchener, J.A.; Matijevic, E.; Mysels, K.J.; Overbeek, J.T.G.; Corkill, J.M.; Goodman, J.F.; Fowkes, F.M.; et al. General Discussion. *Discuss. Faraday Soc.* **1966**, *42*, 60–68. [CrossRef]
44. Iler, R.K. Coagulation of Colloidal Silica by Calcium Ions, Mechanism, and Effect of Particle Size. *J. Coll. Interface Sci.* **1975**, *53*, 476–488. [CrossRef]
45. Baxter, S.; Bryant, K.C. 578. Silica Sols. Part III. Accelerated Gelation, and Particle Size. *J. Chem. Soc.* **1952**, 3024–3027. [CrossRef]
46. Tadros, T.F.; Lyklema, J. The Electrical Double Layer on Silica in the Presence of Bivalent Counter-Ions. *J. Electroanal. Chem. Interfacial Electrochem.* **1969**, *22*, 1–7. [CrossRef]
47. Bhattacharjee, S. DLS and Zeta Potential–What They Are and What They Are Not? *J. Control. Release* **2016**, *235*, 337–351. [CrossRef] [PubMed]
48. Ball, S. Colloidal Gold: The Gold Standard for Drug Delivery? *Drug Dev. Deliv.* **2015**, *15*, 32–35.
49. Carpenter, D.K. Dynamic Light Scattering with Applications to Chemistry, Biology, and Physics. *J. Chem. Educ.* **1977**, *54*, A430. [CrossRef]
50. Koppel, D.E. Analysis of Macromolecular Polydispersity in Intensity Correlation Spectroscopy: The Method of Cumulants. *J. Chem. Phys.* **1972**, *57*, 4814–4820. [CrossRef]
51. Nägele, E. The Zeta-Potential of Cement. *Cem. Concr. Res.* **1985**, *15*, 453–462. [CrossRef]
52. Malvern Instruments Ltd. *Zetasizer Nano Series User Manual*; MAN0317; Malvern Instruments Ltd.: Malvern, UK, 2004.
53. Kitamura, R.; Pilon, L.; Jonasz, M. Optical Constants of Silica Glass from Extreme Ultraviolet to Far Infrared at near Room Temperature. *Appl. Opt.* **2007**, *46*, 8118–8133. [CrossRef]
54. Schröfl, C.; Gruber, M.; Plank, J. Preferential Adsorption of Polycarboxylate Superplasticizers on Cement and Silica Fume in Ultra-High Performance Concrete (UHPC). *Cem. Concr. Res.* **2012**, *42*, 1401–1408. [CrossRef]
55. Plank, J.; Schroefl, C.; Gruber, M.; Lesti, M.; Sieber, R. Effectiveness of Polycarboxylate Superplasticizers in Ultra-High Strength Concrete: The Importance of PCE Compatibility with Silica Fume. *J. Adv. Concr. Technol.* **2009**, *7*, 5–12. [CrossRef]

Article

Influence of Na_2O Content and Ms (SiO_2/Na_2O) of Alkaline Activator on Workability and Setting of Alkali-Activated Slag Paste

Sung Choi and Kwang-Myong Lee *

Department of Civil, Architectural, and Environmental Systems Engineering, Sungkyunkwan University, Jangan-Gu, Suwon 16419, Korea

* Correspondence: leekm79@skku.edu; Tel.: +82-10-8207-7516

Received: 17 May 2019; Accepted: 26 June 2019; Published: 27 June 2019

Abstract: The performance of alkali-activated slag (AAS) paste using activators of strong alkali components is affected by the type, composition, and dosage of the alkaline activators. Promoting the reaction of ground granulated blast furnace slag (GGBFS) by alkaline activators can produce high-strength AAS concrete, but the workability might be drastically reduced. This study is aimed to experimentally investigate the heat release, workability, and setting time of AAS pastes and the compressive strength of AAS mortars considering the Na_2O content and the ratio of Na_2O to SiO_2 (Ms) of binary alkaline activators blended with sodium hydroxide and sodium silicate. The test results indicated that the AAS mortars exhibited a high strength of 25 MPa at 24 h, even at ambient temperature, even though the pastes with an Na_2O content of ≥6% and an Ms of ≥1.0 exhibited an abrupt decrease in flowability and rapid setting.

Keywords: alkali-activator; GGBFS; Na_2O content; Ms (SiO_2/Na_2O); workability; setting time

1. Introduction

Environmental imperatives such as the reduction of CO_2 and conservation of natural resources are becoming issues worldwide. However, because the cement industry emits large amounts of CO_2 and consumes much energy in the supply of raw materials and production processes, efforts are needed to reduce CO_2 emission and energy consumption by replacing cement with supplementary cementitious materials (SCMs) such as ground granulated blast furnace slag (GGBFS) and fly ash (FA). Thus, in order to totally replace cement with GGBFS and FA attempts have been made to create zero-cement concrete by the alkali activation of GGBFS and FA [1–5].

Alkali-activated slag (AAS) concrete, which is produced by using an alkali activator on GGBFS, is a typical zero cement concrete. As GGBFS is calcium aluminosilicate vitreous, it has strong latent hydraulic properties. However, alkaline (pH = 12) activators are needed to stimulate GGBFS. After the GGBFS is stimulated, Si^{4+}, Al^{3+}, Ca^{2+}, Mg^{2+}, and Na^{+} ions are eluted from GGBFS and the reaction of GGBFS proceeds. It should be noted that GGBFS activated in this way has very fast reaction rates due to chemical ionic reactions [6]. Alkaline activators largely impact the properties of AAS concrete and thus, their characteristics such as the alkali concentration, the dosage, and the water-to-solid ratio should be examined prior to the mixture design of AAS concrete [7,8]. In general, soluble alkali or alkali salts can be used as alkaline activators. When considering the strength development and economic efficiency of AAS concrete, the most commonly used activators are Na_2SiO_3 (sodium silicate) and NaOH (sodium hydroxide). A better understanding of the effects of such alkaline activators on the reaction mechanisms of alkali-activated slag could indicate ways to optimize the use of alkaline activators.

Recently, research has been conducted to improve the performance of AAS concrete by blending two types of alkaline activators [9]. When such a binary alkaline activator is used, the characteristics possessed by each alkaline activator can be complementarily utilized [10,11]. NaOH promoted initial reaction of AAS, and Na_2SiO_3 was effective in increasing the strength of AAS concrete. Chang [12] evaluated the setting characteristics of silica-activated pastes depending on the type of alkaline activator used and concluded that setting time significantly reduced as the amount of alkaline activator increased. Zuo et al. [13] showed that though the reaction degree increased as the amount of alkaline activator increased, the reaction degree no longer increased when the amount of alkaline activator was greater than 6 wt.% of cement. Krizan and Zivanovic [14] reported that high strength could be expected when the Ms (SiO_2/Na_2O) were in the range of 0.6–1.5. As described above, as the amount of alkaline activator increases, the reaction of GGBFS increases and the strength of AAS concrete also increases. Therefore, it is necessary to determine an appropriate range for the usage amount of the activator for the mix design of AAS concrete. In particular, the effects of Na_2O and SiO_2 concentrations in alkaline activators on the strength development of AAS concrete should be examined in order to effectively use the alkaline activators.

AAS concrete, which stimulates GGBFS using alkaline activators, rapidly increases in strength in the early ages and exhibits high strength of 40 MPa or more at 28 days [15,16]. However, because the initial reaction rate of GGBFS by alkaline activators is extremely high, several problems with AAS, such as the flowability being lost initially or rapid setting occurring, must be improved for it to be used in a practical setting [17–19]. To address this, the effects of the alkaline activator on the fresh properties of AAS concrete such as workability and setting should be examined, even though the AAS activation mechanism at early ages has not been clearly identified.

In this study, in order to identify the characteristics of AAS pastes, a binary alkaline activator that was a blend of NaOH and Na_2SiO_3 was used. Nine types of AAS paste mixture, varying by their Na_2O content and the ratio of Na_2O to SiO_2 (Ms), were tested. First, the reactivity of the AAS was evaluated for 24 h through the calorimetric measurement of the AAS pastes, and the workability and setting time were measured. Furthermore, to quantitatively analyze the instant at which flowability is lost and setting occurs in the AAS pastes, the viscosity and ultrasonic pulse velocity of the pastes were measured. Lastly, the compressive strength of the AAS mortars was measured up to 24 h. Consequently, it was found that the workability and setting of AAS pastes and the initial strength development of the AAS mortars were significantly affected by Na_2O content and Ms.

2. Materials and Method

2.1. Materials and Alkaline Activator

GGBFS obtained from POSCO (Pohang, Korea) was used as a binder. The chemical composition of the GGBFS determined with X-ray fluorescence (XRF, BRUKER S8 Tiger, Billerica, MA, USA) is given in Table 1. The GGBFS was composed of 44.0% CaO, 33.7% SiO_2, 13.8% Al_2O_3, and 5.2% MgO and thus, its basicity coefficient ($Kb = (CaO + MgO)/(SiO_2 + Al_2O_3)$) was 1.04, which is similar to the neutral value of 1.0 for the ideal alkali activation [12]. The hydration modulus ($HM = (CaO + MgO + Al_2O_3)/SiO_2$) was 1.87, which was 33.5% higher than the required value of 1.4 for sound hydration properties of GGBFS [12,19]. Table 2 shows the physical properties of GGBFS. The density and Blaine air permeability of the GGBFS were 2.90 g/cm^3 and 4253 cm^2/g, respectively. Figure 1 shows the particle size distribution of GGBFS measured with a laser particle size analyzer (PSA, BECKMAN COULTER LS 13 320, Brea, CA, USA). The medium particle size (d_{50}), which represents the particle size of a cluster of particles, was 10.2 μm, and d_{10} and d_{90} were 1.4 μm and 33.2 μm, respectively. The dissolution of GGBFS is dominated by small particles. That is, particles >20 μm react slowly, while particles <2 μm react completely after 24 h in alkali-activated binders [20,21].

The physical properties of the fine aggregates that were used and the mortar production were measured complying with ASTM C 128-15 [22] and ASTM C 33-13 [23]. The fine aggregates were the crushed sands of granite, and their density and absorption rate were 2.62 g/cm^3 and 1.04%, respectively.

Figure 2 shows the particle size distribution of fine aggregates. The max. size of the fine aggregate was 4.76 mm, and the fineness modulus was 2.74.

Two types of alkaline activator, sodium hydroxide (98% purity, Duksan, Ansan, Korea) and sodium silicate (Type 3 industrial water glass, Na_2SiO_3, Ganachem, Ulsan, Korea) were used to activate the GGBFS. The chemical composition of sodium silicate solution was SiO_2 = 28.3 wt.%, Na_2O = 9.3 wt.%, and H_2O = 62.4 wt.%. A 2 M sodium hydroxide solution was prepared by mixing with deionized water. Furthermore, because NaOH would release heat when mixed with water, in order to cool down to ambient temperature, the alkaline activator was stored in a chamber at 20 °C and the relative humidity (RH) of 60% for 24 h prior to casting AAS pastes. The pH of the activators was ranged 13.2–13.6, as measured using a pH meter (HORIBA, LAQUA 74bW, Kyoto, Japan).

Table 1. Chemical composition of ground granulated blast furnace slag (GGBFS) (wt.%).

Type	CaO	SiO_2	Al_2O_3	MgO	SO_3	TiO_2	K_2O	M_nO	Fe_2O_3
GGBFS	44.0	33.7	13.8	5.2	1.2	0.7	0.5	0.2	0.1

Table 2. Physical properties of GGBFS.

Type	Density (g/cm^3)	Blaine (cm^2/g)	Particle Size Distribution (μm)			
			Mean	d_{10}	d_{50}	d_{90}
GGBFS	2.90	4253	14.2	1.4	10.2	33.2

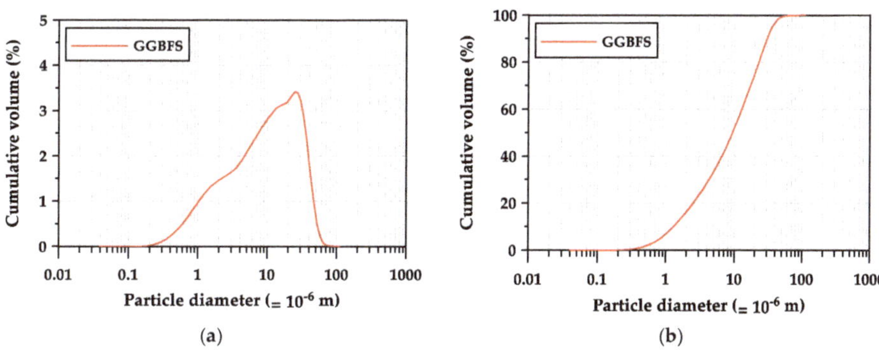

Figure 1. Particle size characteristics of GGBFS: (**a**) particle size distribution, (**b**) cumulative particle size distribution.

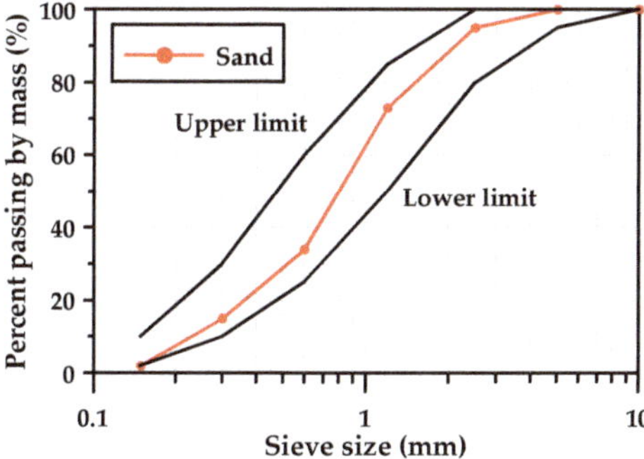

Figure 2. Particle size distribution curve of fine aggregates.

2.2. Mixture Design

Table 3 shows the mixture design of the AAS pastes used to evaluate the workability and setting of the AAS pastes. In this study, a liquid-to-binder ratio (l/b, binder = GGBFS content) was fixed as 0.42 and the Na_2O content and Ms of alkaline activator were varied. It should be noted that the water content of Na_2SiO_3 was included in the required water content. Also, NaOH was added as an alkali activator in order to adjust Ms according to the dosage of Na_2SiO_3. The Na_2O content was set to 4%, 6%, and 8% according to the weight of the GGBFS, and the three levels of Ms were considered at 0.75, 1.00, and 1.25. Thus, nine AAS paste mixtures were designed, and the amount of alkaline activators for each mixture was determined, as shown as Table 3. In the mixture section in the table, the first and second components represent Na_2O content and Ms, respectively. To produce the AAS mortar mixture, twice as much fine aggregate as GGBFS was added to the AAS paste mixture.

Table 3. Mixture design of alkali-activated slag (AAS) pastes with respect to 100 g of binder.

Mixture	Na_2O	SiO_2	Alkaline ($Na_2O + SiO_2$)	Ms (SiO_2/Na_2O)	Alkali Activator		l/b
					NaOH	Na_2SiO_3	
4-0.75	4.0	3.0	7.0	0.75	3.89	10.60	
4-1.00	4.0	4.0	8.0	1.00	3.47	14.13	
4-1.25	4.0	5.0	9.0	1.25	3.04	17.67	
6-0.75	6.0	4.5	10.5	0.75	5.83	15.90	
6-1.00	6.0	6.0	12.0	1.00	5.20	21.20	0.42
6-1.25	6.0	7.5	13.5	1.25	4.56	26.50	
8-0.75	8.0	6.0	14.0	0.75	7.78	21.20	
8-1.00	8.0	8.0	16.0	1.00	6.93	28.27	
8-1.25	8.0	10.0	18.0	1.25	6.08	35.34	

2.3. Test Method

2.3.1. Calorimetric Measurement

GGBFS generates heat during reacting with alkaline activator. To measure the heat, GGBFS and water were stored for 24 h in the calorimeter equipment at 20 °C. The AAS paste mixtures were prepared by mixing the GGBFS, water, and alkaline activators using an internal propeller for 5 min. The mixture was placed into the calorimeter (Tokyo-Riko, Three-point multi-purpose conduction calorimeter, Tokyo, Japan). The temperature range of the equipment was 5–60 °C, and the temperature stability was 5×10^{-3} °C/5 °C. The amount of heat release was measured immediately after mixing the AAS paste up to 24 h.

2.3.2. Workability

The workability of AAS paste mixtures was tested using the mini slump test method. Figure 3 shows a mini slump setup and flow measurement of AAS paste. The mold for the mini slump test was made of brass, and its height was 50 mm, the top opening was 70 mm, and the bottom opening was 100 mm. After filling the mold with the paste mixture, the mold was lifted and the spread length of the mixture was measured. The mini slump of the AAS paste was tested a total 13 times for each paste, from 0 min through 60 min at 5 min intervals after casting the paste complying with ASTM C 1437-15 [24]. In order to eliminate the possibility of measurement errors in the test due to moisture evaporation from the surface of the paste, the paste was remixed for 30 s before mini slump test.

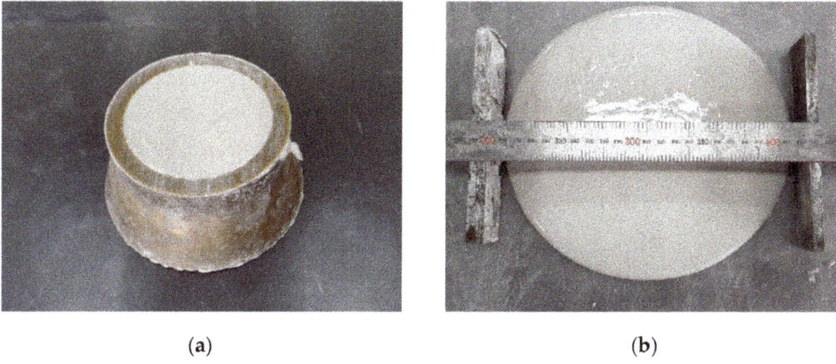

(a) (b)

Figure 3. A mini slump test: (**a**) test setup, (**b**) flow measurement of AAS paste.

2.3.3. Setting Time

The Vicat test (per ASTM C 191-18a [25]) test was carried out to measure the setting time of the AAS pastes inside a room chamber at 23 °C and 50% (RH). The Vicat initial time of setting is defined as the elapsed time between the initial contact of cement and water and the penetration of 25 mm. The Vicat final time of setting end point is defined as the first penetration measurement that did not mark the specimen surface with a complete circular impression.

2.3.4. Rheology and Ultrasonic Pulse Velocity (UPV)

The rheology and UPV tests were adopted to evaluate the timing at which the AAS paste lost its flowability, by measuring the viscosity and ultrasonic pulse velocity of the AAS paste according to the elapsed time. The rheology test measured viscosity through a paste viscometer using the friction resistance between the spindle and the paste rotating at a speed of 5 rpm. To determine the moment that flowability was lost in the AAS paste, the time to reach maximum measurable viscosity was recorded. The UPV test was carried out by placing the oscillator and receiver on acrylic boxes (thickness: 10 mm; ultrasonic velocity of acryl: 2740 m/s) placed 20 mm apart, filling the container with paste, and measuring the transmission time of ultrasonic waves through the oscillator and receiver at intervals of 30 s with the measurement frequency of 5 kHz.

2.3.5. Compressive Strength

To investigate the influence of Na_2O content and Ms on the compressive strength development of AAS mortars up to 24 h, uniaxial compression test was conducted. A mortar mixer was used for mixing 2-liter batches. After premixing GGBFS and fine aggregates for 30 s, the alkaline activator and water were added at low speed. Mixing continued at low speed for 90 s and at high speed for 1 min. Cube mortar specimens ($50 \times 50 \times 50$ mm^3) were prepared complying with ASTM C109-16a [26] and cured in a chamber at 20 ± 2 °C and the relative humidity (RH) of 60 ± 2%, for curing at the environmental condition similar to general ambient condition. Compressive strength of the mortar specimens was measured at 12 h, 18 h, and 24 h after casting the AAS paste. The compressive strength was obtained as an average value of three AAS mortar specimens.

3. Results and Discussion

3.1. Heat Release of Pastes

Figure 4 shows the calorimetric curves for the rate of heat release according to Na_2O content. The AAS pastes, excluding the paste with 4% Na_2O, had two peaks of heat [13]. When Na_2O content was greater than 6% of binder weight, the first peak of heat occurred at 30–60 min, which resulted mainly from the wetting and partial dissolution of GGBFS by NaOH and partly from the reaction of

GGBFS by Na_2SiO_3. The second peak was observed between 6 h and 12 h, which corresponds mainly to the formation of reaction products [27]. That is, when Na_2O content was fixed to be 6% or 8%, the first peak of heat increased as Ms increased, in other words, SiO_2 content increased. In particular, the paste with 6% Na_2O content showed a relatively high first peak and a clear secondary peak. Ravikumar and Neithalath [28] reported a similar result, stating that at constant Na_2O content, the Ms was a main factor in increasing both the speed and the magnitude of the first peak of heat. For the pastes with alkaline activators of 4% Na_2O content, the first peak of heat was relatively low and the secondary peak occurred after the induction since the amount of alkaline activators was less. This is a similar tendency to the rate of heat release curve when only Na_2SiO_3 was used [27].

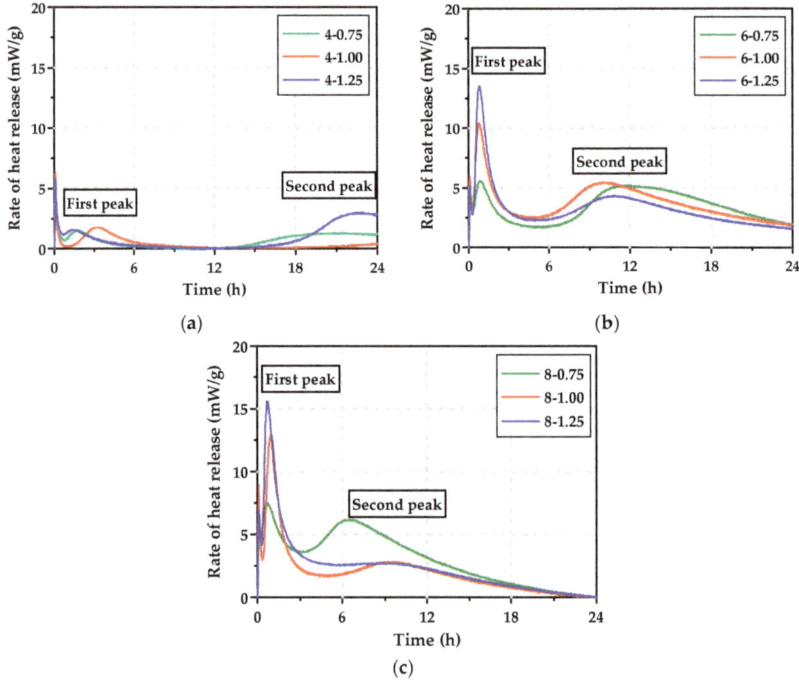

Figure 4. Rate of heat release curves of AAS pastes for 24 h: (**a**) Na_2O content = 4%, (**b**) Na_2O content = 6%, (**c**) Na_2O content = 8%.

3.2. Workability

Figure 5 shows the mini slump test results for AAS pastes with a l/b of 0.42. The paste without an alkaline activator demonstrated an initial flow of 182 mm, while the pastes with alkaline activators showed an initial flow of 205–350 mm. All of the AAS pastes with alkaline activators of 6% and 8% Na_2O content lost their flowability within 15 min. The flowability decreased more rapidly as Ms increased; specifically, for the 8-1.25 paste, with the highest Na_2O content and Ms, hardening occurred just after 5 min after mixing. As mentioned in Section 3.1, this is because the activation and reaction of GGBFS by the alkali activators occurred rapidly at the early ages. Once the AAS paste lost its flowability, the pastes began to harden quickly. These tendencies appeared as Ms increased because more reaction products of the GGBFS were produced by the sufficient supply of SiO_2 [14].

The AAS paste with alkaline activators of 4% Na_2O content maintained a flow of >200 mm for 1 h after mixing. Similarly, Puertas et al. [29] reported that the analysis of the flow performance of AAS pastes with alkaline activators of less than 5% Na_2O content showed that when Ms was lower than 0.8, the flowability of the paste was maintained for 60 min.

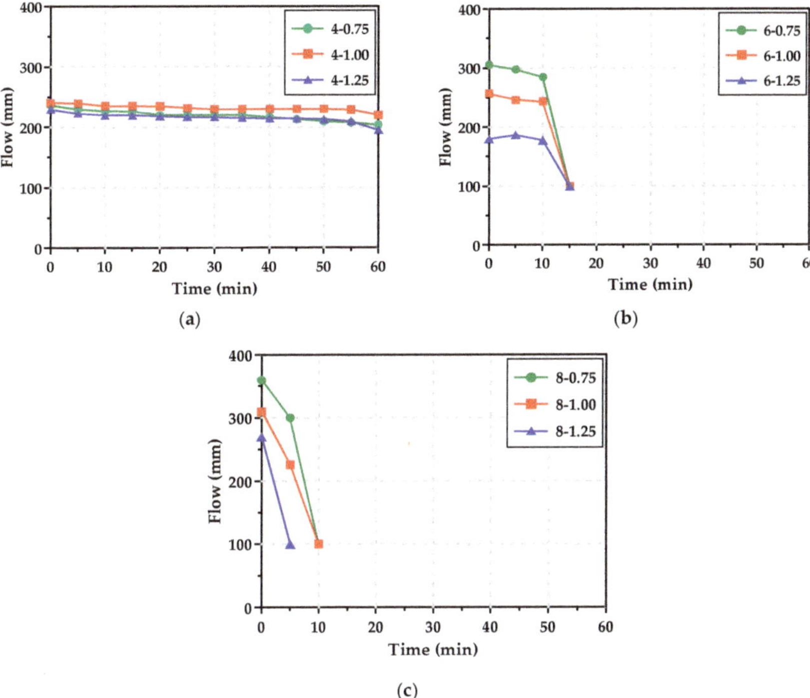

Figure 5. Mini slump curves of AAS pastes: (**a**) Na_2O content = 4%, (**b**) Na_2O content = 6%, (**c**) Na_2O content = 8%.

The mini slump test results of the AAS pastes with l/b of 0.45 or 0.48 are shown in Figure 6. The AAS pastes with l/b of 0.45 and 0.48 lost flowability at 20 min and 25 min, respectively, while the AAS pastes with l/b of 0.42 lost flowability within 15 min. When l/b increased from 0.42 to 0.48, flowability was enhanced by just 10 min.

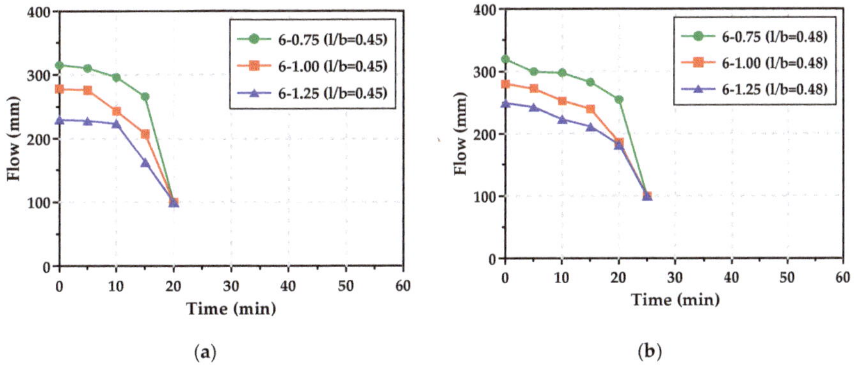

Figure 6. Mini slump curves of AAS pastes: (**a**) l/b = 0.45, (**b**) l/b = 0.48.

It was also observed that the higher the Ms, the lower the initial flow of the AAS paste. This is because as the Ms increased, workability was reduced due to the increase of the reaction products of the GGBFS [14,30,31]. However, the times at which the AAS pastes lost their workability were nearly identical regardless of the Ms.

3.3. Setting Time

Table 4 shows the Vicat test results of the AAS pastes on the initial and final setting time. The setting of the AAS paste was accelerated by the increase of activation and reaction of the GGBFS when the dosage of alkaline increased. Similarly, Chang [12] reported that the setting time of the AAS pastes correlated with the sum of the Na_2O (from NaOH and Na_2SiO_3) and SiO_2 (from Na_2SiO_3) concentrations of the alkali activator. For AAS pastes with less than 8 g of alkaline (per 100 g of GGBFS), setting time significantly increased as the dosage of alkaline was low. For AAS pastes with more than 10 g of dosage of alkaline, setting was accelerated; initial setting occurred within 1 h, as shown in Figure 7. When the dosage of alkaline increased to more than 14 g, final setting occurred within 30 min.

Table 4. Vicat test results of AAS pastes.

Mixture	Initial Setting Time (min)	Final Setting Time (min)
4-0.75	280	355
4-1.00	65	112
4-1.25	33	53
6-0.75	46	60
6-1.00	34	45
6-1.25	24	40
8-0.75	19	28
8-1.00	17	27
8-1.25	16	25

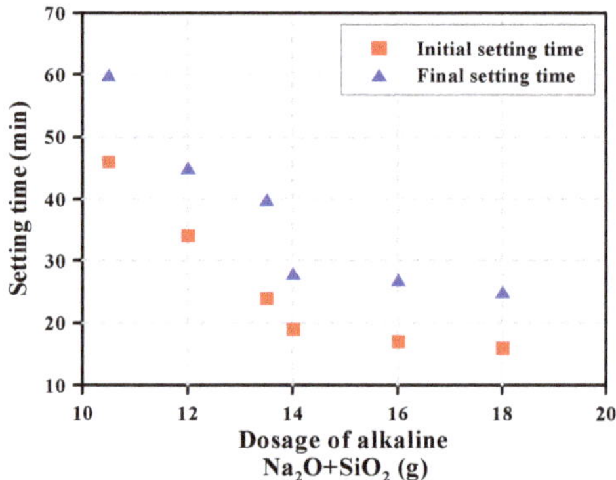

Figure 7. Setting time of AAS paste by the dosage of alkaline.

3.4. Viscosity Measurement Limit and Ultrasonic Pulse Velocity

Figure 8 shows the rheology test results of the AAS pastes with Na_2O content and Ms. The viscosity measurement limit (VML) refers to the point at which flowability is lost in the fluid state. As described in Section 3.2, the flow of the paste showed a tendency for the flowability to decrease abruptly immediately before the loss of flowability, and the shear stress rapidly increased 3 to 5 min before reaching VML. As shown in Table 4, the min. and max. VML times of the AAS paste with ≥6% Na_2O content were 540 s for 8-1.25 and 1040 s for 6-0.75, respectively. These VMLs were much correlated with the time at which AAS pastes lost their flowability, as shown in Figure 5b,c, since the GGBFS reaction was quickened by tens of minutes after the mixing.

The AAS paste with 4% Na_2O retained its flowability for up to 1 h in the mini slump test, but its VML was only 15–20 min. This was because the AAS paste was remixed for 30 s before the measurement of the flow, while in the rheology test, the shear stress of the AAS paste was measured continuously for 30 min without remixing.

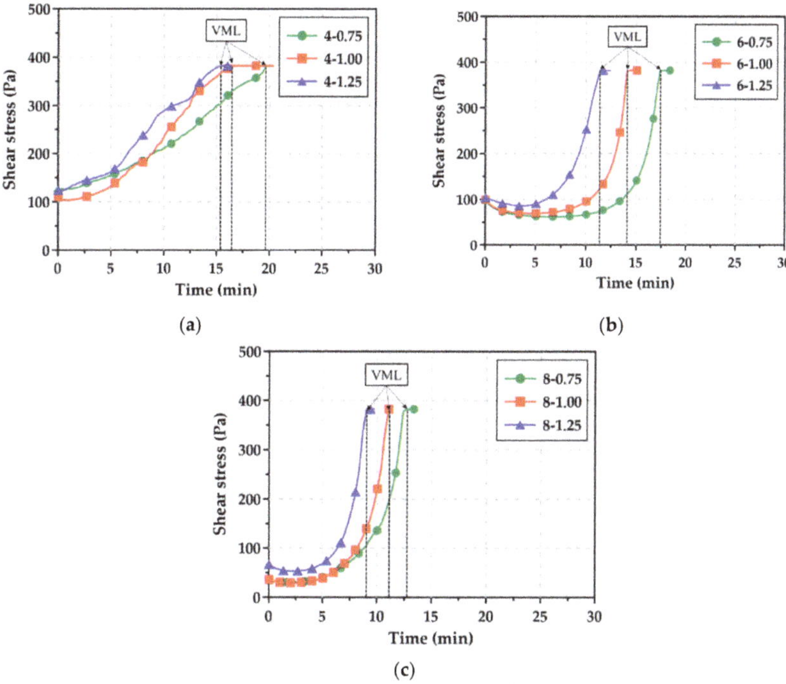

Figure 8. Rheology test curves for AAS pastes: (**a**) Na_2O content = 4%, (**b**) Na_2O content = 6%, (**c**) Na_2O content = 8%.

Figure 9 shows the ultrasonic pulse velocity (UVP) measurements of the AAS pastes; the initial ultrasonic velocity of the paste was 302.5–405.0 m/s regardless of the mixture type [32,33]. This indicates that the paste was in a fluid state. Afterwards, as the paste lost its flowability, the UPV increased sharply, and during certain periods of time before VML was reached, the UPV increased intermittently rather than continuously. Table 5 summarizes the UPVs of the time at which the VML was reached. The UPVs of the AAS pastes increased gradually during the reaction of the GGBFS [34]. The rheology test results showed that when the AAS paste reached the VML, the UPVs were confirmed to be 900–1000 m/s. After the VML, the UPVs increased steadily, and then, the increase rate of UPVs differed in accordance with the Na_2O content and Ms. It is observed in Figure 9 that after the VML, UPVs of the AAS pastes with an Na_2O content of ≥6% and Ms of ≥1.00 rapidly increased, but UPVs of the AAS pastes with 4% Na_2O content or Ms of 0.75 modestly increased.

Table 5. Ultrasonic pulse velocities (UPVs) and viscosity measurement limits (VMLs) of AAS pastes.

Mixture	VML (min:s)	UPV (m/s) at VML
4-0.75	19:40	990.75
4-1.00	16:20	929.75
4-1.25	15:20	939.00
6-0.75	17:20	967.75
6-1.00	14:00	909.50
6-1.25	11:20	911.75
8-0.75	12:20	936.50
8-1.00	11:00	909.50
8-1.25	9:00	972.50

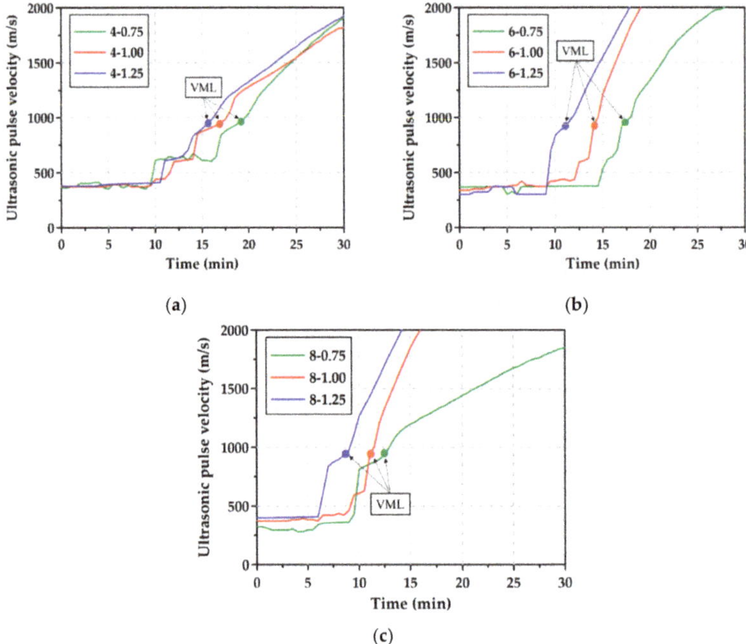

Figure 9. UPV curves for AAS pastes: (**a**) Na_2O content = 4%, (**b**) Na_2O content = 6%, (**c**) Na_2O content = 8%.

3.5. Compressive Strength

Figure 10 shows the compressive strength development of the AAS mortars. The AAS mortars with a 4% Na_2O content had a compressive strength of <1 MPa up to 24 h regardless of Ms, implying that the increase of SiO_2 content did not significantly affect the compressive strength development. This is consistent with the very low heat release for the initial 24 h, as shown in Figure 4a. On the other hand, although the mortar with ≥6% Na_2O showed an increase in strength as time elapsed, the compressive strength differed in accordance with the Ms. When Ms remained constant, the compressive strength development of the AAS mortars with 8% Na_2O was very similar to that of the AAS mortars with a 6% Na_2O. This was in accordance with the results of Zuo et al. [13], who reported that when Na_2O was greater than 6%, the reaction degree did not increase significantly with time, which in turn kept compressive strength relatively constant.

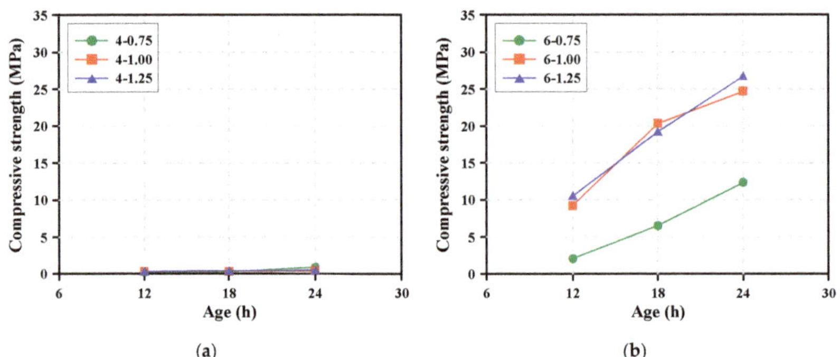

Figure 10. *Cont.*

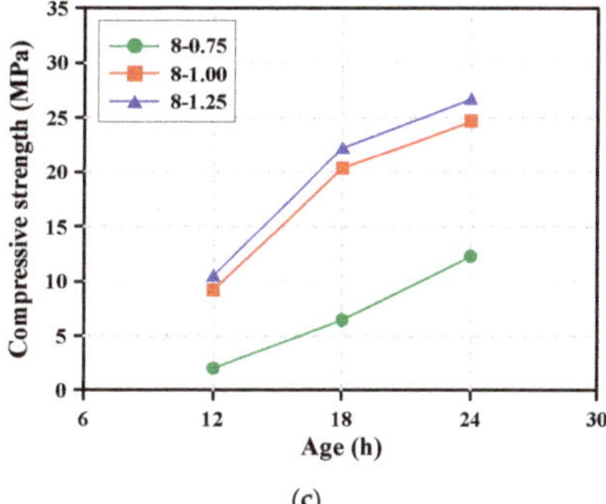

(c)

Figure 10. Compressive strength development of AAS mortar specimens cured in a chamber at 20 °C: (**a**) Na_2O content = 4%, (**b**) Na_2O content = 6%, (**c**) Na_2O content = 8%.

4. Conclusions

The objective of this study was to investigate the effects of Na_2O content and Ms of the alkali activator on the workability and setting of AAS pastes. To evaluate the fresh properties of the pastes, mini slump, Vicat test, and rheology/UPV test were applied. Based on the experimental results, the following conclusions can be drawn:

1. The Na_2O content and Ms strongly affected the rate of heat release, workability, and setting time of alkali-activated pastes. In particular, when alkaline activators with Na_2O content of ≥6% (Ms = 0.75–1.25) were used, the activation and reaction of the GGBFS were enhanced because of the effects of high dosage of alkaline (Na_2O + SiO_2), which resulted in workability rapidly decreasing and setting quickly occurring. However, the Na_2O content and Ms did not significantly affect the compressive strength development of the AAS mortars when the Na_2O content increased from 6% to 8%. On the other hand, for the pastes with Na_2O content = 4%, the rate of heat release was quite low for the initial 24 h and setting was delayed up to 4 h and 40 min, and even the workability was maintained for more than 1 h.
2. In AAS pastes with high Na_2O content, setting occurred within a few minutes after flowability was lost, and thus, it was not easy to measure the flowability with the existing mini slump test. However, since the changes in viscosity of the paste can be continuously monitored over time using the rheology/UPV test, the measurement of VML and UPV could allow the quantitative assessment of the time at which flowability was lost and consequently, to predict the initial setting time of the paste.
3. For the production of AAS concrete with high early strength of 25 MPa at 24 h, it is suggested from the test results that the Na_2O content and Ms of the alkaline activator should be as much as 6% and 1.0, respectively. However, to secure the workability of AAS paste, chemical admixtures such as superplasticizers should be used to adequately control the workability and setting in highly alkaline environments.

Author Contributions: S.C. & K.-M.L. designed the study; S.C. performed the experimental work and analyzed the data; K.-M.L. reviewed and edited the article; S.C. wrote the article.

Funding: This research was supported by a grant (19SCIP-B103706-05) from the Construction Technology Research Program funded by the Ministry of Land, Infrastructure, and Transport of the Korean government.

Conflicts of Interest: The authors declare no conflict of interest.

References

1. Provis, J.L. Alkali-activated materials. *Cem. Concr. Res.* **2018**, *114*, 40–48. [CrossRef]
2. Provis, J.L.; Palomo, A.; Shi, C. Advances in understanding alkali-activated materials. *Cem. Concr. Res.* **2015**, *78*, 110–125. [CrossRef]
3. Shi, C. Early microstructure development of activated lime-fly ash pastes. *Cem. Concr. Res.* **1996**, *26*, 1351–1359. [CrossRef]
4. Poon, C.S.; Kou, S.C.; Lam, L.; Lin, Z.S. Activation of fly ash/cement systems using calcium sulfate anhydrite ($CaSO_4$). *Cem. Concr. Res.* **2001**, *31*, 873–881. [CrossRef]
5. Bellum, R.R.; Nerella, R.; Madduru, S.R.C.; Indukuri, C.S.R. Mix Design and Mechanical Properties of Fly Ash and GGBFS-Synthesized Alkali-Activated Concrete (AAC). *Infrastructures* **2019**, *4*, 20. [CrossRef]
6. Li, C.; Sun, H.; Li, M.L. A Review: The Composition mbetween Alkali -Activated Slag m(Si + Ca) and Metakaolin(Si + Al) Cements. *Cem. Concr. Res.* **2010**, *40*, 341–1349. [CrossRef]
7. Shi, C.; Roy, D.; Krivenko, P. *Alkali-Activated Cements and Concretes*; CRC Press: Boca Raton, FL, USA, 2003.
8. Pacheco, T.F.; Castro, G.J.; Jalali, S. Alkali-activated binders: A review: Part 1. Historical background, terminology, reaction mechanisms and hydration products. *Constr. Build. Mater.* **2008**, *22*, 1305–1314. [CrossRef]
9. Pacheco, T.F.; Castro, G.J.; Jalali, S. Alkali-activated binders: A review. Part 2. About materials and binders manufacture. *Constr. Build. Mater.* **2008**, *22*, 1315–1322. [CrossRef]
10. Mangat, P.S.; Ojedokun, O.O. Influence of curing on pore properties and strength of alkali activated mortars. *Constr. Build. Mater.* **2018**, *188*, 337–348. [CrossRef]
11. Haha, M.B.; Le Saout, G.; Winnefeld, F.; Lothenbach, B. Influence of activator type on hydration kinetics, hydrate assemblage and microstructural development of alkali activated blast-furnace slags. *Cem. Concr. Res.* **2011**, *41*, 301–310. [CrossRef]
12. Chang, J.J. A study on the setting characteristics of sodium silicate-activated slag pastes. *Cem. Concr. Res.* **2003**, *33*, 1005–1011. [CrossRef]
13. Zuo, Y.; Nedeljković, M.; Ye, G. Coupled thermodynamic modelling and experimental study of sodium hydroxide activated slag. *Constr. Build. Mater.* **2008**, *188*, 262–279. [CrossRef]
14. Krizan, D.; Zivanovic, B. Effects of dosage and modulus of water glass on early hydration of alkali-slag cements. *Cem. Concr. Res.* **2002**, *32*, 1181–1188. [CrossRef]
15. Collins, F.G.; Sanjayan, J.G. Workability and mechanical properties of alkali activated slag concrete. *Cem. Concr. Res.* **1999**, *29*, 455–458. [CrossRef]
16. Bakharev, T.; Sanjayan, J.G.; Cheng, Y.B. Sulfate attack on alkali-activated slag concrete. *Cem. Concr. Res.* **2002**, *32*, 211–216. [CrossRef]
17. Neto, A.A.M.; Cincotto, M.A.; Repette, W. Drying and autogenous shrinkage of pastes and mortars with activated slag cement. *Cem. Concr. Res.* **2008**, *38*, 565–574. [CrossRef]
18. Bakharev, T.; Sanjayan, J.G.; Cheng, Y.B. Effect of admixtures on properties of alkali-activated slag concrete. *Cem. Concr. Res.* **2000**, *30*, 1367–1374. [CrossRef]
19. Nedeljković, M.; Li, Z.; Ye, G. Setting, strength, and autogenous shrinkage of alkali-activated fly ash and slag pastes: Effect of slag content. *Materials* **2018**, *11*, 2121. [CrossRef]
20. Wan, H.; Shui, Z.; Lin, Z. Analysis of geometric characteristics of GGBS particles and their influences on cement properties. *Cem. Concr. Res.* **2004**, *34*, 133–137. [CrossRef]
21. Wang, P.; Trettin, R.; Rudert, V. Effect of fineness and particle size distribution of granulated blast-furnace slag on the hydraulic reactivity in cement systems. *Adv. Cem. Res.* **2005**, *17*, 161–167. [CrossRef]
22. ASTM C128-15. *Standard Test Method for Relative Density (Specific Gravity) and Absorption of Fine Aggregate*; ASTM International: West Conshohocken, PA, USA, 2015.
23. ASTM C33-13. *Specification for Concrete Aggregates*; ASTM International: West Conshohocken, PA, USA, 2013.
24. ASTM, C1437-15. *Standard Test Method for Flow of Hydraulic Cement Mortar*; ASTM International: West Conshohocken, PA, USA, 2015.
25. ASTM, C191-18a. *Standard Test Methods for Time of Setting Hydraulic Cement by Vicat Needle*; ASTM International: West Conshohocken, PA, USA, 2018.
26. ASTM C109-16a. *Standard Test Method for Compressive Strength of Hydraulic Cement Mortars (Using 2-in. or [50-mm] Cube Specimens)*; ASTM International: West Conshohocken, PA, USA, 2016.

27. Shi, C.; Day, R.L. A calorimetric study of early hydration of alkali-slag cements. *Cem. Concr. Res.* **1995**, *25*, 1333–1346. [CrossRef]
28. Ravikumar, D.; Neithalath, N. Reaction kinetics in sodium silicate powder and liquid activated slag binders evaluated using isothermal calorimetry. *Thermochim. Acta* **2012**, *546*, 32–43. [CrossRef]
29. Puertas, F.; Varga, C.; Alonso, M.M. Rheology of alkali-activated slag pastes. Effect of the nature and concentration of the activating solution. *Cem. Concr. Compos.* **2014**, *53*, 279–288. [CrossRef]
30. Huanhai, Z.; Xuequan, W.; Zhongzi, X.; Mingshu, T. Kinetic study on hydration of alkali-activated slag. *Cem. Concr. Res.* **1993**, *23*, 1253–1258. [CrossRef]
31. Altan, E.; Erdoğan, S.T. Alkali activation of a slag at ambient and elevated temperatures. Cement and Concrete Composites. *Cem. Concr. Compos.* **2012**, *34*, 131–139. [CrossRef]
32. Wei, S.; Yunsheng, Z.; Jones, M.R. Using the ultrasonic wave transmission method to study the setting behavior of foamed concrete. *Constr. Build. Mater.* **2014**, *51*, 62–74. [CrossRef]
33. Zhang, S.; Zhang, Y.; Li, Z. Ultrasonic monitoring of setting and hardening of slag blended cement under different curing temperatures by using embedded piezoelectric transducers. *Constr. Build. Mater.* **2018**, *159*, 553–560. [CrossRef]
34. Lee, H.K.; Lee, K.M.; Kim, Y.H.; Yim, H.; Bae, D.B. Ultrasonic in-situ monitoring of setting process of high-performance concrete. *Cem. Concr. Res.* **2004**, *34*, 631–640. [CrossRef]

 materials

Article

The Influence of Steel Fiber Tensile Strengths and Aspect Ratios on the Fracture Properties of High-Strength Concrete

Won-Chang Choi [1], Kwon-Young Jung [2], Seok-Joon Jang [3] and Hyun-Do Yun [4,*]

1. Department of Architectural Engineering, Gachon University, Gyeonggi-do 461-701, Korea
2. Korea Land & Housing Corporation, Gyeongsangnam-do 52852, Korea
3. Korea Infrastructure Safety Corporation, Gyeongsangnam-do 52856, Korea
4. Department of Architectural Engineering, Chungnam National University, Daejeon 305-764, Korea

* Correspondence: wiseroad@cnu.ac.kr; Tel.: +82-42-821-5622

Received: 11 June 2019; Accepted: 28 June 2019; Published: 30 June 2019

Abstract: Steel fiber embedded in concrete serves to reduce crack development and prevent crack growth at the macroscopic level of the concrete matrix. Steel fiber-reinforced concrete (SFRC) with high compressive concrete strength is affected primarily by the dimensions, shape, content, aspect ratio, and tensile strength of the embedded steel fiber. In this study, double-ended hook steel fiber was used in SFRC with a concrete compressive strength of 80 MPa. This fiber was used for the study variables with two aspect ratios (64, 80) and tensile strength values up to 1600 MPa. The flexural performance of the SFRC specimens was evaluated using crack mouth open displacement tests, and the test results were compared with code provisions. A modified reinforcement index was also used to quantify the flexural performance based on comparisons with fracture energy.

Keywords: steel fiber; fiber content; aspect ratio; toughness index; high-strength concrete

1. Introduction

The addition of steel fiber into concrete mixtures mitigates brittle failures in the concrete matrix. Specifically, the bridging effect of the steel fiber in the concrete mixture, which occurs after cracking, enhances the mixture's structural behavior in terms of shear strength, flexural strength, ductility, impact resistance, and fatigue [1–3]. The distribution of stress throughout steel fiber-reinforced concrete (SFRC) after cracks appear causes the cracks to widen significantly at the macroscopic level of the concrete matrix so that local crack propagation can be controlled. However, an excessive amount of added steel fiber leads to a reduction of the workability and segregation of the mix [4]. Experimental studies found in the literature [5] are limited to steel fiber aspect ratios of 70 and tensile strength values of 1000 MPa. Therefore, for test purposes, fiber content alone as a test variable is not sufficient. The fiber aspect ratio and tensile strength must also be considered. A wider range of design variables, including various aspect ratios and high tensile strengths, is needed.

Gao et al. (1997) [6] investigated the flexural behavior of high-strength concrete with a range of fiber contents (0%–2.0%) and aspect ratios (46–70). As the fiber content and aspect ratio were increased in the Gao et al. study, the flexural strength improved by 9.6% to 90%, and the experimental results indicated that a fiber content of 1.0% to 1.5% is needed to enhance the flexural behavior of high-strength concrete. Yazıcı et al. (2007) [7] derived similar results to Gao et al. (1997) [6], who showed that the flexural strength increased from 3% to 81% with an increase in the steel fiber content and aspect ratio in SFRC mixtures. The mechanical properties of SFRC with the compressive concrete strength of 50 MPa and variables of different steel fiber contents (0%–1.5%) and aspect ratios (45, 65, 80).

Köksall et al. (2008) [8] also evaluated the flexural performance of concrete with compressive strength values of 40 MPa to 70 MPa and two different fiber aspect ratios (65, 80), as well as fiber contents (0.5%, 1.0%). As in the earlier studies by Gao et al. (1997) [6] and Yazıcı et al. (2007) [7], the flexural strength increased with an increase in fiber content and aspect ratio. However, the flexural strength decreased with an increase in the compressive strength of the concrete. The result was a higher bond strength of the concrete matrix than the tensile strength of the fiber, which caused the fibers in the concrete mixture to rupture. Therefore, Köksall et al. (2012) [9] continued to study the mechanical properties of SFRC with respect to concrete compressive strength and fiber tensile strength. They found that variations of the mechanical properties of SFRC were insignificant when using normal concrete. However, the mechanical properties were highly affected in SFRC with high-strength concrete.

In short, the high tensile strength of the steel fiber used in high-strength SFRC contributes to the mitigation of cracking and to energy dissipation. Related research conducted by the author indicates the necessity for high-performance steel fiber to improve the toughness and flexural strength of high-strength concrete [10]. Studies associated with the mechanical characteristics of high-strength SFRC, which focus on fiber content, aspect ratio, and tensile strength, are needed to ensure good performance without loss of workability due to the excessive fiber content in high-strength SFRC. In this study, we investigated the flexural performance of SFRC by measuring crack mouth opening displacement (CMOD).

2. Experimental Program

2.1. Material Preparation and Fabrication

In this study, the design compressive concrete strength (f_{ck}) was 80 MPa, with a water-to-binder ratio of 25% and steel fiber contents of 0.5% and 1.0% (percentage by volume) as variables. Table 1 shows the mix design of the SFRC used in this study. Typical Portland cement Type I (Density: 3.15 g/cm^3), Crushed aggregate, Fly-ash Density: 2.12g/cm^3, Blaine: 2976 cm^2/g)

Table 1. Mix design of study steel fiber-reinforced concrete.

f_{ck} (MPa)	W/B (%)	Air (%)	Unit Weight (kg/m^3)						
			W	C	SF	FA	S	G	Steel Fiber
80	25	4	165	462	66	132	643	813	0
									39.3
									78.5

Notes: f_{ck}: compressive concrete strength, W/B: water-to-binder ratio, W: water, C: Portland cement Type I (Density: 3.15 g/cm^3), SF: silica fume, FA: fly ash (2.12g/cm^3, Blaine: 2,976 cm^2/g), S: sand (Density: 2.56 g/cm^3, Absorption Ratio: 1.18%), G: gravel (Density: 2.65 g/cm^3).

Figure 1 presents the different dimensions of the double-ended hook steel fiber used in this study. The literature has shown that double-ended hook steel fiber provides excellent performance [11]. In this study, the main variables were the steel fiber aspect ratios and tensile strength values. For the specimens with aspect ratios of 64, two tensile strengths of 1200 MPa and 1600 MPa were used, and for the specimens with the aspect ratio of 80, two tensile strengths of 1000 MPa and 1600 MPa were used.

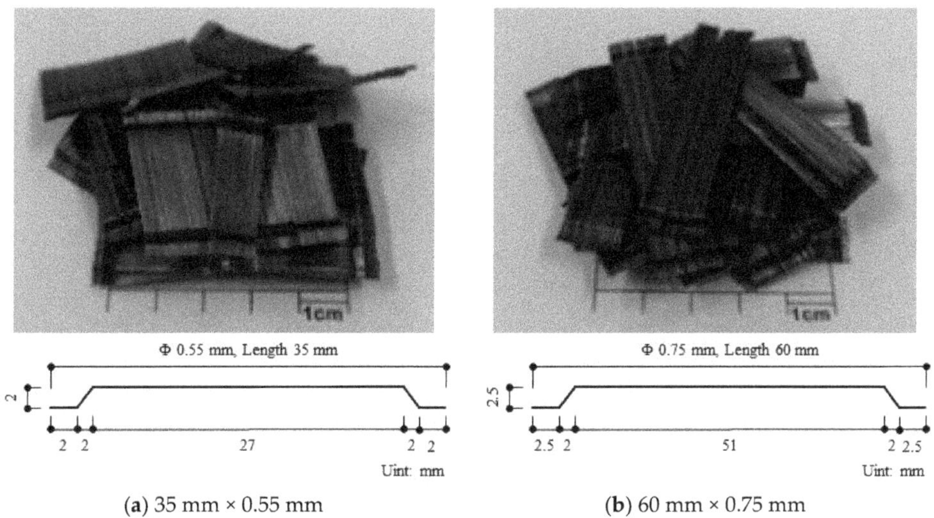

Figure 1. Fibers used in this study.

In order to evaluate the effects of the properties of the steel fiber on the flexural behavior of the SFRC, nine beam specimens, 150 mm × 150 mm × 550 mm, were prepared according to the European Standard EN-14651 standard [12]. These flexural test specimens were cured in a steel mold for 24 h and then, after removal from the mold, were cured in water at 20 °C ± 1 °C until testing. Prior to testing, the specimens were also sawed to create a 5 mm wide and 25 mm deep notch on the bottom of each specimen, as shown in Figure 2.

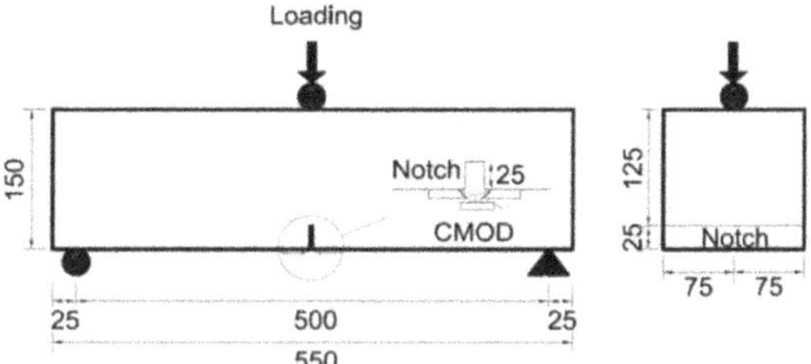

Figure 2. Dimensions of the beam specimen (unit: mm).

The specimen designations used in this study reflect the aspect ratio of the steel fiber, the tensile strength of the steel fiber, and the steel fiber content (its volume fraction). For example, 64-NTS-0.5 indicates that the specimen has a fiber aspect ratio of 64, a normal tensile strength (NTS) (1200 MPa), and 0.5% fiber.

2.2. Test Set-Up

Flexural tests were conducted using a 200 kN universal testing machine (Heungjin, Kimpo, Republic of Korea). Figure 3 shows the test set-up for a CMOD specimen. Center-point loading was applied with a displacement rate of 0.3 mm/min. LVDTs (linear variable differential transformers) were installed to measure the CMOD.

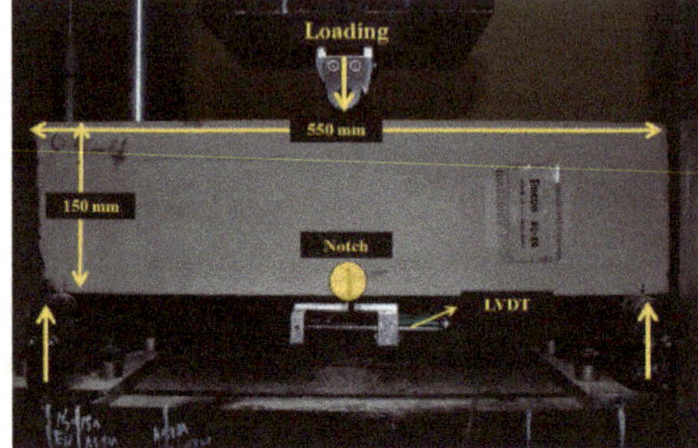

Figure 3. Test set-up for crack mouth opening displacement (CMOD) tests of steel fiber-reinforced concrete (SFRC) specimens.

3. Results and Discussion

3.1. Flexural Stress CMOD Curves for SFRC Specimens

Figure 4 presents the flexural stress-CMOD curves of the SFRC mixtures with respect to their different steel fiber contents and the tensile strengths of the steel fiber.

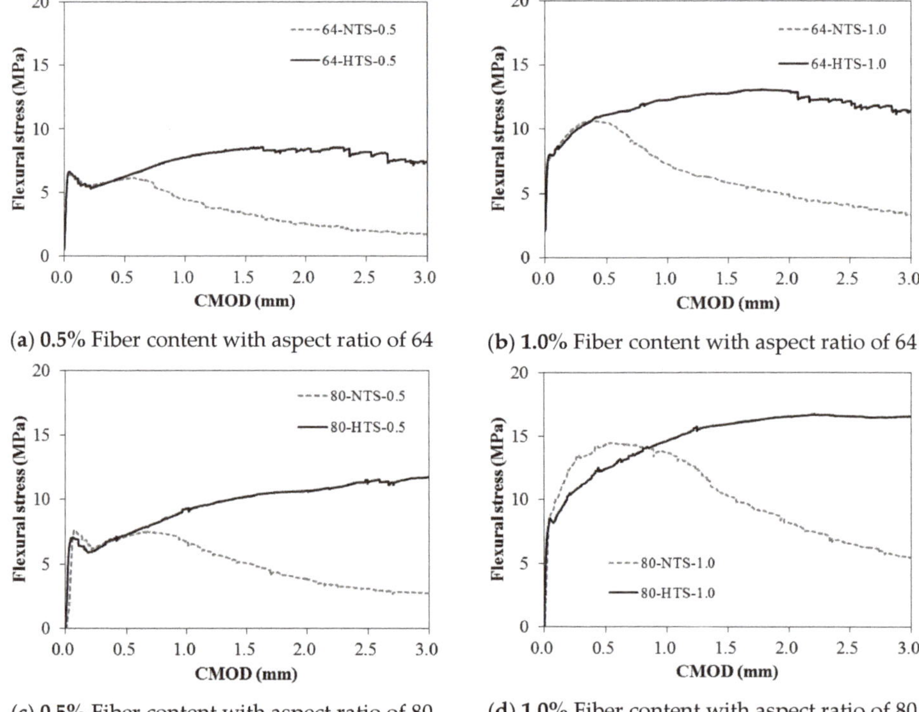

Figure 4. Flexural stress versus CMOD curves for high-strength SFRC specimens with different fiber contents and aspect ratios.

The flexural performance of the SFRC was evaluated in accordance with the EN-14651 standard [12]. The bending stress in the linear region and residual bending stress were calculated using Equations (1) and (2), respectively:

$$f_{LOP} = \frac{3F_{LOP}L}{2bh_{sp}{}^2} \quad (1)$$

$$f_{Rj} = \frac{3F_jL}{2bh_{sp}{}^2} \quad (2)$$

where f_{LOP} is the flexural stress that corresponds to the LOP (N/mm^2), F_{LOP} is the load that corresponds to the limit of proportionality (LOP) (N), L is the span length (mm), b is the width of the specimen (mm), h_{sp} is the distance between the the lowest point of the notch and the top of the specimen (mm), $f_{R,j}$ is the residual flexural tensile strength that corresponds to the specified CMOD, and F_j is the flexural load that corresponds to the specified CMOD.

Figure 5 shows typical load-CMOD curves according to EN-14651 standard. F_{LOP} can be defined in two ways, as represented by Case 1 and Case 2 in the figure. The flexural load F_{LOP} can be determined at the CMOD at 0.05 mm (Case 1 in the figure) or by taking the highest load values in a range within 0.05 mm (Case 2).

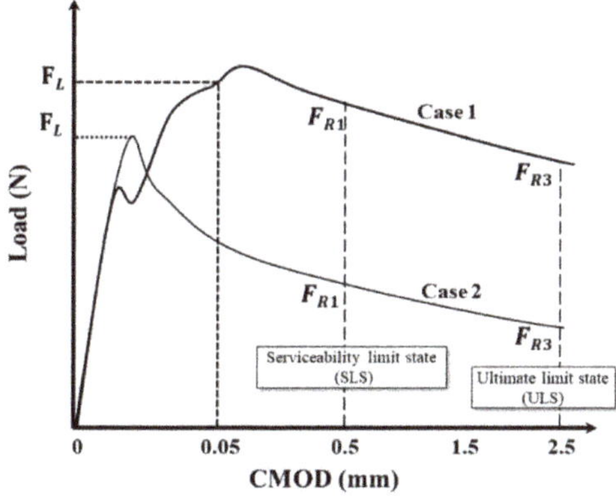

Figure 5. Typical load-CMOD curves (EN-14651).

Table 2 summarizes the test results of the CMOD curves presented in Figure 4.

Table 2. Test Results for Flexural Performance of High-Strength SFRC Mixtures.

Specimen	f_{LOP} (MPa)	f_R (MPa)	f_{R1} (MPa)	f_{R3} (MPa)	f_{R1}/f_{LOP}	f_{R3}/f_{R1}	G_f (N/m)
CON	7.02 (±0.29)	7.02 (±0.29)	-	-	-	-	-
64-NTS-0.5	6.43 (±1.30)	6.87 (±0.57)	6.07 (±3.64)	2.05 (±1.13)	0.94 (±0.18)	0.34 (±0.01)	1950.70 (±370.87)
64-NTS-1.0	7.83 (±0.50)	10.69 (±0.28)	10.50 (±0.58)	4.09 (±2.18)	1.34 (±0.07)	0.39 (±0.07)	4651.73 (±1088.41)
64-HTS-0.5	6.57 (±0.30)	8.89 (±0.62)	6.26 (±1.56)	8.18 (±1.79)	0.96 (±0.10)	1.31 (±0.06)	10,317.59 (±618.37)
64-HTS-1.0	7.98 (±2.18)	13.06 (±1.53)	11.09 (±3.82)	12.18 (±2.40)	1.39 (±0.19)	1.10 (±0.01)	13,673.07 (±1127.82)

Table 2. *Cont.*

Specimen	f_{LOP} (MPa)	f_R (MPa)	f_{R1} (MPa)	f_{R3} (MPa)	f_{R1}/f_{LOP}	f_{R3}/f_{R1}	G_f (N/m)
80-NTS-0.5	7.86 (±0.16)	8.20 (±0.08)	7.13 (±1.22)	3.12 (±0.06)	0.88 (±0.04)	0.44 (±0.03)	4372.64 (±74.61)
80-NTS-1.0	8.81 (±0.77)	14.50 (±0.49)	14.36 (±1.88)	6.55 (±2.63)	1.63 (±0.10)	0.45 (±0.04)	6755.17 (±344.22)
80-HTS-0.5	7.00 (±0.41)	11.85 (±0.41)	7.27 (±0.15)	11.30 (±0.61)	1.04 (±0.08)	1.56 (±0.12)	23,765.02 (±1041.55)
80-HTS-1.0	8.39 (±0.30)	16.86 (±0.39)	12.46 (±0.13)	16.55 (±0.53)	1.49 (±0.04)	1.33 (±0.03)	32,813.33 (±994.72)

Note: CON: control specimen; f_{LOP} is the flexural stress that corresponds to the LOP (N/mm^2).

Strain-softening characteristics were observed for the 64-NTS-0.5 after initial cracking. However, Strain-hardening was shown to occur continuously for the 64-NTS-1.0 specimen with an increase in CMOD after initial cracking. Strain-hardening characteristics were also observed for both 64-HTS-0.5 and 64-HTS-1.0. Furthermore, no abrupt load drop occurred after the specimens reached their maximum flexural strengths.

Depending on the fiber content, the flexural behavior of the specimens with an aspect ratio of 80 and a tensile strength of 1100 MPa was similar to that of specimens with an aspect ratio of 64 and a tensile strength of 1200 MPa. At the fiber content of 0.5%, the load decreased as the CMOD increased after initial cracking. At a fiber content of 1.0%, the load increased continuously after the initial crack and then decreased. The specimens with aspect ratio of 80 and tensile strength of 1600 MPa showed the best flexural behavior in this study. The specimens with 0.5% and 1.0% fiber contents maintained over 3 mm CMOD with no decrease in flexural loading.

The maximum flexural stress of the specimens with 0.5% steel fiber and a tensile strength of 1600 MPa was 20.2%–63.1% lower than that of the other specimens with a mix ratio of 1.0 percent fiber. However, a high level of residual bending stress remained after initial cracking. Therefore, the use of steel fiber with a high tensile strength, and a high aspect ratio is expected to provide excellent flexural behavior for high-strength concrete, even if the concrete has a low fiber content.

Figure 6 shows the failure surface of the CMOD specimens. Two failure modes were observed with the pull-out and/or rupture of fibers. For the specimen 80-HTS-1.0, when increasing the fiber content and tensile strength of the fiber, the majority of the fiber was pulled out from the matrix.

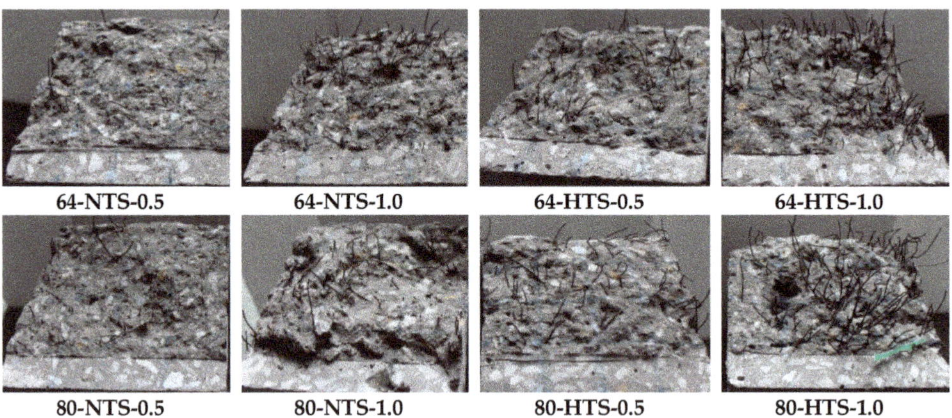

Figure 6. Failure modes of steel fibers in the specimens.

3.2. Evaluation of Residual Flexural Strength of SFRC

The flexural performance of the SFRC specimens was evaluated in accordance with EN-14561 in the International Federation for Structural Concrete (fib) model code 2010 (fib MC 2010) [13] for this study. The steel fiber can be substituted for tensile reinforcement for all and/or partial reinforcement when Equations (3) and (4) are satisfied:

$$\frac{f_{R1}}{f_{LOP}} \geq 0.4 \quad (3)$$

$$\frac{f_{R3}}{f_{R1}} \geq 0.5 \quad (4)$$

where f_{R1} is the residual flexural tensile strength in the serviceability limit state that corresponds to the CMOD at 0.5 mm (N/mm^2), and f_{R3} is the ultimate residual flexural tensile strength that corresponds to the CMOD at 2.5 mm.

The 80-NTS-1.0 specimen has the highest f_{R1} value with 14.36 MPa, and the 80-HTS-1.0 specimen has the highest f_{R3} value with 16.55 MPa. The fiber aspect ratio affects the residual flexural tensile stress up to the service load stage, and then the tensile strength of the steel fiber becomes the dominant parameter to determine the residual flexural tensile stress at the ultimate loading stage.

Even if the number of crosslinking fibers is increased as the crack propagates, the fibers are pulled out or broken when the fiber tensile strength is low. However, the crack will continue to develop without failure when high tensile strength fiber is used.

Figure 7 shows the effects of fiber content on the residual flexural performance of the SFRC specimens for the evaluation that includes the experimental results obtained from this study compared to those found in the literature. Concrete with strength values higher than 60 MPa was considered a high-strength concrete. For the conventional normal-strength concrete, some specimens with low fiber content (0.25 percent) did not satisfy Equation (3), and for the high-strength concrete, some specimens with 1.0% fiber did not satisfy Equation (4). The fiber tensile strength is lower than the adhesion force between the materials, even if the fiber content is high, so the flexural performance is limited due to the rupture of the fibers.

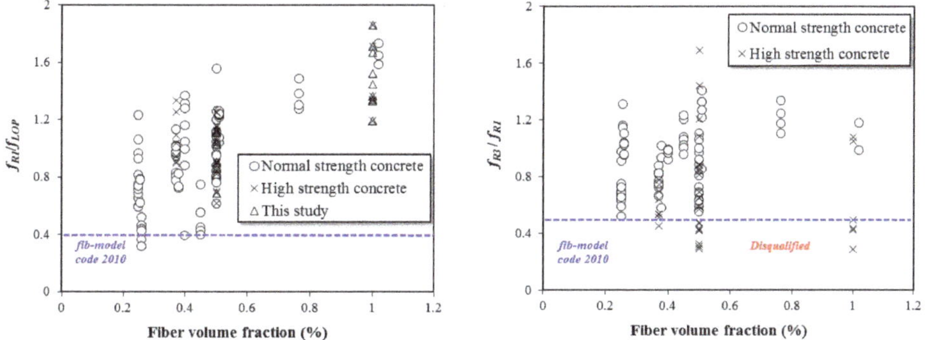

Figure 7. Effects of steel fiber content on the residual flexural performance of SFRC specimens.

Figure 8 shows the effects of the fiber aspect ratio on the flexural performance of the SFRC specimens. In the literature, the fiber aspect ratios range from 50 to 95. The specimens having fiber with an aspect ratio of 50 did not satisfy the criteria according to Equation (3), and the specimens with fiber aspect ratios of 60 to 80 did not satisfy Equation (4).

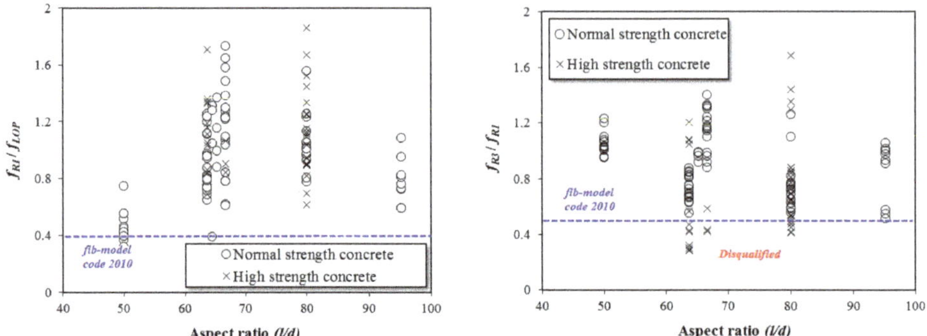

Figure 8. Effects of steel fiber aspect ratio on the residual flexural performance of SFRC specimens.

Figure 9 presents the effects of the tensile strength of the steel fiber on the residual flexural performance of the SFRC specimens. In the literature, tensile strength values of steel fiber range from 1000 MPa to 1200 MPa, and only a limited number of studies have used high-strength steel fiber with over 1600 MPa. In this study, when the tensile strength of the steel fiber was below 1200 MPa, many specimens were disqualified based on Equations (3) and (4). However, all the specimens that contained high tensile strength steel fiber satisfied both Equations (3) and (4). For the specimens with steel fiber tensile strength values below 1200 MPa, the ratio f_{R3}/f_{R1} tended to decrease compared to f_{R1}/f_{LOP}.

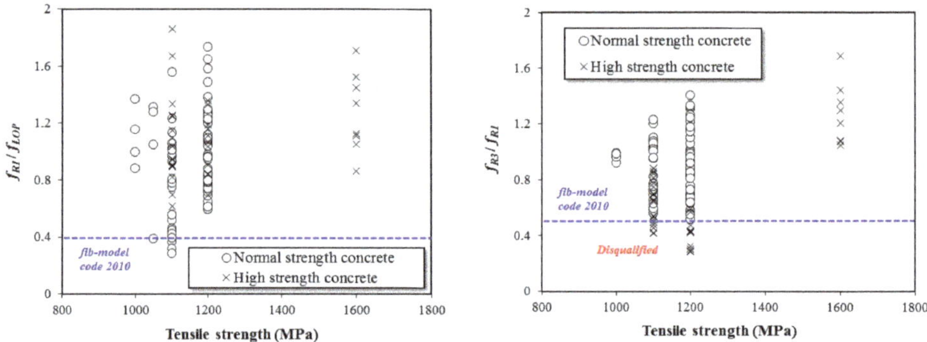

Figure 9. Effects of steel fiber tensile strength on the flexural performance of high-strength SFRC specimens.

Figure 10 presents the flexural performance of the SFRC specimens according to the concrete's compressive strength and fiber content. In the case of the compressive strength of concrete at 40 MPa, most of the specimens are shown to satisfy Equation (3), except for the specimens containing 0.25% steel fiber. The f_{R1}/f_{LOP} of the SFRC in Figure 10a tends to increase with an increase in compressive strength. However, as the compressive strength increases, Equation (4) cannot be satisfied by some of the specimens, and the f_{R3}/f_{R1} value of the SFRC in Figure 10b tends to decrease.

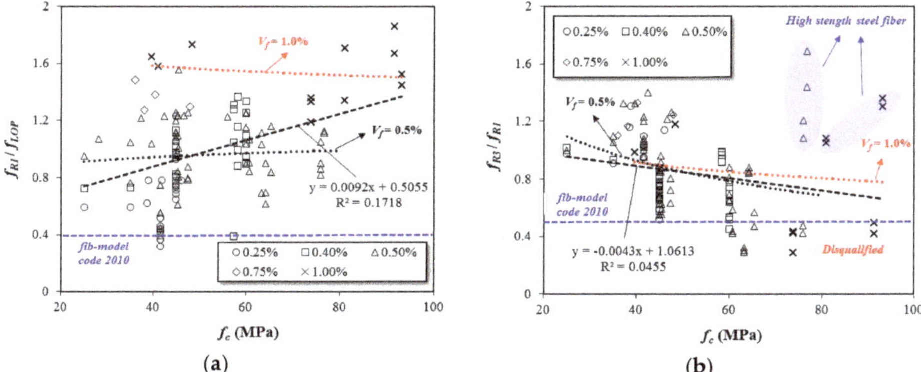

Figure 10. Effects of compressive strength on flexural performance of SFRC specimens, (**a**): Performance index of f_{R1}/f_{LOP} versus compressive strength of concrete (**b**): Performance index of f_{R3}/f_{R1} versus compressive strength of concrete.

The experimental results indicate that all the test specimens satisfy the f_{R1}/f_{LOP} value regardless of the tensile strength of the steel fiber. However, the f_{R3}/f_{R1} values range from 0.34 to 1.56, which do not satisfy the criteria, except for the specimens with a fiber tensile strength of 1600 MPa. Therefore, the results indicate that steel fiber with a tensile strength of 1600 MPa can be used as a feasible substitute for tensile reinforcement in high-strength concrete.

In the literature, a reinforcing index (RI) has been used to quantify the flexural performance of SFRC. The RI is computed by multiplying the fiber content (V_f) and aspect ratio (l/d). In this study, the tensile strength of steel fiber also was determined to be an important parameter for the flexural performance of SFRC. Therefore, a modified RI that includes the tensile strength of steel fiber is proposed in this study. The modified RI value is determined by multiplying $V_f \cdot (l/d) \cdot f_{ft}/100$.

Figure 11 shows the relationship between the flexural performance of the SFRC specimens and modified RI values. The SFRC performance index f_{R1}/f_{LOP} in Figure 11a tends to increase with the newly proposed RI value, but the increment in the computed value (f_{R3}/f_{R1}) in Figure 11b is reduced and tends to be irregular.

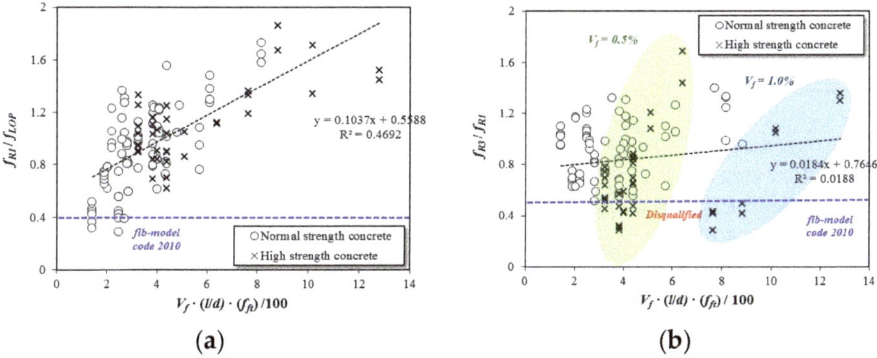

Figure 11. Reinforcing index versus flexural performance values. (**a**): Performance index of f_{R1}/f_{LOP} versus modified RI. (**b**): Performance index of f_{R3}/f_{R1} versus modified RI.

3.3. Determination of Fracture Energy of SFRC

The fracture energy of high-strength concrete can be computed using Equations (5) and (6), as proposed by the Japan Concrete Institute (JCI-S-001-2003) [14]:

$$G(f) = \frac{0.75\ W_0 + W_1}{A_{lig}} \tag{5}$$

$$W_1 = 0.75\left(\frac{S}{L}m_1\right)g \cdot CMOD_c \tag{6}$$

where $G(f)$ is the fracture energy (kN/m), W_1 is the work done by the applied load and self-weight of the specimen (kN·m), W_0 is the energy represented by the area of the load-CMOD curves (kN·m), S is the span length (mm), L is the overall length of the specimen (mm), m_1 is the mass of the specimen (kg), g is the acceleration due to gravity (m/s^2), $CMODc$ is the CMOD value at failure (mm), and A_{lig} is the cross-sectional area of the specimen (mm^2).

Fracture energy refers to the energy at which the material causes a fracture. Köksall et al. (2012) evaluated the fracture energy of high-strength concrete according to steel fiber content and tensile strength. Figure 12 shows the relationship between the fracture energy of the SFRC specimens used in this study and the modified RI in terms of steel fiber characteristics and fiber content. The results indicate that the fracture energy increases with an increase in fiber content and tensile strength.

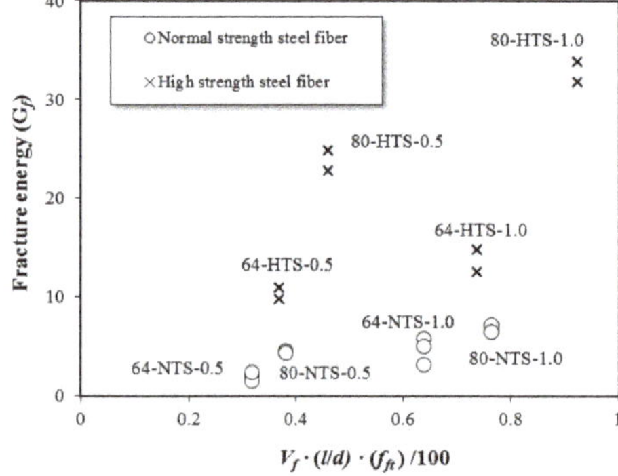

Figure 12. Relationship between fracture energy and the modified reinforcing index.

As shown, the fracture energy increased by 32.5%~138.5%, with an increase in fiber content. As the fiber tensile strength and aspect ratio increased, the fracture energy increased by 194.0%~443.5% and 45.2%~140.0%, respectively. The fracture energy was influenced predominantly by the fiber's tensile strength. In short, as described for the residual bending stress and flexural behavior evaluations, when the tensile strength of the steel fiber is sufficient, the steel fiber does not break, even if cracking occurs. The steel fiber provides high fracture energy until failure due to its continuous bridging effect.

4. Conclusions

This study evaluated the effects of the aspect ratio and mechanical properties of steel fiber on the flexural performance of high-strength concrete in accordance with the EN-14651 standard and JCI code. The conclusions are based on limited experimental results and are summarized as follows.

The flexural behavior in LOP was found to be influenced by the fiber content and aspect ratio. The maximum value was 8.81 MPa for the 80-NTS-1.0 specimen.

In this study, the maximum flexural stress of the SFRC specimens was found to be affected by the fiber properties in the order of fiber content, tensile strength, and aspect ratio. The maximum flexural stress was the highest at 16.86 MPa in specimens with an aspect ratio of 80 and a tensile strength of 1600 MPa with 1.0% steel fiber.

The tensile strength of steel fiber is a dominant parameter that can be used to determine the flexural performance of SFRC. The proposed modified RI indicates that high tensile strength steel fiber can be used in concrete with high compressive strength.

Author Contributions: Formal Analysis, W.-C.C., & S.-J.J.; Data Curation, S.-J.J.; Writing-Original Draft Preparation, W.-C.C.; Writing-Review & Editing, W.-C.C., & K.-Y.J.; Supervision, H.-D.Y.; Project Administration, H.-D.Y.

Funding: This research was funded by National Research Foundation of Korea (2016R1D1A3B02008179).

Conflicts of Interest: The authors declare that there is no conflict of interests regarding the publication of this article.

References

1. Almusallam, T.; Ibrahim, S.; Al-Salloum, Y.; Abadel, A.; Abbas, H. Analytical and experimental investigations on the fracture behavior of hybrid fiber reinforced concrete. *Cem. Concr. Compos.* **2016**, *74*, 201–217. [CrossRef]
2. Lee, J.-H. Influence of concrete strength combined with fiber content in the residual flexural strengths of fiber reinforced concrete. *Compos. Struct.* **2017**, *168*, 216–225. [CrossRef]
3. Yoo, D.-Y.; Yoon, Y.-S.; Banthia, N. Flexural response of steel-fiber-reinforced concrete beams: Effects of strength, fiber content, and strain-rate. *Cem. Concr. Compos.* **2015**, *64*, 84–92. [CrossRef]
4. Koh, K.T.; Kang, S.T.; Park, J.J.; Ryu, G.S. A Study on the Improvement of Workability of High Strength Steel Fiber Reinforced Cementitious Composites. *J. Korea Inst. Struct. Maint. Insp.* **2014**, *8*, 141–148. (In Korean).
5. Caggiano, A.; Folino, P.; Lima, C.; Martinelli, E.; Pepe, M. On the mechanical response of Hybrid Fiber Reinforced Concrete with Recycled and Industrial Steel Fibers. *Constr. Build. Mater.* **2017**, *147*, 286–295. [CrossRef]
6. Gao, J.; Sun, W.; Morino, K. Mechanical properties of steel fiber-reinforced, high-strength, lightweight concrete. *Cem. Concr. Compos.* **1997**, *19*, 307–313. [CrossRef]
7. Yazıcı, Ş.; Inan, G.; Tabak, V. Effect of aspect ratio and volume fraction of steel fiber on the mechanical properties of SFRC. *Constr. Build. Mater.* **2007**, *21*, 1250–1253.
8. Koksal, F.; Altun, F.; Yiğit, I.; Şahin, Y.; Yiğit, I. Combined effect of silica fume and steel fiber on the mechanical properties of high strength concretes. *Constr. Build. Mater.* **2008**, *22*, 1874–1880. [CrossRef]
9. Köksal, F.; Sahin, Y.; Sahin, M. Effect of Steel Fiber Tensile Strength on Mechanical Properties of Steel Fiber Reinforced Concretes. *ACI SP* **2012**, *289*, 129–143.
10. Jang, S.J.; Yun, H.D. Effects of Curing Age and Fiber Volume Fraction on Flexural Behavior of High-Strength Steel Fiber-Reinforced Concrete. *J. Korean Soc. Hazard Mitig.* **2016**, *16*, 15–21. (in Korean). [CrossRef]
11. Soroushian, P.; Bayasi, Z. Fiber type effects on the performance of steel fiber reinforced concrete. *Mater. J.* **1991**, *88*, 129–134.
12. EN 14651. Test Method for Metallic Fibered Concrete–Measuring the Flexural Tensile Strength (Limit of Proportionality (LOP), Residual). 2005. Available online: http://auber.com.vn/upload/download/445669281373.pdf (accessed on 29 June 2019).
13. Du Béton, F.I. Model Code 2010—First Complete Draft. fib Bulletin, (55), 318. 2010. Available online: https://www.fib-international.org/publications/fib-bulletins/model-code-2010-first-complete-draft,-vol-1-pdf-146-detail.html (accessed on 29 June 2019).
14. JCI-S-001-2003. Method of Test for Fracture Energy of Concrete by Use of Notched Beam. 2003. Available online: http://www.jci-net.or.jp/j/jci/study/jci_standard/JCI-S-001-2003-e.pdf (accessed on 29 June 2019).

Article

Effects of a Short Heat Treatment Period on the Pullout Resistance of Shape Memory Alloy Fibers in Mortar

Min Kyoung Kim [1], Dong Joo Kim [1,*], Young-Soo Chung [2] and Eunsoo Choi [3,*]

1 Department of Civil and Environmental Engineering, Sejong University, 2019, Neungdong-ro, Gwangjin-Gu, Seoul 05006, Korea
2 Department of Civil Engineering, Chung-Ang University, 221 Heuksuk-dong, Dongjak-gu, Seoul 156-756, Korea
3 Department of Civil Engineering, Hongik University, 72-1 Sangsu-dong, Mapo, Seoul 121-791, Korea
* Correspondence: djkim75@sejong.ac.kr (D.J.K.); eunsoochoi@hongik.ac.kr (E.C.); Tel.: +82-2-3408-3820 (D.J.K.); +82-2-320-3060 (E.C.); Fax: +82-2-3408-4332 (D.J.K.); +82-2-322-1244 (E.C.)

Received: 15 June 2019; Accepted: 14 July 2019; Published: 16 July 2019

Abstract: The feasibility of the crack closure of cementitious composites reinforced with shape memory alloy (SMA) fibers was investigated by performing single-fiber pullout tests. To demonstrate the fast crack closing ability, in this study, a heat treatment (300 °C) was applied for a short time (10 min). A short heat treatment was applied for 10 min, after the slip reached 0.5 mm, to activate the shape memory effects of cold-drawn SMA fibers. Two types of alloys were investigated, NiTi and NiTiNb, with two geometries, either smooth or dog-bone-shaped. During the heat treatment, the pullout stress of the SMA fibers initially decreased due to thermal extension, and then increased after heating for 1–3 min, resulting from the shape memory effects. However, their pullout stress recovery during and after the heat treatment was different for the different alloys and fiber geometries. The NiTi fibers generally produced a higher and faster recovery in terms of their pullout stress than the NiTiNb fibers, while the dog-bone-shaped fibers showed a faster pullout stress recovery than the smooth fibers.

Keywords: fibers; smart materials; fiber/matrix bond; physical properties; heat treatment

---

1. Introduction

Considerable research has been performed on extending the service life of civil infrastructures through the prevention of early concrete deterioration. As an early deterioration countermeasure, various high-performance construction materials, such as ultra-high-performance concretes (UHPCs), high-performance fiber-reinforced cementitious composites (HPFRCCs), and self-healing concretes (SHCs), have been developed by numerous researchers [1–6]. However, even the high-performance construction materials are unable to close existing concrete cracks in a short time for a quick (or instantaneous) repair of concrete infrastructure. Although both the UHPCs and HPFRCCs have been shown to produce a significantly high tensile strength and cracking control capacity by generating multiple microcracks during tensile strain hardening [1,7], as well as a larger redistribution capacity of stresses [8], they remain unable to close existing cracks. Although the SHCs can heal (or fill) cracks by precipitating crystalline calcium carbonates within the cracks, they require a water supply and a minimum of a few days to heal the cracks [5,9]. Therefore, the aforementioned high-performance construction materials cannot be applied for urgent repairs, even though they have shown a superior prevention of early concrete deterioration.

Due to the issues noted above, the authors of this study propose the development of short shape memory alloy (SMA) fiber-reinforced cement composites with a fast crack closing capacity.

Kim et al. [10,11] have already demonstrated that an 8 h long-term heat treatment of cold-drawn SMA fibers significantly enhanced the pullout resistance of fibers embedded in a mortar matrix due to the diameter recovery in the cold-drawn SMA fibers. However, it is quite difficult to maintain a heat treatment for a long time on site. Kim et al. [12] demonstrated the pre-stressing effect of shape memory alloy fiber-reinforced cementitious composites (SMA-FRCCs) under direct tension after a short (10 min) heat treatment. They reported that the shape memory effect can be activated by short-term heat treatment for 10 min. Consequently, in this study, we investigated whether a short (10 min) heat treatment period could activate the shape memory effect and eventually close existing cracks.

This research is aimed at developing SMA-FRCCs with a crack closing capacity, as well as pre-stressing effects by utilizing the shape memory effect. The detailed objectives are (1) to investigate the geometry change in the SMA fibers during short heating periods, (2) to investigate the heat treatment effects on the pullout resistance of the SMA fibers, and (3) to discover the influence of different SMA fiber geometries and compositions on the pullout resistance recovery during and after the short heat treatment period.

Shape Memory Alloy Fiber-Reinforced Cementitious Composites (SMA-FRCCs)

Considerable research has been performed on the application of SMA wires or bars to civil infrastructures and buildings by activating their shape memory effects and utilizing their super-elastic characteristics. Researchers have reported on the effectiveness of wrapping SMA wires around concrete cylinders or columns to enhance ductility and strength by utilizing the confinement effects induced from shape memory effects. Choi et al. [13] proposed a new jacketing method to confine concrete cylinders and/or reinforced concrete columns using SMA wires, demonstrating that the shape memory effect of SMA wires was considerably more effective than the steel jacketing method for generating confinement effects [14,15]. Tran et al. [16] also reported that the shape memory-induced confinement effect of SMA wires significantly increased the strength and ductility of concrete cylinder samples. Furthermore, SMA wires have also been used to reinforce cement-based matrices in order to generate pre-stressing effects in the cement composites; Sawaguchi et al. [17] applied Fe-Mn-Si-based shape memory alloys containing NbC to generate pre-stressing effects.

In addition to the confinement and pre-stressing effects obtained through SMA wire application, the self-crack closing behavior of composites reinforced with SMA wire has also been investigated. SMA wires added to epoxy or cement mortar were applied to generate a crack closing capability in a beam [18–20]. In order to create the crack closing ability, it is important to have a strong interfacial bond strength between the fiber or wire and the surrounding matrix. Accordingly, Umezaki [18] investigated the pullout resistance of SMA wires, reporting that spiral SMA wires produced a higher pullout resistance than smooth SMA wires. Wang et al. [21] investigated the internal stress distribution to prevent debonding at the SMA fiber and epoxy matrix interface, in order to eventually utilize the shape memory effect more efficiently. Watanabe et al. [22] also reported that the interfacial bond strength of Fe-based SMA wire was notably enhanced as the amount of pre-strain in the SMA wires increased. Kuang and Ou [23] developed a self-healing concrete beam by utilizing the super-elastic behavior of SMAs and the cohering characteristic of the repairing adhesive.

Although several researchers have studied the behavior of SMA wire-reinforced composites in an attempt to utilize their shape memory effect, the majority of their SMA applications have been limited to continuous SMA wires rather than short SMA fibers. The implementation of short SMA fibers has rarely been attempted, and the behavior of short SMA fiber-reinforced cementitious composites is yet to be discovered. In this study, to develop short SMA fiber-reinforced cementitious composites with a self-crack closing ability, the pullout resistance of these fibers embedded in a mortar matrix was systematically investigated by applying a short heat treatment period.

2. Experiments

An experimental program was designed to investigate the feasibility of crack closure of SMA-FRCCs by applying a short heat treatment period (10 min). The pullout stress recovery of SMA fibers embedded in a mortar matrix was investigated during this heat treatment with single-fiber pullout tests. Two types of SMAs (NiTi and NiTiNb) were used, while two types of fiber geometries (smooth and dog-bone-shaped) were investigated experimentally. First, the geometric changes in the diameter and length of cold-drawn SMA fibers during the short heat treatment time were investigated by analyzing stereoscopic microscope images taken every minute of the heat treatment. Second, the pullout stress versus slip (and time) responses of the SMA fibers in a mortar matrix were investigated by performing single-fiber pullout tests. The conditions of the single-fiber pullout tests are provided in Table 1. During the pullout tests, a short heat treatment period was applied while maintaining a constant slip (0.5 mm) for 10 min once the fiber slip reached 0.5 mm. Then, the fiber was pulled further until complete fiber pullout was achieved.

Table 1. Conditions of the single-fiber pullout tests.

Alloy	Fiber Diameter (mm) [a]	Shape	Heat Treatment	Notation
NiTi	1.0 → 0.96	Smooth	No	NT_S_N
			Yes	NT_S_H
		Dog-bone shaped	No	NT_D_N
			Yes	NT_D_H
NiTiNb	1.12 → 1.08	Smooth	No	NTN_S_N
			Yes	NTN_S_H
		Dog-bone shaped	No	NTN_D_N
			Yes	NTN_D_H

[a] → refers to the diameter change after the cold-drawing process.

2.1. Materials and Sample Preparation

The composition and strength of the mortar matrix are provided in Table 2; the compressive strength of the mortar was 55 MPa. The properties of the SMA fibers investigated in this study are summarized in Table 3. Four types of SMA fibers (NT_S, NT_D, NTN_S, and NTN_D, described in Table 1) were used and are shown in Figure 1. The NiTi (NT) alloy contained 50% nickel by atomic composition, whereas the NiTiNb (NTN) alloy contained 41% nickel, 50% titanium, and 9% niobium. The initial 1.0 mm diameter of the NT wires was reduced to 0.96 mm after the cold drawing process.

Table 2. Matrix mixture composition by weight ratio and compressive strength.

Cement (Type 3)	Fly Ash	Silica Sand	High-Range Water-Reducing Admixture	Water	f'_{ck} (MPa)
1.00	0.15	1.00	0.009	0.35	55

Table 3. Properties of the shape memory alloy (SMA) fibers.

Notation	Length (mm)	Diameter (mm)	Young's Modulus (GPa)	Tensile Strength (MPa)	Transformation Temperature (°C) [a]			
					M_f	M_s	A_s	A_f
NT_S NT_D	30	0.96	41	973	0.7	36.7	43.2	62.8
NTN_S NTN_D	30	1.08	21	1270	-	−100≥	−100≤	-

[a] M_f is the temperature at the end of martensitic transformation, M_s is the temperature at the start of martensitic transformation, A_s is the temperature at the start of austenitic transformation, and A_f is the temperature at the end of austenitic transformation.

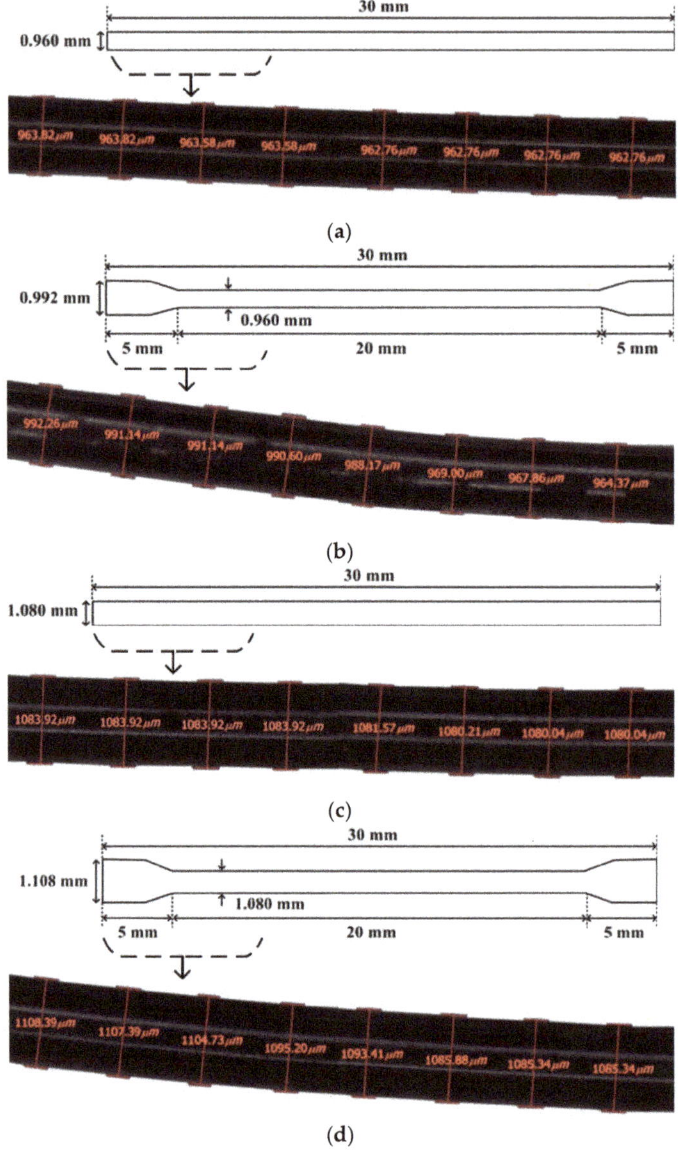

Figure 1. Shape memory alloy (SMA) fiber geometries, for (**a**) NiTi smooth (NT_S), (**b**) NiTi dog-bone-shaped (NT_D), (**c**) NiTiNb smooth (NTN_S), and (**d**) NiTiNb dog-bone-shaped (NTN_D).

In comparison, the NTN wires with a 1.12 mm initial diameter were also cold-drawn to a 1.08 mm reduced diameter. The transformation temperatures of both cold-drawn NT and NTN fibers are summarized in Table 3.

Six samples of each type of SMA fiber were prepared, with three samples pulled out with the heat treatment, whereas the others were tested with no heat treatment. The notation NT_S_N in Table 1 represents the pullout samples of the NT smooth (S) fiber with no (N) heat treatment, while samples of the NTN dog-bone-shaped (D) fiber with the short heat (H) treatment periods were designated as

NTN_D_H. The fibers were first installed in fiber holding devices to maintain the embedment length (15 mm) and inclination angle (90°) of the fiber within a mortar matrix. The devices holding the SMA fibers were then placed in molds to produce bell-shaped pullout samples. The mortar matrix was prepared using a Hobart-type laboratory mixer with a 20 L capacity. A detailed procedure on mixing and curing this type of sample can be found in Kim et al. [10,11]. The pullout samples, after curing in water for 14 days, were dried and tested after 16 days at room temperature in the laboratory.

2.2. Test Setup and Procedure

The length and diameter of the SMA fibers could be measured by analyzing images taken by a stereoscopic microscope with a Huvitz Lusis HC-30MU camera (Huvitz, Gyeonggi-do Province, Korea). A short period of heat treatment was applied using a heat gun (DeWalt D26411) (DeWalt, Towson, MD, USA). The operation temperature of the heat gun was set at 300 °C and the airflow speed was 250 L/min. The heating gun was installed at a 35 mm distance from the SMA fibers during the heat treatment. The heat treatment in the experiment was performed using a commercial heating gun for the practical application, even in structural members. The temperature applied by using a heat gun was about 300 °C, so that it would not generate any significant damage on the cement-based matrix [24]. Therefore, it was applied to raise the temperature of SMA fibers (over A_f) embedded within the mortar matrix. Moreover, heat treatment at 300 °C for 10 min does not significantly affect the tensile strength of SMA-FRCCs (reduced about 10%) [12], unlike SFRCs, which exhibit about a 40% reduction in tensile strength by heating at 300 °C for 2 h prior to testing [25]. During the pullout resistance investigation of the SMA fibers, a universal test machine with a 500 kgf capacity was used and the displacement speed was maintained at 1 mm/min during the fiber pullout. To avoid any slip of fiber in the grip system, the other part of embedded fiber was fully held with a sufficient pressure in the grip, as can be seen in Figure 2. The test procedure is as follows: the fiber was initially pulled out to a 0.5 mm slip with a velocity of 1 mm/min; then, the 0.5 mm slip was kept constant for 10 min; and finally, the fiber was further pulled with the same 1 mm/min velocity until complete pullout was achieved. The test set-up for the single-fiber pullout test is shown in Figure 2.

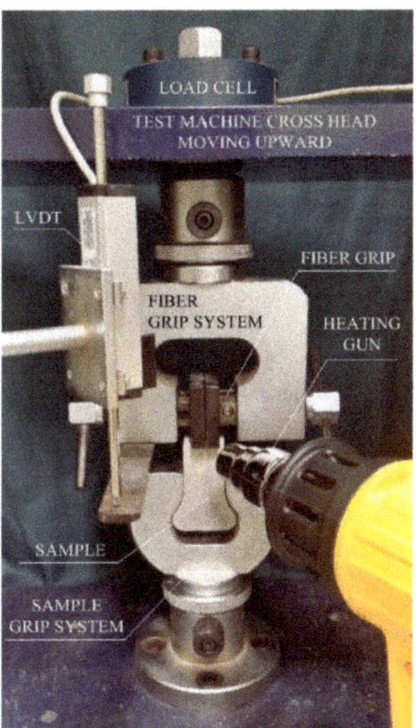

Figure 2. Test setup for the single-fiber pullout test.

3. Results and Discussion

3.1. Length and Diameter of Cold-Drawn SMA Fibers during the Short Heat Treatment Period

The changes in the diameter and length of the cold-drawn SMA fibers during the short heat treatment time are provided in Figure 3; Figure 3a–d shows the changes in the length and diameter of the NT_S, NT_D, NTN_S, and NTN_D samples, respectively. During the 10 min heat treatment, the lengths of the cold-drawn SMA fibers clearly decreased, whereas their diameters, which were originally reduced by the cold-drawing process, increased due to the shape memory effect.

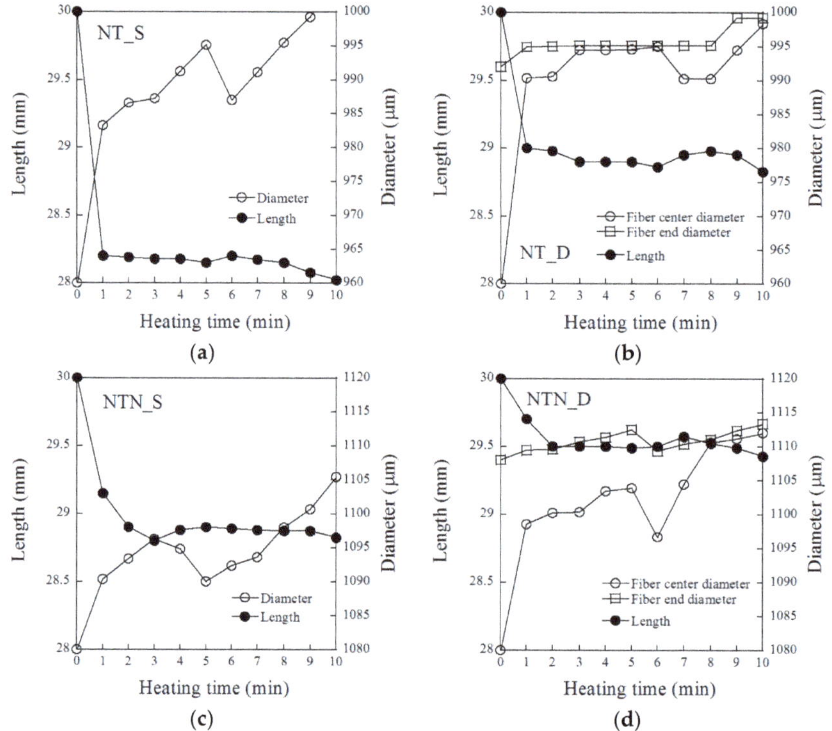

Figure 3. Effects of heating time on the length and diameter of the cold-drawn shape memory alloy (SMA) fibers for (**a**) NT_S, (**b**) NT_D, (**c**) NTN_S, and (**d**) NTN_D.

Although all SMA fibers had noticeable changes in their diameter and length, the changes were different, based on the type of alloy and geometry. During the heat treatment, the NT_S length was shortened from 30 to 28 mm (−6.67%), while that of the NTN_S decreased from 30 to 28.8 mm (−4.0%), as shown in Figure 3a,c, respectively. On the other hand, the NT_S diameter expanded from 960 to 1000 μm (+4.17%), while the NTN_S increased from 1080 to 1105 μm (+2.31%). In addition, the NT_D also presented a greater length reduction (from 30 to 28.8 mm) and wider expansion of the diameter (from 960 to 998 μm) than the NTN_D, as shown in Figure 3b,d. Therefore, the NT series generally showed higher shape memory effects, for both the diameter expansion and length reduction, than the NTN series. Moreover, the S geometry of the SMA fibers produced greater changes in both length and diameter than the D-shaped geometry, as shown in Figure 3. For example, the length of the NT_D decreased from 30 to 28.8 mm, whereas the NT_S changed from 30 to 28 mm. In addition, the length and diameter of all SMA fibers changed significantly within 1 to 3 min of heat treatment. Based on this observation, a short-term heat treatment of only a few minutes could successfully activate the shape memory effect of cold-drawn SMA fibers. Therefore, in this study, the heat treatment period was determined to be 10 min during the pullout test.

3.2. Pullout Stress Versus Slip Response

The pullout stress versus slip curves of the NT fibers are provided in Figure 4, while those of the NTN fibers are shown in Figure 5. The pullout responses of the SMA fibers clearly changed relative to the different fiber alloys and geometries, and if the heat treatment was applied. The dog-bone-shaped SMA fibers largely produced a better pullout resistance than the smooth SMA fibers because the former utilized a mechanical interaction between the bulged end of the fiber and mortar matrix, unlike the smooth fibers, which only used the frictional bond resistance at the interface. In addition, for the short heat treatment samples, after the 0.5 mm initial fiber pullout, there was a noticeable increase in the pullout stress, even though the slip was a constant 0.5 mm, as shown in Figure 4b,d and Figure 5b,d. These figures additionally provide close-up graphs of the pullout response from the 0 to 1.0 mm slip. During the initial pullout to the 0.5 mm slip, the pullout stress reached its first peak point at a slip between 0.1 and 0.3 mm, and then decreased until the slip was 0.5 mm. When the slip reached 0.5 mm, it was kept constant for 10 min. For the test series with the heat treatment, the pullout stress at 0.5 mm slip immediately increased during the heat treatment above the initial peak stress. However, for the samples without a heat treatment, there was no change in the pullout stress at the 0.5 mm slip. Furthermore, among the dog-bone-shaped fibers, the NTN consistently showed a higher pullout resistance than the NT fibers, regardless of heat treatment. The higher pullout resistance of the NTN_D fibers over the NT_D fibers likely originated from the higher strength of the NTN.

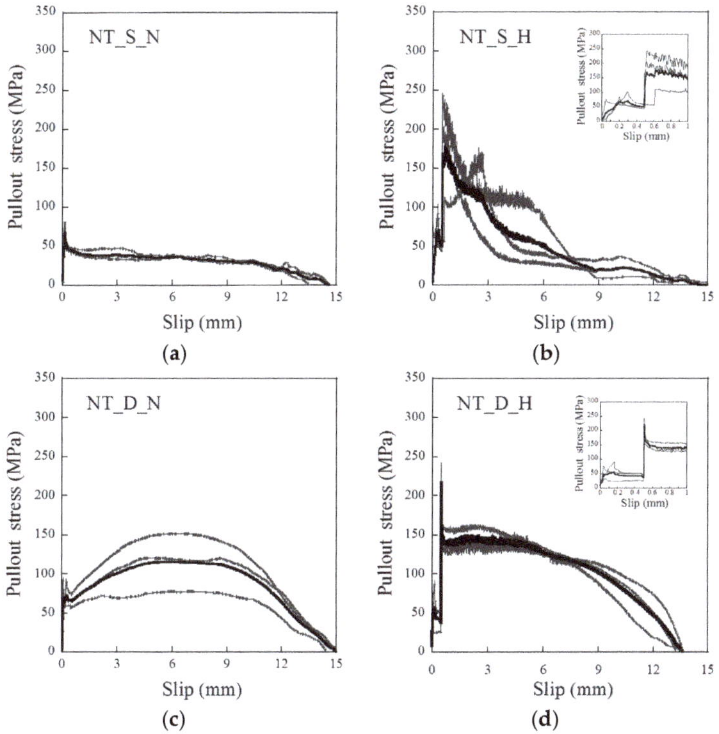

Figure 4. Pullout stress versus slip response of the NiTi fibers for (**a**) NT_S_N, (**b**) NT_S_H, (**c**) NT_D_N, and (**d**) NT_D_H. The insets in (**b**) and (**d**) are close-ups of the 0 to 1 mm slip regions for each graph.

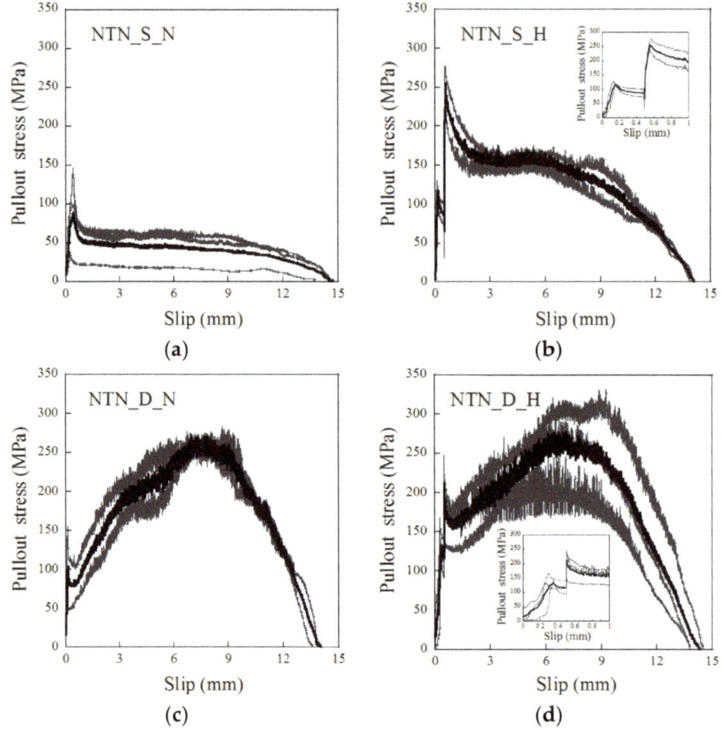

Figure 5. Pullout stress versus slip response of the NiTiNb fibers for (**a**) NTN_S_N, (**b**) NTN_S_H, (**c**) NTN_D_N, and (**d**) NTN_D_H. The insets in (**b**) and (**d**) are close-ups of the 0 to 1 mm slip regions for each graph.

Kim et al. [10] investigated the effects of heat treatment on the pullout resistance of SMA fibers embedded in the mortar matrix. They applied the pre-heat treatment (at 200 °C for 8 h) prior to the pullout test and then additional heat treatment (200 °C) was applied during the pullout test. The heat treatment of SMA fibers considerably recovered their reduced diameter after cold-drawing and consequently increased their pullout resistance. Moreover, Kim et al. [11] applied the pre-heat treatment (80 °C for 7 h) prior to the pullout test and then additional heat treatment (100 °C) during the pullout test. They reported that the heat treatment generally increased the pullout resistance regardless of the fiber compositions. Comprehensively, the enhanced pullout resistance of SMA fibers after heat treatment demonstrates their crack closing potentials in fiber-reinforced cementitious composites. However, current research requires not only a long time for pre-heat treatment of at least 7 h, but also continuous heat treatment during the pullout test. Therefore, it is necessary to reduce the time for heat treatment. Consequently, the pullout resistance was found to be enhanced by applying the heat treatment (300 °C) for only a short time (10 min) during the pullout test without pre-heat treatment.

3.3. Pullout Resistance during the Short Heat Treatment Period

The changes in the SMA fiber pullout stress during the short heat treatment period are provided in Figures 6 and 7 for the NT and NTN fibers, respectively. For the NT fibers, as shown in Figure 6a,c, if there was no heat treatment, the pullout stress was a constant value as the slip amount was kept at 0.5 mm for 10 min. However, for the series with the heat treatment, the pullout stress decreased at the beginning of the heat treatment, and then started to increase significantly, after 1–3 min, to a value greater than the residual pullout stress ($P_{0.5,re}$) after 5 min of heating. After 10 min of heating, the pullout stress reached the recovered pullout stress ($P_{0.5,rc}$), as illustrated in Figure 8, and the $P_{0.5,rc}$ of the NT fibers was higher than the $P_{0.5,re}$, regardless of the fiber geometry. When the fiber pullout was restarted after the 10 min heat treatment, the pullout stress of the fiber instantly increased to the re-pullout stress ($P_{0.5,rp}$), as shown in Figure 6b,d and Figure 8. The $P_{0.5,rp}$ was significantly higher than the $P_{0.5,rc}$. The pullout resistance of the NT fibers initially decreased at the beginning of the heat treatment, owing to the thermal extension (or elongation) of the fibers. Then, the NT fibers began recovering the pullout resistance after 1–3 min, by activating their shape memory effects, i.e., shortening the length and expanding the diameter of the fibers; these eventually produced a higher $P_{0.5,rc}$ than the $P_{0.5,re}$. In Figure 8, the NTN fibers showed a similar response to the heat treatment as the NT fibers, with both maintaining a constant value for 10 min after the 0.5 mm initial slip when there was no heat treatment, whereas their values varied noticeably during the short heat treatment period. However, the $P_{0.5,rc}$ of the NTN fibers was lower than their $P_{0.5,re}$, unlike the NT fibers.

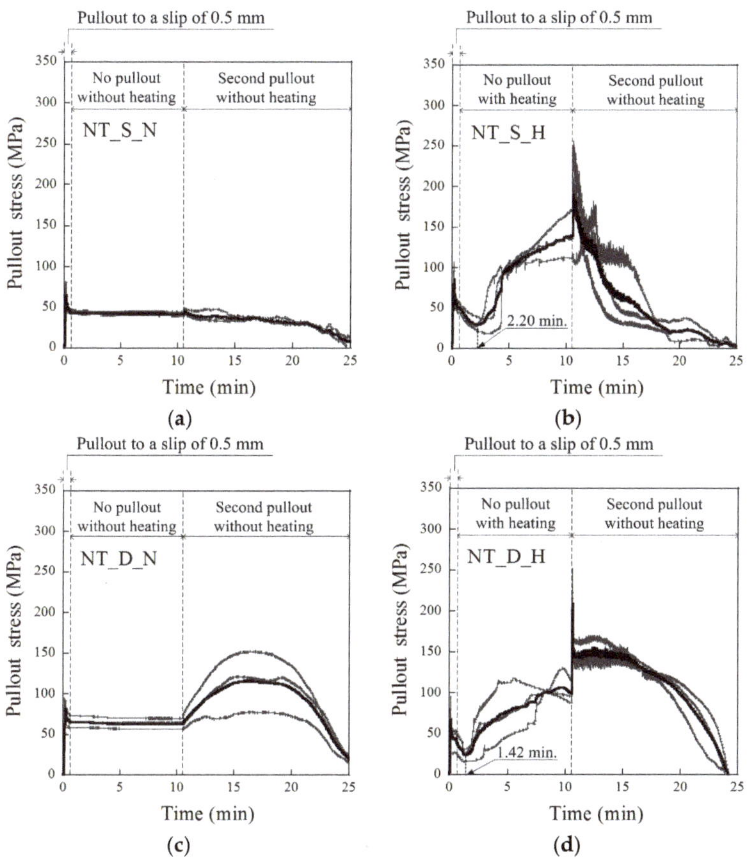

Figure 6. History of the pullout stress versus time for the NiTi fibers of (**a**) NT_S_N, (**b**) NT_S_H, (**c**) NT_D_N, and (**d**) NT_D_H.

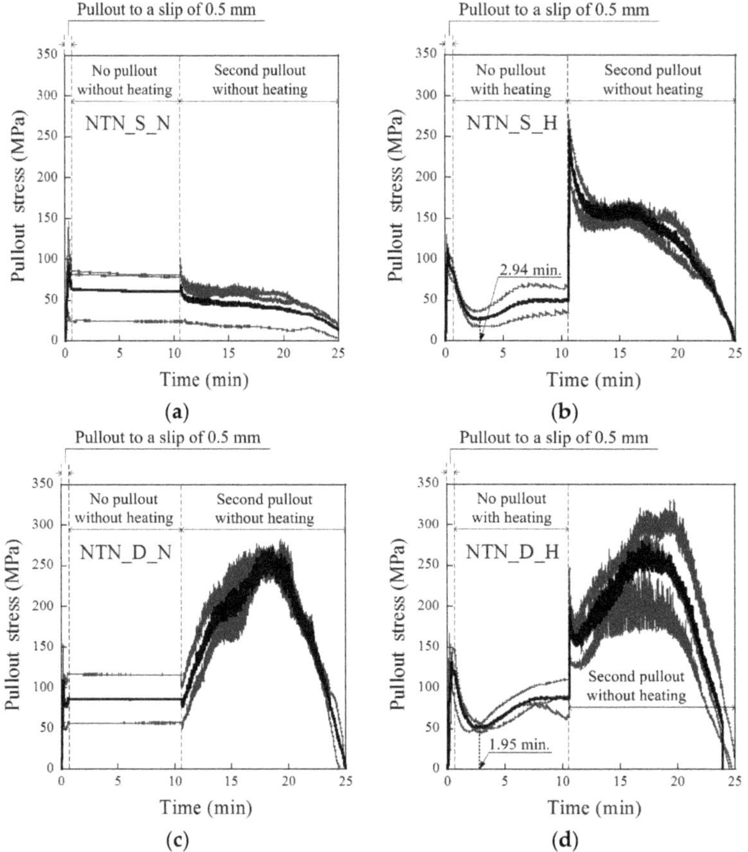

Figure 7. History of the pullout stress versus time for the NiTiNb fibers of (**a**) NTN_S_N, (**b**) NTN_S_H, (**c**) NTN_D_N, and (**d**) NTN_D_H.

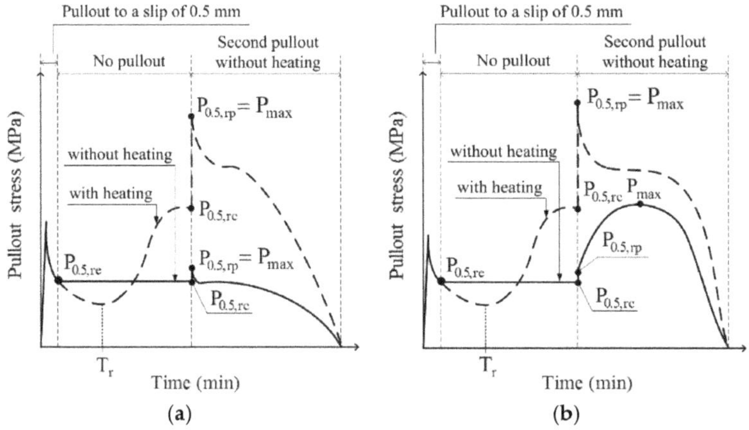

Figure 8. *Cont.*

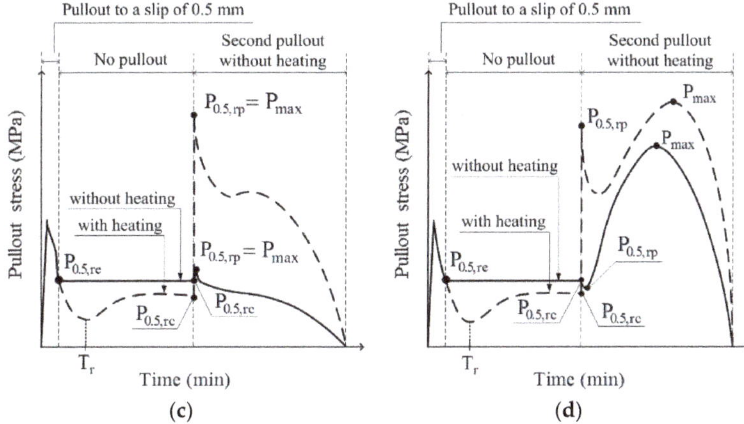

Figure 8. Typical pullout stress versus time curves of the shape memory alloy (SMA) fibers, showing the pullout parameters and pullout stress recovery for (**a**) NT_S, (**b**) NT_D, (**c**) NTN_S, and (**d**) NTN_D for samples with heating (solid lines) or without heating (dashed lines).

To quantitatively compare the pullout resistances of all test series, several parameters describing the SMA fiber pullout behavior are summarized in Table 4, including the $P_{0.5,re}$, $P_{0.5,rc}$, $P_{0.5,rp}$, maximum pullout stress ($P_{0.5,max}$), and pullout energy after the 0.5 mm slip to complete pullout (PE). After the short heat treatment time, the PE values were noticeably enhanced, as presented in Table 4, although the degree of enhancement varied according to the geometry and alloy of the fibers. The PE values of the dog-bone-shaped SMA fibers were clearly higher than those of the smooth SMA fibers: the value of NT_D_H was 1063.2 MPa-m, whereas that of NT_S_H was 525.5 MPa-m.

Table 4. Pullout test results.

Notation	Sample	$P_{0.5,re}$ (MPa)	$P_{0.5,rc}$ (MPa)	$P_{0.5,rp}$ (MPa)	P_{max} (MPa)	PE (MPa-mm)
NT_S_N	SP1	41.5	39.5	43.6	43.6	272.2
	SP2	46.2	44.2	48.8	48.8	335.2
	SP3	43.2	42.1	46.1	46.1	317.1
	Aver. [a]	**43.7**	**41.9**	**45.7**	**45.7**	**308.2**
	STD [b]	1.9	1.9	2.1	2.1	26.5
NT_S_H	SP1	56.6	112.5	181.9	181.9	471.3
	SP2	44.5	171.2	214.5	214.5	382.9
	SP3	49.9	136.1	254.9	254.9	658.0
	Aver.	**50.3**	**139.9**	**190.6**	**190.6**	**525.5**
	STD	4.9	24.1	29.9	29.9	114.7
NT_D_N	SP1	58.7	56.0	58.1	94.4	578.7
	SP2	73.5	69.1	47.9	152.3	1137.0
	SP3	64.2	64.2	65.8	121.2	959.1
	Aver.	**65.6**	**63.1**	**65.7**	**116.4**	**904.0**
	STD	6.4	5.4	7.3	23.7	232.9
NT_D_H	SP1	51.2	95.0	251.0	251.0	1123.9
	SP2	53.2	87.6	232.3	232.3	959.7
	SP3	26.3	115.2	193.7	193.7	1106.7
	Aver.	**43.6**	**99.3**	**208.8**	**208.8**	**1063.2**
	STD	12.2	11.7	23.9	23.9	73.7

Table 4. *Cont.*

Notation	Sample	$P_{0.5,re}$ (MPa)	$P_{0.5,rc}$ (MPa)	$P_{0.5,rp}$ (MPa)	P_{max} (MPa)	PE (MPa-mm)
NTN_S_N	SP1	23.9	23.9	26.3	26.3	181.5
	SP2	85.9	81.0	85.5	85.5	641.8
	SP3	81.6	78.3	99.8	99.8	652.3
	Aver.	**63.8**	**61.1**	**69.3**	**69.3**	**494.3**
	STD	28.3	26.3	31.8	31.8	219.5
NTN_S_H	SP1	73.1	33.7	239.4	239.4	1472.1
	SP2	95.9	66.4	276.6	276.6	1780.3
	Aver.	**84.5**	**50.0**	**257.5**	**257.5**	**1625.9**
	STD	11.4	16.4	18.6	18.6	154.1
NTN_D_N	SP1	57.2	57.1	57.4	273.8	2073.3
	SP2	115.8	115.8	116.3	282.2	2358.4
	Aver.	**84.8**	**84.5**	**86.7**	**271.8**	**2375.9**
	STD	29.3	29.4	29.5	4.2	142.6
NTN_D_H	SP1	148.0	63.8	247.0	296.1	2600.0
	SP2	92.8	109.9	154.7	330.4	2865.0
	SP3	120.9	115.1	236.2	269.0	2031.1
	Aver.	**120.3**	**96.6**	**198.3**	**280.1**	**2850.8**
	STD	22.5	23.1	41.2	25.1	347.9

[a] Aver.: average values; [b] STD: standard deviation.

3.4. Pullout Stress Recovery Starting Time (T_r) during the Heat Treatment

As observed in the pullout stress versus time curves of the SMA fibers during the short-term heat treatment in both Figures 6 and 7, their pullout stress first decreased at the beginning of the heat treatment and then began increasing to the end of the heat treatment. The pullout stress recovery starting time was quite different based on the type of SMA fiber geometry and alloy. Table 5 provides the recovery starting times (T_r) of the pullout stress relative to the test series during the short heat treatment period. The average T_r values of the test series (NT_S_H, NT_D_H, NTN_S_H, and NTN_D_H) were 2.20, 1.42, 2.94, and 1.95 min, respectively. The NT fibers usually produced faster T_r than the NTN fibers, while the D-shaped fibers produced faster T_r than the S fibers. The earlier pullout stress recovery of the dog-bone-shaped SMA fibers is thought to originate from their higher PE values, indicating a superior pullout resistance. Accordingly, the SMA fibers with a deformed geometry generating a larger PE are favorable for generating a faster crack closing ability. However, the parameters influencing the recovery start time (T_r) require further investigation.

Table 5. Pullout stress recovery starting time during heating (T_r).

Notation	Sample	T_r (min.)
NT_S_H	SP1	3.35
	SP2	1.93
	SP3	3.17
	Average	**2.20**
NT_D_H	SP1	1.47
	SP2	1.51
	SP3	1.42
	Average	**1.42**
NTN_S_H	SP1	4.20
	SP2	2.77
	Average	**2.94**
NTN_D_H	SP1	3.14
	SP2	1.93
	SP3	1.95
	Average	**1.95**

3.5. Pullout Stress Recovery Ratios

The ratios R_1 and R_2, between the pullout stresses, $P_{0.5,re}$, $P_{0.5,rc}$, and $P_{0.5,rp}$, were calculated to quantify the crack closing ability of the SMA fibers during and after the short-term heat treatment. The ratio between the $P_{0.5,rc}$ and $P_{0.5,re}$ is denoted as R_1, which represents the potential crack closing ability. The ratio between the $P_{0.5,rp}$ and $P_{0.5,re}$ was additionally analyzed and denoted as R_2, which is thought to correlate with the ability to prevent crack reopening. Both R_1 and R_2 ratios are provided in Figure 9b. The NT fibers generally produce higher recovery ratios than the NTN fibers, as shown in Figure 9b; the values of R_1 were 2.78, 2.28, 0.59, and 0.80 for NT_S_H, NT_D_H, NTN_S_H, and NTN_D_H, respectively, while those of R_2 were 3.79, 4.79, 3.05, and 1.65 for NT_S_H, NT_D_H, NTN_S_H, and NTN_D_H, respectively. The R_1 ratios of the NT fibers were higher than 1.0, whereas those of the NTN fibers were less than 1.0. Therefore, the pullout stress recovery during the short heat treatment period is higher for the NT fibers than the NTN fibers. Additionally, R_1 values below 1.0 are not sufficient for closing cracks, whereas those higher than 1.0 are indicative of a satisfactory or strong crack closing capacity. The R_2 ratios of the NT fibers were also higher than the NTN fibers. Therefore, the NT fibers after the heat treatment displayed a higher pullout resistance during the second pullout process.

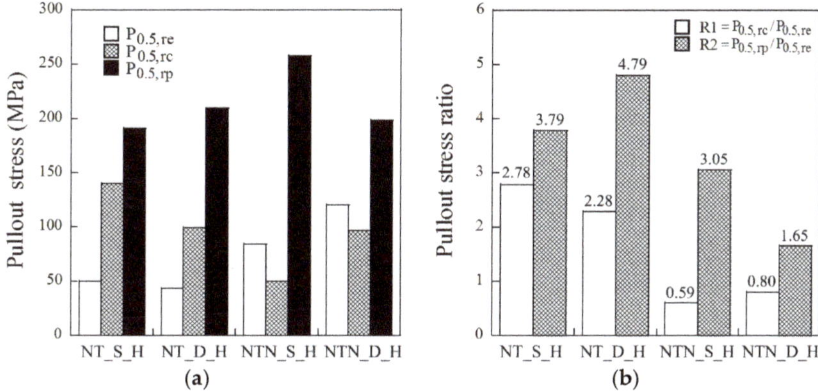

Figure 9. (**a**) Pullout stress and (**b**) pullout stress ratios owing to the heat treatment, for the different samples.

3.6. Pullout Energy Ratios

The short heat treatment period also noticeably increased the amount of pullout energy after the 0.5 mm initial slip to complete fiber pullout, as provided in Table 4 and Figure 10a.

The enhanced SMA fiber pullout energy after the heat treatment originated from the lateral recovery of the fiber diameter due to shape memory effects. To quantitatively compare the heat treatment effect on the PE, the pullout energy ratios between samples with and without heat treatment were estimated and are shown in Figure 10. The pullout energy ratios of the NT_S, NT_D, NTN_S, and NTN_D were 1.71, 1.18, 3.29, and 1.20, respectively. The higher pullout energy ratio is thought to be favorable for resisting crack reopening. The NTN_S fiber produced the highest pullout energy ratio, while the NTN_D fiber produced the highest pullout energy. In addition, it was also evident that the smooth fibers generated higher pullout energy ratios than the dog-bone-shaped fibers because the pullout energy of the D fiber was much higher than that of the S fiber for the series without heat treatment. The *PE* value of NTN_S_N was 494.3 MPa-mm, while that of NTN_D_N was 2375 MPa-mm.

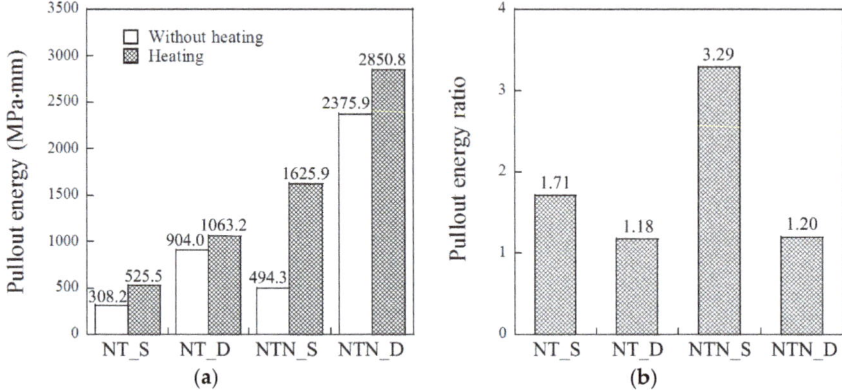

Figure 10. (**a**) Pullout energy (for samples with heating (grey) or without heating (white)), and (**b**) pullout energy ratio owing to the heating treatment.

4. Conclusions

This study investigated the effects of applying a short heat treatment period (10 min) to shape memory alloy (SMA) fibers on their geometry and pullout resistance. The short heat treatment time clearly activated the shape memory effects, eventually generating the pullout stress recovery of SMA fibers in mortar. The conclusions below can be drawn from this experimental study:

- During the short heat treatment period, the length of the cold-drawn SMA fibers clearly decreased, whereas their diameter expanded due to the shape memory effect. The NiTi fibers generally showed greater shape memory effects in both diameter and length than the NiTiNb fibers, while the smooth geometry SMA fibers had greater shape memory effects than the dog-bone-shaped geometry fibers;
- SMA fibers with a dog-bone-shaped geometry generally showed higher pullout resistances than those with a smooth geometry;
- The short heat treatment period noticeably increased the pullout stress, although the degree of enhancement varied, relative to the SMA fiber alloy and geometry: (1) the NiTi fibers generally produced a faster pullout stress recovery than the NiTiNb fibers, while the SMA fibers with dog-bone-shaped geometries showed a faster recovery than those with a smooth geometry; (2) the NiTi fibers revealed higher pullout stress ratios during and after the heat treatment than the NiTiNb fibers; and (3) the SMA fibers with dog-bone-shaped geometries generated a larger amount of pullout energy (*PE*), which is favorable for faster crack closing.

Self-healing concrete can close (heal or fill) cracks; however it requires a few days minimum to heal the cracks. Consequently, in this study, we found that the SMA-FRCCs have a fast crack closing capacity (just only 10 min). The crack-closing behavior of SMA-FRCCs under a load is now under investigation and the parameters influencing the recovery start time require further investigation. Moreover, it is also necessary to investigate the residual stress of SMA-FRCCs [26].

Author Contributions: M.K.K., D.J.K., Y.-S.C., and E.C. conceived, designed, and wrote the manuscript; M.K.K. performed the experiments.

Funding: This research was supported by the Basic Science Research Program through the National Research Foundation of Korea (NRF) funded by the Ministry of Education, Science and Technology (Project No. 2019-R1A2C2008542).

Conflicts of Interest: The authors declare no conflicts of interest.

References

1. Naaman, A.E.; Reinhardt, H.W. Characterization of high performance fiber reinforced cementitious composites-HPFRCC. In Proceedings of the Second International Workshop on High Performance Fiber Reinforced Cement Composites (HPFRCC2), Ann Arbor, MI, USA, 11–14 June 1995; pp. 1–24.
2. Orange, G.; Acker, P.; Vernet, C. A new generation of UHPConcrete: DUCTAL damage resistance and micro mechanical analysis. In Proceedings of the Third International Workshop on High Performance Fiber Reinforced Cement Composites (HPFRCC3), Mainz, Germany, 16–19 May 1999; pp. 101–112.
3. Rossi, P. High performance multimodal fiber reinforced cement composites (HPMFRCC): The LCPC experience. *ACI Mater. J.* **1997**, *94*, 478–483.
4. Rossi, P.; Arca, A.; Parant, E.; Fakhri, P. Bending and compressive behavior of a new cement composite. *Cem. Concr. Res.* **2005**, *35*, 27–33. [CrossRef]
5. Ahn, T.H.; Kishi, T. Crack self-healing behavior of cementitious composites incorporating various mineral admixtures. *J. Adv. Concr. Technol.* **2010**, *8*, 171–186. [CrossRef]
6. Li, M.; Li, V.C. Cracking and healing of engineered cementitious composites under chloride environment. *ACI Mater. J.* **2011**, *108*, 333–340.
7. Kim, D.J.; El-Tawil, S.; Naaman, A.E. Rate-dependent tensile behavior of high performance fiber reinforced cementitious composites. *Mat. Struct.* **2009**, *42*, 399–414. [CrossRef]
8. Blanco, A.; Pujadas, P.; Fuente, D.; Cavalaro, S.H.P.; Aguado, A. Influence of the type of fiber on the structural response and design of FRC slabs. *J. Struct. Eng.* **2016**, *142*, 04016054. [CrossRef]
9. Edvardsen, C. Water permeability and autogenous healing of cracks in concrete. *ACI Mater. J.* **1999**, *96*, 448–455.
10. Kim, D.J.; Kim, H.A.; Chung, Y.-S.; Choi, E. Pullout resistance of straight NiTi SMA fibers in cement mortar after cold drawing process and heat treatment. *Compos. Part B Eng.* **2014**, *67*, 588–594. [CrossRef]
11. Kim, D.J.; Kim, H.A.; Chung, Y.-S.; Choi, E. Pullout resistance of deformed shape memory alloy fibers embedded in cement mortar. *J. Intell. Mat. Syst. Str.* **2014**, *27*, 1–12. [CrossRef]
12. Kim, M.K.; Kim, D.J.; Chung, Y.-S.; Choi, E. Direct tensile behavior of shape-memory alloy fiber-reinforced cement composites. *Constr. Build. Mater.* **2016**, *102*, 462–470. [CrossRef]
13. Choi, E.; Nam, T.-H.; Cho, S.-C.; Chung, Y.-S.; Park, T. The behavior of concrete cylinders confined by shape memory alloy wires. *Smart Mater. Struct.* **2008**, *17*, 065032. [CrossRef]
14. Choi, E.; Chung, Y.-S.; Choi, J.-H.; Kim, H.-T.; Lee, H. The confining effectiveness of NiTiNb and NiTi SMA wire jackets for concrete. *Smart Mater. Struct.* **2010**, *19*, 035024. [CrossRef]
15. Choi, E.; Kim, Y.-W.; Chung, Y.-S.; Yang, K.-T. Bond strength of concrete confined by SMA wire jackets. *Phys. Procedia* **2010**, *10*, 210–215. [CrossRef]
16. Tran, H.; Balandraud, X.; Destrebecq, J.F. Improvement of the mechanical performances of concrete cylinders confined actively or passively by means of SMA wires. *Arch. Civ. Mech. Eng.* **2015**, *15*, 292–299. [CrossRef]
17. Sawaguchi, T.; Kikuchi, T.; Ogawa, K.; Kajiwara, S.; Ikeo, Y.; Kojima, M.; Ogawa, T. Development of prestressed concrete using Fe-Mn-Si-based shape memory alloys containing NbC. *Mater. Trans.* **2006**, *41*, 580–583. [CrossRef]
18. Umezaki, E. Improvement in separation of SMA from matrix in SMA embedded smart structures. *Mat. Sci. Eng. A Struct.* **2000**, *285*, 363–369. [CrossRef]
19. Lee, K.-J.; Lee, J.-H.; Jung, C.-Y.; Choi, E. Crack-closing performance of NiTi and NiTiNb fibers in cement mortar beams using shape memory effects. *Compos. Struct.* **2018**, *202*, 710–718. [CrossRef]
20. Lee, J.-H.; Lee, K.-J.; Choi, E. Flexural capacity and crack-closing performance of NiTi and NiTiNb shape-memory alloy fibers randomly distributed in mortar beams. *Compos. Part B Eng.* **2018**, *153*, 264–276. [CrossRef]
21. Wang, Y.; Zhou, L.; Wang, Z.; Huang, H.; Ye, L. Analysis of internal stresses induced by strain recovery in a single SMA fiber-matrix composite. *Comp. Part B Eng.* **2011**, *42*, 1135–1143. [CrossRef]
22. Watanabe, Y.; Miyazaki, E.; Okada, H. Enhanced mechanical properties of Fe-Mn-Si-Cr shape memory fiber/plaster smart composite. *Mater. Trans.* **2002**, *43*, 974–983. [CrossRef]
23. Kuang, Y.; Ou, J. Self-repairing performance of concrete beams strengthened using superelastic SMA wires in combination with adhesives released from hollow fiber. *Smart Mater. Struct.* **2008**, *17*, 025020. [CrossRef]

24. Kodur, V.K.R.; Sultan, M.A. Effect of temperature on thermal properties of high strength concrete. *J. Mater. Civil Eng.* **2003**, *15*, 101–107. [CrossRef]
25. Kim, J.; Lee, G.-P.; Moon, D.Y. Evaluation of mechanical properties of steel-fibre-reinforced concrete exposed to high temperatures by double punch test. *Constr. Build. Mater.* **2015**, *79*, 182–191. [CrossRef]
26. Pujadas, P.; Blanco, A.; Cavalaro, S.H.P.; Fuente, A.; Aguado, A. Multidirectional double punch test to assess the post-cracking behaviour and fibre orientation of FRC. *Constr. Build. Mater.* **2014**, *58*, 214–224. [CrossRef]

Article

Experimental Study on the Shrinkage Behavior and Mechanical Properties of AAM Mortar Mixed with CSA Expansive Additive

Sung Choi [1], Gum-Sung Ryu [2,*], Kyeong-Taek Koh [2], Gi-Hong An [2] and Hyeong-Yeol Kim [2]

1 Department of Civil, Architectural, and Environmental Systems Engineering, Sungkyunkwan University (SKKU), 2066, Seobu-Ro, Jangan-Gu, Suwon-Si, Gyeonggi-Do 16419, Korea; csomy1113@naver.com

2 Department of Infrastructure Safety Research, Korea Institute of Civil Engineering and Building Technology, 283, Goyangdae-Ro, Ilsanseo-Gu, Goyang-Si 10223, Korea; ktgo@kict.re.kr (K.-T.K.); agh0530@kict.re.kr (G.-H.A.); hykim1@kict.re.kr (H.-Y.K.)

* Correspondence: ryu0505@kict.re.kr; Tel.: +82-31-910-0050

Received: 5 September 2019; Accepted: 28 September 2019; Published: 11 October 2019

Abstract: In this study, a calcium sulfoaluminate-based expansive additive (0%, 2.5%, 5.0%, and 7.5% by the mass of the binder) was added to compensate for the shrinkage of alkali-activated material (AAM) mortar. Modulus of elasticity curves based on the ACI 209 model were derived for the AAM mortar mixed with the additive by measuring the compressive strength and modulus of elasticity. Moreover, autogenous shrinkage and total shrinkage were measured for 150 days, and drying shrinkage was calculated by excluding autogenous shrinkage from total shrinkage. For the autogenous and drying shrinkage of AAM mortar, shrinkage curves by age were obtained by deriving material constants using the exponential function model. Finally, shrinkage stress was calculated using the modulus of elasticity of the AAM mortar and the curves obtained using the shrinkage model. The results showed that the calcium sulfoaluminate-based expansive additive had an excellent compensation effect on the drying shrinkage of AAM mortar, but the effect was observed only at early ages when the modulus of elasticity was low. From a long-term perspective, the shrinkage compensation effect was low when the modulus of elasticity was high, and thus, shrinkage stress could not be reduced.

Keywords: alkali-activated material; calcium sulfoaluminate-based expansive additive; concrete shrinkage; modulus of elasticity; shrinkage stress

---

1. Introduction

Cement is an excellent and economical construction material, and to date, no construction material that can perfectly replace cement exists. As environmental problems emerge worldwide, however, efforts are being made to identify a suitable substitute for cement because it emits CO_2 gas in large quantities [1,2]. It is difficult to completely replace cement in the construction industry because of its widespread usage, but some proportion may be replaceable provided binders suitable for different structural purposes are developed. Various materials that may serve as substitutes for cement have been investigated by many researchers [3–8]. Among them, ground granulated blast furnace slag (GGBFS) and fly ash (FA), which have been used to partially replace cement, can exhibit performances equal to those of ordinary Portland cement (OPC) if alkali-activated materials (AAMs) are used. AAMs are eco-friendly materials that can improve the performance of concrete because they have high initial strength and excellent durability. Thus, many researchers have studied the performance of AAM concrete in this regard [9]. Most studies on AAMs, however, have focused on the reactivity and physical performances of binders; the shrinkage characteristics of AAMs have hardly been investigated. Cartwright et al. [10], however, reported that AAMs exhibited 3–6 times more shrinkage than OPC. Various previous studies have also reported that AAMs can cause serious problems when used in structures because their shrinkage is significantly higher than that of OPC [11,12]. However, the mechanisms and inferences relating to the high shrinkage of AAMs have not been comprehensibly summarized.

AAMs cause microcracks because they cause much shrinkage [13–16], as such microcracks may degrade the strength and durability of concrete [17–21], various methods have been proposed to reduce the shrinkage of AAMs. According to Chatterji [22], the use of the expansive additives containing alkali metal components can compensate for shrinkage because they generate many expansive hydrates, such as ettringite ($3CaO{\cdot}3Al_2O_3{\cdot}CaSO_4{\cdot}32H_2O$) and calcium hydroxide ($Ca(OH)_2$), and this expansion effect is affected by the mixing and curing conditions of concrete [23,24]. Palacios and Puertas [15] conducted research with various shrinkage-reducing agents to reduce the shrinkage of AAMs. In addition to these methods, which are materials-specific, other techniques for reducing shrinkage have been studied. Thomas et al. [25] proposed a curing method to reduce drying shrinkage, and Sakulich and Bentz [26] reported that use of lightweight aggregates subjected to pre-wetting can reduce autogenous shrinkage due to the internal curing effect. Summarizing the results of these previous studies reveals that the shrinkage caused by AAMs is determined by specific parameters, including the types and mix proportions of AAMs as well as the curing conditions [27–29].

With regard to the shrinkage of AAM mortar, structural problems can be caused simply by the generated shrinkage, but most issues are caused by shrinkage cracking, which occurs when the stress caused by shrinkage is higher than the mortar strength. The shrinkage stress acting on a structure increases with the amount of shrinkage and the modulus of elasticity. In general, the shrinkage of concrete is high at early ages and decreases over time, but the modulus of elasticity increases over time. Thus, in the long term, the shrinkage is low, but the shrinkage stress acting on the structure can be evaluated differently. In particular, AAM mortars show higher initial shrinkage than OPC and are also subject to continuous shrinkage over the long term. Therefore, it is necessary to accurately predict the shrinkage stress generated in AAM mortars by measuring the modulus of elasticity by age and to thereafter apply appropriate shrinkage-reducing technologies accordingly.

In this study, a calcium sulfoaluminate-based (CSA) expansive additive was used to compensate for the shrinkage of AAM mortar. The shrinkage characteristics of the AAM mortar were analyzed by age by measuring its autogenous and total shrinkage for 150 days according to the content of the CSA expansive additive, and a shrinkage model was proposed based on the results. In addition, the shrinkage stress of the AAM mortar was calculated by measuring its modulus of elasticity, and the shrinkage stress compensation effect of the mortar mixed with the CSA expansive additive was analyzed.

2. Materials and Method

2.1. Materials and Mixture Proportions of AAM Mortar

The AAM mortar used in this study contained a two-component binder, wherein GGBFS and FA were mixed in the ratio 7:3. Table 1 shows the analysis results of the major chemical components of the GGBFS and FA. The GGBFS was procured from Sampyo Cement Corp. (Dangjin, Korea). It had a density of 2.91 g/cm^3 and a fineness of 4683 cm^2/g. The GGBFS was composed of 41.9% CaO, 33.4% SiO_2, 13.8% Al_2O_3, and 4.9% MgO, and thus, its basicity coefficient (Kb = (CaO + MgO)/(SiO_2 + Al_2O_3)) was 1.00, which is similar to the neutral value of 1.0 for ideal alkali activation [30]. The hydration modulus of GGBFS according to a formula proposed in Ref. [31] (HM = (CaO + MgO + Al_2O_3)/SiO_2) was 1.82. This value was higher than the required value of 1.4, which indicates good hydration properties of the GGBFS [31].

The FA was type 2 fly ash (KS L 5405) procured from Sampyo Cement Corp. (Boryeong, Korea) [30]. It had a density of 2.20 g/cm^3 and a fineness of 3216 cm^2/g. The FA was composed of 60.3% SiO_2, 24.2% Al_2O_3, and 7.3% Fe_2O_3. Moreover, SiO_2, Al_2O_3, and Fe_2O_3 accounted for 91.8% of the total, whereas M_2O (K_2O + Na_2O) accounted for 1.9%. The alkali-activator was used to accelerate the reaction of the binder. The alkali-activator was in the form of white powder with a specific density of 1.026 g/cm^3 and a molar ratio of 0.95. In addition, alkali-activators are manufactured separately by adjusting the chemical components. The SiO_2/Na_2O ratio of the alkali-activator (SiO_2: 46.17%, Na_2O: 50.18%) used in this study was 0.92.

To compensate for the shrinkage of the AAM mortar, a powdered CSA expansive additive was used. The CSA expansive additive contained lime, gypsum, and bauxite as its major components. The specific density of the CSA expansive additive was 2.86 g/cm^3, and its Blaine fineness was 3754 cm^2/g. River sand with a density of 2.53 g/cm^3 and a water absorption rate of 1.08% was used as a fine aggregate. The maximum size of the fine aggregate was 4.76 mm, and the fineness modulus was 2.77.

Table 1. Chemical composition of ground granulated blast furnace slag (GGBFS), fly ash (FA), and calcium sulfoaluminate-based (CSA) expansion agent.

Type	CaO	SiO_2	Al_2O_3	Fe_2O_3	SO_3	MgO	K_2O	Na_2O
GGBFS	41.9	33.4	13.8	0.6	4.8	4.9	0.5	0.2
FA	3.7	60.3	24.2	7.3	0.5	1.9	1.2	0.8
CSA	36.4	30.2	24.2	1.7	5.3	1.4	0.5	0.3

Table 2 summarizes the mix proportions of the AAM mortar. The water-to-binder (W/B) ratio was 45.1% and the sand-to-binder (S/B) ratio was 1.2. The activator-to-water ratio was 24.0%. Then, 0.0%, 2.5%, 5.0%, and 7.5% of CSA expansive additive based on the amount of the binder (GGBFS:FA = 7:3) was added. The AAM mortar was dry mixed for 30 s after inserting the binder as well as the powdered AAM and CSA expansive additive. Water was then added, and the mortar was mixed at a low speed (15 rpm) for 10 min. After inserting the fine aggregate, the mortar was mixed for 90 s at a speed of 30 rpm.

Table 2. Mix proportions of alkali-activated material (AAM) mortar.

Type	W/B	S/B	Water (g)	Binder (g)	Activator (g)	Sand (Fine Aggregate) (g)	CSA Expansive Additive (g)
EA-0.0	0.451	1.2	451	1000	108	1200	0
EA-2.5	0.451	1.2	451	1000	108	1200	25
EA-5.0	0.451	1.2	451	1000	108	1200	50
EA-7.5	0.451	1.2	451	1000	108	1200	75

2.2. Test Methods

Cube mortar specimens of dimensions 50 mm × 50 mm × 50 mm complying with ASTM C109-16a [32] were prepared. AAM mortar was poured into a cubic mold, cured for 1 d, and demolded. It was then cured in a chamber with constant temperature (20 ± 2 °C) and relative humidity (90 ± 2%). The compressive strength test was conducted at 1, 7, and 28 d of age. The modulus of elasticity of the concrete was calculated by applying loads up to 40% of the ultimate load at a rate of 0.25 MPa/s and obtaining the deformation values for the loads using an interpolation method in accordance with ASTM C469M-14 [33].

An embedded gauge (PMFL-50-2LT, Tokyo Sokki Kenkyujo Co., Ltd., Tokyo, Japan) was used to measure the shrinkage of the AAM mortar. Figure 1 shows the method for measuring the length change of the mortar due to shrinkage. An embedded gauge for the length change rate and a temperature sensor (Thermocouple t type, Tokyo Sokki Kenkyujo Co., Ltd., Tokyo, Japan) were installed at the center of 100 mm × 100 mm × 400 mm specimens. Before pouring the mortar, a Teflon sheet and polystyrene board were placed on the inner surface of the mold to minimize friction with the mold and provide restraint in the length direction. After pouring the mortar, a polyester film was installed on the specimen surface to prevent the evaporation and absorption of moisture on the surface. The AAM mortar specimens were cured for 1 d and demolded. The autogenous shrinkage specimen was sealed using a polyester film to control drying shrinkage. The specimen for measuring total shrinkage, namely the sum of drying shrinkage and autogenous shrinkage, was not sealed after demolding. Two specimens were prepared for the length change rate test to measure total shrinkage and autogenous shrinkage, and the measurement results were averaged. The test on the length change rate of the AAM mortar was conducted in a chamber with constant temperature (20 °C) and relative humidity (60%).

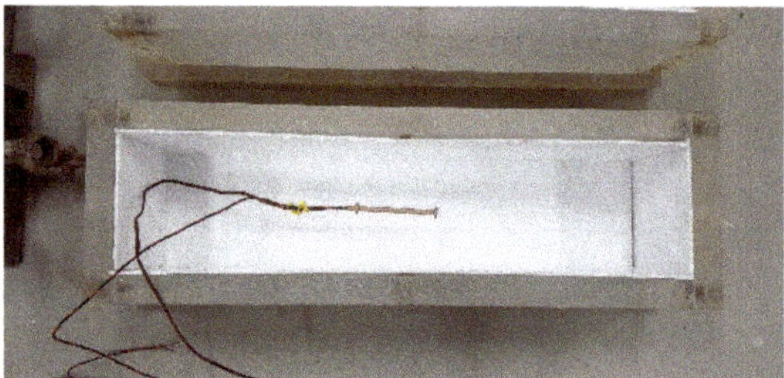

Figure 1. Measurement of length change.

3. Results and Discussion

3.1. Compressive Strength and Modulus of Elasticity

Table 3 shows the tests results of the compressive strength and modulus of elasticity of the AAM mortar according to the content of the expansive additive. The target strength of the AAM mortar was 40 MPa, and all the mixtures met the target strength at 28 d of age. The compressive strength results at 1 d of age showed that the compressive strength of Expansive Additive (EA)-0.0 was 3.16 MPa, but the AAM mortar specimens mixed with the expansive additive exhibited compressive strengths exceeding 5 MPa, indicating a strength increase of more than 2 MPa compared to that for EA-0.0. In particular, EA-5.0 exhibited a compressive strength of 5.91 MPa, the highest observed strength, at 1 d of age. At 28 d of age, the strengths of EA-2.5 and EA-5.0 were respectively 12.6% and 13.7% higher than that of EA-0.0. The strength of EA-7.5, however, was only 8.4% higher. Therefore, it was found that AAM mortar exhibited the highest strength when the content of the expansive additive was 5%.

When the modulus of elasticity of the AAM mortar specimens mixed with the expansive additive were compared, it was found that the initial modulus of elasticity increased as the content of expansive additive increased. While the modulus of elasticity of EA-0.0 at 1 d of age was 1.21 GPa, that of EA-7.5 with the highest expansive additive content was 2.46 GPa, which was approximately two times higher. As the age increased, however, the effect of the addition of the expansive additive on the modulus of elasticity decreased. At 28 d of age, the modulus of elasticity of the AAM mortar ranged from 20.07 to 21.17 GPa, showing that the influence of the expansive additive was not significant.

Table 3. Compressive strength and modulus of elasticity of AAM mortar.

Type	Compressive Strength (MPa)			Modulus of Elasticity (GPa)		
	1 d	**7 d**	**28 d**	**1 d**	**7 d**	**28 d**
EA-0.0	3.16	33.96	43.59	1.21	15.49	20.53
EA-2.5	5.18	37.70	49.09	2.13	15.67	20.07
EA-5.0	5.91	35.22	49.56	2.07	15.84	20.16
EA-7.5	5.68	34.98	47.26	2.46	16.15	21.17

AAM mortar with the CSA expansive additive exhibited an increase in the modulus of elasticity at early ages, but there was no significant difference in the modulus of elasticity at 28 d of age. This means that the rate of increase of the modulus of elasticity may vary depending on the content of the expansive additive. Various prediction equations on the modulus of elasticity were applied to compare the moduli of elasticity of the AAM mortar according to the content of the expansive additive, but the model on the modulus of elasticity proposed by American Concrete Institute (ACI) 209, which can predict the modulus of elasticity of the AAM mortar most accurately, was used [34]. As a result, R^2 between the modulus of elasticity measured from the experiment and that obtained by the ACI 209 model for AAM mortar was 98.89% or higher, indicating a high correlation.

$$E_{cmt} = E_{cm28}\sqrt{\frac{t}{a+bt}} \tag{1}$$

where E_{cmt} is the modulus of elasticity of the AAM mortar at t days of age, and E_{cm28} is the compressive strength at 28 d of age. a and b are material constants related to the compressive strength. Figure 2 shows the experimental values of the modulus of elasticity of the AAM mortars and the prediction curves for the modulus of elasticity as per the ACI 209 model. Table 4 shows the results of the derivation of the values of a and b according to the content of the expansive additive by applying the ACI 209 model. Nagataki and Gomi [35] reported that if the content of CSA expansive additive exceeds a certain value, the strength may decrease but the modulus of elasticity and creep increase continuously. The experiment results also showed that the material constants of the modulus of elasticity did not change considerably when the content of the expansive additive was less than 5%, but they exhibited a different tendency when the content was 7.5% because the increment in the modulus of elasticity increased. When the content of the expansive additive was less than 5%, the value of a, which represents the increment in the modulus of elasticity, decreased, whereas the value of b, which denotes the rate of increase in the modulus of elasticity, showed a tendency to slowly increase as the content of the expansive additive increased at 1 d of age. This means that when the expansive additive is added, the initial modulus of elasticity is high and the modulus of elasticity rapidly increases, but there is no significant difference in the final modulus of elasticity. As a result, the prediction curves for the modulus of elasticity as per the ACI 209 model showed a tendency similar to the actually measured modulus of elasticity. Regarding the curves for the modulus of elasticity predicted by the ACI 209 model, EA-7.5 exhibited a somewhat high modulus of elasticity in the long term, but the AAM mortar specimens with the expansive additive content of less than 5% exhibited no significant difference in the modulus of elasticity.

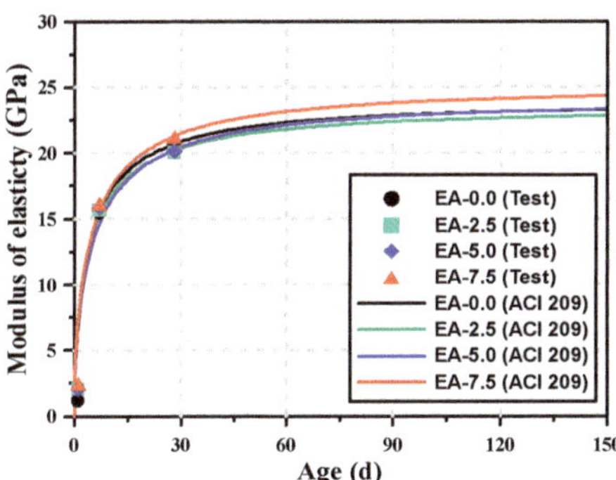

Figure 2. Comparison of modulus of elasticity between the test results and the results of the ACI 209 model.

Table 4. Material constants of the modulus of elasticity of AAM mortar.

Type	EA-0.0	EA-2.5	EA-5.0	EA-7.5
a	8.4	7.4	7.3	8.0
b	0.689	0.722	0.727	0.703
R^2	0.9915	0.9895	0.9889	0.9914

3.2. Shrinkage

Figure 3 shows the shrinkage test results of the AAM mortars at 3 and 150 d of age according to the content of the expansive additive. In the AAM mortar shrinkage test, measurement was started based on the final setting time. According to the results of various studies on AAMs, OPC-based mortar has large initial autogenous shrinkage, which tends to decrease over time. However, AAM mortar has higher initial autogenous shrinkage than OPC mortar and involves high shrinkage in the long term [11,36]. In this experiment, the shrinkage test results of the AAM mortars also showed that rapid shrinkage occurred until approximately 0.5 d and continuous shrinkage occurred until 150 d. EA-0.0 exhibited the highest total shrinkage and autogenous shrinkage because it had no shrinkage compensation effect caused by the expansive additive. As the content of the expansive additive increased, the shrinkage of the AAM mortar decreased. The shrinkage compensation effect of the expansive additive was most clearly observed within 1 d of age. Moreover, the shrinkage curves for up to 150 d of age showed that the shrinkage compensation effect of the expansive additive was not significant for autogenous shrinkage, but the opposite was true for total shrinkage.

The total shrinkage of AAM mortar is the sum of autogenous and drying shrinkage. Thus, drying shrinkage can be calculated using the measured total and autogenous shrinkage. Table 5 summarizes the autogenous, total, and drying shrinkage of the AAM mortars at 1 and 150 d of age. When the shrinkage of AAM mortars was analyzed at 1 d of age, the autogenous shrinkage reduction rates of EA-2.5, EA-5.0, and EA-7.5 were 23.3%, 27.0%, and 35.3% respectively compared to the autogenous shrinkage of EA-0.0, and their corresponding drying shrinkage reduction rates were 65.0%, 65.9%, and 85.1%. This indicates that the addition of expansive additive to AAM mortar reduces both autogenous and drying shrinkage at 1 d of age, but the influence on drying shrinkage is higher. When the shrinkage of the AAM mortars was analyzed at 150 d of age, the autogenous shrinkage reduction rates of EA-2.5, EA-5.0, and EA-7.5 were 1.8%, 3.9%, and 7.5% respectively compared to the autogenous shrinkage of EA-0.0; thus, the reduction rates were lower compared to those at 1 d of age. The drying shrinkage reduction rates of EA-2.5, EA-5.0, and EA-7.5 were 38.0%, 67.4%, and 71.8% respectively compared to the drying shrinkage of EA-0.0, indicating that the reduction rates were lower compared to those at 1 d of age even though shrinkage was reduced by the expansive additive in the long term. This result can be attributed to the initial shrinkage compensation effect of the CSA expansive additive, and the shrinkage reduction rate decreased in the long term because the shrinkage compensation effect of the expansive additive was not significant after 3 d of age.

Table 5. ε_{ash*}, ε_{tsh**}, ε_{dsh***} of AAM mortar at 1 and 150 d of age.

Type	ε_{ash} ($\mu\varepsilon$)		ε_{tsh} ($\mu\varepsilon$)		$\varepsilon_{dsh} = \varepsilon_{tsh} - \varepsilon_{ash}$ ($\mu\varepsilon$)	
	1 d	150 d	1 d	150 d	1 d	150 d
EA-0.0	−1096	−2330	−1559	−3096	−463	−766
EA-2.5	−859	−2290	−1003	−2765	−162	−475
EA-5.0	−800	−2211	−958	−2461	−158	−250
EA-7.5	−709	−2165	−778	−2381	−69	−216

* ε_{ash}: Autogenous shrinkage, ** ε_{tsh}: Total shrinkage, *** ε_{dsh}: Drying shrinkage.

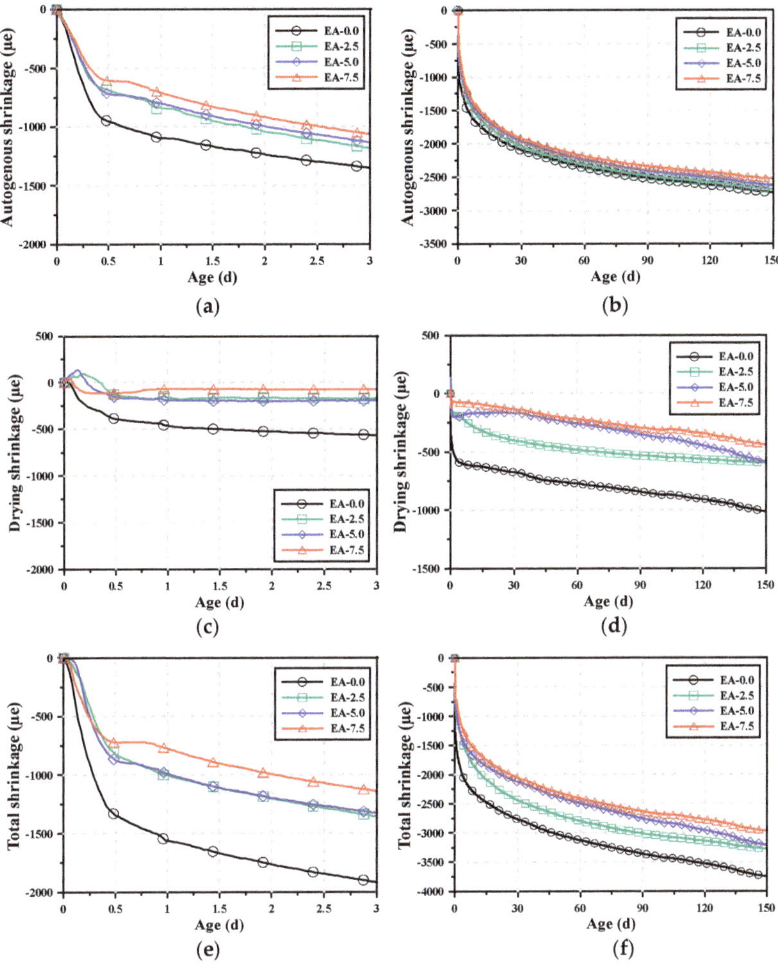

Figure 3. Autogenous shrinkage curves of AAM mortars: (**a**) autogenous shrinkage for 3 d, (**b**) autogenous shrinkage for 150 d, (**c**) drying shrinkage for 3 d, (**d**) drying shrinkage for 150 d, (**e**) total shrinkage for 3 d, and (**f**) total shrinkage for 150 d.

3.3. Shrinkage Modeling

The shrinkage of concrete and mortar is affected by various mixing and curing conditions, and it must be measured for a long period of time. Therefore, various models have been proposed to predict shrinkage in advance. The reaction mechanisms of the AAMs, however, are different from those of the existing OPC-based mortar and concrete. They are also characterized by different binder reaction times and reaction rates. Therefore, a model that reflects the effects of materials must be selected to predict the shrinkage of the AAM mortar using a model. Hu et al. [37] applied various models to predict the autogenous and drying shrinkage of alkali-activated slag mortar and examined their suitability. They also reported that the exponential function model is suitable for predicting the autogenous and drying shrinkage of AAM mortar. In the exponential function model, the effects of materials and mixing can be reflected through the values of the constants. Thus, an exponential function model was applied to analyze the shrinkage of the AAM mortars mixed with the expansive additive, as follows:

$$\varepsilon_{(t)} = \varepsilon_s[1 - \alpha \cdot \exp(-bt)] \tag{2}$$

In this exponential model, $\varepsilon_{(t)}$ is the shrinkage at t days of age and ε_{150} is the shrinkage at 150 d of age. Both α and b are material constants, where α denotes the generated amount of shrinkage and b is the slope of generated shrinkage over time, and subscripts $_a$ and $_d$ refer to the autogenous shrinkage and drying shrinkage respectively. Table 6 summarizes the material constants and amount of shrinkage at 150 d of age for the AAM mortars mixed with the expansive additive using the exponential model. When the material constants for the autogenous shrinkage of the AAM mortars were analyzed, the value of α_a for EA-0.0 was −0.499, while those for EA-2.5, EA-5.0, and EA-7.5, which used an expansive additive, ranged from −0.553 to −0.572. Thus, the values of α_a for the AAM mortar specimens that used the expansive additive were lower. The values of b_a, however, showed no significant difference despite a slight increase when the expansive additive was added. Moreover, when the material constants for the drying shrinkage of the AAM mortars were analyzed, it was found that the value of α_d for EA-0.0 was −0.485, but those for the AAM mortar specimens that used the expansive additive ranged from −0.700 to −0.944. Thus, the use of the expansive additive significantly decreased the value of α_d. The value of b_d, however, exhibited no significant difference, similar to the case of autogenous shrinkage.

Figure 4 shows the relationships between the expansive additive content and the coefficients of the shrinkage model. While the material constants of autogenous shrinkage did not exhibit significant changes even when the expansive additive content increased, α_d significantly decreased as the expansive additive content increased. This indicates that the CSA expansive additive has a larger impact on drying shrinkage than on autogenous shrinkage. An increase in the expansive additive content can compensate for and reduce shrinkage, but this mostly results from the expansion effect at early ages. The value of b was not affected by the expansive additive content. This is because the shrinkage reduction effect of the expansive additive was not significant in a long term.

Table 6. Material constants of the exponential function model for autogenous and drying shrinkage.

Type	Autogenous Shrinkage				Drying Shrinkage			
	α_a	b_a	$\varepsilon_{as,150}$	R^2	α_d	b_d	$\varepsilon_{ds,150}$	R^2
EA-0.0	−0.499	0.023	3745.2	0.9574	−0.485	0.0127	−1016.9	0.9977
EA-2.5	−0.553	0.025	3270.1	0.9566	−0.700	0.0136	−591.3	0.9752
EA-5.0	−0.560	0.024	3207.8	0.9605	−0.874	0.0097	−587.2	0.9575
EA-7.5	−0.572	0.026	2963.5	0.9565	−0.944	0.0121	−441.0	0.9521

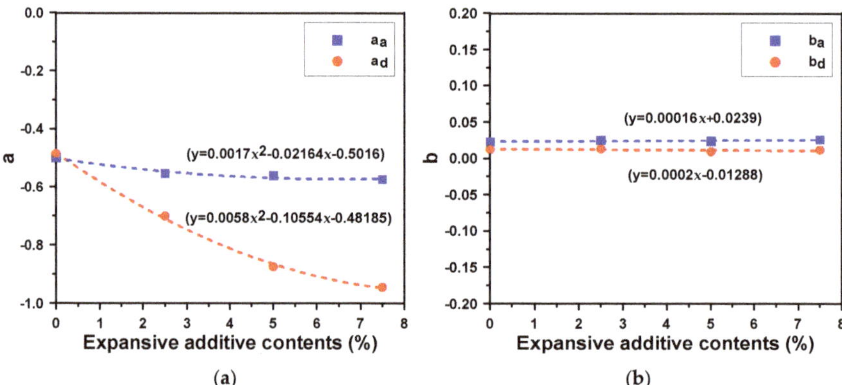

Figure 4. Relationships between the expansive additive content and material coefficients of the shrinkage model: (**a**) for coefficient "α", and (**b**) for coefficient "b".

3.4. Shrinkage Stress

To calculate the shrinkage stress of the AAM mortars, the prediction curves for the modulus of elasticity obtained by the ACI 209 model and the total shrinkage prediction curves of the AAM mortars obtained by the exponential function were used. The shrinkage stress of the AAM mortars ($f_{(sh,\,t+\Delta t)}$) was calculated by adding the stress caused by the shrinkage generated per unit time (Δf_{sh}) to the shrinkage stress acting on the AAM mortar ($f_{(sh,\,t)}$).

$$f_{(sh,\,t+\Delta t)} = f_{(sh,\,t)} + \Delta f_{sh}, \quad (3)$$

$$\Delta f_{sh} = E \cdot \Delta \varepsilon_{sh}. \quad (4)$$

Figure 5 shows the stress generated by shrinkage at 30 min intervals (Δf_{sh}) and the shrinkage stress accumulated in the AAM mortars (f_{sh}) over time. The shrinkage stress generated per unit time in Figure 5a shows that a large amount shrinkage occurred until 1 d of age, and thus, Δf_{sh} was also high even though the modulus of elasticity was low in that period. Δf_{sh} rapidly decreased after 1 d of age, but it showed a tendency to slowly increase as the age increased. This is because the modulus of elasticity increased with the age. For the AAM mortars, shrinkage stress continuously occurred until 150 d as total shrinkage increased. Δf_{sh} mostly ranged from 0.001–0.003 MPa, but it showed a higher range (0.004–0.006 MPa) within 60 d of age. Large shrinkage stress was observed in some sections because some differences were observed in $\Delta \varepsilon_{sh}$ when the shrinkage obtained by the data logger was divided into 30 min intervals. The increase in the expansive additive content decreased the maximum value of Δf_{sh} at 1 d of age, but the expansive additive content could not significantly affect Δf_{sh} after 1 d of age.

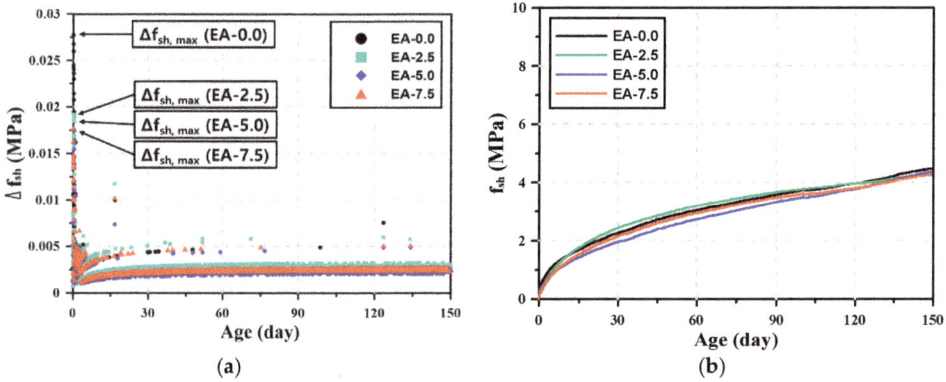

Figure 5. Stress generated by shrinkage. (**a**) Stress generated per unit time (Δf_{sh}), and (**b**) shrinkage stress (f_{sh}).

As shown in Figure 5b, the shrinkage stress accumulated in the AAM mortars (f_{sh}) was hardly affected by the expansive additive content. At early ages, the shrinkage stress decreased as the expansive additive content increased. After 6 d of age, however, such tendency was not observed. For the AAM mortars, the shrinkage stress continuously increased until 150 d of age. Table 7 summarizes the cumulative shrinkage stress at 1, 28, and 150 d of age. The shrinkage stress at 1 d of age ranged from 0.54 to 0.26 MPa, which was approximately 12.1–6.1% of the shrinkage stress at 150 d of age. The shrinkage stress at 28 d of age ranged from 2.37 to 1.90 MPa, which was approximately 55.5–43.5% of the shrinkage stress at 150 d of age. These results show that relatively larger shrinkage stress occurred at early ages. After 28 d of age, however, continuous shrinkage stress was observed in the AAM mortars. This value increased until 150 d without reduction even when the expansive additive was added. When the shrinkage stress values according to the expansive additive content at 1 and 150 d of age were compared, it was found that the shrinkage stress of the AAM mortars reduced by 29.6–51.9% at 1 d of age depending on the expansive additive content. However, the shrinkage stress reduction rate decreased to less than 5% at 150 d of age. This is because the expansive additive could not reduce the shrinkage stress in the long term as it reacted at early ages, resulting in an expansion, and could not generate further expansion thereafter.

Table 7. Shrinkage stress by age according to expansive additive content.

Type	f_{sh} (MPa)			Increase/Decrease Rate Compared to the Reference (EA-0.0) (%)		
	1 d	28 d	150 d	1 d	28 d	150 d
EA-0.0	0.54	2.20	4.48	0.0	0.0	0.0
EA-2.5	0.38	2.37	4.27	−29.6	+7.7	−4.7
EA-5.0	0.37	1.90	4.37	−31.5	−13.6	−2.5
EA-7.5	0.26	2.08	4.27	−51.9	−5.5	−4.7

4. Conclusions

The compressive strength, moduli of elasticity, and shrinkage test results of AAM mortars were obtained according to the content of the CSA expansive additive, and a shrinkage model was derived based on the results. In addition, the following conclusions were drawn from calculations of the shrinkage stress.

1. The addition of CSA expansive additive increased the compressive strength and modulus of elasticity of the AAM mortars at 1 d of age, and this tendency increased as the CSA expansive additive content increased. After 1 d of age, however, the addition of CSA expansive additive could hardly improve the compressive strength and moduli of elasticity of the AAM mortars.
2. The AAM mortars exhibited autogenous shrinkage to a greater extent than drying shrinkage, but the CSA expansive additive had a larger impact on reducing drying shrinkage. Therefore, limitations exist with regard to controlling the shrinkage of AAM mortar with the CSA expansive additive alone, and it is necessary to apply additional methods for controlling autogenous shrinkage.
3. A model capable of predicting the modulus of elasticity and shrinkage of AAM mortar by age was proposed, and the shrinkage stress of the AAM mortar mixed with the CSA expansive additive at 150 d of age was calculated using the proposed model. The result showed that the CSA expansive additive was suitable for controlling the cracks caused by the shrinkage of the AAM mortar at early ages, but it caused no improvement with regard to shrinkage generated in the long term. Therefore, to ensure effective control of the shrinkage stress of AAM mortar, it is necessary to accurately quantify the stress through long-term shrinkage stress monitoring and to apply techniques capable of controlling shrinkage stress properly depending on the time of shrinkage occurrence.
4. The results of this study in predicting cracks caused by the shrinkage of AAM mortar will be used for material improvement and curing management methods to minimize the shrinkage of AAM.

Author Contributions: All authors have contributed equally to the authorship of the article.

Funding: This research was supported by International Collaborative Research and Development Program (No. N051000001) funded by the Ministry of Trade, Industry and Energy (MOTIE, Korea).

Conflicts of Interest: The authors declare no conflict of interest.

References

1. Anand, S.; Vrat, P.; Dahiya, R.P. Application of a system dynamics approach for assessment and mitigation of CO_2 emissions from the cement industry. *J. Environ. Manag.* **2006**, *79*, 383–398. [CrossRef] [PubMed]
2. Rosković, R.; Bjegović, D. Role of mineral additions in reducing CO_2 emission. *Cem. Concr. Res.* **2005**, *35*, 974–978. [CrossRef]
3. Coppola, L.; Coffetti, D.; Crotti, E.; Pastore, T. CSA-based Portland-free binders to manufacture sustainable concretes for jointless slabs on ground. *Constr. Build. Mater.* **2018**, *187*, 691–698. [CrossRef]
4. Khale, D.; Chaudhary, R. Mechanism of geopolymerization and factors influencing its development: A review. *J. Mater. Sci.* **2007**, *42*, 729–746. [CrossRef]
5. Pacheco-T, F.; Castro-Gomes, J.; Jalali, S. Alkali-activated binders: A review: Part 1. Historical background, terminology, reaction mechanisms and hydration products. *Constr. Build. Mater.* **2008**, *22*, 1305–1314. [CrossRef]
6. Duxson, P.; Provis, J.L.; Lukey, G.C.; Van Deventer, J.S. The role of inorganic polymer technology in the development of 'green concrete'. *Cem. Concr. Res.* **2007**, *37*, 1590–1597. [CrossRef]
7. Duxson, P.; Fernández-Jiménez, A.; Provis, J.L.; Lukey, G.C.; Palomo, A.; van Deventer, J.S. Geopolymer technology: The current state of the art. *J. Mater. Sci.* **2007**, *42*, 2917–2933. [CrossRef]
8. Choi, S.; Lee, K.M. Influence of Na_2O Content and Ms (SiO_2/Na_2O) of Alkaline Activator on Workability and Setting of Alkali-Activated Slag Paste. *Materials* **2009**, *12*, 2072. [CrossRef]
9. Coppola, L.; Coffetti, D.; Crotti, E.; Marini, A.; Passoni, C.; Pastore, T. Lightweight cement-free alkali-activated slag plaster for the structural retrofit and energy upgrading of poor quality masonry walls. *Cem. Concr. Compos.* **2019**, *104*, 103341. [CrossRef]
10. Cartwright, C.; Rajabipour, F.; Radlińska, A. Shrinkage characteristics of alkali-activated slag cements. *J. Mater. Civ. Eng.* **2014**, *27*, B4014007. [CrossRef]
11. Kumarappa, D.B.; Peethamparan, S.; Ngami, M. Autogenous shrinkage of alkali activated slag mortars: Basic mechanisms and mitigation methods. *Cem. Concr. Res.* **2018**, *109*, 1–9. [CrossRef]
12. Atiş, C.D.; Bilim, C.; Çelik, Ö.; Karahan, O. Influence of activator on the strength and drying shrinkage of alkali-activated slag mortar. *Constr. Build. Mater.* **2009**, *23*, 548–555. [CrossRef]
13. Coppola, L.; Coffetti, D.; Crotti, E. Pre-packed alkali activated cement-free mortars for repair of existing masonry buildings and concrete structures. *Constr. Build. Mater.* **2018**, *173*, 111–117. [CrossRef]
14. Neto, A.A.M.; Cincotto, M.A.; Repette, W. Drying and autogenous shrinkage of pastes and mortars with activated slag cement. *Cem. Concr. Res.* **2008**, *38*, 565–574. [CrossRef]
15. Palacios, M.; Puertas, F. Effect of shrinkage-reducing admixtures on the properties of alkali-activated slag mortars and pastes. *Cem. Concr. Res.* **2007**, *37*, 691–702. [CrossRef]
16. Ye, H.; Cartwright, C.; Rajabipour, F.; Radlińska, A. Understanding the drying shrinkage performance of alkali-activated slag mortars. *Cem. Concr. Res.* **2017**, *76*, 13–24. [CrossRef]
17. Puertas, F.; Palacios, M.; Vázquez, T. Carbonation process of alkali-activated slag mortars. *J. Mater. Sci.* **2006**, *41*, 3071–3082. [CrossRef]
18. Marjanović, N.; Komljenović, M.; Baščarević, Z.; Nikolić, V. Comparison of two alkali-activated systems: Mechanically activated fly ash and fly ash-blast furnace slag blends. *Procedia Eng.* **2015**, *108*, 231–238. [CrossRef]
19. Bilek, V.; Hurta, J.; Done, P.; Zidek, L. Development of alkali-activated concrete for structures—Mechanical properties and durability. *Perspect. Sci.* **2016**, *7*, 190–194. [CrossRef]
20. Torres-Carrasco, M.; Tognonvi, M.T.; Tagnit-Hamou, A.; Puertas, F. Durability of alkali-activated slag concretes prepared using waste glass as alternative activator. *ACI Mater. J.* **2015**, *112*, 791–800. [CrossRef]
21. Al-Otaibi, S. Durability of concrete incorporating GGBS activated by water-glass. *Constr. Build. Mater.* **2008**, *22*, 2059–2067. [CrossRef]
22. Chatterji, S. Mechanism of expansion of concrete due to the presence of dead burnt CaO and MgO. *Cem. Concr. Res.* **1995**, *25*, 51–56. [CrossRef]
23. Collepardi, M. *Scienza e Tecnologia Del Calcestruzzo*, 2nd ed.; Hoepli Editore: Milano, Italy, 1987; p. 230.
24. Maltese, C.; Pistolesi, C.; Lolli, A.; Bravo, A.; Cerulli, T.; Salvioni, D. Combined effect of expansive and shrinkage reducing admixtures to obtain stable and durable mortars. *Cem. Concr. Res.* **2005**, *35*, 2244–2251. [CrossRef]

25. Thomas, R.J.; Lezama, D.; Peethamparan, S. On drying shrinkage in alkali-activated concrete: Improving dimensional stability by aging or heat-curing. *Cem. Concr. Res.* **2017**, *91*, 13–23. [CrossRef]
26. Sakulich, A.R.; Bentz, D.P. Mitigation of autogenous shrinkage in alkali activated slag mortars by internal curing. *Mater. Struct.* **2013**, *46*, 1355–1367. [CrossRef]
27. Collins, F.; Sanjayan, J.G. Microcracking and strength development of alkali activated slag concrete. *Cem. Concr. Compos.* **2001**, *23*, 345–352. [CrossRef]
28. Collins, F.; Sanjayan, J.G. Strength and shrinkage properties of alkali-activated slag concrete containing porous coarse aggregate. *Cem. Concr. Res.* **1999**, *29*, 607–610. [CrossRef]
29. Bakharev, T.; Sanjayan, J.G.; Cheng, Y.B. Alkali activation of Australian slag cements. *Cem. Concr. Res.* **1999**, *29*, 113–120. [CrossRef]
30. Korean Industrial Standards. *Fly Ash [KS L 5405]*; Korean Agency for Technology and Standards: Seoul, Korean, 2016.
31. Chang, J.J. A study on the setting characteristics of sodium silicate-activated slag pastes. *Cem. Concr. Res.* **2003**, *33*, 1005–1011. [CrossRef]
32. *Standard Test Method for Compressive Strength of Hydraulic Cement Mortars (Using 2-in. or [50-mm] Cube Specimens)*; ASTM C109/C109M-16a; ASTM International: West Conshohocken, PA, USA, 2016.
33. *Standard Test Method for Static Modulus of Elasticity and Poisson's Ratio of Concrete in Compression*; ASTM C469M-14; ASTM International: West Conshohocken, PA, USA, 2014.
34. Videla, C.; Carreira, D.J.; Garner, N. *Guide for modeling and calculating shrinkage and creep in hardened concrete*; ACI Report 209; American Concrete Institute: Farmington Hills, MI, USA, 2008.
35. Nagataki, S.; Gomi, H. Expansive admixtures (mainly ettringite). *Cem. Concr. Compos.* **1998**, *20*, 163–170. [CrossRef]
36. Lee, K.M.; Lee, H.K.; Lee, S.H.; Kim, G.Y. Autogenous shrinkage of concrete containing granulated blast-furnace slag. *Cem. Concr. Res.* **2006**, *36*, 1279–1285. [CrossRef]
37. Hu, X.; Shi, C.; Zhang, Z.; Hu, Z. Autogenous and drying shrinkage of alkali-activated slag mortars. *J. Am. Ceram. Soc.* **2019**, *102*, 4963–4975. [CrossRef]

Article

Tensile Behavior Characteristics of High-Performance Slurry-Infiltrated Fiber-Reinforced Cementitious Composite with Respect to Fiber Volume Fraction

Seungwon Kim [1,*], Dong Joo Kim [2], Sung-Wook Kim [3] and Cheolwoo Park [1,*]

1 Department of Civil Engineering, Kangwon National University, 346 Jungang-ro, Samcheok 25913, Korea
2 Department of Civil and Environmental Engineering, Sejong University, 209 Neungdong-ro, Gwangjin-gu, Seoul 05006, Korea; djkim75@sejong.ac.kr
3 Department of Infrastructure Safety Research, Korea Institute of Civil Engineering and Building Technology 283, Goyang-daero, Ilsanseo-gu, Goyang 10223, Korea; swkim@kict.re.kr
* Correspondence: inncoms@kangwon.ac.kr (S.K.); tigerpark@kangwon.ac.kr (C.P.); Tel.: +82-33-570-6518 (S.K.); +82-33-570-6515 (C.P.)

Received: 19 September 2019; Accepted: 11 October 2019; Published: 13 October 2019

Abstract: Concrete has high compressive strength, but low tensile strength, bending strength, toughness, low resistance to cracking, and brittle fracture characteristics. To overcome these problems, fiber-reinforced concrete, in which the strength of concrete is improved by inserting fibers, is being used. Recently, high-performance fiber-reinforced cementitious composites (HPFRCCs) have been extensively researched. The disadvantages of conventional concrete such as low tensile stress, strain capacity, and energy absorption capacity, have been overcome using HPFRCCs, but they have a weakness in that the fiber reinforcement has only 2% fiber volume fraction. In this study, slurry infiltrated fiber reinforced cementitious composites (SIFRCCs), which can maximize the fiber volume fraction (up to 8%), was developed, and an experimental study on the tensile behavior of SIFRCCs with varying fiber volume fractions (4%, 5%, and 6%) was carried out through direct tensile tests. The results showed that the specimen with high fiber volume fraction exhibited high direct tensile strength and improved brittleness. As per the results, the direct tensile strength is approximately 15.5 MPa, and the energy absorption capacity was excellent. Furthermore, the bridging effect of steel fibers induced strain hardening behavior and multiple cracks, which increased the direct tensile strength and energy absorption capacity.

Keywords: SIFRCC; fiber volume fraction; direct tensile strength; energy absorption capacity; direct tensile test

---

1. Introduction

Concrete is widely used in architectural structures and social infrastructure facilities because it is economical and has high compressive strength and durability. However, concrete is characterized by brittle fracture due to low bending and tensile strengths and weak crack resistance compared to the high compressive strength [1–6]. Recently, studies have been conducted to develop high-performance construction materials with excellent performance by improving the disadvantages or maximizing the advantages of concrete. Attempts are being made to mechanically increase strength or improve ductility, because strength and ductility have opposite properties [7].

With the development of construction technology, the construction of high-rise of buildings and long structures, the use of 100 MPa or higher ultra-high-strength concrete is increasing [8]. However, although ultra-high-strength concrete has high compressive strength, it has low tensile strength, bending strength, and toughness, and weak resistance to cracks [8]. Above all, ultra-high-strength concrete has the intrinsic problem of brittle fracture to the peak stress. To overcome this problem, active studies are being conducted on high-performance fiber-reinforced cementitious composites (HPFRCCs), in which brittle fracture to ductile fracture are induced [8].

HPFRCCs are characterized by improved tensile force, strain capacity, and energy absorption capacity, which are the weaknesses of conventional concrete [9,10]. However, the maximum fiber volume fraction of conventional HPFRCCs and fiber-reinforced concrete is limited to 2.0% due to fiber balling [9,10]. A majority of fiber balling occurs during the fiber addition process due to excessive fibers during the mixing of HPFRCCCs and fiber-reinforced concrete. Thus, the limited fiber volume fraction has been one of the biggest disadvantages. Studies are being actively conducted to understand the direct tensile behavior characteristics of HPFRCCs [9,10]. To overcome these disadvantages, an increase in the steel fiber volume fraction may improve tensile strength and energy absorption capacity. Therefore, HPFRCCs reported thus far have exhibited deflection hardening behavior under flexural tensile load, rather than direct tensile behavior and strain hardening behavior with multiple micro-cracks [9,10]. Therefore, it is very difficult to obtain a tensile stress-strain curve under direct tensile load to acquire information on multiple cracks [9,10].

To maximize the mechanical properties of HPFRCCs and overcome the limitation of fiber volume fraction, this study developed slurry-infiltrated fiber-reinforced cementitious composites (SIFRCCs), which can incorporate a high volume of steel fibers. The SIFRCCs can incorporate up to 8% fiber volume fraction, thus maximizing tensile strength, energy absorption capacity, and strain capacity, which are shortcomings of the conventional concrete and fiber-reinforced concrete. An experimental research on the tensile behavior characteristics was conducted with respect to the fiber volume fraction of high-performance SIFRCCs through a direct tensile test.

2. Existing Works Related to Direct Tensile Test

In a study on the flexural tensile strength of fiber-reinforced concrete members, the ductile behavior improved after cracking at 80 MPa or lower compressive strength. However, studies on structural behavior analysis of members of 150 MPa or higher compressive strength are relatively insufficient and predictions of the bending strength are limited [8,11]. Flexural tensile tests of fiber-reinforced concrete involve many difficulties, but flexural tensile test are mainly conducted for direct tensile tests because of the problems of slip phenomenon of specimens in the drawing process [8]. However, attempts at direct tensile tests are being made continuously because reliability can be compromised due to many assumption conditions in the process of estimating tensile strength through flexural tensile tests. To address this problem, French regulations have presented a method of performing tests by directly installing notches in the specimens [12].

According to a research report on ultra high performance concrete (UHPC), crack review appears to be unnecessary for UHPC considering its higher tensile strength than conventional concrete due to the action of steel fibers, and the characteristic of low crack width relative to the tensile load also appears to be unnecessary [8,13,14]. The report mentioned that even though the crack width is small, it is necessary to examine crack behavior under various load conditions [8,15]. It also stated that because the crack behavior can vary under the working load and extreme load due to the ductile tensile behavior, it is necessary to validate the crack examination of UHPC with the existing design standards through direct tensile test [8,15]. Furthermore, unlike the direct tensile test, indirect tensile tests, such as tensile test through bending, are burdensome because they require inverse analysis using numerical analysis [8,15]. Although direct tensile tests are not widely used in the measurement of the tensile behavior of concrete, they have been considered to be the most direct and proper method owing to the characteristics of UHPC that exhibits ductile behavior after cracking [8,15].

3. Experiment Overview

3.1. SIFRCCs

The SIFRCCs developed in this study can maximize the steel fiber volume fraction of the existing fiber-reinforced concrete and HPFRCCs. With its high fiber volume ratio, high tensile strength, energy absorption capacity, and strain capacity can be expected. Unlike fiber-reinforced concrete and HPFRCCs, SIFRCCs can be produced by the following steps. First, the mold is filled with steel fibers in advance. Second, high-performance slurry is prepared after mixing. This slurry should be poured to avoid mixing the concrete matrix and steel fibers. The high-performance slurry should be pour from one end to the other within the cluster of steel fibers such that there are no bubbles. This is done to avoid voids. SIFRCCs are characterized by the omission of coarse aggregates for the high-performance slurry to achieve sufficient filling performance between the steel fibers [9,16].

3.2. Experiment Method

To analyze the compressive strength of SIFRCCs, fiber volume fractions of 4%, 5%, and 6% were considered. The compressive strength test was performed in accordance with Korea Standards (KS F 2405) [17]. The cylinder specimens used had a diameter of 100 mm and a height of 200 mm.

To analyze the tensile behavior characteristics with respect to the fiber volume fraction of SIFRCCs, the characteristics of direct tensile behavior were experimented. In the case of direct tensile test, the test method is not clearly defined. Therefore, we conducted an experimental study on the tensile behavior characteristics such as energy absorption capacity through the direct tensile strength, which indicates the maximum tensile stress, strain capacity at the direct tensile strength, and stress-strain curve based on literature review.

To analyze the tensile behavior characteristics of SIFRCCs, fiber volume fractions of 4%, 5%, and 6% were considered. The direct tensile test was performed by displacement control method at the rate of 1 mm/min. A direct tensile test specimen appropriate for the dedicated tensile jig was fabricated as shown in Figure 1. The tensile behavior characteristics were analyzed using the 300-ton class universal testing machine shown in Figure 2. The cross-section of the direct tensile test specimen has a width of 50 mm and a height of 25 mm. The scope of gauge length for tensile performance measurement was set to 50 mm. To induce multiple cracks of SIFRCCs within the range of gauge length, a wire mesh was installed outside the gauge length and used facilitate crack inducement. Furthermore, to derive the tensile stress-strain curve of the SIFRCCs, a linear variable differential transformer (LVDT) (Tokyo Sokki, Tokyo, Japan) that can take measurements up to 25 mm was installed on either side of the tensile jig. The direct tensile strength was calculated using the following Equation (1) [7]:

$$f = \frac{P_{max}}{bh} \tag{1}$$

where P_{max} is the maximum load (N), b is the width (mm) at the gauge length, and h. is the height at the gauge length (mm).

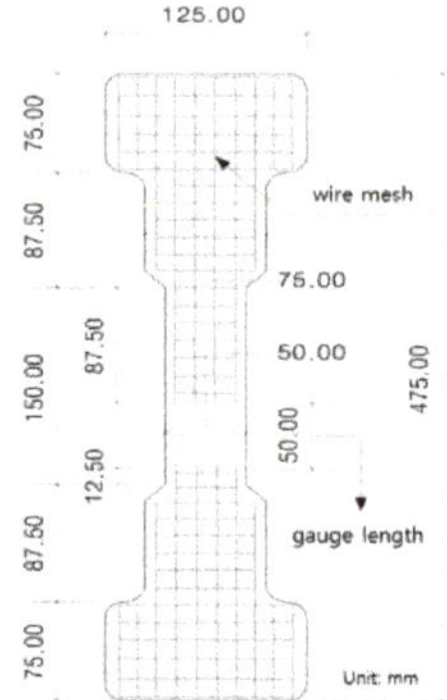

Figure 1. Schematic of the specimen.

Figure 2. Experimental setup.

3.3. Materials

This study used type 1 ordinary Portland cement. Table 1 lists the physical and chemical properties of the used cement. In this study, silica fume was used to achieve high-performance and high strength of the slurry. Table 2 lists the physical and chemical properties of the used silica fume.

Table 1. Physical and chemical properties of the used cement.

Physical Properties					
Specific Gravity	**Fineness (cm^2/g)**	**Stability (%)**	**Setting Time (min)**		**LOI (%)**
			Initial	**Final**	
3.15	3400	0.10	230	410	2.58
Chemical compositions (%, mass)					
SiO_2	CaO	MgO	SO_3	Al_2O_3	
21.95	60.12	3.32	2.11	6.59	

Table 2. Physical and chemical properties of silica fume.

Physical Properties				
Specific Gravity			**Fineness (cm^2/g)**	
2.10			200,000	
Chemical compositions (%, mass)				
SiO_2	CaO	MgO	SO_3	Al_2O_3
96.00	0.38	0.10	-	0.25

Furthermore, fine aggregates with a diameter of 0.5 mm or less were used to improve the filling performance of the high-performance slurry and to reduce material separation. Coarse aggregates were not used to secure filling performance. To improve the filling performance of the slurry, a high-performance polycarboxylic acid water-reducing agent was used. The admixture used in this experiment has high strength and high flow characteristics and has excellent unit water quantity reduction property and material separation resistance.

For steel fibers, double hook steel fibers for conventional concrete with a diameter of 0.75 mm, a length of 60 mm, and an aspect ratio of 80 were used. Regarding physical properties, the steel fibers have a density of 7.8 g/cm^3 and a tensile strength of 1200 MPa. Figure 3 shows the shape of the used steel fibers.

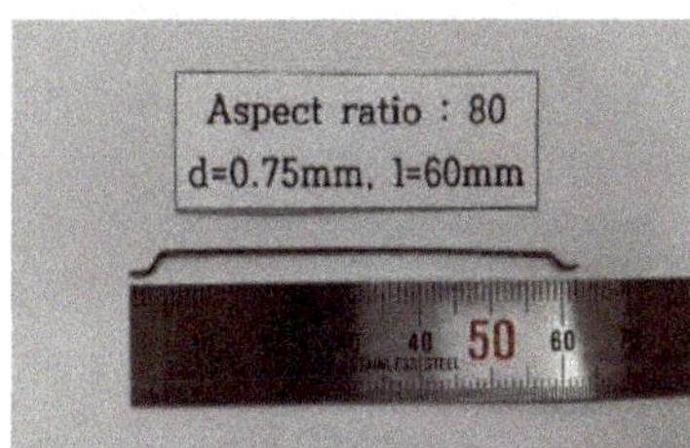

Figure 3. Shape of the used steel fibers.

3.4. Mixing and Fabrication of Specimens

To mix the SIFRCCs, the water-binder ratio was fixed to 0.35 to achieve the filling performance of the high-performance slurry for filling the inner space of the steel fibers that were placed in advance. The amount of the high range water reducing (HRWR) agent was set to 2.5% of the binder weight. To reduce material separation and achieve the required strength, fine aggregates were added for 0.5% of binder weight and the silica fume was added for 15% of the cement weight. Table 3 shows the SIFRCC mixing formula. The fiber volume fraction variables were set to 4%, 5%, and 6%.

Table 3. Slurry-Infiltrated Fiber-Reinforced Cementitious Composites (SIFRCCs) mixing formula.

Variables	W/B (%)	Unit Material Quantity (kg/m^3)					
		W	**C**	**Fine Aggregate**	**Silica Fume**	**HRWR**	**Steel Fibers**
4%	35	407.4	962.8	566.4	169.9	28.3	312
5%							390
6%							468

To analyze the tensile behavior characteristics of the SIFRCCs with respect to the fiber volume fraction of 4%, 5%, and 6% in the direct tensile test, direct tensile test specimens were fabricated with the mixing ratio of each variable in Table 3.

4. Results and Analysis

4.1. Compressive Strength

Figure 4 shows the compressive strength test results with respect to the fiber volume fraction of SIFRCCs. In the case of 6% fiber volume fraction, the average compressive strength was analyzed to be approximately 83 MPa. The average compressive strength of 5% fiber volume fraction was approximately 75 MPa, lower by approximately 10%. Furthermore, the average compressive strength of 4% fiber volume fraction was approximately 66 MPa lower by approximately 12% compared to the 5% fiber volume fraction and by approximately 21% compared to the 6% fiber volume fraction. The compressive strength increased in proportion to the fiber volume fraction. The input amount of steel fibers appeared to increase with increasing fiber volume fraction, which generated the restraining effect of the specimen itself, and this affected the increase of compressive strength. Figure 5 shows the result of the compressive stress-strain test with respect to the fiber volume fraction.

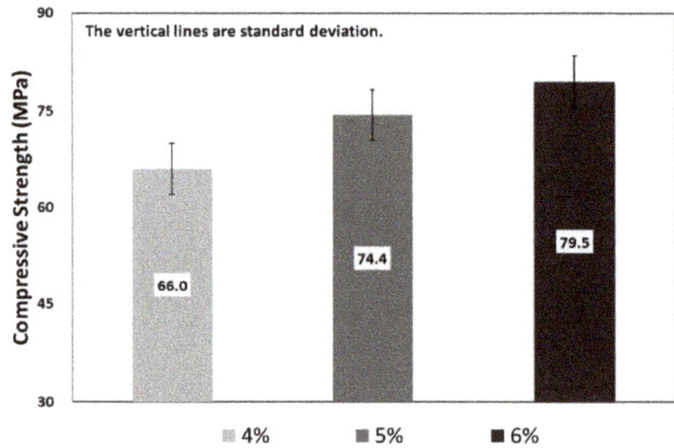

Figure 4. Compressive strength test results with respect to the fiber volume fraction.

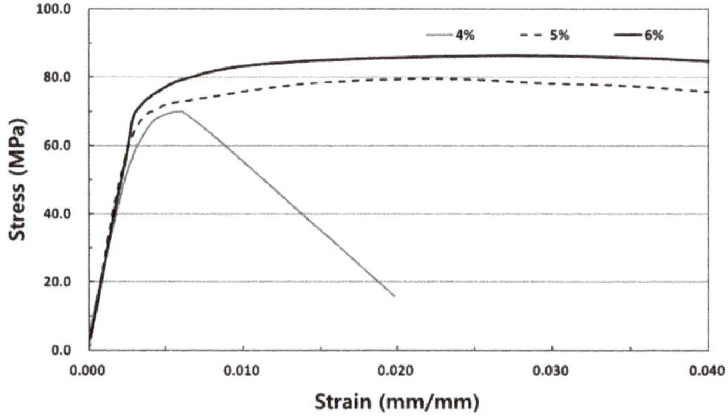

Figure 5. Compressive stress-strain curve with respect to the fiber volume fraction.

4.2. Direct Tensile Strength

Figure 6 shows the results of the direct tensile strength test with respect to the fiber volume fraction of SIFRCCs. The direct tensile strength test result of 6% fiber volume fraction showed a high average direct tensile strength of approximately 15.5 MPa. The average direct tensile strength of 5% fiber volume fraction was approximately 14.2 MPa, lower by approximately 9%. The average direct tensile strength of 4% fiber volume fraction was approximately 11.0 MPa, lower by approximately 23% compared to 5% fiber volume fraction and by approximately 30% compared to 6% fiber volume fraction.

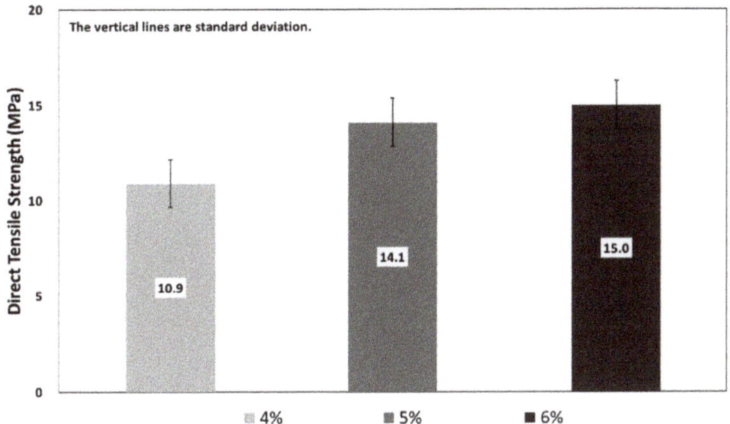

Figure 6. Direct tensile strength with respect to fiber volume fraction.

The direct tensile strength test of the SIFRCCs showed that the cracks gradually spread and lead to fracture after the initial cracking due to the reinforcement of steel fibers. This phenomenon was evident as the fiber volume fraction increased. Furthermore, the direct tensile strength also showed an increasing trend with the increasing fiber volume fraction, similar to the compressive strength test result of the SIFRCCs with respect to the fiber volume fraction.

4.3. *Strain Capacity and Tensile Stress-Strain Curve*

Figure 7 shows the strain capacity test result with respect to the fiber volume fraction at the direct tensile strength. The result of the strain capacity test under the direct tensile strength with respect to the fiber volume fraction of SIFRCCs verified excellent strain capacity of 0.7% (0.007) at 5% fiber volume fraction. Since the direct tensile strength increased with increasing fiber volume fraction, the strain capacity was expected increase as well, but the strain capacity was the lowest at 6% fiber volume fraction. This is because in the case of 6% fiber volume fraction, the steel fibers resist the direct tensile load as the fiber amount increases, and due to the small cross-section size (25×50 mm^2) of the direct tensile test specimen, the adhesion performance of the high-performance slurry matrix and steel fibers decreased. Furthermore, considering that the length of the steel fibers is 60 mm, the size of the specimens is considered to be affected by the specimens because the arrangement of steel fibers was parallel to the tensile load when the direct tensile test specimen was fabricated.

Figure 8 shows the tensile stress-strain curve with respect to the fiber volume fraction of the SIFRCCs, and Figure 9 shows the tensile stress-strain curve of the 6% fiber volume fraction of the SIFRCCs. The strain at the direct tensile strength was analyzed to be 0.53% (0.0053), and the energy absorption capacity was 62.10 kJ/m^3, which was the lowest among all variables. Regarding the compressive stress-strain test result for 6% fiber volume fraction, the post-peak behavior exhibited a strain hardening behavior, but the post-peak behavior of the tensile stress-strain curve test result exhibited a strain softening behavior. Considering that the cross-section size of the gauge length for measurement is 25×50 mm^2, the result was somewhat different from the compression behavior because the adhesion performance between the high-performance slurry matrix and steel fibers was insufficient.

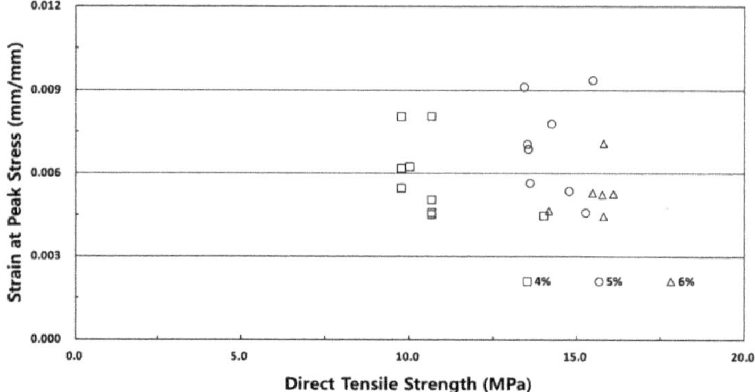

Figure 7. Strain capacity test results at the direct tensile strength.

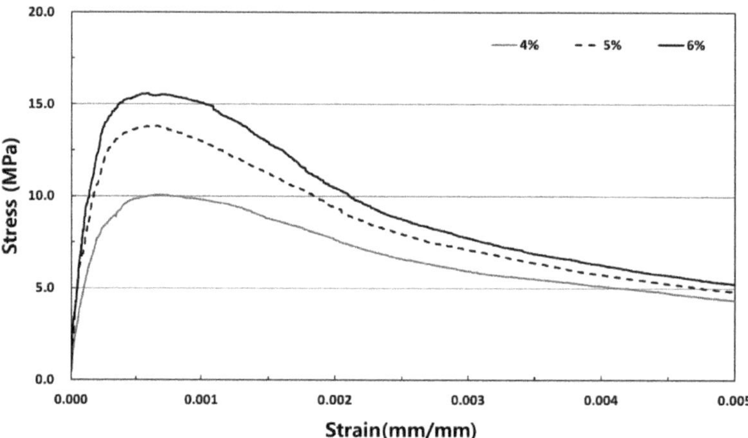

Figure 8. Tensile stress-strain curve with respect to the fiber volume fraction.

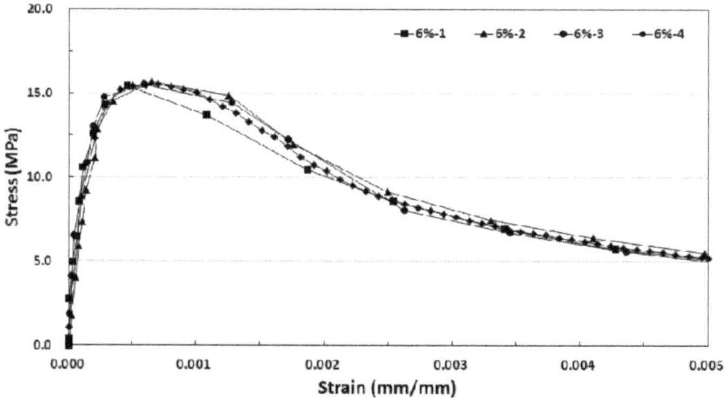

Figure 9. Tensile stress-strain curve of Slurry-Infiltrated Fiber-Reinforced Cementitious Composites (SIFRCCs) with 6% fiber volume fraction.

Figure 10 shows the tensile stress-strain curve of the SIFRCCs with 5% fiber volume fraction. The strain at the direct tensile strength was 0.7% (0.0070), indicating the highest strain capacity among all variables. The energy absorption capacity was also the highest at 88.05 kJ/m^3 on average. Similar to the tensile stress-strain test result for the specimen with 6% fiber volume fraction, the post-peak behavior exhibited a strain softening behavior.

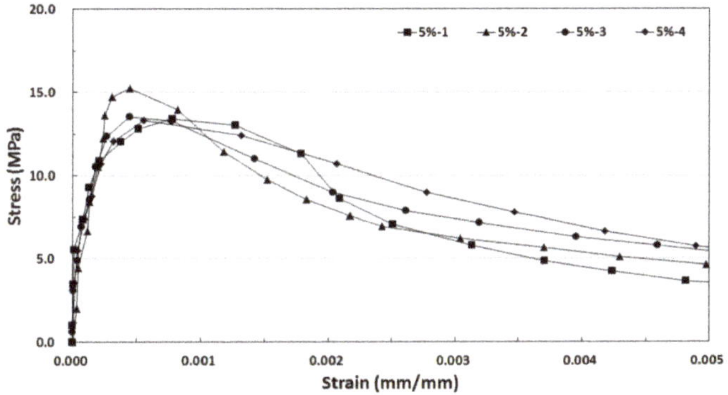

Figure 10. Tensile stress-strain curve of SIFRCCs with 5% fiber volume fraction.

Figure 11 shows the tensile stress-strain curve of SIFRCCs with 4% fiber volume fraction. Unlike the tensile stress-strain test result for 5% and 6% fiber volume fractions, the post-peak behavior characteristics were closest to the strain hardening behavior. The direct tensile strength for 4% fiber volume fraction was lower than those of others, but the energy absorption capacity was 62.27 kJ/m^3, which is higher than that of the 6% fiber volume fraction (the highest direct tensile strength). This is considered to be because the size of the direct tensile test specimen is too small and the interface adhesion property between the high-performance slurry and steel fibers did not reach the maximum, resulting in different tensile behavior characteristics for each fraction.

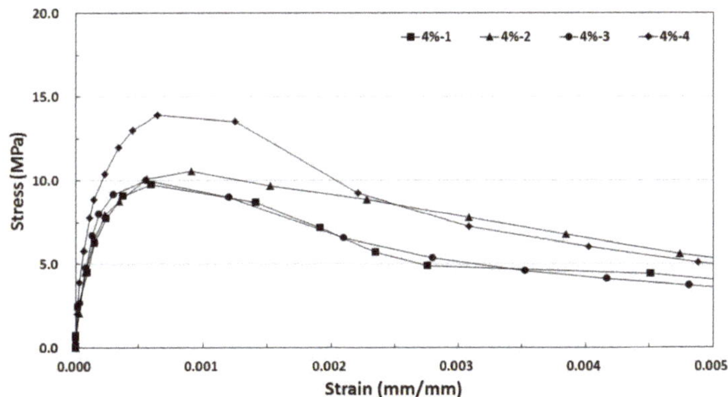

Figure 11. Tensile stress-strain curve of SIFRCCs with 4% fiber volume fraction.

5. Conclusions

To overcome the limited fiber volume fraction of the conventional fiber-reinforced concrete and HPFRCCs, this study developed SIFRCCs that contains a high fiber volume fraction to maximize the tensile strength, energy absorption capacity and strain capacity. An experimental study on tensile behavior characteristics was conducted with respect to different fiber volume fractions of the high-performance SIFRCCs through direct tensile tests. The conclusions of this study are as follows.

(1) The analysis result of the tensile behavior characteristics through direct tensile tests of the SIFRCCs showed a high direct tensile strength, more than 15 MPa (Vf = 6%), which is higher than conventional HPFRCCs and UHPC due to increasing the fiber volume fraction. Also the load of SIFRCCs with respect to the fiber volume fraction continuously increased because of the high fiber volume fraction after the initial crack, and sufficient residual strength was obtained after the maximum strength. This sufficient residual strength is expected to bring about positive effects to the brittle fracture of structures when unexpected loads is applied.

(2) The reinforcement with a high fiber volume fraction improved brittleness, which is a disadvantage of conventional concrete. After the initial cracking, cracks gradually spread and led to fracture. The maximum strain capacity was approximately 0.7%, which showed excellent energy absorption capacity. However, the interface adhesion performance between the high-performance slurry and steel fibers was insufficient due to small cross-section of the direct tensile test specimen. It is expected that the strain capacity and energy absorption capacity can be improved by increasing the cross-section size.

(3) The energy absorption capacity increased but the strain capacity tended to decrease with increasing fiber volume fractions. It is that the size of the direct tensile specimen is too small to exert the maximum interface adhesive characteristics between the high-performance slurry and steel fibers, indicating different direct tensile behavior for each fiber volume fraction.

(4) The bridging effect of steel fibers caused strain hardening behavior and multiple cracks, resulting in the increase of the direct tensile strength and energy absorption capacity. In the case of the strain capacity, it was suppressed by the high-performance slurry restraint due to the increase of the adhesion of the slurry and steel fibers.

Author Contributions: Conceptualization, S.K., C.P., D.J.K., S.-W.K.; methodology, S.K.; writing—original draft preparation, S.K., C.P., D.J.K., S.-W.K.

Funding: This research was supported by Construction Technology Research Project funded by the Ministry of Land, Infrastructure and Transport of Korea government, grant number 18SCIP-B146646-01.

Conflicts of Interest: The authors declare no conflicts of interest.

References

1. Kim, H.-S.; Kim, G.-Y.; Lee, S.-K.; Choe, G.-C.; Nam, J.-S. Direct Tensile Properties of Fiber-Reinforced Cement Based Composites according to the Length and Volume Fraction of Amorphous Metallic Fiber. *J. Korea Inst. Build. Constr.* **2019**, *19*, 201–207.
2. Naaman, A.E.; Shah, S.P. Pull-out mechanism in steel fiber-reinforced concrete. *J. Struct. Div.* **1976**, *102*, 1537–1548.
3. Chan, Y.-W.; Chu, S.-H. Effect of silica fume on steel fiber bond characteristics in reactive powder concrete. *Cem. Concr. Res.* **2004**, *34*, 1167–1172. [CrossRef]
4. Li, V.C.; Wang, Y.; Backer, S. Effect of inclining angle, bundling and surface treatment on synthetic fibre pull-out from a cement matrix. *Composites* **1990**, *21*, 132–140. [CrossRef]
5. Kim, H.-S.; Kim, G.-Y.; Lee, S.-K.; Son, M.-J.; Choe, G.-C.; Nam, J.-S. Strain rate effects on the compressive and tensile behavior of bundle-type polyamide fiber-reinforced cementitious composites. *Compos. Part B Eng.* **2019**, *160*, 50–65. [CrossRef]
6. Nam, J.-S.; Shinohara, Y.; Atou, T.; Kim, H.-S.; Kim, G.-Y. Comparative assessment of failure characteristics on fiber-reinforced cementitious composite panels under high-velocity impact. *Compos. Part B Eng.* **2016**, *99*, 84–97. [CrossRef]
7. Choi, J.-I.; Park, S.-E.; Lee, B.Y.; Kim, Y.Y. Tensile Properties of Polyethylene Fiber-Reinforced Highly Ductile Composite with Compressive Strength of 100 MPa Class. *J. Korea Concr. Inst.* **2018**, *30*, 497–503. [CrossRef]
8. Park, J.W.; Lee, G.C. Effect of Compressive Strength and Curing Condition on the Direct Tensile Strength Properties of Ultra High Performance Concrete. *J. Korea Inst. Build. Constr.* **2017**, *17*, 175–181. [CrossRef]

9. Kim, S.; Jung, H.; Kim, Y.; Park, C. Effect of steel fiber volume fraction and aspect ratio type on the mechanical properties of SIFCON-based HPFRCC. *Struct. Eng. Mech.* **2018**, *65*, 163–171.
10. Kim, S.-W.; Cho, H.-M.; Lee, H.-Y.; Park, C.-W. Flexural Performance Characteristics of High Performance Slurry Infiltrated Fiber Reinforced Cementitious Composite according to Fiber Volume Fraction. *J. Korea Inst. Struct. Maint. Insp.* **2015**, *19*, 109–115.
11. Yang, I.-H.; Kim, K.-C.; Joh, C.-B. Flexural strength of hybrid steel fiber-reinforced ultra-high strength concrete beams. *J. Korea Concr. Inst.* **2015**, *27*, 280–287.
12. AFNOR. *NF P 18–470*; AFNOR: Pairs, France, 2016; p. 96.
13. Lee, J.W.; Joh, C.B. The effect of specimen size & shape on tensile strength of K-UHPC. In Proceedings of the Korea Concrete Institute, Gyeongju, Korea, 2–4 May 2012; pp. 97–98.
14. Park, J.W.; Lee, G.C.; Koh, K.T.; Ryu, G.S. Study on Direct Tensile Properties and Reliability Review of Steel Fiber Reinforced UHPC. *J. Korea Inst. Build. Constr.* **2018**, *18*, 125–132.
15. Korea Institute of Construction Technology (KICT). *Design Specific Technology for Ultra High Performance Concrete*; Report of 4th Year of Super Bridge; Korea Institute of Construction Technology: Goyang, Korea, 2010.
16. Kim, S.; Kim, Y.; Kim, S.-W.; Kim, Y.J.; Park, C. Strengthening Effect of Externally Reinforced RC Structure with SIFRCC on Impact Load. In Proceedings of the 5th International Conference on Protective Structures, Poznan, Poland, 19–23 August 2018; pp. 522–528.
17. Korea Standards. Standard test method for compressive strength of concrete. In *KS F 2405:2010*; Korean Standards Association: Seoul, Korea, 2017.

Review

Recent Progress in Nanomaterials for Modern Concrete Infrastructure: Advantages and Challenges

Karla P. Bautista-Gutierrez [1], Agustín L. Herrera-May [1,2,*], Jesús M. Santamaría-López [1], Antonio Honorato-Moreno [1] and Sergio A. Zamora-Castro [1,*]

1 Maestría en Ingeniería Aplicada, Facultad de Ingeniería de la Construcción y el Hábitat, Universidad Veracruzana, Calzada Ruiz Cortines 455, Boca del Río, Veracruz 94294, Mexico; karlapaty40@gmail.com (K.P.B.-G.); jsantamaria@uv.mx (J.M.S.-L.); antonio.honorato.ing@gmail.com (A.H.-M.)

2 Micro and Nanotechnology Research Center, Universidad Veracruzana, Calzada Ruiz Cortines 455, Boca del Río, Veracruz 94294, Mexico

* Correspondence: leherrera@uv.mx (A.L.H.-M.); szamora@uv.mx (S.A.Z.-C.); Tel.: +55-442-226-6337 (A.L.H.-M.)

Received: 23 August 2019; Accepted: 23 October 2019; Published: 29 October 2019

Abstract: Modern concrete infrastructure requires structural components with higher mechanical strength and greater durability. A solution is the addition of nanomaterials to cement-based materials, which can enhance their mechanical properties. Some such nanomaterials include nano-silica (nano-SiO_2), nano-alumina (nano-Al_2O_3), nano-ferric oxide (nano-Fe_2O_3), nano-titanium oxide (nano-TiO_2), carbon nanotubes (CNTs), graphene and graphene oxide. These nanomaterials can be added to cement with other reinforcement materials such as steel fibers, glass, rice hull powder and fly ash. Optimal dosages of these materials can improve the compressive, tensile and flexural strength of cement-based materials, as well as their water absorption and workability. The use of these nanomaterials can enhance the performance and life cycle of concrete infrastructures. This review presents recent researches about the main effects on performance of cement-based composites caused by the incorporation of nanomaterials. The nanomaterials could decrease the cement porosity, generating a denser interfacial transition zone. In addition, nanomaterials reinforced cement can allow the construction of high-strength concrete structures with greater durability, which will decrease the maintenance requirements or early replacement. Also, the incorporation of nano-TiO_2 and CNTs in cementitious matrices can provide concrete structures with self-cleaning and self-sensing abilities. These advantages could help in the photocatalytic decomposition of pollutants and structural health monitoring of the concrete structures. The nanomaterials have a great potential for applications in smart infrastructure based on high-strength concrete structures.

Keywords: carbon nanotubes; cement-based materials; concrete infrastructure; graphene; graphene oxide; mechanical strength; nanomaterials; nano-Al_2O_3; nano-Fe_2O_3; nano-SiO_2; nano-TiO_2; smart infrastructure

1. Introduction

Construction engineering requires materials that enhance the mechanical properties of the cement-based composites for modern concrete infrastructure. For instance, the compressive, tensile and flexural strength of concrete structures need to be improved. For this, nanomaterials can be mixed with cementitious matrices to obtain concrete with high mechanical strength [1–10]. Nanotechnology can facilitate the development of nanomaterials incorporated into cement-based materials to increase their mechanical strength [11–25], decreasing their environment impact [26]. The CO_2 emissions generated during the production of ordinary Portland cement can represent approximately between

5% and 7% of the world man-made emissions of this gas [27,28]. A main challenge of the cement industry is the reduction of the CO_2 emissions. One alternative solution is the construction of concrete structures with higher mechanical strength and higher durability, which will decrease their maintenance requirements or need for early replacement. Thus, the concrete structures can have thinner sections, which will require less quantity of cement-based composites for their construction.

The cement-based materials can be mixed with nanomaterials such as nano-silica (nano-SiO_2), nano-alumina (nano-Al_2O_3), nano-ferric oxide (nano-Fe_2O_3), nano-titanium oxide (nano-TiO_2), carbon nanotubes (CNTs), graphene and graphene oxide. In recent years, several researchers [29–49] have studied the incorporation of nanomaterials into cement-based materials. The mixture of cementitious composites and nanomaterials can increase the mechanical strength of the resulting concrete structures. Thus, the life cycle of these structures can be extended or they can require smaller amounts of steel reinforcing bars. A common nanomaterial employed in cement-based composites is nano-silica. This material accelerates the cement hydration due to the generation of calcium-silicate-hydrate (C–S–H) and dissolution of tricalcium silicates (C_3S) [50]. In addition, this acceleration of cement hydration is caused by the nano-silica acting as a seed for nucleation of C–S–H [50]. Nano-silica can improve the durability, workability and mechanical properties of cement-based materials [51–59]. On other hand, nano-Al_2O_3 particles can increase the compressive strength of cement-based materials [41,60–63]. Al_2O_3 nanofibers with a dosage of 0.25% by cement weight may enhance the compressive strength of cement-based materials by up to 30% [50]. Another nanomaterial that can be added to cementitious matrices is nano-Fe_2O_3. Optimal values of this nanomaterial improve the compressive strength of concrete specimens [64,65]. The cement added with TiO_2 nanoparticles can be used to build a photocatalytic concrete with self-cleaning and air-purification characteristics [66]. This concrete type can allow effective photocatalytic decomposition of pollutants, including volatile organic compounds, carbon monoxide, chlorophenols and aldehydes generated from automobiles and industrial emissions [66–68]. Also, graphene family nanomaterials can be incorporated into cement composites to enhance their mechanical strength and durability, as well as provide self-sensing abilities [69–72]. Other novel properties of cement-based materials containing nanomaterials are their low electrical resistivity and self-sensing capabilities [73]. For instance, cement-based composites with CNTs have strain-sensing abilities, which may allow the measurement of their electrical resistance under applied loads [74]. It represents an advantage to obtain strain-sensing concrete structure systems for structural health monitoring [75,76].

This review includes recent studies about the effects on the mechanical strength, durability and workability of cement-based composites due to the incorporation of nanomaterials such as nano-SiO_2, nano-Fe_2O_3, nano-TiO_2, nano-Al_2O_3, CNTs, graphene and graphene oxide. In addition, these studies include nanomaterials that provide self-cleaning and self-sensing abilities to concrete structures. Also, the main challenges of using nanomaterials in cement-based materials are discussed.

2. Nanomaterials in Cement-Based Materials

2.1. Nano-Silica (Nano-SiO_2)

Nano-silica is a nanomaterial employed for civil engineering applications that can replace micro-silica and silica fume. Nano-silica reacts with lime during the cement hydration process and it generates a C–S–H gel that may improve the mechanical strength and durability of concrete. A good dispersion of nano-silica into cement-based materials can accelerate the hydration process of cement paste, allowing a denser microstructure. On the other hand, an excessive number of nanoparticles can cause agglomeration due to their high surface energy, which will provide a non-uniform dispersion. Figure 1 shows the scale ranges of several materials used in concrete fabrication [77].

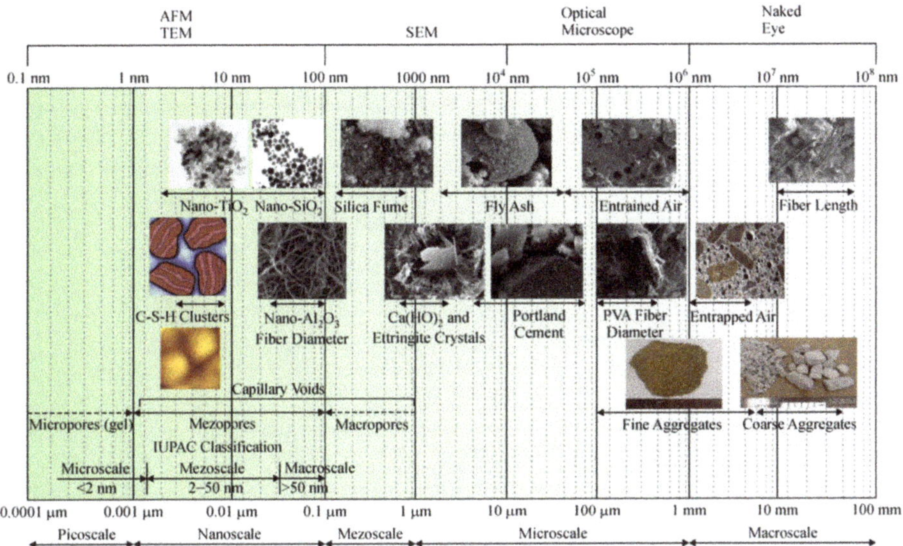

Figure 1. Scale range of several materials used in the concrete fabrication. Reprinted with permission from [77]. Copyright©2017, Higher Education Press and Springer-Verlag, Berlin/Heidelberg.

Flores-Vivian et al. [77] used Portland cement containing nano-silica to modify the rheological performance and improve the durability and strength. They used a nano-silica content of 0.25% by weight of cement-based materials. Other researchers such as Braz de Abreu et al. [22] reported the use of stabilized nano-silica particles (between 3 and 200 nm in size) in Brazilian-type CP V ARI PLUS Portland cement. They fabricated three types of concrete mixes: a reference concrete, a concrete added with stabilized nano-silica and a concrete including stabilized nano-silica with silica fume. After, they studied the results of concrete compressive strength tests at curing ages of 3, 7 and 28 days. The concrete compressive strength with only stabilized nano-silica increased up to 27%, 20% and 11% at 28 days compared with the reference concrete. On the contrary, the concrete with stabilized nano-silica and silica fume registered even higher compressive strength values (i.e., 28%, 37% and 24% at 28 days) compared to the control concrete. Thus, a mixture of nano-silica and silica fume with Portland cement generated a concrete with higher compressive strength.

Heidari and Tavakoli [78] fabricated a mixture using nano-silica and ceramic powder. They investigated the properties of ceramic power based on the ASTM C 618 standard, using 92% as material in the mixture. In this mixture, the cement is replaced with ceramic powder (phase A). In the second phase (phase B), the ceramic powder percentage is reduced, and the nano-silica is added. They employed the binder content as a constant (320 kg/m^3) and a water-cement ratio of 0.5. During phase A, mixtures were made with a ceramic powder percentage of 0%, 10%, 15%, 20%, 25%, 30% and 40% of the cement weight using the same proportion of aggregates and water. During phase B, mixtures are made with 0.5% and 1% of nanosilica and different ceramic powder content of 10%, 15%, 20% and 25% of the cement weight. All concrete mixtures are fabricated considering the ASTM C192 standard. The results of the compressive strength tests of the concrete (phase A) were obtained with different curing ages (7, 28, 56 and 91 days). These results show that the concrete strength proportionally decreases with the amount of ceramic powder added to the concrete. The concrete specimen containing 1% of nano-silica and 10% of ceramic powder improved its compressive strength. The impact on the pozzolanic reaction of nano-SiO$_2$ is more effective at an early age.

Supit and Shaikh [79] determined the durability properties of high-volume fly ash concrete with addition of nano-silica. They used type I Portland cement and different series of mixtures with a water-cement ratio of 0.40. Compressive strength tests for all concrete mixtures were measured at ages of 3, 7, 28, 56 and 90 days. The incorporation of nano-silica into ordinary concrete increased the compressive strength reaching up to 150% more at early ages. For ages of 28, 56 and 90 days, the compressive strength showed an increment between 45% and 75%. The nano-silica accelerated the hydration process and allowed a cementitious matrix with denser microstructure. The 4% nano-silica-modified concrete decreased its water absorption (between two and three times lower) in comparison to concrete without nano-silica. The resistance of chloride penetration was studied at ages of 28 and 90 days, in which the mixture with 2% nano-silica registered the lowest penetration value. Based on microstructure analysis, nano-silica-modified concrete mixtures presented denser microstructures. Thus, nano-silica modified concretes could be classified as low permeability concretes.

In order to reinforce reactive powder concrete (RPC), Han et al. [80] added nano-SiO_2-coated TiO_2 (NSCT) to RPC. These nanomaterials were studied by scanning electron microscopy (SEM), thermogravimetry (TG) analysis and powder X-ray diffraction (XRD). The acceleration of cement hydration due to the effect of the nucleus played a dominant role in the first days. The CH crystals particles size registered a reduction when the content of NSCT was increased (see Figure 2). The flexural and compressive strength of NSCT reinforced RPC (NSCTRRPC) specimens were investigated at curing ages of 3 and 28 days, considering different contents of NSCT (i.e., 1%, 3% and 5% by cement weight). The NSCTRRPC specimens enhanced their flexural and compressive strength in comparison to RPC specimens without NSCT. Figures 3 and 4 depict the flexural and compressive strength of NSCTRRPC specimens at curing ages of 3 and 28 days. For 3% NSCT dosage at a curing age of 3 days, a maximum flexural strength value (9.77 MPa) of the NSCTRRPC specimen was achieved. It represents an increment of 83.3% compared with the RPC without NSCT. For curing age of 28 days and 5% NSCT content, the flexural strength (14.38 MPa) of the NSCTRRPC specimen was increased up to 87% with respect to RPC without NSCT. Thus, NSCT increases the flexural strength of RCP specimens at both early age (3 days) and later age (28 days). The composites with NSCT registered small increments in their compressive strength at curing age of 3 days. On the contrary, maximum levels of the compressive strength of NSCT modified composites were measured at a curing age of 28 days. Thus, the highest compressive strength (111.75 MPa) of NSCTRRPC specimens is obtained with 3% NSCT dosage. This strength value registered an increase of 12.26% in comparison with RPC without NCST. However, the flexural strength of NSCTRRPC specimens had higher increment levels than that of compressive strength for the same test composites. This is caused by the NSCT that significantly enhances the toughness of the RPC [80].

Li et al. [81] examined the properties of ultra-high-performance concrete, which is obtained with particles of nano-limestone (nano-$CaCO_3$) and nano-silica. They used type I Portland cement and fly ash, and silica fume as binding agents. The percentages of nano-silica and nano-limestone by cement weight were of 0.5%, 1.0%, 1.5% and 2.0% and 2.0%, 3.0% and 4.0%, respectively. The mixture workability was reduced with respect the control specimen and it was maintained when the amount of nano-limestone is increased. This is due to the small size of nanoparticles that are found on the surface, leaving less water to contribute towards fluidity. The compressive and tensile strength of concretes including nano-limestone and nano-silica were improved with respect to concretes without any additions. The microstructure with highest values of density and mechanical strength was obtained with content levels of 1% nano-silica and 3% nano-limestone, respectively. The mechanical strength of concrete containing nano-silica and nano-limestone is increased with the reduction of the water-cement ratio.

Sadeghi et al. [82] reported non-destructive compressive strength tests of self-compacting concretes added with steel fibers, polypropylene and nano-silica. They employed the ultrasonic pulse velocity technique in concrete to register mechanical strength of concrete specimens. These concrete specimens ($100 \times 100 \times 100$ mm^3) were fabricated based on II Portland cement at ages of 7, 28 and 90 days. In addition, they used 40 different types of mixtures considering 2%, 4% and 6% of replacement with nano-silica and superplasticizer. In the specimens were measured the wave transmission velocity and compressive strength using the exponential relationship between both parameters. An increment of steel fiber volume above 3% increased the wave pulse transmission velocity in the specimens. The compressive strength and wave pulse transmission velocity increased when the percentage of nano-silica achieved above 4% of cement weight; however, both decreased afterwards.

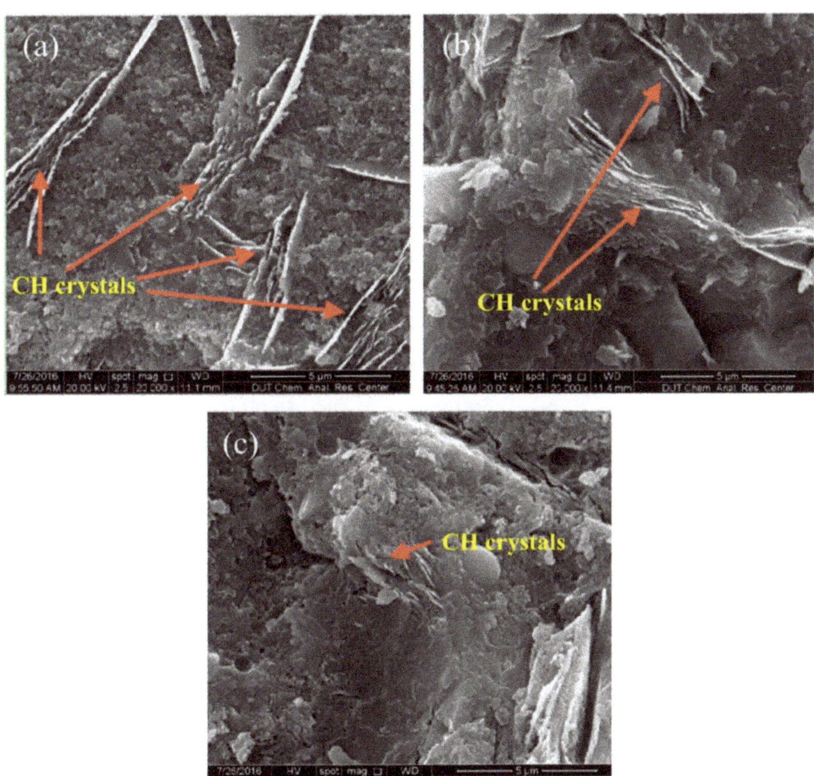

Figure 2. SEM micrographs of CH crystals in concrete with nano-SiO_2-coated TiO_2 at curing age of 28 days (20,000×): (**a**) sample T0; (**b**) sample T3; (**c**) sample T5. Reprinted with permission from [80]. Copyright©2017, Elsevier B.V.

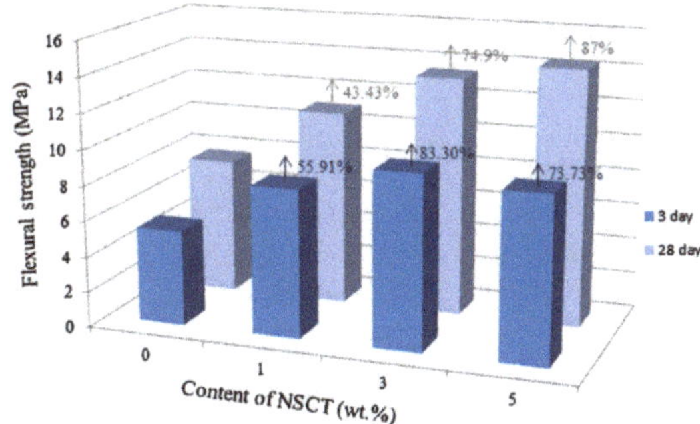

Figure 3. Flexural strength of NSCTRRPC test specimens with different values of NSCT content at curing age of 3 and 28 days. Reprinted with permission from [80]. Copyright©2017, Elsevier B.V.

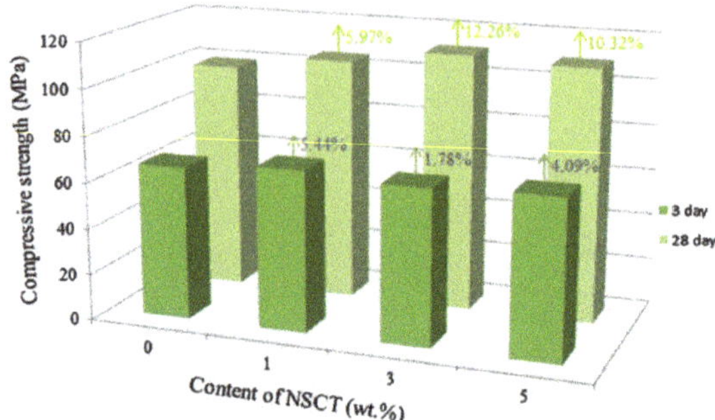

Figure 4. Compressive strength of NSCTRRPC test specimens with different values of NSCT content at curing age of 3 and 28 days. Reprinted with permission from [80]. Copyright©2017, Elsevier B.V.

Najigivi et al. [83] implemented tests using ordinary Portland cement and different nano-silica particles types according to average size. They named each one with the letters N and M, which both particle types reached an amorphous structure with a high pozzolanic reaction. These researchers used a water-cement ratio of 0.40, including nano-silica particles with proportions of 0.5%, 1.0%, 1.5% and 2.0% within the N particles and 2% in the M particles. In all the combinations of these tests, both nano-silica particles types decreased the concrete fluidity. The lime-cure concrete with maximum compressive strength was achieved using 2% nano-silica particles of M type with quicklime solution. This concrete reached the maximum values of compressive strength (40.2 MPa, 53.5 MPa and 57.1 MPa) at curing ages of 7, 28 and 90 days. This increment is due to the calcium hydroxide compounds reacted with nano-silica at a superficial level, generating additional C–S–H gel.

Zhang et al. [84] investigated the durability of concrete specimens containing nano-silica and steel fiber. They used five different contents of nano-silica (1%, 3%, 5%, 7% and 9%) and five-volume levels of steel fiber (0.5%, 1%, 1.5%, 2% and 2.5%). The durability tests of concrete specimens included the carbonation and cracking resistance, and permeability and freezing-thawing resistance. The durability tests are examined considering the carbonation depth of the specimens, total cracking area per unit area of the concrete specimen, cracks number, relative dynamic elastic modulus of the samples obtained after of the freezing-thawing cycles, and permeation depth of the water. For instance, a reduction in both the generated cracks number and water permeation depth of the concrete specimens can improve the concrete durability. Figures 5 and 6 show the total cracking area per unit area and cracks number of concrete samples containing 15% fly ash and five different nano-silica dosages. The cracks number in the concrete specimens decreased when the nano-silica dosages increased from 1% to 7%. The minimum number of cracks is achieved with 7% nano-silica dosage, but this number is increased when the nano-silica content is 9%. In addition, the total cracking area significantly decreased for nano-silica contents between 3% and 5%. Although, a nano-silica dosage of 9% caused an increment of 71.8% of the total cracking area compared with 5% nano-silica content. On the contrary, water permeation depth of the concrete specimens is showed in Figure 7. When the nano-silica content increments between 1% and 5%, the water permeation depth of the concrete specimens is significantly reduced. This improvement level decreases for nano-silica dosages of 7% and 9%, respectively. Based on these results, the nano-silica added concrete specimens enhanced their durability when the nano-silica content is within a certain limit. However, a high content of nano-silica could affect the durability of the concrete.

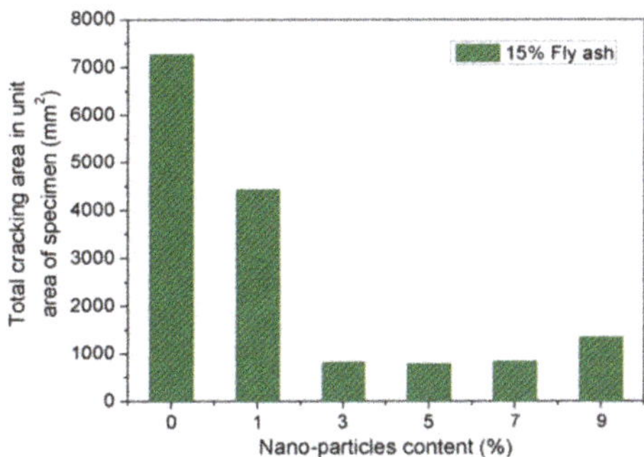

Figure 5. Influence of nano-silica dosage on the total cracking area per area unit of concrete specimens. Reprinted with permission from [84]. Copyright©2019, MDPI AG.

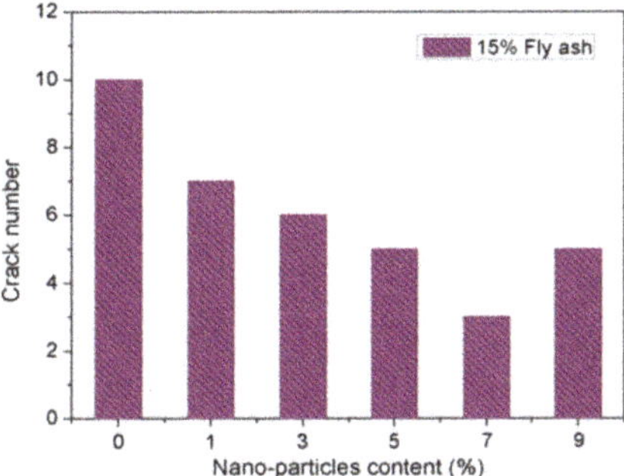

Figure 6. Effect of nano-silica dosage on the cracks number of concrete samples. Reprinted with permission from [84]. Copyright©2019, MDPI AG.

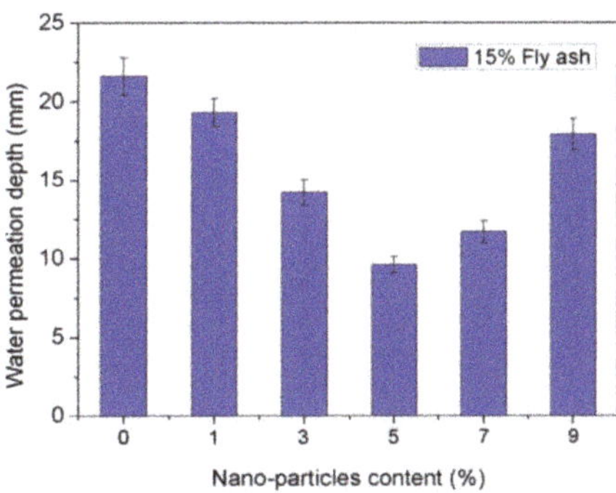

Figure 7. Effect of nano-silica dosage on the water permeation depth of concrete samples. Reprinted with permission from [84]. Copyright©2019, MDPI AG.

Tavakoli et al. [85] reported the effect on the compressive strength caused by addition of silica fume and nano-silica in concrete samples at curing age of 7, 28 and 56 days. They used type II Portland cement with different percentages of nano-silica (0.5% and 1%) and silica fume (5% and 10%) by cement weight. For each case, concrete samples containing nano-silica and silica fume increased their compressive strength compared to control specimen without these materials. The concrete samples achieved the highest compressive strength (52.9 MPa) using 10% of silica fume and 1% of nano-silica particles at curing age of 56 days. This strength value is 42.2% higher than that of the concrete sample without nano-silica and silica fume. More investigations about of nano-silica modified cement were reported by Nazerigivi and Najigivi [86]. They studied the influence on the mechanical strength of concrete specimens caused by incorporation of two different nano-silica sizes (15 nm and 80 nm) with percentages of 0.5%, 1.0%, 1.5% and 2.0% by cement weight. They employed ordinary Portland cement and 16 different concrete samples and one control concrete sample for each mechanical test type, as indicated Table 1. A water-to-binder ratio of 0.40 was used into all the concrete samples. With lime solution, these samples are cured at ages of 7, 28 and 90 days. Tables 1–3 indicate the measurements of the compressive, split tensile and flexural strength of all the concrete samples. For the three curing ages, the nano-silica added concrete samples improved their compressive strength with respect to that of the control specimen. The compressive strength had a gradual increment when the nano-silica dosage was increased up to 2% of 15 nm plus 1.5% of 80 nm; after, it had a small decrease. The generation of C–S–H gel may be accelerated due to ultra-high specific surface and ultra-fine particle size of nano-silica incorporated in concrete samples [82]. The split tensile and flexural strength of all the nano-silica modified concrete specimens were improved with respect to the control sample. Both split tensile and flexural strength registered gradual increments with the incorporation of nano-silica up to 2% of 15 nm plus 1.5% of 80 nm; after, these mechanical strengths had a small reduction. It could be caused because the total quantity of nano-silica is higher than that to obtain the lime-silica hydration reaction [82].

Mastali and Dalvand [87] reported a theoretical and experimental study of the effects on the mechanical properties of concrete samples due to the presence of 1.0% nano-silica and 7% silica fume, respectively. They realized 270 tests with different designs of self-compacting concrete, in which the impact resistance and mechanical properties of concrete samples were enhanced. The incorporation of nano-silica and silica fume in the cement of silica fume and self-compacting concrete increased 70% its impact resistance for the first crack. Fiber reinforced specimens with water-cement ratio of 0.34 and 0.27 registered the highest average of tensile and flexural strength, respectively.

Table 1. Compressive strength of concrete specimens containing several nano-silica contents. Reprinted with permission from [86]. Copyright©2019, Elsevier B.V.

Sample	Nano-Silica		Compressive Strength (MPa)			Improvement of Compressive Strength (%)		
	15 nm Nanoparticle	80 nm Nanoparticle	7 Days	28 Days	90 Days	7 Days	28 Days	90 Days
C0 (control)	0	0	26.3	34.8	40.3	0	0	0
N1	0.5	0.5	28.2	38.4	44.9	7.2	10.3	11.4
N2	0.5	1	30.3	41.3	49.2	15.2	18.7	22.1
N3	0.5	1.5	32.9	44.2	53.0	25.1	27.0	31.5
N4	0.5	2	35.2	46.8	57.6	33.8	34.5	42.9
N5	1	0.5	31	42.5	50.7	17.9	22.1	25.8
N6	1	1	33.4	46.2	54.1	27.0	32.8	34.2
N7	1	1.5	36.3	47.5	58.2	38.0	36.5	44.4
N8	1	2	40.7	50	63.3	54.8	43.7	57.1
N9	1.5	0.5	35.2	47	56.1	33.8	35.1	39.2
N10	1.5	1	37.2	49.1	59.8	41.4	41.1	48.4
N11	1.5	1.5	41.3	52.2	64.7	57.0	50.0	60.5
N12	1.5	2	46.4	58.7	69.3	76.4	68.7	72.0
N13	2	0.5	39	50.3	63.4	48.3	44.5	57.3
N14	2	1	41.9	54	67.2	59.3	55.2	66.7
N15	2	1.5	52.1	63.7	78.1	98.1	83.0	93.8
N16	2	2	50.3	61	75.2	91.3	75.3	86.6

Table 2. Split tensile strength of concrete specimens containing several nano-silica contents. Reprinted with permission from [86]. Copyright©2019, Elsevier B.V.

Sample	Nano-Silica		Split Tensile Strength (MPa)			Improvement of Split Tensile Strength (%)		
	15 nm Nanoparticle	80 nm Nanoparticle	7 Days	28 Days	90 Days	7 Days	28 Days	90 Days
C0 (control)	0	0	1.3	1.5	1.9	0	0	0
N1	0.5	0.5	1.8	2	2.7	38.5	33.3	42.1
N2	0.5	1	2.1	2.4	3	61.5	60.0	57.9
N3	0.5	1.5	2.5	2.9	3.5	92.3	93.3	84.2
N4	0.5	2	3	3.3	3.9	130.8	120.0	105.3
N5	1	0.5	2.2	2.8	3.2	69.2	86.7	68.4
N6	1	1	2.7	3.1	3.8	107.7	106.7	100.0
N7	1	1.5	3.1	3.7	4.2	138.5	146.7	121.1
N8	1	2	3.8	4.2	4.8	192.3	180.0	152.6
N9	1.5	0.5	2.9	3.3	4.1	123.1	120.0	115.8
N10	1.5	1	3.3	3.8	4.6	153.8	153.3	142.1
N11	1.5	1.5	4.1	4.5	5.2	215.4	200.0	173.7
N12	1.5	2	4.4	4.8	5.7	238.5	220.0	200.0
N13	2	0.5	3.6	4	4.9	176.9	166.7	157.9
N14	2	1	4.5	4.8	5.5	246.2	220.0	189.5
N15	2	1.5	4.9	4.3	5.9	276.9	186.7	210.5
N16	2	2	4.3	4.8	5.1	230.8	220.0	168.4

Table 3. Flexural strength of concrete specimens containing several nano-silica contents. Reprinted with permission from [86]. Copyright©2019, Elsevier B.V.

Sample	Nano-Silica		Flexural Strength (MPa)			Improvement of Flexural Strength (%)		
	15 nm Nanoparticle	80 nm Nanoparticle	7 Days	28 Days	90 Days	7 Days	28 Days	90 Days
C0 (control)	0	0	4	4.2	4.5	0	0	0
N1	0.5	0.5	4.4	4.5	5	10.0	7.1	11.1
N2	0.5	1	4.6	4.9	5.3	15.0	16.7	17.8
N3	0.5	1.5	5.1	5.2	5.7	27.5	23.8	26.7
N4	0.5	2	5.4	5.6	6	35.0	33.3	33.3
N5	1	0.5	4.6	5.1	5.4	15.0	21.4	20.0
N6	1	1	5.3	5.5	5.9	32.5	31.0	31.1
N7	1	1.5	5.5	5.8	6.3	37.5	38.1	40.0
N8	1	2	5.9	6.2	6.7	47.5	47.6	48.9
N9	1.5	0.5	5.4	5.7	6.1	35.0	35.7	35.6
N10	1.5	1	5.6	6	6.6	40.0	42.9	46.7
N11	1.5	1.5	6.2	6.5	7	55.0	54.8	55.6
N12	1.5	2	6.5	6.8	7.3	62.5	61.9	62.2
N13	2	0.5	5.8	6.2	6.8	45.0	47.6	51.1
N14	2	1	6.5	6.7	7.2	62.5	59.5	60.0
N15	2	1.5	7	7.3	7.8	75.0	73.8	73.3
N16	2	2	6.8	7	7.2	70.0	66.7	60.0

Mohammed et al. [88] evaluated the influence on the properties of concrete due to the nano-silica inclusion. This nano-silica incorporation caused a reduction of 13% in the pore amount of the cementitious paste. They studied the relationships that improved the compression strength of the concrete. The workability was modified negatively, which was not affected with the incorporation of superplasticizer to the concrete paste. When the nano-silica inclusion was increased in the experimentation, the permeability and infiltration rate were reduced based on the SEM results.

The incorporation of nano-silica optimal dosage in concrete samples may improve their compressive, tensile and flexural strength. The nano-silica added in cement with other materials such as polypropylene, glass and steel fibers with fixed proportions can increase the mechanical properties of the concrete. Concretes with nano-silica absorbed $Ca(OH)_2$ crystals, filling the voids of the C–S–H structure, leading to a denser microstructure.

2.2. Nano-Ferric Oxide (Nano-Fe_2O_3)

The optimal addition of nano-Fe_2O_3 in concrete specimens may improve their compressive strength. In addition, the volume electrical resistance of cement mortars with inclusion of nano-Fe_2O_3 can be altered through the applied load, allowing the measure of compressive stress [73]. It can be used for structural heath monitoring of concrete structures without require additional sensors.

Fang et al. [89] measured the mechanical properties of cement samples with different additions of nano-Fe_2O_3 (3%, 5% and 10% by cement weight) at ages of 7, 14 and 28 days. Figure 8 shows the SEM image of cement specimens with different additions of nano-Fe_2O_3. When the nano-Fe_2O_3 content increases then the surface morphology is denser. For all the measurements, the addition of nano-Fe_2O_3 in cement mortars increased their compressive strength compared to control mortar. Maximum values of the compressive strength of the cement samples were achieved using 10% of nano-Fe_2O_3 content. For this nano-Fe_2O_3 dosage at curing ages of 7, 14 and 28 days, the compressive strength of the cement mortar was increased up to 66.81%, 69.76% and 25.20%, respectively.

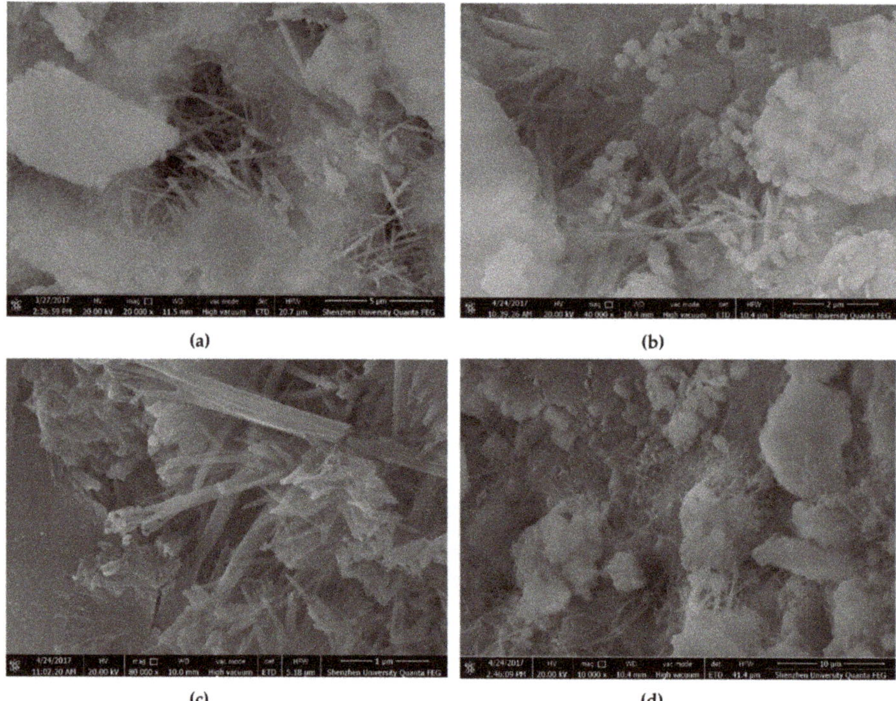

Figure 8. SEM image of concrete specimens with addition of nano-Fe_2O_3 (**a**) 0%; (**b**) 3%; (**c**) 5%; (**d**) 10%. Reprinted with permission from [89]. Copyright©2018, Atlantis Press.

Rashad [90] presented a review the effects of nano-Fe_2O_3, nano-Al_2O_3, nano-Fe_3O_4 and nanoclay on some properties of cement composites. These properties were the mechanical strength, hydration heat, water absorption, workability, setting time and durability. For instance, the inclusion of nano-Fe_2O_3 in the cementitious matrix decreased the water absorption and heat rate values as well as accelerated the peak times. Moreover, the workability of the composite was reduced when the nano-Fe_2O_3 content was increased. On the other hand, nano-Fe_2O_3 (0.5%–5% in concretes and 0.5%–10% in mortars) added into the cementitious matrix improved the compressive strength. Nazari et al. [91] also studied the workability of concrete including nano-Fe_2O_3. For this case, cement was partially substituted with nano-Fe_2O_3 (i.e., 0%, 0.5%, 1%, 1.5% and 2% by cement weight) and a water to binder ratio of 0.4 was employed. The workability of concrete is decreased when the nano-Fe_2O_3 dosage is increased. In addition, Nazari and Riahi [92] developed two models using genetic programming and artificial neural networks to predict the percentage of water absorption and split tensile strength of concrete samples containing nano-Fe_2O_3.

Khoshakhlagh et al. [93] studied the changes of the concrete properties achieved by adding different percentages (1%–5% by cement weight) of nano-Fe_2O_3 and superplasticizer. The flexural, compressive and tensile strength, and the water permeability of the concrete specimens were improved with the incorporation of nano-Fe_2O_3 up to 4% by cement weight. The content of nano-Fe_2O_3 up to 4wt.% of the concrete specimens increased the coefficient of water absorption. The concrete specimens with nano-Fe_2O_3 enhanced their hydration heat, workability and the compressive, flexural and tensile strength.

2.3. Nano-Titanium Oxide (Nano-TiO_2)

The addition of nano-TiO_2 in concrete specimens can provide self-cleaning properties to the concrete. The concrete containing these nanoparticles can allow a photocatalytic degradation of pollutants (e.g., VOCs, CO, NO_x, aldehydes and chlorophenols) from industrial and automobile emissions. However, this effect is less efficient with aging due to carbonatation [94,95].

(a)

Figure 9. *Cont.*

(b)

Figure 9. Examples of photocatalytic cement-based coatings that contain TiO_2 thin film: (**a**) parking lot view (Phoenix, AZ, USA) and (**b**) bike lane (Brooklyn, NY, USA). Reprinted with permission from [23]. Copyright©2016, Higher Education Press and Springer-Verlag Berlin/Heidelberg.

Figure 9 depicts two applications of photocatalytic cement-based coatings, which allows a self-cleaning effect in function the decomposition of gases and organic pollutions [23]. This is due to a TiO_2 thin film on the concrete surface that can provide active oxygen under UV light present in sunlight. Thus, it catalyzes the degradation of organic matters located at the nano-TiO_2 coated concrete surface [27]. The concrete surface is cleaned with the rainwater, which can prevent the buildup of dirt. Another important characteristic of nano-TiO_2 is the chemical stability and low price in comparison with other materials. Moreover, nano-TiO_2 can enhance the resistance to water permeability of cement-based structures [25]. Figure 10 depicts the SEM image of fracture surfaces of cementitious composites considering two sizes of nano-TiO_2 [96]. Wang et al. [97] investigated the mechanical and physical properties of cement mortar specimens considering different contents of nano-TiO_2 under curing temperatures of 0, 5, 10 and 20 °C. They used natural river sand, Portland cement (type I ordinary), and TiO_2 nanoparticles with size of 15 nm. In the experimental tests were used nano-TiO_2 dosages of 1%, 2%, 3%, 4% and 5% by cement weight, respectively. In the fabrication of the specimens, the nano-TiO_2 was dispersed in water through ultrasonication. After, cement and sand are mixed during 1 minute. Then, the well-dispersed nano-TiO_2 was added and mixed during 60 seconds and after water was incorporated. In the following stage, the mortars are collocated into molds and cured using different temperatures. For the specimens were used a water to binder ratio of 0.5. Figure 11 depicts the SEM images of cement pastes including 2 wt.% nano-TiO_2, at curing age of 28 days under different temperatures. The compressive strength characterization is determined according to ASTM C109 [98] employing a hydraulic testing machine under a controlled load of 1350 N/s. The flexural strength test was evaluated regarding the ASTM C293 [99]. This characterization is determined at curing ages of 3, 7, 28 and 56 days. Figure 11 depicts the results of the hydration degree of the mortar specimens. First, hydration degree of the mortar specimens enhanced through the increment of the nano-TiO_2 dosage. TiO_2 nanoparticles can supply an extra space for the precipitation of hydration products. Figures 12 and 13 depict the response of the compressive and flexural strength of the cement mortar samples. Both compressive and flexural strength registered downward trend at low curing temperature. On contrary, flexural and compressive strength of mortar specimens containing nano-TiO_2 had fast increment with respect to ordinary mortar until that the nano-TiO_2 content achieved up to 2 wt.%. This increase slowed down for TiO_2 nanoparticles dosages higher than 2 wt.%. The enhanced strength of mortar samples is caused by TiO_2 nanoparticles that facilitated the cement hydration and filled the pores in C–S–H gels [97]. These nanoparticles present large surface area to volume ratio, allowing an extra surface area to precipitate hydration products. In addition, TiO_2 nanoparticles form a bond between them self and C–S–H gel that improves their strength [97].

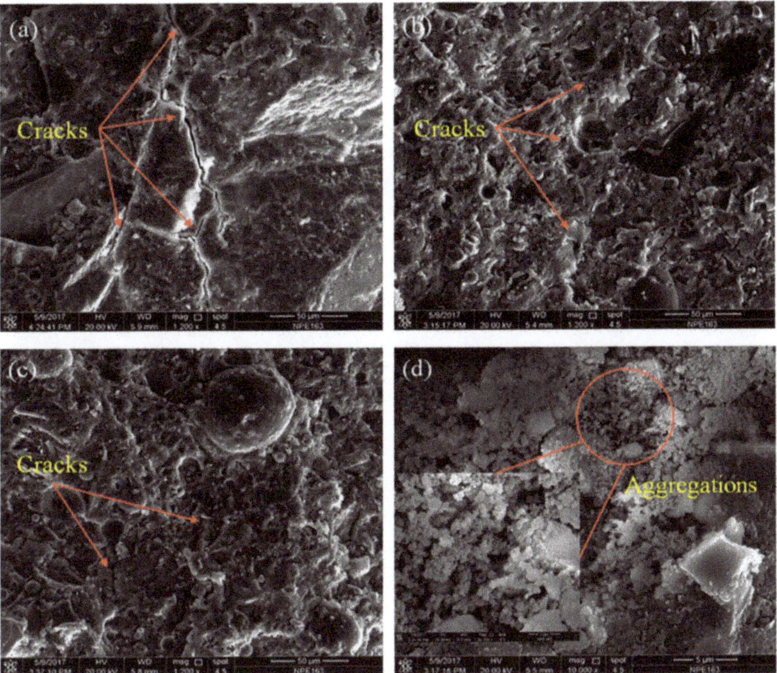

Figure 10. SEM image of the fracture surfaces considering: (**a**) control cementitious composites (1200×); (**b**) nano-TiO_2 (50 nm size) modified cementitious composites (1200×); (**c**) nano-TiO_2 (500 nm size) modified cementitious composites (1200×); (**d**) aggregations of nano-TiO_2 (50 nm size) in cementitious composites (1200×). Reprinted with permission from [96]. Copyright©2019, Elsevier B.V.

Feng et al. [100] examined the microstructures of concrete matrices incorporating nano-TiO_2 as well as the mechanical properties of the cement pastes. Figure 14 is a SEM image of TiO_2 nanoparticles and their selected area electron diffraction. The incorporation of nano-TiO_2 (0.1%, 0.5%, 1.0% and 1.5% by cement weight) in cement paste using a water-cement ratio of 0.4 improved the flexural strength (4.52%, 8.00%, 8.26% and 6.71%) at 28 days age.

Jalal et al. [101] studied the characteristics of high resistance self-compacting concrete containing fly ash and nano-TiO_2. They used Portland cement that was replaced up to 15% weight of waste ash and up to 5% weight of nano-TiO_2. The addition of nano-TiO_2 in the concrete improved the consistency and reduced the segregation probability. Considering the water absorption and capillarity, a significant decrease was obtained due to the nano-TiO_2.

Figure 11. SEM images of cement samples with addition of 2 wt.% nano-TiO_2 cured under temperatures of (**a**) 0 °C, (**b**) 5 °C, (**c**) 10 °C, and (**d**) 20 °C at 28 days. Reprinted with permission from [97]. Copyright©2018, Hindawi.

The weight losses in concrete samples were caused by the rapid formation of hydrated products. The self-compacting concrete with nano-TiO_2 registered a microstructure more refined, which enhanced the resistance to mechanic failures. Other researchers, Yu et al. [102] reported the improvement of concrete microstructure incorporating nano-TiO_2, which increased its mechanical strength. The TiO_2 nanoparticles catalyze the decomposition of harmful gases in the air. In addition, the concrete with nano-TiO_2 achieved a maximum compressive strength that was 7% higher in comparison with the non-added nanoparticle concrete. In addition, Yu et al. [102] investigated the changes of temperatures that can induce cracks and accelerate the hydration reaction.

Chunping et al. [103] investigated the durability of ultra-high performance concrete due to the incorporation of nano-TiO_2. This concrete added with 1% nano-TiO_2 improved its mechanical properties. They investigated the effects on the dry shrinkage, carbonation resistance, freeze-thaw resistance and resistance to chloride ingress. The addition of nano-TiO_2 in concrete could allow it a self-cleaning and photocatalytic behavior. In addition, the normal concrete containing nano-TiO_2 could decrease the capillary porosity.

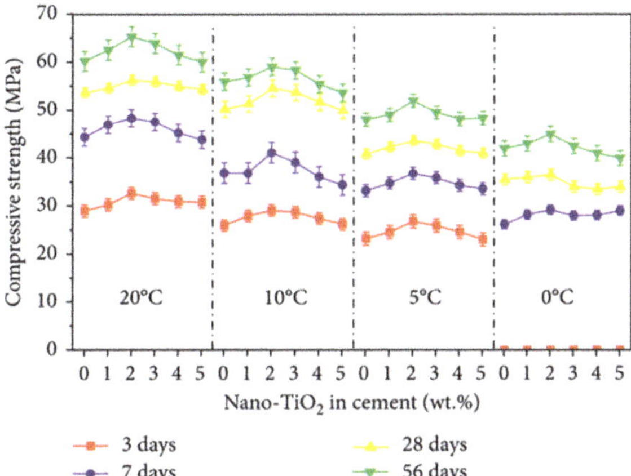

Figure 12. Compressive strength of cement mortar samples incorporating different nano-TiO_2 dosages. Reprinted with permission from [97]. Copyright©2018, Hindawi.

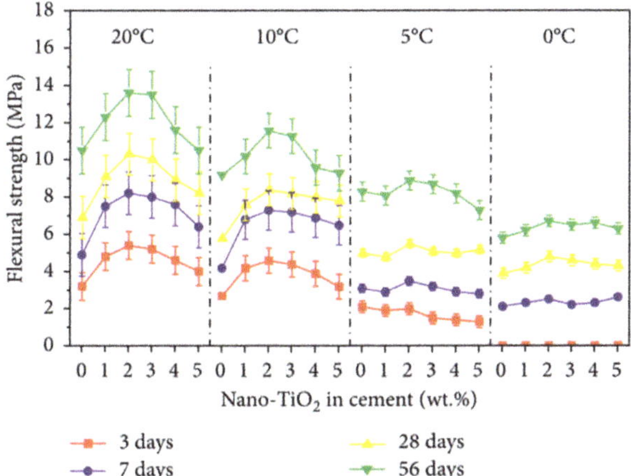

Figure 13. Flexural strength of cement mortar samples containing different nano-TiO_2 dosages. Reprinted with permission from [97]. Copyright©2018, Hindawi.

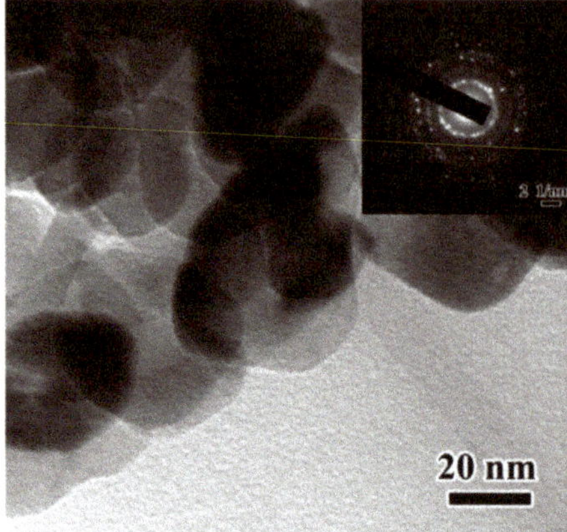

Figure 14. TEM image of the morphology of the TiO_2 nanoparticles and their selected area electron diffraction (SAED). Reprinted with permission from [100]. Copyright©2013, American Chemical Society.

2.4. Nano-Alumina (Al_2O_3)

The use of nano-Al_2O_3 can accelerate the formation process of C–S–H gel, especially at early-ages, which enhances the strength of composites [104]. For instance, Muzenski et al. [50] fabricated ultra-high strength cement-based materials using Al_2O_3 nanofibers with a content of 0.25% by weight of cementitious materials, which improved the compressive strength up to 200 MPa. This represents an increment of 30% in comparison to material strength with only 1% of silica fume. This high compressive strength was achieved with low amount of silica fume. This improved performance is caused by the nanofibers that act as a seed to generate hydration products and contribute the reinforcement for the C–S–H formations, which decrease the number of micro-cracks. In addition, to reach the maximum mechanical performance of the cement-based materials is necessary a suitable dispersion of the Al_2O_3 nanofibers. A longer dispersion time could reduce the fibers agglomeration, allowing the enhance of mechanical performance. For instance, the compressive strength at 28 days age achieved higher values for specimens with Al_2O_3 nanofibers dispersed for 3 h. Nevertheless, higher quantities of Al_2O_3 nanofibers and supplementary cementitious materials did not increase the mechanical behavior of the cement-based materials. Figure 15a,b depicts SEM images of the Al_2O_3 nanofibers diluted in cement pastes.

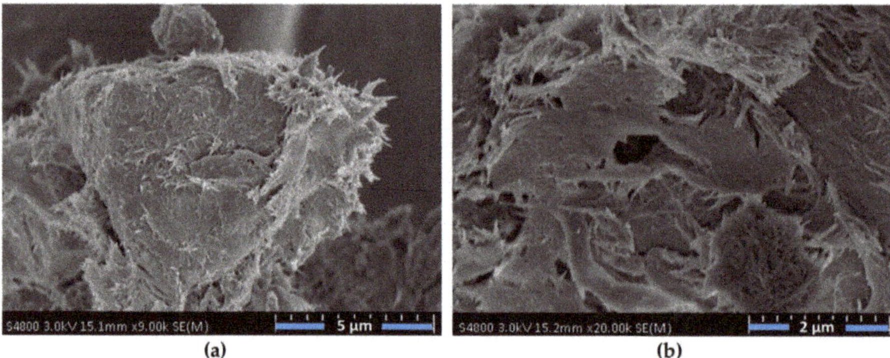

Figure 15. SEM image of the Al_2O_3 nanofibers diluted in cement pastes at (**a**) 9000× magnification and (**b**) 20,000× magnification. Reprinted with permission from [50]. Copyright© 2019, Elsevier B.V.

Yang et al. [105] investigated the effect of nano-Al_2O_3 on the chloride-binding capacity of cement paste samples. These samples were prepared with nano-Al_2O_3 dosages of 0.5%, 1.0%, 3.0% and 5%. The chloride-binding capacity was examined using conventional equilibrium tests, in which the samples were exposed with a NaCl solution at 0.05 mol/L, 0.1 mol/L, 0.3 mol/L, 0.5 mol/L and 1.0 mol/L, respectively. Based on the experimental results, the bound chloride content had an increase of 37.2% at NaCl solution (0.05 mol/L) by adding 5.0% of nano-Al_2O_3. Thus, an appropriate adding of nano-Al_2O_3 improved the chloride-binding of cement paste samples.

Mohseni at al. [106] studied the effects of nano-alumina and rice husk ash (RHA) in polypropylene fiber (PPF)-reinforced cement mortars. The RHA is an agricultural waste material, which can be recycled to obtain economic and environmental benefits. Figure 16 shows the SEM images of nano-alumina and RHA. The compressive strength of the mortar samples is increased up to 18% and 20% due to the addition of 3% nano-Al_2O_3 with 20% RHA at 28 and 90 days. The flexural strength of the mortar samples increased up to 34% and 41% by adding 3% nano-Al_2O_3 with 10% RHA. This addition of nano-Al_2O_3 generated a denser microstructure in the mortar samples.

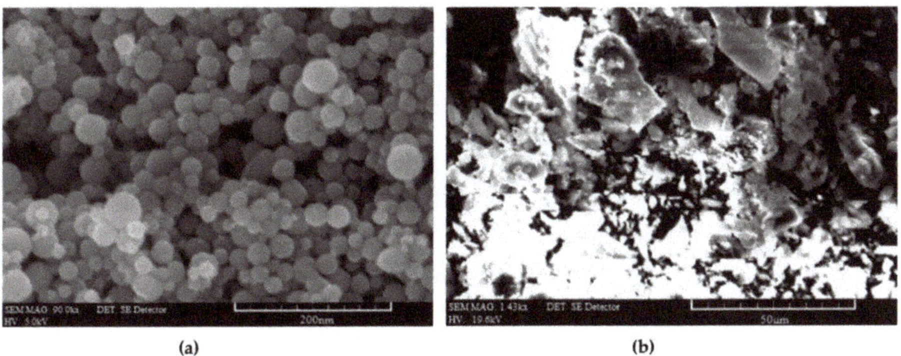

(a) (b)

Figure 16. SEM image of the (**a**) nano-Al_2O_3 and (**b**) rice husk ash. Reprinted with permission from [106]. Copyright© 2016, Elsevier B.V.

Barbhuiya et al. [107] examined the influence of the incorporation of nano-Al_2O_3 on the microstructural properties of the cement paste hydrated at 7 days age. Ordinary Portland cement is substituted with nano-Al_2O_3 powder with 2% and 4% by cement weight and the water-cement ratio is fixed to 0.4. In this early-age, they did not note changes at the compressive strength of the cement specimen at early age.

Based on the XRD analysis, Barbhuiya et al. [107] did not find a new crystalline phase developed by adding nano-Al_2O_3 within 7 days of curing. They reported the generation of dense microstructure with larger crystal of portlandite within the cement matrix due to the nano-Al_2O_3 addition, as shown in Figures 17 and 18. Gowda et al. [108] reported the influence of nano-Al_2O_3 in the water absorption and electrical resistivity of cement mortars. They used 1%, 3% and 5% of nano-Al_2O_3 by cement weight. The water absorption had a small reduction with the addition of 1% and 3% nano-Al_2O_3. However, the water absorption registered a small increment with the addition of 5% nano-Al_2O_3. The highest electrical resistivity of the cement mortar is achieved with 5% nano-Al_2O_3.

Figure 17. SEM image of 2% nano-Al_2O_3 by weight of cement paste hydrated up to 7 days. Reprinted with permission from [107]. Copyright© 2014, Elsevier Ltd.

Figure 18. SEM image of 4% nano-Al_2O_3 by weight of cement paste hydrated up to 7 days. Reprinted with permission from [107]. Copyright© 2014, Elsevier B.V.

2.5. Carbon Nanotubes (CNTs)

Recently, several researchers [109–123] have reported the effects of CNTs on the electrical and mechanical properties of concrete samples. For instance, CNTs can decrease the formation and growth of micro-cracks in concrete. The CNTs have important mechanical and electrical properties, including their high strength and high conductivity. For instance, CNTs have high mechanical performance with high aspect ratios (length to diameter ratio) that may generate stronger cement composites [27]. The CNTs cement-based composites have strain-sensing behavior that can measure their electrical parameters under applied loads [124]. This behavior can allow the development of strain-sensing systems of concrete structures for potential applications of damage detection and structural health monitoring [75,125].

García-Macías et al. [124] developed a micromechanics model to determine the piezoresistive behavior of cement-based nanocomposites incorporating CNTs and considering the waviness and non-uniform distributions of nanoinclusions. In order to validate the theoretical model, they tested cement-based samples that were doped with multi-walled CNTs (MWCNTS) and exposed to uniaxial compression. These samples were fabricated of concrete, mortar and composite cement paste. Figure 19 depicts the SEM pictures of the MWCNTs dispersion in water solution after sonication and in a cement mortar sample. For the compression loads on the MWCNTs reinforced cement-based composites, they used an equipment of servo-controlled pneumatic universal testing with load capacity of 14 kN, as shown in Figure 20a,b. For the cement paste, mortar and concrete samples are incorporated

MWCNTs with electrical conductivity between 10^1 and 10^4 S/m. Figure 21a–c illustrates the response of the electrical conductivity of different cement-based composites using the theoretical model and experimental setup. The cement paste, mortar and concrete specimens used filler concentrations of 1%, 0.75% and 0.75% by cement weight. The proposed analytical model may predict the electrical resistance performance of MWCNTs reinforced cement-based materials under compression loads. Ruan et al. [118] reported the influence of different types and dosages of MWCNTs on the mechanical properties of RPC under water or heat curing. They fabricated RPC including four types of MWCNTs with dosages of 0%, 0.25% and 0.50% with water/heat curing, respectively. The mechanical performance of the MWCNTs filled RFC specimens were examined. This mechanical performance considered the flexural strength, fracture energy, compressive/ toughness and flexural strength to compressive strength ratio. The fabrication of the RPC specimens included MWCNTs, water, water reducer, fly ash, quartz sand, cement and silica fume. The cement, silica fume and quartz sand ratio was 1:0.25:1.1. In addition, 20% of cement was substituted by fly ash to enhance the mobility of the mixtures and decrease the cement amount. The four types of MWCNTs used were classified as T1 (functionalized MWCNTs with carboxyl groups), T2 (functionalized MWCNTs with hydroxyl groups), T3 (helical MWCNTs through catalytic cracking) and T4 (nickel-coated MWCNTs).

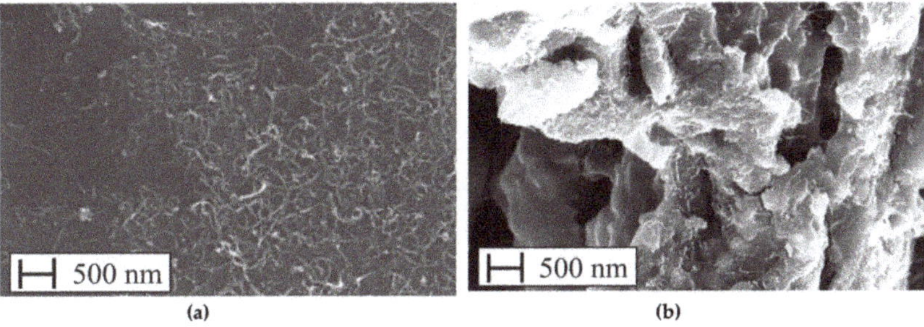

Figure 19. SEM images of the MWCNTs dispersion in (**a**) water suspensions after sonication and (**b**) in a mortar specimen after curing. Reprinted with permission from [124]. Copyright© 2017, Elsevier B.V.

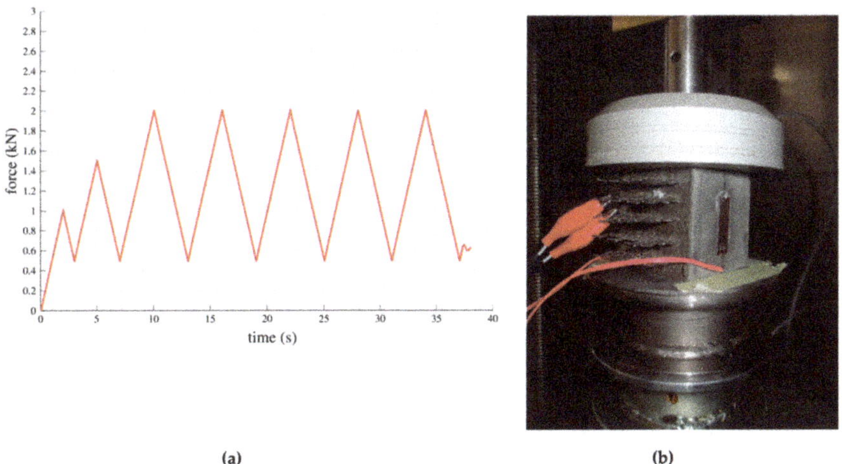

Figure 20. (**a**) Compression load versus time and (**b**) uniaxial testing machine used in the MWCNTs reinforced cement-based specimens. Reprinted with permission from [124]. Copyright© 2017, Elsevier B.V.

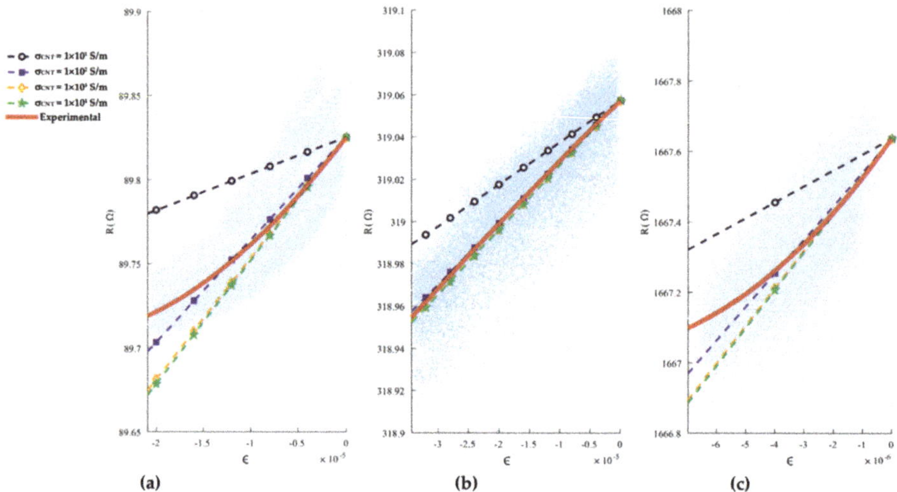

Figure 21. Results of the theoretical and experimental electrical resistance versus applied mechanical strain for (**a**) cement paste, (**b**) mortar and (**c**) concrete samples. Reprinted with permission from [124]. Copyright© 2017, Elsevier B.V.

Figure 22 illustrates the flexural strengths of the MWCNTs filled RPC under water curing. With exception of the specimen filled with 0.5% MWCNTs dosage of T3, all the others specimens filled with dosages of 0.25% and 0.50% MWCNTs showed enhanced flexural strength. The specimen T2 with 0.25% MWCNTs content had the maximum increase (27.2%) of the flexural strength. However, the specimen with 0.50% MWCNTs content registered a decrease (3.8%) of the flexural strength. For the specimens T1, T2 and T3, the flexural strength had better results for low dosage of MWCNTs than that by high dosage of MWCNTs. In addition, the incorporation of the four types of MWCNTs with dosages of 0.25% and 0.50% improved the compressive strength of the RPC specimens under water curing. The compressive strength increased 18.1% with the incorporation of 0.50% MWCNTs content, compared with the RPC without MWCNTs. The dosage of MWCNTs improved the compressive toughness of all the RPC specimens with water curing, as shown Figure 23. The highest compressive toughness was measured in the specimen T2 with 0.25% MWCNTs content, which represented an increase of 39.2% in comparison to the RPC without MWCNTs. Figure 24a,b depicts SEM images of the wide distribution network of MWCNTs in RPC, which can improve the mechanical properties of RPC.

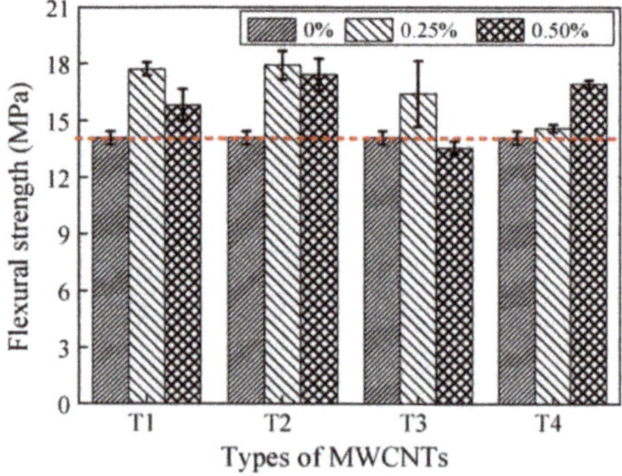

Figure 22. Flexural strengths of the MWCNTs reinforced RPC specimens under water curing. Reprinted with permission from [118]. Copyright© 2018, Elsevier B.V.

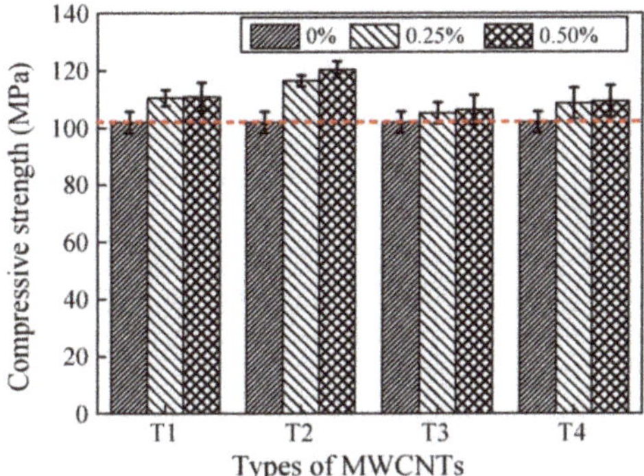

Figure 23. Compressive strengths of the MWCNTs reinforced RPC specimens under water curing. Reprinted with permission from [118]. Copyright© 2018, Elsevier B.V.

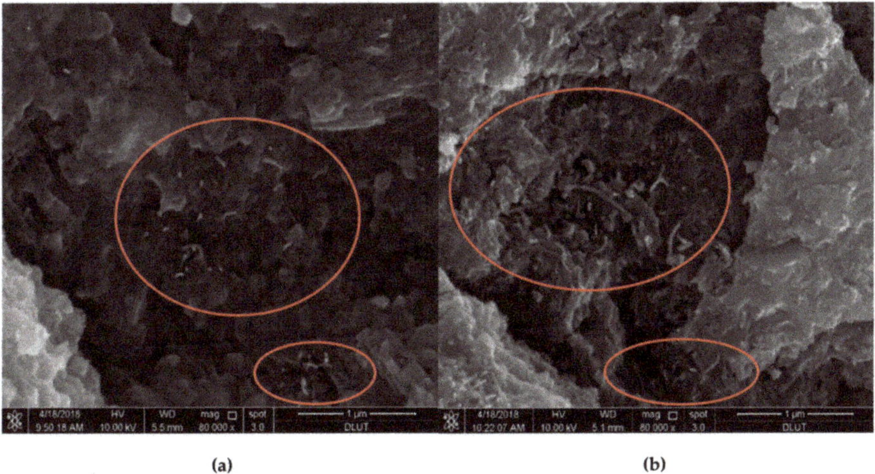

(a) **(b)**

Figure 24. (**a**,**b**) SEM images of the extensive distribution of MWCNTs in RPC. Reprinted with permission from [118]. Copyright© 2018, Elsevier B.V.

Lushnikova and Zaoui [120] studied the effect of different types of CNTs incorporated into cement specimens. They used molecular dynamics simulations to determine the influence of CNTs on the mechanical properties of C–S–H such as shear modulus, bulk modulus, elastic constants and Poisson ratio. The results of these simulations registered an improvement of all the studied mechanical properties. Thus, the CNTs are nanomaterials that could enhance the mechanical properties of concrete. Moreover, Sedaghatdoost and Behfarnia [121] examined the influence on the mechanical properties of the Portland cement caused by addition of MWCNTs at the ratios between 0 and 0.15% by weight of cement specimens. These specimens were heated using high temperatures (200–800 °C). The incorporation of 0.1% MWCNT by cement weight improved the compressive, tensile and flexural strength by 35%, 8%, and 11.2%, respectively. In addition, the cement paste was more stable and denser due to the addition of MWCNT. Also, Hawreen and Bogas [122] studied the effects on the long-term creep and shrinkage of concrete due to the incorporation of different types of CNTs. They used concretes with 0.05%–0.5% of unfunctionalized and functionalized CNTs and water to cement ratios of 0.35–0.55. The compressive strength of the concrete with CNTs was increased up to 21%. The addition of CNTs caused a reduction in the early and long-term shrinkage of concrete up to 54% and 15%, respectively. The concrete with addition of CNTs had 17%–18% lower long-term creep in comparison to the concrete without CNTs. Carbon nanotubes are innovative materials for the construction industry that can decrease the formation of nano-cracks. Moreover, the inclusion of CNTs in concrete may increase the compressive and flexural strength of the concrete.

2.6. Graphene-Based Nanomaterials

Recent studies [126–147] have examined the performance of cement-based materials incorporating graphene family nanomaterials (GFN) such as graphene, graphene oxide (GO), reduced graphene oxide (rGO), and graphene nanosheets (GNS). These nanomaterials have extraordinary electrical, mechanical, chemical and thermal properties. Thus, GFN reinforced cement-based materials can improve their structural strength and durability, as well as allow self-cleaning surfaces and self-sensing abilities [148–158].

Hu et al. [158] fabricated cement composite including nano-silica coated GO, which enhanced its mechanical properties. The compressive strengths of cement composites containing GO and nano-silica coated GO were studied at curing ages of 1, 3, 7 and 28 days. The nano-silica coated GO reinforced cement composites increased their compressive strength up to 120.6%, 124.1%, 126.7% and 133% compared to plain cement with curing ages of 1, 3, 7 and 28 days, respectively. For the GO reinforced cement composites without nano-silica, their compressive strengths improved up to 106.0%, 106.7%, 112.2% and 113.6% with respect to plain cement at curing time of 1 day, 3 days, 7 days and 28 days. The coated nano-silica on GO allowed a finer surface structure and better dispersion, which helped to eliminate the agglomeration of GO in pore solution. The nano-silica coated GO promoted the deposition and growth of C–S–H, refining the cement composite microstructure and improving their macro-mechanical properties. Moreover, Hu et al. [159] functionalized GO via triethanolamine (TEA), which was added into oil well cement (OWC) to enhance its mechanical behavior. The incorporation of TEA-GO and GO was at 0.3 wt.%, keeping a fixed water-to-cement ratio of 0.44. First, they mixed GO/TEA-GO power with water and after added cement within 15 s. In the second stage, the cement was collocated in the cup of Waring blender at 4000 rpm and after was mixing during 15 s at 120,000 rpm. Finally, the cement composites are cured at 60 °C for 1, 3, 7 and 28 days. Both GO and TEA-GO had good dispersion in water; although, TEA-GO with smaller size reported better uniformity. The TEA-GO incorporated to cement allows nucleation sites and acts as seeds to provide the cement hydration. In order to evalue the influence of GO/TEA-GO on the mechanical properties of cement composite, compressive and flexural strength were obtained through mechanical hydraulic pressure testing machine. For the compressive strength tests, the TEA-GO reinforced samples (50.8 $\times$ 50.8 $\times$ 50.8 mm^3) were characterized a loading rate of 1.2 KN/s. On the other hand, the flexural strength tests used TEA-GO reinforced samples (160 $\times$ 40 $\times$ 40 mm^3) under a loading rate of 0.2 KN/s. Figure 25 illustrates the average compressive and flexural strengths of GO and TEA-GO reinforced cement samples. The TEA-GO reinforced cement samples presented higher increments (9.4%–31%) of the compressive strength than that of GO reinforced cement (4.1%–17.2%). This is due to that TEA-GO modified cement provides crystals more mature and fewer pores and cracks compared to blank cement and GO reinforced cement samples (see Figure 26). Therefore, the TEA-GO significantly refines the microstructure of cement specimens. On the contrary, the TEA-GO reinforced cement samples achieved the higher increments (8.1%–36.7%) with respect to those of cement incorporating GO (7.8%–20%). The TEA-GO enhanced the mechanical performance of cement due to the increment of hydration degree and limitation to crack propagation. Figure 27 depicts SEM images of the cracks types in three different OWC composites. In the blank cement sample, the cracks penetrate in a straight-through form. For GO reinforced cement sample, the cracks are thinner with random branches. Finally, the cement containing TEA-GO registers fewer cracks with thinner dimensions.

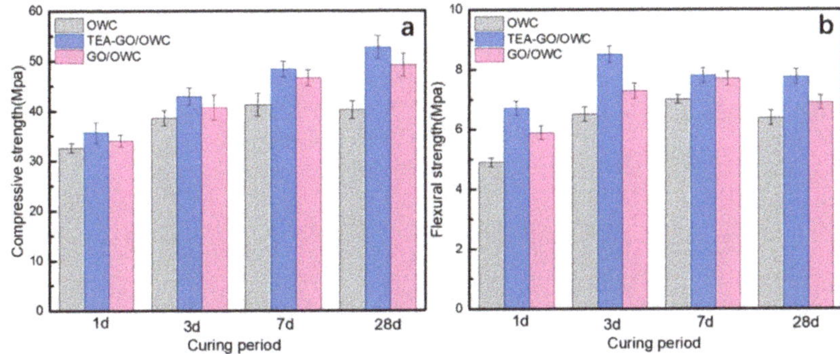

Figure 25. Response of the (**a**) compressive and (**b**) flexural strength of OWC composites containing GO and TEA-GO. Reprinted with permission from [159]. Copyright© 2019, Elsevier B.V.

Figure 26. SEM image of the (**a**) blank OWC sample, (**b**) GO reinforced OWC specimen and (**c**) TEA-GO reinforced OWC specimen after compressive strength test at age 28 days. Reprinted with permission from [159]. Copyright© 2019, Elsevier B.V.

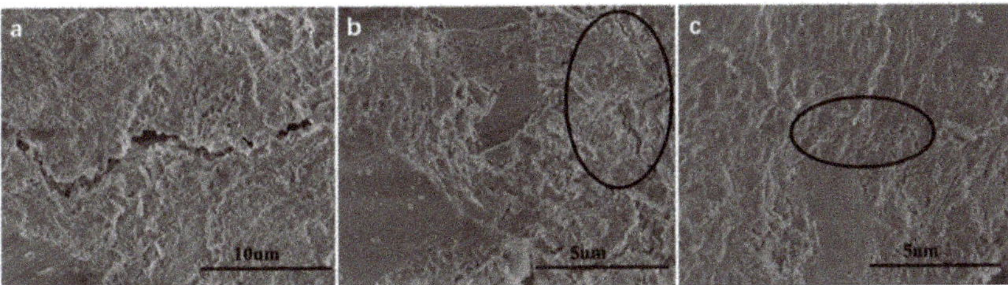

Figure 27. SEM images of the (**a**) plain OWC matrix including a straight-through crack, (**b**) GO/OWC specimen containing fine cracks and (**c**) TEA-GO/OWC specimen. Reprinted with permission from [159]. Copyright© 2019, Elsevier B.V.

Tao et al. [160] investigated the influence of graphene nanoplatelets (GNPs) on the microstructure, pore structure, piezoresistive and mechanical behavior of cement mortar. They quantitatively examined the piezoresistive performance of GNP-reinforced cement mortars exposed to cyclic compressive loads. A PI 42.5 cement is used as binder and natural quartz sands are employed as aggregates. Different dosages (M0 = 0%, M1 = 0.05%, M2 = 0.1%, M3 = 0.5% and M4 = 1%) of GNPs by cement weight were incorporated in cement matrix. The compressive and flexural strength of the GNP-reinforced cement mortars were characterized through a 25 KN high-performance fatigue testing machine. For the compressive and flexural tests were employed loading rates of 144 kN/min and 3 kN/min, respectively. The four-probe method is employed to measure the piezoresistive properties of the cement mortars including GNPs. The cement mortars are dried at 80 °C during 24 h to eliminate the capillary water, which affects the piezoresistive response [161]. Figure 28 illustrates the fracture surfaces of cement mortars containing different GNPs dosages. Later, the cement mortars specimens, regarding the probes and cables, are examined using the mechanical testing machine (see Figure 29). First, the initial resistance of the specimens is determined at stable voltage. Then, the external loads are applied to specimens using constant loading rate of 0.5 kN/s and the piezoresistive properties are obtained through an Instron actuator. Figure 30 depicts the compressive and flexural strengths of the cement mortars. Both strengths firstly increment their values and after decrease when the GNP dosage increases.

Figure 28. SEM images of the cement mortars added with different GNP content: (**a**) M0 (0%), (**b**) M1 (0.05%), (**c**) M2 (0.1%), (**d**) M3 (0.5) and (**e**) M4 (1%). Reprinted with permission from [160]. Copyright© 2019, Elsevier B.V.

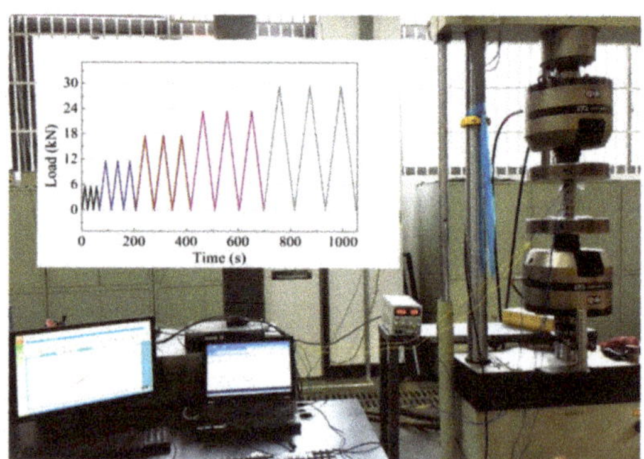

Figure 29. Measurement setup of piezoresistive test of the GNP-added cement mortars under cyclic compressions. Reprinted with permission from [160]. Copyright© 2019, Elsevier B.V.

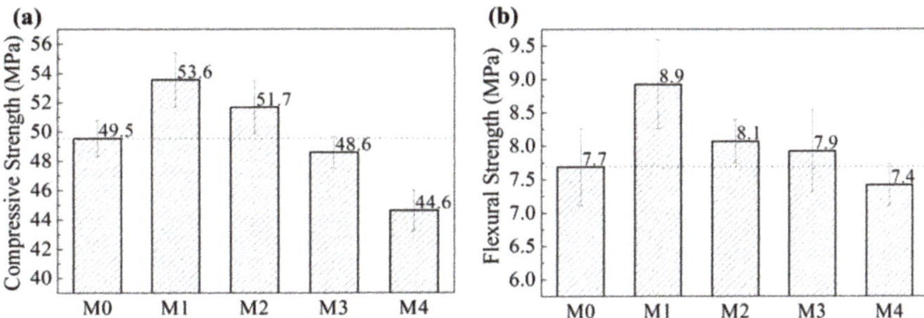

Figure 30. Experimental results of the (**a**) compressive and (**b**) flexural strengths of the GNP reinforced cement mortars. Reprinted with permission from [160]. Copyright© 2019, Elsevier B.V.

Higher magnitudes of compressive and flexural strengths (53.6 MPa and 8.9 MPa, respectively) are reached in the cement mortar with GNP dosage of 0.25% (M1) by cement weight. These values represent increments of 8.3% and 15.6% compared with the compressive and flexural strengths of the cement mortar without GNP. However, the values of both strengths decrease when the GNP dosages exceed 0.05%. Thus, cement hydrates with homogeneous spatial distribution could be obtained with appropriate values of GNPs dispersed in cement matrix.

Qureshi and Panesar [162] characterized the influence of GO and rGO on the performance of cement-based composite. They investigated the microstructural properties of GO and rGO using X-ray diffraction (XRD), optical microscope, Fourier-transform infrared spectroscopy (FTIR), SEM, Energy dispersive X-ray (EDS) and Raman spectroscopy techniques. The average C:O ratio of 54.46 and 82.18 in GO and rGO, respectively, were employed in the cement-based materials. To enhance the dispersibility of rGO in water, rGO was processed with superplasticizer. The dosages of both GO and rGO were of 0.02%, 0.04% and 0.06% of cement weight. To reach uniform mixture and efficient dispersion of both GO and rGO in the cement specimens, a water to cement ratio of 0.45 is implemented. In comparison with the control cement sample without GO and rGO, the final setting time and workability of GO reinforced cement specimens gradually decreased using higher GO content up to 0.06% of cement weight. This is caused by the dominant oxygen functional groups and hydrophilic behavior of GO. On the other hand, final setting time and workability of rGO-added cement specimens increased with respect to the control cement sample. It is due to the superplasticizer content and the almost hydrophobic behavior of rGO. The GO composites had greater dosage of C–S–H and $Ca(OH)_2$ than the rGO composites at ages of 1, 7 and 28 days. In addition, the GO composites showed micropores filled with crystalline compounds and C–S–H gel. For the rGO composites was found random pore filling nature with ettringite elements. Figure 31 depicts the SEM-EDX results of the microstructure of GO composite pores. Based on the EDX and SEM results of GO and rGO, these nanomaterials present suitable compatibility with cement hydration products, reinforcing the microstructure of the cement composites. Figure 32 depicts the response of the flexural strength and compressive of cement specimens added with GO and rGO at curing age of 28 days. In comparison to control cement sample, cement composites incorporated with 0.02%, 0.04% and 0.06% GO and rGO dosages had an increase of 10.2%, 7.8% and 10.6 %, and 9.6%, 13.3% and 14.9%, respectively. This is due to the high number of functional groups of GO in chemical bonding with cement hydration products and the high mechanical strength of rGO.

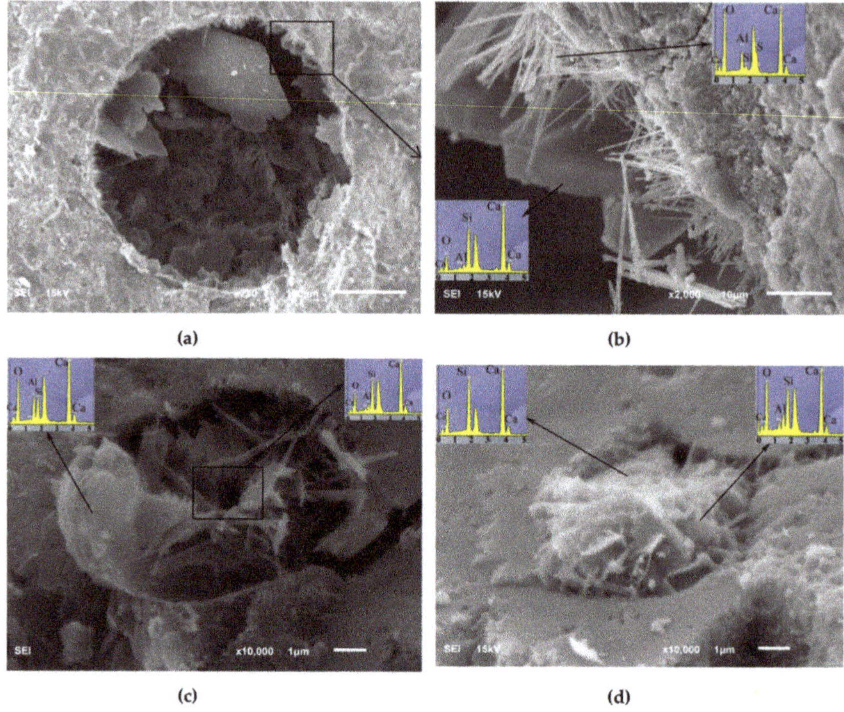

Figure 31. SEM-EDX results of the microstructure of GO composite pores in: (**a**) 0.06% GO composite, (**b**) amplification of (a) showing a surface growth nature of ettringite and flakes structures, (**c**) a small pore of 0.02% GO and (**d**) a small pore of 0.04% GO. Reprinted with permission from [162]. Copyright© 2019, Elsevier B.V.

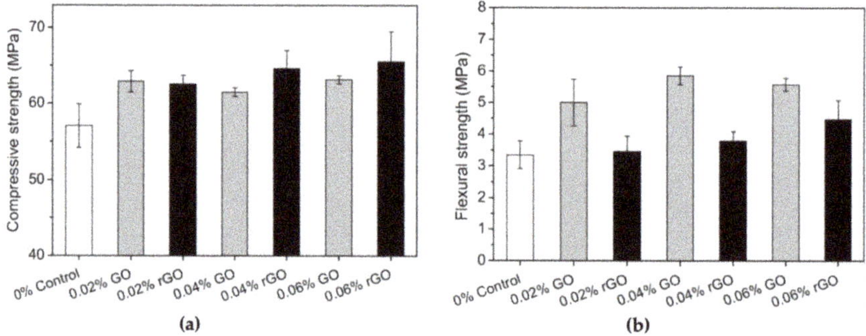

Figure 32. Experimental response of the (**a**) compressive and (**b**) flexural strength of GO and rGO reinforced cement specimens at curing age of 28 days. Reprinted with permission from [162]. Copyright© 2019, Elsevier B.V.

Krystet et al. [163] studied the mechanical properties and microstructure of cementitious materials with addition of electrochemically exfoliated graphene (EEG). EEG enhanced the mechanical properties, microstructure and workability of cementitious materials. EEG did not provide aggregate in alkaline environment and the cement mortars incorporating EEG did not decrease its workability and fluidity. The mixture of 0.05 wt.% of graphene with ordinary Portland cement improved the compressive and tensile strength of the cement material up to 79% and 8%, respectively. EEG contributes to hydration reactions of calcium silicates, allowing an intense generation of C–S–H phase and a compact microstructure.

Kaur and Kothiyal [164] compared the effect of polycarboxylate superplasticizer (PCE-SP) added GO and functionalized CNT (SP@GO and SP@FCNT) on the mechanical properties of cement nanocomposites (CNCs). They used two types of SPs to alter the GO and FCNTs structural features, and to enhance the dispersion of these nanomaterials in aqueous solution and cement matrix. The stabilized GO and FCNT allowed to enhance the mechanical strength of the CNC specimens. After, they fabricated three cubes of CNC specimens (70.6 mm × 70.6 mm × 70.6 mm^3) containing SP@GO and SP@FCNT with different dosages (i.e., 0.02%, 0.04%, 0.08% and 0.16% by cement weight). These specimens were water curing at ages of 7, 14 and 28 days to characterize their mechanical strength. The mechanical tests were done using universal testing machine, applying load at the rate of 3.8 kN/s and 0.5 kN/s, respectively. Figures 33 and 34 show the measurements of average compressive and flexural strength of the SP@GO and SP@FCNT modified CNC specimens. With respect to cement specimen, the maximum values of compressive and tensile strengths of CNC specimens were improved up to 23.2% and 38.5% due to addition of 0.02% and 0.08% SP@GO by cement weight, respectively. On the other hand, addition of 0.08% and 0.04% of SP@FCNT by cement weight enhanced the compressive and tensile strengths of the CNC specimens by 16.5% and 35.8%, respectively. Figure 35 depicts FE-SEM images of CNC specimens containing different SP@GO dosages.

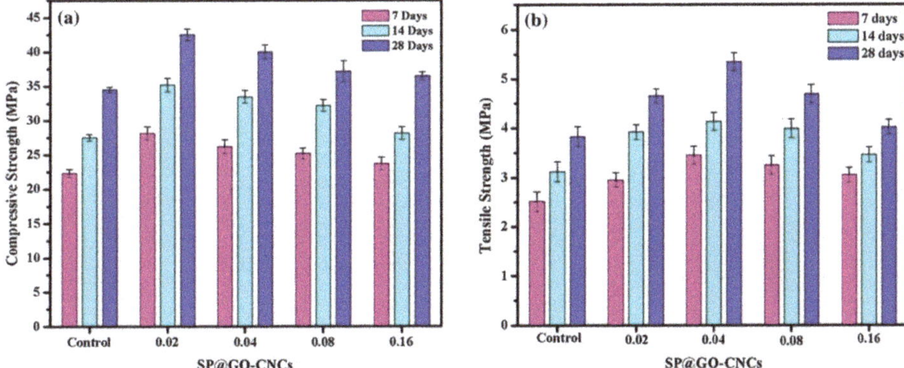

Figure 33. Experimental results of the (**a**) compressive and (**b**) flexural strength of SP@GO modified CNC specimens at curing ages of 7, 14 and 28 days. Reprinted with permission from [164]. Copyright© 2019, Elsevier B.V.

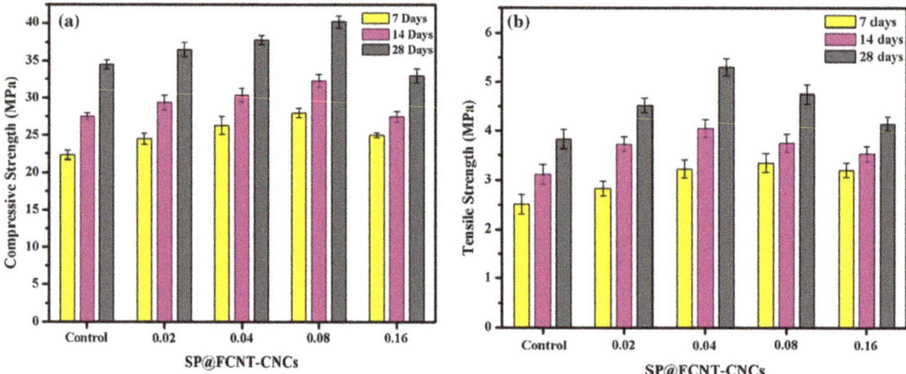

Figure 34. Experimental results of the (**a**) compressive and (**b**) flexural strength of SP@FCNT modified CNC specimens at curing ages of 7, 14 and 28 days. Reprinted with permission from [164]. Copyright© 2019, Elsevier B.V.

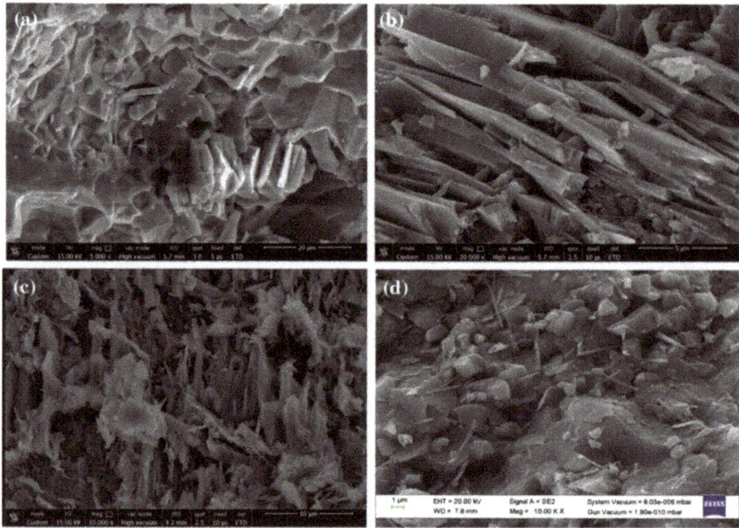

Figure 35. FE-SEM images of CNC specimens incorporating (**a**) 0.02% SP@GO, (**b**) 0.04% SP@GO, (**c**) 0.08% SP@GO and (**d**) 0.16% SP@GO at curing age of 28 days. Reprinted with permission from [164]. Copyright© 2019, Elsevier B.V.

The addition of graphene family nanomaterials in cement composites can enhance their mechanical strength. This will allow the construction of lighter concrete components with extended durability, thus, the consumption of concrete components could be decreased. This will help with the reduction of the gas pollutants resulting from concrete production.

3. Challenges

The nanotechnology has allowed the fabrication of nanomaterials that can be incorporated in cement-based materials to generate higher mechanical properties of the concrete structures. The effect of the nanomaterials on the performance of cement-based materials includes the enhance of their compressive, tensile and flexural strength, reduction of the total porosity (i.e., refinement of the microstructure), acceleration of C–S–H gel generation and increment of Young modulus. Furthermore, incorporation of nanomaterials such as nano-TiO_2 and CNTs can provide self-cleaning and self-sensing properties, respectively, of the products obtained with cement-based materials. To achieve the optimal mechanical properties of the cement-based materials it is very important to mix a suitable dosage of nanomaterial with the cement-based materials. For instance, excessive quantities of nanomaterials added to cement can result in lower compressive, tensile and flexural strength of the cement-based structures. This is caused by nonhomogeneous dispersion of nanomaterials in the cement paste. Thus, the mechanical properties of the nanomaterial reinforced cement-based materials depend of several factors such as the dosage and type of nanomaterial, dispersion method, curing days and curing method. Between these factors, the dispersion method can have a significantly effect on the performance of the nanomaterial reinforced concrete.

An important challenge for the application of nanomaterials in the construction industry is the development of efficient methods for the dispersion of nanomaterials in cement samples. An alternative solution to incorporate the nanomaterial into cement-based materials consists in the dispersion of the nanomaterial in water before of incorporating it to the dry components of the cement-based materials [165]. For this case, ultrasonic dispersion can be employed as an effective method for the dispersion of the nanomaterials, although, this method requires electrical energy that increases its cost. A bad dispersion of the nanomaterials into cement specimens and the formation of great amount of agglomerates may alter the kinetics of the hydration process, modifying the properties of the cement specimens. For instance, Singh et al. [166] reported that the method used to incorporate nano-silica into cement composites can affect the porosity and mechanical properties of the composites. A bad nano-silica dispersion in cement-based materials may generate voids and weak zones, altering the mechanical properties of the materials. Surfactants such plasticizers and superplasticizers can be used to improve the dispersion of nanomaterials in cement-based materials [51,167,168]. Thus, surface active agents can enhance the homogeneity of dispersion due to the generation of aggregates around nanoparticles [169]. This good dispersion is achieved because of both hydrophobic and hydrophilic groups. The nanomaterials interact with the hydrophobic groups and the hydrophilic groups decrease the water-surface tension, increasing the dispersion of the nanomaterials [167]. Nevertheless, several surfactants (e.g., polymeric matrices) employed for the dispersion of nanoparticles can affect the cement hydration kinetics. Figure 36 depicts a nanomaterial dispersion process employed to obtain a cement-based composite.

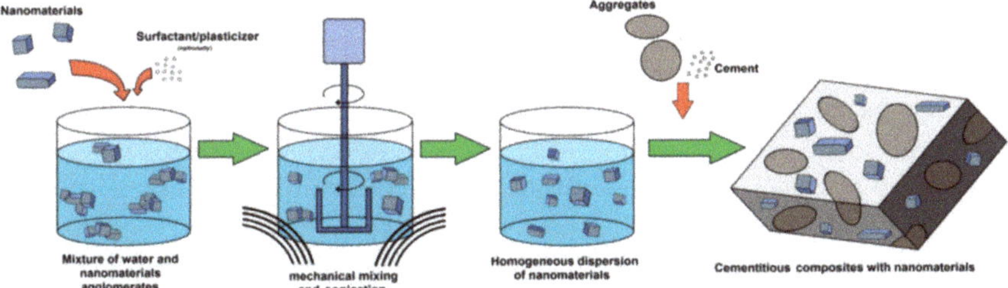

Figure 36. Schematic view of a common nanomaterial dispersion process employed to fabricate a cement-based composite. Reprinted with permission from [167]. Copyright©2018, MDPI AG.

To develop the large scale production of these modified cement-based materials it will be necessary to develop an efficient nanomaterial dispersion method that allows a stable and satisfactory dispersion in cement-based materials [167]. However, the re-agglomeration of the nanoparticles may change their size, which could affect their behavior of nanomaterials in cement-based materials [170].

The application nanomaterials in cement-based composites is attractive due to that enhances their mechanical properties only using small dosages of nanomaterials. However, the high cost of these nanomaterials is a limitation to achieve their commercial application in cementitious materials. In the case of graphene and graphene oxide, their fabrication scale is small and relatively expensive. In the future, an important challenge is the reduction of fabrication costs of the nanomaterials. For instance, SiO_2 nanoparticles may be generated with low cost from hydrothermal solutions generated due the magmatic ore intrusion [77]. Thus, nano-silica and other minerals can be recovered when the steam is condensed in water during the operation of a geothermal power plant [77].

Moreover, another challenge is the optimal dosage determination of the nanomaterial added in the cement paste to obtain the higher mechanical strength and greater durability. To meet this goal, more studies about the effect of different types of nanomaterials on the mechanical properties of cement-based materials are required. These studies must include the effect of the combination of nanomaterials with other cementitious materials. Thus, the optimal quantities of the combination of these materials must be examined to find the best mechanical behavior of the cement-based materials. The main challenge of the use of carbon nanotubes (CNTs)/nanofibers in cement paste is the dispersion due to their strong self-attraction and high hydrophobicity [171,172]. This poor dispersion may cause defects zones in cement/CNTs composites, which constrain the use of CNTs in cementitious matrices [27]. Thus, more investigations must be made with respect to developing methods to improve the dispersion of CNTs in cement-based composites. For example, some researchers [173,174] have studied chemical and surface modification methods for carbon nanotubes to improve their dispersion and bonding between carbon nanotubes and the matrix. Other studies [175–177] had examined the mechanical properties of cement samples, which were prepared with different combinations of carbon nanotubes and nano-silica. In addition, future researches must consider prediction models of the relationship between external mechanical deformations and electrical resistivity of cement-based composites incorporating CNTs.

In addition, the future applications of nanomaterials in the construction industry will require one to consider the local environmental conditions (e.g., elements of the local environmental dust). These conditions could damage the performance of the concrete, reducing its durability and increasing the cost of maintenance. For instance, The Jubilee Church in Rome (2003) was one of the first buildings that used self-cleaning and reinforced concrete [178–180]. This construction had three iconic shells constructed from 2001 and 2002, which employed precast panels with photocatalytic nano-TiO_2 particles. Thus, the nano-TiO_2 particles could absorb energy from light and employ it to achieve a photocatalytic degradation of pollutions. In 2019, Cardellicchio [179,180] reported a study about premature evidence of decay of the three shells that showed failure of their self-cleaning performance. This study considered the material pathologies and their possible damage sources. The surfaces of the shells still contain nano-TiO_2 particles in the form of anatine, which was detected by this study through chemical analysis. Nonetheless, the self-cleaning properties of the shells with nano-TiO_2 are only activated when both the sunlight allows the redox of pollutant and the photo-induced hydrophilicity permits the cleaning of the shells [180]. For this case, the hydrophilic characteristic is limited by two main conditions. One condition is linked with common composition of the pozzolanic powder in Rome that cannot be oxidized by titanium dioxide. Another condition is the abrasive effect of the rainwater on the surface of the shells that improves the superficial roughness, increases the bond between powder particles and concrete [180]. The erosive action occurs on the convex surface, which was registered by a colorimetric analysis showing a tendency towards whitish-grey hue caused by the scattering of the sunlight. These two conditions generate a patina which decreases the photocatalytic effect of the surface of the concrete shells (see Figure 37). The efficiency of the nano-TiO_2 particles incorporated in concrete is affected by the porosity and roughness of the concrete surface. The porosity of the concrete surface may allow the water retention and its roughness may help the adhesion of powder on the surface. The chemical and abrasive characteristics of pozzolanic powder may decrease the efficiency of the self-cleaning of the nano-TiO_2 added concrete. Therefore, future buildings that use concrete with self-cleaning properties may be affected by powder of volcanic origins or precipitations incorporating desert dust [180].

Figure 37. View of the damages on the surface of the self-cleaning concrete shells of Jubilee Church in Rome, which were caused by the local environmental dust. Reprinted with permission from [180]. Copyright©2019, Taylor & Francis Group.

More investigations about new nanomaterials to improve the mechanical properties of cement-based composites are required. In addition, to achieve the commercial application of these nanomaterials reinforced cement-based composites is necessary to know the effect of the nanomaterials on the mechanical properties of these composites. Therefore, future researches must include the development of novel theoretical models that can predict the mechanical properties of the cement-based materials as function of dosage level of the nanomaterials. Also, more researches about the impact of the nanomaterials in the public health and environment must be developed. There are few studies about the effects of the nanomaterials used in cement-based materials on the public health and environment. For instance, Lee et al. [181] reported some effects of nanoparticles used in the construction industry on environment health and safety. Lam et al. [182] presented a review about the carbon nanotube toxicity, which could produce pulmonary inflammation and cardiac toxicity. Moreover, nano-TiO_2 could generate inflammation in mammalian cells [183]. Thus, many researches must be done to reduce the negative impacts of the nanomaterials on the health and environment. In addition, more studies about the bioavailability and environment mobility of nanomaterials required to be done [182].

4. Conclusions

The incorporation of nanomaterials in concrete can improve their compressive, tensile and flexural strength. Recent investigations had considered nanomaterials such as nano-silica, nano-titania, nano-ferric oxide, nano-alumina, CNT, graphene and GO. The addition of these nanomaterials in concrete can achieved denser microstructures, decreasing the water absorption. The workability of the concrete could be improved by adding these nanomaterials. The nano-TiO_2 modified concrete can provide it self-cleaning properties and other benefits to help the environment clean. In addition, nano-TiO_2 added in concrete can allow the photocatalytic degradation of pollutants (e.g.; NO_x, VOCs, CO, chlorophenols, and aldehydes) from automobile and industrial emissions. The CNTs reinforced cement-based composites can have self-sensing abilities for applications of structural health monitoring or damage detection. In addition, graphene and GO added in cement-based materials can increase their mechanical strength and durability, as well as develop self-cleaning surfaces and self-sensing abilities.

In the construction industry, the fabrication of cement-based composites can generate high levels of CO_2 gas. To address this problem, one solution is the addition of nanomaterials to cement-based composites, which can provide structural components with high mechanical strength and great durability. Thus, the maintenance requirements and replacement frequency of the cement-based structural components can be decreased. These advantages can allow the reduction of the percentage of cement used in the construction industry. This in turn will decrease the CO_2 emissions caused by the cement fabrication process.

The application of nanotechnology in cement-based materials is still in a research stage. The results of experimental tests of nanomaterials-reinforced cement specimens have demonstrated that they can enhance the mechanical strength and durability of the resultng concretes. Moreover, these nanomaterials can allow a novel generation of smart cement-based composites with strain-sensing abilities for damage inspection and structural health monitoring.

Author Contributions: K.P.B.-G. and A.H.-M. contributed with the sub-sections of nano-silica, nano-Fe_2O_3 and nano-Al_2O_3; J.M.S.-L. supervised all the sections of the manuscript; S.A.Z.-C. reviewed the sections of introduction and challenges, and A.L.H.-M. wrote the sections of nano-TiO_2, carbon nanotubes and graphene-based nanomaterials.

Funding: This research was supported by the project "PFCE 2019 DES Técnica Veracruz 30MSU0940B-21".

Conflicts of Interest: The authors declare no conflict of interest.

References

1. Aslani, F. Nanoparticles in self-compacting concrete. A review. *Mag. Concr. Res.* **2015**, *67*, 1084–1100. [CrossRef]
2. Kawashima, S.; Hou, P.; Corr, D.J.; Shah, S.P. Modification of cement-based materials with nanoparticles. *Cem. Concr. Compos.* **2013**, *36*, 8–15. [CrossRef]
3. Chuah, S.; Pan, Z.; Sanjayan, J.G.; Wang, C.M.; Duan, W.H. Nano reinforced cement and concrete composites and new perspective from graphene oxide. *Constr. Build. Mater.* **2014**, *73*, 113–124. [CrossRef]
4. Lazaro, A.; Yu, Q.L.; Brouwers, H.J.H. Nanotechnologies for sustainable construction. In *Sustainability of Construction Materials*, 2nd ed.; Khatib, J.M., Ed.; Woodhead Publishing: Duxford, UK, 2016; pp. 55–78. [CrossRef]
5. Hanus, M.J.; Harris, A.T. Nanotechnology innovations for the construction industry. *Prog. Mater. Sci.* **2013**, *58*, 1056–1102. [CrossRef]
6. Saleh, H.M.; El-Sheikh, S.M.; Elshereafy, E.E.; Essa, A.K. Performance of cement-slag-titanate nanofibers composite immobilized radioactive waste solution through frost and flooding events. *Constr. Build. Mater.* **2019**, *223*, 221–232. [CrossRef]
7. Jassam, T.H.; Kien-Woh, K.; Yang-Zhi, J.N.; Lau, B.; Yaseer, M.M.M. Novel cement curing technique by using controlled release of carbon dioxide coupled with nanosilica. *Constr. Build. Mater.* **2019**, *223*, 692–704. [CrossRef]
8. Du, H. Properties of ultra-lightweight cement composites with nano-silica. *Constr. Build. Mater.* **2019**, *199*, 696–704. [CrossRef]
9. Xu, J.; Shen, W.; Corr, D.J.; Shan, S.P. Effects of nanosilica on cement grain/C–S–H gel interfacial properties quantified by modulus mapping and nanoscratch. *Mater. Res. Express* **2019**, *6*, 045061. [CrossRef]
10. Evangelista, A.C.J.; de Morais, J.F.; Tam, V.; Soomro, M.; Di Gregorio, L.T.; Haddad, A.N. Evaluation of Carbon Nanotube Incorporation in Cementitious Composite Materials. *Materials* **2019**, *12*, 1504. [CrossRef]
11. Srikanth, M.; Asmatulu, R. Nanotechnology safety in the construction and infrastructure industries. In *Nanotechnology Safety in the Construction and Infrastructure Industries*, 1st ed.; Asmatulu, R., Ed.; Elsevier: Burlington, MA, USA, 2013; pp. 99–1013. [CrossRef]
12. Konsta-Gdoutos, M.S.; Aza, C.A. Self-sensing carbon nanotube (CNT) and nanofiber (CNF) cementitious composites for teal time damage assessment in smart structures. *Cem. Concr. Compos.* **2014**, *53*, 162–169. [CrossRef]
13. Carriço, A.; Bogas, J.A.; Hawreen, A.; Guedes, M. Durability of multi-walles carbon nanotube reinforced concrete. *Constr. Build. Mater.* **2018**, *164*, 121–133. [CrossRef]
14. Papanikolaou, I.; Arena, N.; Al-Tabbaa, A. Graphene nanoplatelet reinforced concrete for self-sensing structures—A lifecycle assessment perspective. *J. Clean. Prod.* **2019**, *240*, 118202. [CrossRef]

15. Mahdikhani, M.; Bamshad, O.; Shirvani, M.F. Mechanical properties and durability of concrete specimens containing nano silica in sulfuric acid rain condition. *Constr. Build. Mater.* **2018**, *167*, 929–935. [CrossRef]
16. Zabihi, N.; Ozkul, M.H. The fresh properties of nano silica incorporating polymer-modified cement pastes. *Constr. Build. Mater.* **2018**, *168*, 570–579. [CrossRef]
17. Luo, Z.; Li, W.; Wang, K.; Shah, S.P. Research progress in advanced nanomechanical characterization of cement-based materials. *Cem. Concr. Compos.* **2018**, *94*, 277–295. [CrossRef]
18. Mijowska, E.; Horszczaruk, E.; Sikora, P.; Cendrowski, K. The effect of nanomaterials on thermal resistance of cement-based composites exposed to elevated temperature. *Mater. Today Proc.* **2018**, *5*, 15968–15975. [CrossRef]
19. Rai, S.; Tiwari, S. Nano Silica in Cement Hydration. *Mater. Today Proc.* **2018**, *5*, 9196–9202. [CrossRef]
20. El-Gamal, S.M.A.; Hashem, F.S.; Amin, M.S. Influence of carbon nanotubes, nanosilica and nanometakaolin on some morphological-mechanical properties of oil well cement pastes subjected to elevated water curing temperature and regular room air curing temperature. *Constr. Build. Mater.* **2017**, *146*, 531–546. [CrossRef]
21. Li, L.G.; Huang, Z.H.; Zhu, J.; Kwan, A.K.H.; Chen, H.Y. Synergistic effects of micro-silica and nano-silica on strength and microstructure of mortar. *Constr. Build. Mater.* **2017**, *140*, 229–238. [CrossRef]
22. Braz de Abreu, G.B.; Marques-Costa, S.M.; Gumieri, A.G.; Fonseca-Calixto, J.M.; França, F.C.; Silva, C.; Delgado-Quinõnes, A. Mechanical properties and microstructure of high performance concrete containing stabilized nano-silica. *Matéria* **2017**, *22*, e11824. [CrossRef]
23. Sobolev, K. Modern developments related to nanotechnology and nanoengineering of concrete. *Front. Struct. Civ. Eng.* **2016**, *10*, 131–141. [CrossRef]
24. Yang, Q.; Liu, P.; Ge, Z.; Wang, D. Self-Sensing Carbon Nanotube-Cement Composite Material for Structural Health Monitoring of Pavements. *J. Test. Eval.* **2020**, 48. [CrossRef]
25. Silvestre, J.; Silvestre, N.; de Brito, J. Review on concrete nanotechnology. *Eur. J. Environ. Civ. Eng.* **2016**, *20*, 455–485. [CrossRef]
26. Huseien, G.F.; Shah, K.W.; Sam, A.R.M. Sustainability of nanomaterials based self-healing concrete: An all-inclusive insight. *J. Build. Eng.* **2019**, *23*, 155–171. [CrossRef]
27. Singh, N.B.; Kalra, M.; Saxena, S.K. Nanoscience of Cement and Concrete. *Mater. Today Proc.* **2017**, *4*, 5478–5487. [CrossRef]
28. Sikora, P.; Horszczaruk, E.; Cendrowski, K.; Mijowska, E. The influence of nano-Fe_3O_4 on the microstructure and mechanical properties of cementitious composites. *Nanoscale Res. Lett.* **2016**, *11*, 182. [CrossRef]
29. An, J.; Nam, B.H.; Alharbi, Y.; Cho, B.H.; Khawaji, M. Edge-oxidized graphene oxide (EOGO) in cement composites: Cement hydration and microstructure. *Compos. Part B Eng.* **2019**, *173*, 106795. [CrossRef]
30. Xiao, H.; Zhang, F.; Liu, R.; Zhang, R.; Liu, Z.; Liu, H. Effects of pozzolanic and non-pozzolanic nanomaterials on cement-based materials. *Constr. Build. Mater.* **2019**, *213*, 1–9. [CrossRef]
31. He, R.; Huang, X.; Zhang, J.; Geng, Y.; Guo, H. Preparation and evaluation of exhaust-purifying cement concrete employing titanium dioxide. *Materials* **2019**, *12*, 2182. [CrossRef]
32. Sun, J.; She, X.; Tan, G.; Tanner, J.E. Modification effects of nano-SiO_2 on early compressive strength and hydration characteristics of high-volume fly ash concrete. *J. Mater. Civ. Eng.* **2019**, *31*, 04019057. [CrossRef]
33. Wang, J.; Wang, W.; Wu, X.; Zhou, Y. Mechanical properties of cement asphalt mortar under low temperature condition. *J. Test. Eval.* **2019**, *47*, 1995–2009. [CrossRef]
34. Tiong, M.; Gholami, R.; Rahman, M.E. Cement degradation in CO_2 storage sites: A review on potential applications of nanomaterials. *J. Pet. Explor. Prod. Technol.* **2019**, *9*, 329. [CrossRef]
35. Yoo, D.-Y.; You, I.; Zi, G.; Lee, S.-J. Effects of carbon nanomaterial type and amount on self-sensing capacity of cement paste. *Measurement* **2019**, *134*, 750–761. [CrossRef]
36. Sivasankaran, U.; Raman, S.; Nallusamy, S. Experimental analysis of mechanical properties on concrete with nano silica additive. *J. Nano Res.* **2019**, *57*, 93–104. [CrossRef]
37. Saleh, H.M.; El-Saied, F.A.; Salaheldin, T.A.; Hezo, A.A. Macro- and nanomaterials for improvement of mechanical and physical properties of cement kiln dust-based composite materials. *J. Clean. Prod.* **2018**, *204*, 532–541. [CrossRef]
38. Paul, S.C.; van Rooyen, A.S.; van Zijl, G.P.A.G.; Petrik, L.F. Properties of cement-based composites using nanoparticles: A comprehensive review. *Constr. Build. Mater.* **2018**, *189*, 1019–1034. [CrossRef]

39. Da Rocha Segundo, I.G.; Lages Dias, E.A.; Pereira Fernandes, F.D.; de Freitas, E.F.; Costa, M.F.; Oliveira Carneiro, J. Photocatalytic asphalt pavement: The physicochemical and rheological impact of TiO_2 nano/microparticles and ZnO microparticles onto the bitumen. *Road Mater. Pavement Des.* **2019**, *20*, 1452–1467. [CrossRef]
40. Indukuri, C.S.R.; Nerella, R.; Madduru, S.R.C. Effect of graphene oxide on microstructure and strengthened properties of fly ash and silica fume based cement composites. *Constr. Build. Mater.* **2019**, *229*, 116863. [CrossRef]
41. Khooshechin, M.; Tanzadeh, J. Experimental and mechanical performance of shotcrete made with nanomaterials and fiber reinforcement. *Constr. Build. Mater.* **2018**, *165*, 199–205. [CrossRef]
42. Staub de Melo, J.V.; Trichês, G. Study of the influence of nano-TiO_2 on the properties of Portland cement concrete for application on road surfaces. *Road Mater. Pavement Des.* **2018**, *19*, 1011–1026. [CrossRef]
43. Diamond, S.A.; Kennedy, A.J.; Melby, N.L.; Moser, R.D.; Poda, A.R.; Weiss, C.A.; Brame, J.A. Assessment of the potential hazard of nano-scale TiO_2 in photocatalytic cement: Application of a tiered assessment framework. *NanoImpact* **2017**, *8*, 11–19. [CrossRef]
44. Sikora, P.; Cendrowski, K.; Markowska-Szczupak, A.; Horszczaruk, E.; Mijowska, E. The effects of silica/titania nanocomposite on the mechanical and bactericidal properties of cement mortars. *Constr. Build. Mater.* **2017**, *150*, 738–746. [CrossRef]
45. Balopoulos, V.D.; Archontas, N.; Pantazopoulou, S.J. Model of the mechanical behavior of cementitious matrices reinforced with nanomaterials. *J. Eng.* **2017**, *2017*, 7329540. [CrossRef]
46. Bossa, N.; Chaurand, P.; Levard, C.; Borschneck, D.; Miche, H.; Vicente, J.; Geantet, C.; Aguerre-Chariol, O.; Michel, F.M.; Rose, J. Environmental exposure to TiO2 nanomaterials incorporated in building material. *Environ. Pollut.* **2017**, *220*, 1160–1170. [CrossRef]
47. Bastos, G.; Patiño-Barbeito, F.; Patiño-Cambeiro, F.; Armesto, J. Nano-inclusions applied in cement-matrix composites: A review. *Materials* **2016**, *9*, 1015. [CrossRef]
48. Liew, K.M.; Kai, M.F.; Zhang, L.W. Carbon nanotube reinforced cementitious composites: An overview. *Compos. Part A Appl. Sci. Manuf.* **2016**, *91*, 301–323. [CrossRef]
49. Wu, Z.; Shi, C.; Khayat, K.H.; Wan, S. Effects of different nanomaterials on hardening and performance of ultra-high strength concrete (UHSC). *Cem. Concr. Compos.* **2016**, *70*, 24–34. [CrossRef]
50. Muzenski, S.; Flores-Vivian, I.; Sobolev, K. Ultra-high strength cement-based composites designed with aluminum oxide nano-fibers. *Constr. Build. Mater.* **2019**, *220*, 177–186. [CrossRef]
51. Hendrix, D.; McKeon, J.; Wille, K. Behavior of colloidal nanosilica in an ultrahigh performance concrete environment using dynamic light scattering. *Materials* **2019**, *12*, 1976. [CrossRef]
52. Varghese, L.; Kanta Rao, V.V.L.; Parameswaran, L. Effect of nanosilica and microsilica on bond and flexural behaviour of reinforced concrete. In Recent Advances in Structural Engineering. In *Lecture Notes in Civil Engineering*; Rao, A., Ramanjaneyulu, K., Eds.; Springer: Singapore, 2019; Volume 2, pp. 825–839. [CrossRef]
53. Alhawat, M.; Ashour, A.; El-Khoja, A. Properties of concrete incorporating different nano silica particles. *Mater. Res. Innov.* **2019**. [CrossRef]
54. Varghese, L.; Rao, V.K.; Parameswaran, L. Improvement of early-age strength of high-volume siliceous fly ash concrete with nanosilica—A review. *Adv. Civ. Eng. Mater.* **2018**, *7*, 599–615. [CrossRef]
55. Hela, R.; Bodnarova, L.; Rundt, L. Development of ultra high performance concrete and reactive powder concrete with nanosilica. *IOP Conf. Ser. Mater. Sci. Eng.* **2018**, *371*, 012017. [CrossRef]
56. Setiati, N.R. Effects of additional nanosilica of compressive strength on mortar. *IOP Conf. Ser. Mater. Sci. Eng.* **2017**, *223*, 012065. [CrossRef]
57. Khaloo, A.; Mobini, M.H.; Hosseini, P. Influence of different types of nano-SiO_2 particles on properties of high-performance concrete. *Constr. Build. Mater.* **2016**, *113*, 188–201. [CrossRef]
58. Isfahani, F.T.; Redaelli, E.; Lollini, F.; Li, W.; Bertolini, L. Effects of nanosilica on compressive strength and durability properties of concrete with different water to binder ratios. *Adv. Mater. Sci. Eng.* **2016**, *2016*, 8453567. [CrossRef]
59. Moon, J.; Taha, M.M.R.; Youm, K.-S.; Kim, J.J. Investigation of pozzolanic reaction in nanosilica-cement blended pastes based on solid-state kinetic models and ^{29}Si MAS NMR. *Materials* **2016**, *9*, 99. [CrossRef]
60. Oltulu, M.; Şahin, R. Single and combined effects of nano-SiO_2, nano-Al_2O_3 and nano-Fe_2O_3 powders on compressive strength and capillary permeability of cement mortar containing silica fume. *Mater. Sci. Eng. A* **2011**, *528*, 7012–7019. [CrossRef]

61. Shaikh, F.U.A.; Hosan, A. Effect of Nano Alumina on Compressive Strength and Microstructure of High Volume Slag and Slag-Fly Ash Blended Pastes. *Front. Mater.* **2019**, *6*, 90. [CrossRef]
62. Zhan, B.J.; Xuan, D.X.; Poon, C.S. The effect of nanoalumina on early hydration and mechanical properties of cement pastes. *Constr. Build. Mater.* **2019**, *202*, 169–176. [CrossRef]
63. Jaishankar, P.; Karthikeyan, C. Characteristics of cement concrete with nano alumina particles. *IOP Conf. Ser. Earth Environ. Sci.* **2017**, *80*, 012005. [CrossRef]
64. Xing, X.; Xu, J.; Bai, E.; Zhu, J.; Wang, Y. Response surface research of the preparation of nano-Fe_2O_3 cement-based composite. *Mater. Rep.* **2018**, *32*, 1367–1372. [CrossRef]
65. Mutuk, H.; Mutuk, T.; Gumus, H.; Mesci Oktay, B. Shielding behaviors and analysis of mechanical treatment of cement containing nanosized powders. *Acta Phys. Pol. A* **2016**, *130*, 172–174. [CrossRef]
66. Faraldos, M.; Kropp, R.; Anderson, M.A.; Sobolev, K. Photocatalytic hydrophobic concrete coatings to combat air pollution. *Catal. Today* **2016**, *259*, 228–236. [CrossRef]
67. Chen, J.; Poon, C.-S. Photocatalytic construction and building materials: From fundamentals to applications. *Build. Environ.* **2009**, *44*, 1899–1906. [CrossRef]
68. Kamaruddin, S.; Stephan, D. Quartz–titania composites for the photocatalytical modification of construction materials. *Cem. Concr. Compos.* **2013**, *36*, 109–115. [CrossRef]
69. Zhao, L.; Guo, X.; Liu, Y.; Zhao, Y.; Chen, Z.; Zhang, Y.; Guo, L.; Shu, X.; Liu, J. Hydration kinetics, pore structure, 3D network calcium silicate hydrate, and mechanical behavior of graphene oxide reinforced cement composites. *Constr. Build. Mater.* **2018**, *190*, 150–163. [CrossRef]
70. Chen, Z.-S.; Zhou, X.; Wang, X.; Guo, P. Mechanical behavior of multilayer GO carbon-fiber cement composites. *Constr. Build. Mater.* **2018**, *159*, 205–212. [CrossRef]
71. Lv, S.; Hu, H.; Zhang, J.; Luo, X.; Lei, Y.; Sun, L. Fabrication of GO/cement composites by incorporation of few-layered GO nanosheets and characterization of their crystal/chemical structure and properties. *Nanomaterials* **2017**, *7*, 457. [CrossRef]
72. Wang, B.; Jiang, R.; Wu, Z. Investigation of the mechanical properties and microstructure of graphene nanoplatelet-cement composite. *Nanomaterials* **2016**, *6*, 200. [CrossRef]
73. Sanchez, F.; Sobolev, K. Nanotechnology in concrete—A review. *Constr. Build. Mater.* **2010**, *24*, 2060–2071. [CrossRef]
74. Tian, Z.; Li, Y.; Zheng, J.; Wang, S. A state-of-the-art on self-sensing concrete: Materials, fabrication and properties. *Comp. Part B: Eng.* **2019**, *177*, 107437. [CrossRef]
75. Pisello, A.L.; Alessandro, A.D.; Sambuco, S.; Rallini, M.; Ubertini, F.; Asdrubali, F.; Materazzi, A.L.; Cotana, F. Multipurpose experimental characterization of smart nanocomposite cement-based materials for thermal-energy efficiency and strain-sensing capability. *Sol. Energy Mater. Sol. Cells* **2017**, *161*, 77–88. [CrossRef]
76. García-Macías, E.; Downey, A.; D'Alessandro, A.; Castro-Triguero, R.; Laflamme, S.; Ubertini, F. Enhanced lumped circuit model for smart nanocomposite cement-based sensors under dynamic compressive loading conditions. *Sens. Actuators A* **2017**, *260*, 45–57. [CrossRef]
77. Flores-Vivian, I.; Pradoto, R.G.K.; Moini, M.; Kozhukhova, M.; Potapov, V.; Sobolev, K. The effect of SiO_2 nanoparticles derived from hydrothermal solutions on the performance of portland cement based materials. *Front. Struct. Civ. Eng.* **2017**, *11*, 436–445. [CrossRef]
78. Heidari, A.; Tavakoli, D. A study of the mechanical properties of ground ceramic powder concrete incorporating nano-SiO_2 particles. *Constr. Build. Mater.* **2013**, *38*, 255–264. [CrossRef]
79. Supit, S.W.M.; Shaikh, F.U.A. Durability properties of high-volume fly ash concrete containing nano-silica. *Mater. Struct.* **2015**, *48*, 2431–2445. [CrossRef]
80. Han, B.; Li, Z.; Zhang, L.; Zeng, S.; Yu, X.; Han, B.; Ou, J. Reactive powder concrete reinforced with nano SiO_2-coated TiO_2. *Constr. Build. Mater.* **2017**, *148*, 104–112. [CrossRef]
81. Li, W.; Huang, Z.; Cao, F.; Sun, Z.; Shah, S. Effects of nano-silica and nano-limestone on flowability and mechanical properties of ultra-high-performance concrete matrix. *Constr. Build. Mater.* **2015**, *95*, 366–374. [CrossRef]
82. Sadeghi Nik, A.; Lotfi Omran, O. Estimation of compressive strength of self-compacted concrete with fibers consisting nano-SiO_2 using ultrasonic pulse velocity. *Constr. Build. Mater.* **2013**, *44*, 654–662. [CrossRef]

83. Najigivi, A.; Khaloo, A.; Iraji Zad, A.; Abdul Rashid, S. Investigating the effects of using different types of SiO_2 nanoparticles on the mechanical properties of binary blended concrete. *Compos. Part B Eng.* **2013**, *54*, 52–58. [CrossRef]
84. Zhang, P.; Li, Q.; Chen, Y.; Shi, Y.; Ling, Y.-F. Durability of steel fiber-reinforced concrete containing SiO_2 Nano-Particles. *Materials* **2019**, *12*, 2184. [CrossRef] [PubMed]
85. Tavakoli, D.; Heidari, A. Properties of concrete incorporating silica fume and nano-SiO_2. *Indian J. Sci. Technol.* **2013**, *6*, 108–112.
86. Nazarigivi, A.; Najigivi, A. Study on mechanical properties of ternary blended concrete containing two different sizes of nano SiO_2. *Compos. Part B Eng.* **2019**, *167*, 20–254. [CrossRef]
87. Mastali, M.; Dalvand, A. The impact resistance and mechanical properties of fiber reinforced self-compacting concrete (SCC) containing nano-SiO2 and silica fume. *Eur. J. Environ. Civ. Eng.* **2016**, 1–27. [CrossRef]
88. Mohammed, B.S.; Liew, M.S.; Alaloul, W.S.; Khed, V.C.; Hoong, C.Y.; Adamu, M. Properties of nano-silica modified pervious concrete. *Case Stud. Constr. Mater.* **2018**, *8*, 409–422. [CrossRef]
89. Fang, Y.; Sun, Y.; Lu, M.; Xing, F.; Li, W. Mechanical and pressure-sensitive properties of cement mortar containing nano Fe_2O_3. *Adv. Eng. Res.* **2018**, *146*, 206–210. [CrossRef]
90. Rashad, A.M. A synopsis about the effect of nano-Al_2O_3, nano-Fe_2O_3, nano-Fe_3O_4 and nano-clay on some properties of cementitious materials—A short guide for Civil Engineer. *Mater. Des.* **2013**, *52*, 143–157. [CrossRef]
91. Nazari, A.; Riahi, S.; Riahi, S.; Shamekhi, S.F.; Khademno, A. Benefits of Fe_2O_3 nanoparticles in concrete mixing matrix. *J. Am. Sci.* **2010**, *6*, 102–106.
92. Nazari, A.; Riahi, S. Computer-aided design of the effects of Fe_2O_3 nanoparticles on split tensile strength and water permeability of high strength concrete. *Mater. Des.* **2011**, *32*, 3966–3979. [CrossRef]
93. Khoshakhlagh, A.; Nazari, A.; Khalaj, G. Effects of Fe_2O_3 nanoparticles on water permeability and strength assessment of high strength self-compacting concrete. *J. Mater. Sci. Technol.* **2012**, *28*, 73–82. [CrossRef]
94. Zhang, R.; Cheng, X.; Hou, P.; Ye, Z. Influences of nano-TiO_2 on the properties of cement-based materials: Hydration and drying shrinkage. *Constr. Build. Mater.* **2015**, *81*, 35–41. [CrossRef]
95. Chen, Y. A review on the effects of nanoparticles on properties of self-compacting concrete. *IOP Conf. Ser. Mater. Sci. Eng.* **2018**, *452*, 022134. [CrossRef]
96. Li, Z.; Han, B.; Yu, X.; Zheng, Q.; Wang, Y. Comparison of the mechanical property and microstructures of cementitious composites with nano- and micro-rutile phase TiO_2. *Arch. Civ. Mech. Eng.* **2019**, *19*, 615–626. [CrossRef]
97. Wang, L.; Zhang, H.; Gao, Y. Effect of TiO_2 nanoparticles on physical and mechanical properties of cement at low temperatures. *Adv. Mater. Sci. Eng.* **2018**, *2018*, 8934689. [CrossRef]
98. ASTM. *ASTM C109-93: Standard Specification for Compressive Strength of Mortars*; American Society for Testing and Materials: West Conshohocken, PA, USA, 2007.
99. ASTM. *ASTM C293/C293M–10: Standard Test Method for Flexural Strength of Concrete*; American Society for Testing and Materials: West Conshohocken, PA, USA, 2007.
100. Feng, D.; Xie, N.; Gong, C.; Leng, Z.; Xiao, H.; Li, H.; Shi, X. Portland Cement Paste Modified by TiO_2 Nanoparticles: A Microstructure Perspective. *Ind. Eng. Chem. Res.* **2013**, *52*, 11575–11582. [CrossRef]
101. Jalal, M.; Fathi, M.; Farzad, M. Effects of fly ash and TiO_2 nanoparticles on rheological, mechanical, microstructural and thermal properties oh high strength self-compacting concrete. *Mech. Mater.* **2013**, *61*, 11–27. [CrossRef]
102. Yu, X.; Kang, S.; Long, X. Compressive strength of concrete reinforced by TiO_2 nanoparticles. *AIP Conf. Proc.* **2018**, *2036*, 030006. [CrossRef]
103. Chunping, G.; Qiannan, W.; Jintao, L.; Wei, S. The effect of nano TiO_2 on the durability of ultra-high-performance concrete with and without a flexural load. *Ceram-Silikáty* **2018**, *62*, 374–381. [CrossRef]
104. Liu, J.; Li, Q.; Xu, S. Influence of nanoparticles on fluidity and mechanical properties of cement mortar. *Constr. Build. Mater.* **2015**, *101*, 892–901. [CrossRef]
105. Yang, Z.; Gao, Y.; Mu, S.; Chang, H.; Sun, W.; Jiang, J. Improving the chloride binding capacity of cement paste by adding nano-Al_2O_3. *Constr. Build. Mater.* **2019**, *195*, 415–422. [CrossRef]

106. Mohseni, E.; Mehrinejad Khotbehsara, M.; Naseri, F.; Monazami, M.; Sarker, P. Polypropylene fiber reinforced cement mortars containing rice husk ash and nano-alumina. *Constr. Build. Mater.* **2016**, *111*, 429–439. [CrossRef]
107. Barbhuiya, S.; Mukherjee, S.; Nikraz, H. Effects of nano-Al_2O_3 on early-age microstructural properties of cement paste. *Constr. Build. Mater.* **2014**, *52*, 189–193. [CrossRef]
108. Gowda, R.; Narendra, H.; Nagabushan, B.; Rangappa, D.; Prabhakara, R. Investigation of nano alumina on the effect of durability and micro-structural properties of the cement mortar. *Mater. Today Proc.* **2017**, *4*, 12191–12197. [CrossRef]
109. Rashad, A.M. Effect of carbon nanotubes (CNTs) on the properties of traditional cementitious materials. *Constr. Build. Mater.* **2017**, *153*, 81–101. [CrossRef]
110. Parvaneh, V.; Khiabani, S.H. Mechanical and piezoresistive properties of self-sensing smart concretes reinforced by carbon nanotubes. *Mech. Adv. Mater. Struct.* **2019**, *25*, 993–1000. [CrossRef]
111. Ding, S.; Ruan, Y.; Yu, X.; Han, B.; Ni, Y.-Q. Self-monitoring of smart concrete column incorporating CNT/NCB composite fillers modified cementitious sensors. *Constr. Build. Mater.* **2019**, *201*, 127–137. [CrossRef]
112. Ramezani, M.; Kim, Y.H.; Sun, Z. Modeling the mechanical properties of cementitious materials containing CNTs. *Cem. Concr. Compos.* **2019**, *104*, 103347. [CrossRef]
113. Douba, A.; Emiroglu, M.; Kandil, U.F.; Reda Taha, M.M. Very ductile polymer concrete using carbon nanotubes. *Constr. Build. Mater.* **2019**, *196*, 468–477. [CrossRef]
114. Hawreen, A.; Bogas, J.A. Influence of carbon nanotubes on steel–concrete bond strength. *Mater. Struct.* **2018**, *51*, 155. [CrossRef]
115. Baloch, W.L.; Khushnood, R.A. Wasim Khaliq. Influence of multi-walled carbon nanotubes on the residual performance of concrete exposed to high temperatures. *Constr. Build. Mater.* **2018**, *185*, 44–56. [CrossRef]
116. Irshidat, M.R.; Al-Shannaq, A. Using textile reinforced mortar modified with carbon nano tubes to improve flexural performance of RC beams. *Compos. Struct.* **2018**, *200*, 127–134. [CrossRef]
117. Qissab, M.A.; Abbas, S.T. Behaviour of reinforced concrete beams with multiwall carbon nanotubes under monotonic loading. *Eur. J. Environ. Civ. Eng.* **2018**, *22*, 1111–1130. [CrossRef]
118. Ruan, Y.; Han, B.; Yu, X.; Zhang, W.; Wang, D. Carbon nanotubes reinforced reactive powder concrete. *Compos. Part A Appl. Sci. Manuf.* **2018**, *112*, 371–382. [CrossRef]
119. Yoo, D.Y.; Kim, S.; Lee, S.H. Self-sensing capability of ultra-high-performance concrete containing steel fibers and carbon nanotubes under tension. *Sens. Actuators A Phys.* **2018**, *276*, 125–136. [CrossRef]
120. Lushnikova, A.; Zaoui, A. Improving mechanical properties of C–S–H from inserted carbon nanotubes. *J. Phys. Chem. Solids* **2017**, *105*, 72–80. [CrossRef]
121. Sedaghatdoost, A.; Behfarnia, K. Mechanical properties of Portland cement mortar containing multi-walled carbon nanotubes at elevated temperatures. *Constr. Build. Mater.* **2018**, *176*, 482–489. [CrossRef]
122. Hawreen, A.; Bojas, J.A. Creep, shrinkage and mechanical properties of concrete reinforced with different types of carbon nanotubes. *Constr. Build. Mater.* **2019**, *198*, 70–81. [CrossRef]
123. García-Macías, E.; Castro-Triguero, R.; Sáez, A.; Ubertini, F. 3D mixed micromechanics-FEM modeling of piezoresistive carbon nanotube smart concrete. *Comput. Methods Appl. Mech. Eng.* **2018**, *340*, 396–423. [CrossRef]
124. García-Macías, E.; D'Alessandro, A.; Castro-Triguero, R.; Pérez-Mira, D.; Ubertini, F. Micromechanics modeling of the uniaxial strain-sensing property of carbon nanotube cement-matrix composites for SHM applications. *Compos. Struct.* **2017**, *163*, 195–215. [CrossRef]
125. Horszczaruk, E.; Sikora, P.; Łukowski, P. Application of nanomaterials in production of self-sensing concretes: Contemporary developments and prospects. *Arch. Civ. Eng.* **2016**, *62*, 61–74. [CrossRef]
126. Gao, Y.; Jing, H.; Zhou, Z.; Chen, W.; Du, M.; Du, Y. Reinforced impermeability of cementitious composites using graphene oxide-carbon nanotube hybrid under different water-to-cement ratios. *Constr. Build. Mater.* **2019**, *222*, 610–621. [CrossRef]
127. Phrompet, C.; Sriwong, C.; Ruttanapun, C. Mechanical, dielectric, thermal and antibacterial properties of reduced graphene oxide (rGO)-nanosized C3AH6 cement nanocomposites for smart cement-based materials. *Compos. Part B Eng.* **2019**, *175*, 107128. [CrossRef]
128. Tragazikis, I.K.; Dassios, K.G.; Dalla, P.T.; Exarchos, D.A.; Matikas, T.E. Acoustic emission investigation of the effect of graphene on the fracture behavior of cement mortars. *Eng. Fract. Mech.* **2019**, *210*, 444–451. [CrossRef]

129. Liu, J.; Fu, J.; Ni, T.; Yang, Y. Fracture toughness improvement of multi-wall carbon nanotubes/graphene sheets reinforced cement paste. *Constr. Build. Mater.* **2019**, *200*, 530–538. [CrossRef]
130. Liu, Y.; Wang, M.; Wang, W. Electric induced curing of graphene/cement-based composites for structural strength formation in deep-freeze low temperature. *Mater. Des.* **2018**, *160*, 783–793. [CrossRef]
131. Lv, S.; Hu, H.; Hou, Y.; Lei, Y.; Sun, L.; Zhang, J.; Liu, L. Investigation of the effects of polymer dispersants on dispersion of GO nanosheets in cement composites and relative microstructures/performances. *Nanomaterials* **2018**, *8*, 964. [CrossRef]
132. Lu, L.; Zhao, P.; Lu, Z. A short discussion on how to effectively use graphene oxide to reinforce cementitious composites. *Constr. Build. Mater.* **2018**, *189*, 33–41. [CrossRef]
133. Roy, R.; Mitra, A.; Ganesh, A.T.; Sairam, V. Effect of graphene oxide nanosheets dispersion in cement mortar composites incorporating metakaolin and silica fume. *Constr. Build. Mater.* **2018**, *186*, 514–524. [CrossRef]
134. Long, W.J.; Wei, J.J.; Xing, F.; Khayat, K.H. Enhanced dynamic mechanical properties of cement paste modified with graphene oxide nanosheets and its reinforcing mechanism. *Cem. Concr. Compos.* **2018**, *93*, 127–139. [CrossRef]
135. Alharbi, Y.; An, J.; Cho, B.H.; Khawaji, M.; Chung, W.; Nam, B.H. Mechanical and Sorptivity Characteristics of Edge-Oxidized Graphene Oxide (EOGO)-Cement Composites: Dry- and Wet-Mix Design Methods. *Nanomaterials* **2018**, *8*, 718. [CrossRef]
136. Duan, Z.; Zhang, L.; Lin, Z.; Fan, D.; Saafi, M.; Castro Gomes, J.; Yang, S. Experimental test and analytical modeling of mechanical properties of graphene-oxide cement composites. *J. Compos. Mater.* **2018**, *52*, 3027–3037. [CrossRef]
137. Ghazizadeh, S.; Duffour, P.; Skipper, N.T.; Bai, Y. Understanding the behaviour of graphene oxide in Portland cement paste. *Cem. Concr. Res.* **2018**, *111*, 169–182. [CrossRef]
138. Jiang, W.; Li, X.; Lv, Y.; Zhou, M.; Liu, Z.; Ren, Z.; Yu, Z. Cement-based materials containing graphene oxide and polyvinyl alcohol fiber: Mechanical properties, durability, and microstructure. *Nanomaterials* **2018**, *8*, 638. [CrossRef] [PubMed]
139. Long, W.-J.; Fang, C.; Wei, J.; Li, H. Stability of GO Modified by different dispersants in cement paste and its related mechanism. *Materials* **2018**, *11*, 834. [CrossRef]
140. Huang, P.; Lv, L.; Liao, W.; Lu, C.; Xu, Z. Microstructural Properties of Cement Paste and Mortar Modified by Low Cost Nanoplatelets Sourced from Natural Materials. *Materials* **2018**, *11*, 783. [CrossRef]
141. An, J.; McInnis, M.; Chung, W.; Nam, B.H. Feasibility of using graphene oxide nanoflake (GONF) as additive of cement composite. *Appl. Sci.* **2018**, *8*, 419. [CrossRef]
142. Bai, S.; Jiang, L.; Xu, N.; Jin, M.; Jiang, S. Enhancement of mechanical and electrical properties of graphene/cement composite due to improved dispersion of graphene by addition of silica fume. *Constr. Build. Mater.* **2018**, *164*, 433–441. [CrossRef]
143. Chuah, S.; Li, W.; Chen, S.J.; Sanjayan, J.G.; Duan, W.H. Investigation on dispersion of graphene oxide in cement composite using different surfactant treatments. *Constr. Build. Mater.* **2018**, *161*, 519–527. [CrossRef]
144. Zhao, L.; Guo, X.; Liu, Y.; Ge, C.; Chen, Z.; Guo, L.; Shu, X.; Liu, J. Investigation of dispersion behavior of GO modified by different water reducing agents in cement pore solution. *Carbon* **2018**, *127*, 255–269. [CrossRef]
145. Wang, L.; Zhang, S.; Zheng, D.; Yang, H.; Cui, H.; Tang, W.; Li, D. Effect of graphene oxide (GO) on the morphology and microstructure of cement hydration products. *Nanomaterials* **2017**, *7*, 429. [CrossRef]
146. Long, W.-J.; Wei, J.-J.; Ma, H.; Xing, F. Dynamic mechanical properties and microstructure of graphene oxide nanosheets reinforced cement composites. *Nanomaterials* **2017**, *7*, 407. [CrossRef] [PubMed]
147. Lu, L.; Ouyang, D. Properties of cement mortar and ultra-high strength concrete incorporating graphene oxide nanosheets. *Nanomaterials* **2017**, *7*, 187. [CrossRef] [PubMed]
148. Jintao, L.; Qinghua, L.; Shilang, X. Reinforcing mechanism of graphene oxide sheets on cement-based materials. *J. Mater. Civ. Eng.* **2019**, *31*, 04019014. [CrossRef]
149. Birenboim, M.; Nadiv, R.; Alatawna, A.; Buzaglo, M.; Schahar, G.; Lee, J.; Kim, G.; Peled, A.; Regev, O. Reinforcement and workability aspects of graphene-oxide-reinforced cement nanocomposites. *Compos. Part B Eng.* **2019**, *161*, 68–76. [CrossRef]
150. Peng, H.; Ge, Y.; Cai, C.S.; Zhang, Y.; Liu, Z. Mechanical properties and microstructure of graphene oxide cement-based composites. *Constr. Build. Mater.* **2019**, *194*, 102–109. [CrossRef]

151. Belli, A.; Mobili, A.; Bellezze, T.; Tittarelli, F.; Cachim, P. Evaluating the self-sensing ability of cement mortars manufactured with graphene nanoplatelets, virgin or recycled carbon fibers through piezoresistivity tests. *Sustainability* **2018**, *10*, 4013. [CrossRef]
152. Li, G.; Yuan, J.B.; Zhang, Y.H.; Zhang, N.; Liew, K.M. Microstructure and mechanical performance of graphene reinforced cementitious composites. *Compos. Part A Appl. Sci. Manuf.* **2018**, *114*, 188–195. [CrossRef]
153. Shamsaei, E.; de Souza, F.B.; Yao, X.; Benhelal, E.; Akbari, A.; Duan, W. Graphene-based nanosheets for stronger and more durable concrete: A review. *Constr. Build. Mater.* **2018**, *183*, 642–660. [CrossRef]
154. Li, X.; Wang, L.; Liu, Y.; Li, W.; Dong, B.; Duan, W.H. Dispersion of graphene oxide agglomerates in cement paste and its effects on electrical resistivity and flexural strength. *Cem. Concr. Compos.* **2018**, *92*, 145–154. [CrossRef]
155. Sikora, P.; Abd Elrahman, M.; Stephan, D. The Influence of nanomaterials on the thermal resistance of cement-based composites—A review. *Nanomaterials* **2018**, *8*, 465. [CrossRef]
156. Xu, Y.; Zeng, J.; Chen, W.; Jin, R.; Li, B.; Pan, Z. A holistic review of cement composites reinforced with graphene oxide. *Constr. Build. Mater.* **2018**, *171*, 291–302. [CrossRef]
157. Kashif Ur Rehman, S.; Ibrahim, Z.; Memon, S.A.; Aunkor, M.T.H.; Faisal Javed, M.; Mehmood, K.; Shah, S.M.A. Influence of graphene nanosheets on rheology, microstructure, strength development and self-sensing properties of cement based composites. *Sustainability* **2018**, *10*, 822. [CrossRef]
158. Hu, M.; Guo, J.; Li, P.; Chen, D.; Xu, Y.; Feng, Y.; Yu, Y.; Zhang, H. Effect of characteristics of chemical combined of graphene oxide-nanosilica nanocomposite fillers on properties of cement-based materials. *Constr. Build. Mater.* **2019**, *225*, 745–753. [CrossRef]
159. Hu, M.; Guo, J.; Fan, J.; li, P.; Chen, D. Dispersion of triethanolamine-functionalized graphene oxide (TEA-GO) in pore solution and its influence on hydration, mechanical behavior of cement composite. *Constr. Build. Mater.* **2019**, *216*, 128–136. [CrossRef]
160. Tao, J.; Wang, X.; Wang, Z.; Zeng, Q. Graphene nanoplatelets as an effective additive to tune the microstructures and piezoresistive properties of cement-based composites. *Constr. Build. Mater.* **2019**, *209*, 665–678. [CrossRef]
161. Sixuan, H. Multifunctional Graphite Nanoplatelets (GNP) Reinforced Cementitious Composites. Master's Thesis, National University of Singapore, Singapore, 2012.
162. Qureshi, T.S.; Panesar, D.K. Impact of graphene oxide and highly reduced graphene oxide on cement based composites. *Constr. Build. Mater.* **2019**, *206*, 71–83. [CrossRef]
163. Krystek, M.; Pakulski, D.; Patroniak, V.; Górski, M.; Szojda, L.; Ciesielski, A.; Samorì, P. High-Performance Graphene-Based Cementitious Composites. *Adv. Sci.* **2019**, *6*, 1801195. [CrossRef]
164. Kaur, R.; Kothiyal, N.C. Comparative effects of sterically stabilized functionalized carbon nanotubes and graphene oxide as reinforcing agent on physico-mechanical properties and electrical resistivity of cement nanocomposites. *Constr. Build. Mater.* **2019**, *202*, 121–138. [CrossRef]
165. Horszczaruk, E. Properties of cement-Based composites modified with magnetite nanoparticles: A review. *Materials* **2019**, *12*, 326. [CrossRef]
166. Singh, L.P.; Karade, S.R.; Bhattacharyya, S.K.; Yousuf, M.M.; Ahalawat, S. Beneficial role of nanosilica in cement based materials—A review. *Constr. Build. Mater.* **2013**, *47*, 1069–1077. [CrossRef]
167. Sikora, P.; Augustyniak, A.; Cendrowski, K.; Nawrotek, P.; Mijowska, E. Antimicrobial activity of Al_2O_3, CuO, Fe_3O_4, and ZnO nanoparticles in scope of their further application in cement-based building materials. *Nanomaterials* **2018**, *8*, 212. [CrossRef] [PubMed]
168. Martins, R.M.; Bombard, A.J.F. Rheology of fresh cement paste with superplasticizer and nanosilica admixtures studied by response surface methodology. *Mater. Struct.* **2012**, *45*, 905. [CrossRef]
169. Mateos, R.; Vera, S.; Valiente, M.; Díez-Pascual, A.M.; San Andrés, M.P. Comparison of anionic, cationic and nonionic surfactants as dispersing agents for graphene based on the fluorescence of riboflavin. *Nanomaterials* **2017**, *7*, 403. [CrossRef] [PubMed]
170. Stephens, C.; Brown, L.; Sanchez, F. Quantification of the re-agglomeration of carbon nanofiber aqueous dispersion in cement pastes and effect on the early age flexural response. *Carbon* **2016**, *107*, 482–500. [CrossRef]
171. Sanchez, F.; Zhang, L.; Ince, C. Multi-scale Performance and Durability of Carbon Nanofiber/Cement Composites. In *Nanotechnology in Construction 3*; Bittnar, Z., Bartos, P.J.M., Němeček, J., Šmilauer, V., Zeman, J., Eds.; Springer: Berlin/Heidelberg, Germany, 2009.

172. Shah, S.P.; Konsta-Gdoutos, M.S.; Metaxa, Z.S.; Mondal, P. Nanoscale Modification of Cementitious Materials. In *Nanotechnology in Construction 3*; Bittnar, Z., Bartos, P.J.M., Němeček, J., Šmilauer, V., Zeman, J., Eds.; Springer: Berlin/Heidelberg, Germany, 2009.
173. Kang, S.-T.; Seo, J.-Y.; Park, S.-H. The Characteristics of CNT/Cement Composites with Acid-Treated MWCNTs. *Adv. Mater. Sci. Eng.* **2015**, *2015*, 308725. [CrossRef]
174. Stynoski, P.; Mondal, P.; Wotring, E.; Marsh, C. Characterization of silica-functionalized carbon nanotubes dispersed in water. *J. Nanopart. Res.* **2013**, *15*, 1396. [CrossRef]
175. Sikora, P.; Elrahman, M.A.; Chung, S.-Y.; Cendrowski, K.; Mijowska, E.; Stephan, D. Mechanical and microstructural properties of cement pastes containing carbon nanotubes and carbon nanotube-silica core-shell structures, exposed to elevated temperature. *Cem. Concr. Compos.* **2019**, *95*, 193–204. [CrossRef]
176. Mendoza-Reales, O.A.; Sierra-Gallego, G.; Tobón, J.I. The mechanical properties of Portland cement mortars blended with carbon nanotubes and nanosilica: A study by experimental design. *DYNA* **2016**, *83*, 136–141. [CrossRef]
177. Li, W.; Ji, W.; Torabian Isfahani, F.; Wang, Y.; Li, G.; Liu, Y.; Xing, F. Nano-silica sol-gel and carbon nanotube coupling effect on the Performance of Cement-Based Materials. *Nanomaterials* **2017**, *7*, 185. [CrossRef]
178. Richard Meier & Parners Architects LLP. Available online: https://www.richardmeier.com/?projects=jubilee-church-2 (accessed on 11 October 2019).
179. Cardellicchio, L. On conservation issues of contemporary architecture: The technical design development and the ageing process of the Jubilee Church in Rome by Richard Meier. *Front. Archit. Res.* **2018**, *7*, 107–121. [CrossRef]
180. Cardellicchio, L. Self-cleaning and colour-preserving efficiency of photocatalytic concrete: Case study of the Jubilee Church in Rome. *Build. Res. Inf.* **2019**, 1–20. [CrossRef]
181. Lee, J.; Mahendra, S.; Alvarez, P.J.J. Nanomaterials in the construction industry: A review of their applications and environmental health and safety considerations. *ACS Nano* **2010**, *4*, 3580–3590. [CrossRef] [PubMed]
182. Lam, C.W.; James, J.T.; McCluskey, R.; Arepalli, S.; Hunter, R.L. A review of carbon nanotubes toxicity and assessment of potential occupational and environmental health risks. *Crit. Rev. Toxicol.* **2006**, *36*, 189–217. [CrossRef] [PubMed]
183. Handy, R.D.; Henry, T.B.; Scown, T.M.; Johnston, B.D.; Tyler, C.R. Manufactured nanoparticles: Their uptake and effects on fish—A mechanistic analysis. *Ecotoxicology* **2008**, *17*, 396. [CrossRef] [PubMed]

Article

Compressive Behavior Characteristics of High-Performance Slurry-Infiltrated Fiber-Reinforced Cementitious Composites (SIFRCCs) under Uniaxial Compressive Stress

Seungwon Kim [1], Seungyeon Han [2], Cheolwoo Park [1,*] and Kyong-Ku Yun [2,*]

[1] Department of Civil Engineering, Kangwon National University, 346 Jungang-ro, Samcheok 25913, Korea; inncoms@kangwon.ac.kr

[2] KIIT (Kangwon Institute of Inclusive Technology), Kangwon National University, 1 Gangwondaegil, Chuncheon 24341, Korea; syhan8704@kangwon.ac.kr

* Correspondence: tigerpark@kangwon.ac.kr (C.P.); kkyun@kangwon.ac.kr (K.-K.Y.); Tel.: +82-33-570-6515 (C.P.); +82-33-250-6236 (K.-K.Y.)

Received: 11 October 2019; Accepted: 19 December 2019; Published: 1 January 2020

Abstract: The compressive stress of concrete is used as a design variable for reinforced concrete structures in design standards. However, as the performance-based design is being used with increasing varieties and strengths of concrete and reinforcement bars, mechanical properties other than the compressive stress of concrete are sometimes used as major design variables. In particular, the evaluation of the mechanical properties of concrete is crucial when using fiber-reinforced concrete. Studies of high volume fractions in established compressive behavior prediction equations are insufficient compared to studies of conventional fiber-reinforced concrete. Furthermore, existing prediction equations for the mechanical properties of high-performance fiber-reinforced cementitious composite and high-strength concrete have limitations in terms of the strength and characteristics of contained fibers (diameter, length, volume fraction) even though the stress-strain relationship is determined by these factors. Therefore, this study developed a high-performance slurry-infiltrated fiber-reinforced cementitious composite that could prevent the fiber ball phenomenon, a disadvantage of conventional fiber-reinforced concrete, and maximize the fiber volume fraction. Then, the behavior characteristics under compressive stress were analyzed for fiber volume fractions of 4%, 5%, and 6%.

Keywords: slurry-infiltrated fiber-reinforced cementitious composite; high-performance fiber-reinforced cementitious composite; fiber volume fraction; compressive stress; stress-strain relationship; filling slurry matrix

---

1. Introduction

Conventional concrete, which is commonly used as a construction material, does not have an adequate resistance level in terms of collisions or explosive loads; thus, concerns are arising about human casualties from the brittle fracturing when reinforced concrete (RC) structures explode [1,2]. With the increasing threat of global terrorism, as well as safety concerns, there is a need for reinforced concrete structures that can withstand the sudden occurrence of dynamic loads like terrorist impact and blast. The energy absorbing capacity of the material plays an important role in developing protective structures. Fiber-reinforced concrete is being popular due to its greater impact resistance properties [3,4]. The possibility of high-performance fiber-reinforced cementitious composites (HPFRCCs) satisfying blast resistance design requirements for external blasts more effectively than conventionally reinforced concrete. The HPFRCCs have been expected to improve such drawbacks of concrete and improve the impact resistance [5]. From previous research, it has been found that the mechanical properties of the concrete can be improved by adding fibers into the mix. Otter and Naaman reported that the addition of fiber in the concrete had a significant effect on the strength and toughness of concrete [6]. The possibility of HPFRCCs satisfying blast resistance design requirements for external blasts more effectively than conventionally reinforced concrete. HPFRCC has been found to be an economical solution against the blast loadings [7].

As the compressive stress of concrete increases, the elastic area becomes larger, and the load-bearing capacity decreases sharply after the manifestation of the maximum strength [8]. The strengthening mechanism of fiber involves the transfer of stress from the matrix to the fiber by interlocking the fibers and matrices when the fiber surface is deformed. The stress is thus shared by the fibers and matrix in tension until the matrix cracks, and then the total stress is progressively transferred to the fibers [9]. These characteristics of high-strength concrete are determined from the size of strain and the shape and rising and falling curves at the manifestation of the maximum strength; these characteristics are also determined by analyzing compressive stress test results of concrete [8]. We examined established prediction equations and found that studies of high volume fractions remained insufficient compared to studies of conventional fiber-reinforced concrete. Furthermore, the mechanical properties of HPFRCCs and high-strength concrete revealed limitations in each prediction equation for the strength and characteristics of contained fibers (diameter, length, volume fraction) even though the stress-strain relationship was determined by these factors [10–17]. Slurry infiltrated fiber concrete (SIFCON) is one of the HPFRCCs. SIFCON is made by distributing short discrete fibers in the mold to the full volume or designed volume fraction and then infiltrated by a fine liquid cement-based slurry or mortar. The fibers can be sprinkled by hand or by using fiber-dispending units for large sections. [15]. Therefore, this study aimed to develop a high-performance slurry-infiltrated fiber-reinforced cementitious composite (SIFRCCs) that could prevent the fiber balling, a disadvantage of conventional fiber-reinforced concrete, and maximize the fiber volume fraction. Most fiber balling occurred during the fiber addition process due to excessive fibers in the mixing of HPFRCCs or fiber-reinforced concrete. Further, the compressive behavior of the developed SIFRCCs was analyzed for fiber volume fractions of 4%, 5%, and 6%. Cylindrical specimens were fabricated for each variable according to the mixing design of high-performance filling slurry, and an experimental study was conducted on the behavior characteristics under compressive stress to examine the mechanical properties of SIFRCCs with a high volume fraction.

2. Mechanical Properties of Fiber-Reinforced Concrete

Previously proposed stress-strain relationship equations have been modified based on the research results of Propovics [10] and Sargin [11]. Propovics [10] defined the stress-strain relationship through the rate of decrease of the elastic modulus, as shown in Equation (1), and Sargin [11] reflected the properties of a section through the changes of the components and the strength of concrete, as shown in Equation (2).

$$\frac{f_c}{f_{ck}} = \frac{\beta(\varepsilon/\varepsilon_0)}{\beta - 1 + (\varepsilon/\varepsilon_0)^\beta} \tag{1}$$

$$\frac{f_c}{f_{ck}} = \frac{A(\varepsilon/\varepsilon_0) + B(\varepsilon/\varepsilon_0)^2}{1 + C(\varepsilon/\varepsilon_0) + D(\varepsilon/\varepsilon_0)^2} \tag{2}$$

Here, f_c is the stress acting on the concrete; f_{ck}, the compressive stress of concrete; ε, the compressive strain; ε_0, the strain at peak stress; and β, a coefficient that determines the slope and shape of the curve. Furthermore, A~D are coefficients determined by the boundary condition.

A review of existing studies' results revealed that the mechanical properties of concrete, including the maximum strength, elastic modulus, and strain at maximum strength, need to be defined before the stress-strain relationship of concrete under compressive stress can be defined. Of these, the elastic modulus is an important design variable for a concrete structure and has a significant effect on the stress-strain relationship. The elastic modulus tends to increase with the strength of the matrix and is proportional to the square root or cube root of the compressive stress of the matrix [1]. The strain at peak stress is usually fixed at 0.002 or 0.0022 for conventional concrete (average strength) [8]. However, the measured compressive stress of high-strength concrete has been reported to have a value exceeding 0.002, and many studies have proposed prediction equations of the strain at peak stress that reflect this value [8].

Steel fibers are mixed for increasing the tensile strength of concrete; however, they also change the mechanical properties in compressive stress measurements [8]. The mixing of steel fibers, generally, increases the compressive stress and the strain and elastic modulus at peak stress [8].

The strengthening effect of fibers is realized as an algebraic sum or a magnification of the fiber reinforcing effects for the elastic modulus of the matrix alone [8,17]. The prediction equations for the stress-strain relationship that reflect changes in the mechanical properties with the mixing of fibers revealed that the effect of fibers was reflected only for conventional concrete with no fiber reinforcement [8–17].

Furthermore, a prediction equation for estimating the elastic modulus of concrete is generally proposed on the basis of an empirical formula using regression analysis of experimental data of various ranges. The elastic modulus of conventional concrete can be estimated using the compressive stress, unit weight of concrete, etc. [8–11,15]. However, HPFRCCs, such as SIFRCCs, contain steel fibers and other materials in conventional concrete, depending on the mixing conditions, and these added materials can greatly affect the estimation of the elastic modulus. For HPFRCCs, such as SIFRCCs, the volume fraction of each composite material can be calculated according to the composite theory [18–21].

We examined established prediction equations and found that studies of high volume fractions remained insufficient compared to studies of conventional fiber-reinforced concrete. Therefore, we conducted an experimental study of the behavior characteristics under compressive stress to examine the mechanical properties of SIFRCCs with a high volume fraction of fibers.

3. Experimental

3.1. Materials

3.1.1. Cement

This study used type 1 ordinary Portland cement (OPC), the physical and chemical properties of which are listed in Table 1 [21].

Table 1. Physical and chemical properties of the used cement.

Physical Properties					
Specific Gravity	Blaine (cm^2/g)	Stability (%)	Setting Time (min)		Loss on Ignition (%)
			Initial	Final	
3.15	3400	0.10	230	410	2.58
Chemical Composition (%, mass)					
SiO_2	CaO	MgO	SO_3		Al_2O_3
21.95	60.12	3.32	2.11		6.59

3.1.2. Silica Fume

To realize high-performance and high-strength filling slurry, this study used silica fume (Elkem Korea, South Korea), the physical and chemical properties of which are listed in Table 2.

Table 2. Physical and chemical properties of the used silica fume.

Physical Properties				
Specific Gravity			**Blaine (cm^2/g)**	
2.10			200,000	
Chemical Composition (%, mass)				
SiO_2	CaO	MgO	SO_3	Al_2O_3
96.00	0.38	0.10	-	0.25

3.1.3. Aggregates

To improve the filling performance of the slurry and reduce material separation, fine aggregates with diameters of 0.5 mm or smaller were used. No coarse aggregates were used.

3.1.4. Admixture

To improve the filling performance of the slurry, a polycarboxylic acid, the high-performance water reducing agent with high dispersion performance, was used. The admixture used in this experiment had high strength and high flow characteristics, as well as excellent unit water quantity reduction property and material separation resistance. Table 3 lists the characteristics of the used high-performance water-reducing agent.

Table 3. Characteristics of the used high-performance water reducing agent.

Principal Component	Specific Gravity	pH	Alkali Content (%)	Chloride Content (%)
Polycarboxylate	1.05 ± 0.05	5.0 ± 1.5	<0.01	<0.01

3.1.5. Steel Fibers

Double-hook steel fibers for conventional concrete with a diameter of 0.75 mm, length of 60 mm, and an aspect ratio of 80 were used. These steel fibers had a density of 7.8 g/cm^3 and a tensile strength of 1200 MPa. Figure 1 shows the shape of the used steel fibers.

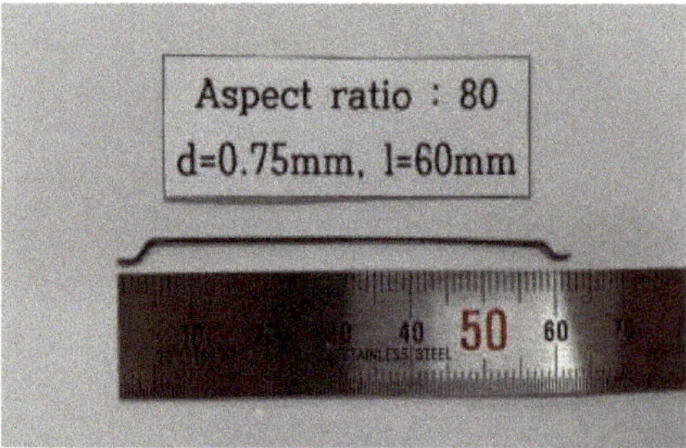

Figure 1. The shape of the used steel fiber.

3.2. SIFRCCs

SIFRCCs that can accommodate a high volume fraction of steel fibers were developed in this study to prevent the fiber ball phenomenon, the main disadvantage of conventional fiber-reinforced concrete, and maximize the fiber volume fraction [22–26]. Unlike conventional fiber-reinforced concrete, SIFRCCs is a type of HPFRCC that can contain a high volume of steel fibers. It is fabricated by filling steel fibers by dispersing them and then filling high-performance slurry. SIFRCCs affords the advantages of preventing the fiber ball phenomenon and allowing a high fiber volume fraction [23,24,26].

3.3. Mixing and Fabrication of Specimens

OPC, fine aggregate, water, superplasticizer, and additional silica fume were mixed to prepare a slurry. The water binder ratio was 0.35. First, the cylindrical mold with diameters of 100 mm and heights of 200 mm were filled with steel fiber with respect to volume fraction. Randomly sprinkled steel fibers in the mold should not overfill the depth of mold and level up as much as possible [25,26]. The slurry as prepared after mixing the contents was poured until no more bubbles were seen to ensure infiltration of slurry into the fibers because the void has negative effects on the strength of the concrete. The amount of high-performance water reducing agent was set to 2.5% of the binder weight [25,26]. To reduce the material separation and achieve the required strength, fine aggregates (50% of binder weight) and silica fume (15% of cement weight) were added [25,26]. Table 4 shows SIFRCCs mixing.

Table 4. SIFRCCs (high-performance slurry-infiltrated fiber-reinforced cementitious composite) mixing proportion.

Fiber (% vol.)	W/B (Water-Binder Ratio)	Unit Material Quantity (kg/m^3)					
		Water	Cement	Fine Aggregate	Silica Fume	Superplasticizer	Steel Fiber
4%	0.35	407.4	962.8	566.4	169.9	28.3	312
5%							390
6%							468

To analyze the compressive behavior characteristics of the SIFRCCs for fiber volume fractions of 4% (312 kg per cubic meters), 5% (390 kg per cubic meters), and 6% (468 kg per cubic meters), cylindrical specimens with diameters of 100 mm and heights of 200 mm were fabricated with respect to the mixing ratio of each variable, as listed in Table 4. Figure 2 shows the fabrication of specimens.

(**a**) Placing steel fibers

(**b**) Filling slurry

Figure 2. Fabrication of specimens.

3.4. Compression Test of SIFRCCs

A universal testing machine (UTM) was used to study the compressive behavior of specimens. The specimens were placed centrally between the two compression plates, such that the center of moving head was vertically above the center of specimen, as shown in Figure 3, then the load was applied on the specimens by moving the movable head. The load and corresponding contraction were measured at different intervals. A review of existing studies on the mechanical properties of high-strength concrete and fiber-reinforced concrete found that there was an applicable strength limit for each prediction equation [8,10–17]. If this strength limit is exceeded, structural design problems can be generated by estimation of the unsafe side. Furthermore, major mechanical properties that define the stress-strain relationship have been found to be determined by the compressive stress of concrete and the fiber volume and shape [8,10–17]. Therefore, this study also used the high fiber volume fraction as a major variable. In addition, an experimental study on the elastic modulus under compressive stress, strain at peak strength, and Poisson's ratio was performed. Figure 3 shows the experimental setup for compression tests. Compressive strength, modulus of elasticity, and Poisson's ratio were determined according to ASTM (American Society for Testing and Materials) specifications (ASTM C873 [27] and ASTM C469 [28], respectively).

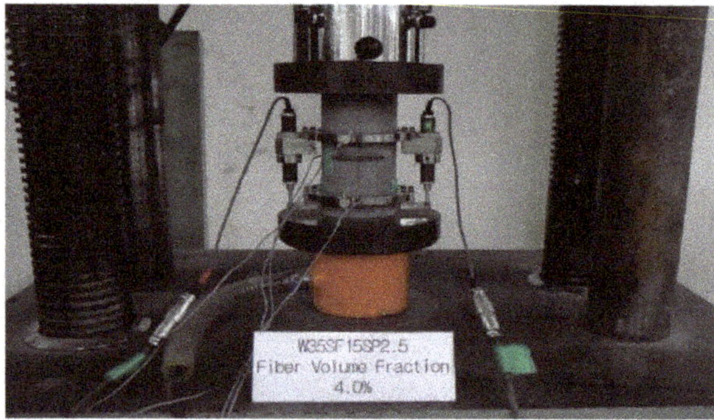

Figure 3. Experimental setup for compressive stress and elastic modulus tests.

4. Experiment Results and Analysis

4.1. Compressive Stress

Figure 4 shows the compressive stress experiment results with respect to the fiber volume fraction for SIFRCCs. For fiber volume fraction of 6%, the average compressive stress was ~83 MPa. For fiber volume fraction of 5%, the average compressive stress was ~75 MPa; this was ~10% lower than that for the fiber volume fraction of 6%. Furthermore, for the fiber volume fraction of 4%, the average compressive stress was ~66 MPa; this was ~12% and ~21% lower than that for fiber volume fractions of 5% and 6%, respectively. The compressive stress tended to increase in proportion to the fiber volume fraction. It seemed that the increased amount of steel fibers with increasing fiber volume fraction produced a restraining effect on the specimen itself, thereby affecting the increase in compressive stress.

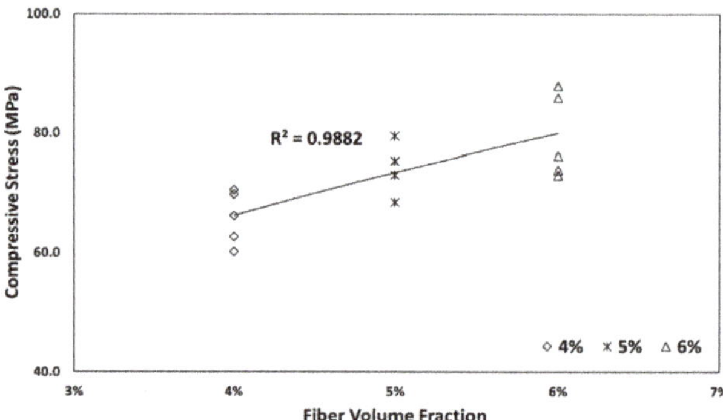

Figure 4. Compressive stress experiment results with respect to fiber volume fraction.

Many studies have shown that the mixing of fibers in fiber-reinforced concrete generally does not influence the compressive stress of the matrix itself. The compression experiment results of Shah and Rangan [29] reported that although the compressive strain at fracture increased significantly when fibers were mixed, the compressive stress did not clearly improve. As the fiber volume fraction increased, the compressive stress gradually increased at first; however, above a certain fraction (4% or more), this tendency changed somewhat. Within a certain mixing level of fibers, the compressive stress increased because the resistance was increased by crack suppression; however, above this level, the compressive stress decreased owing to additional defects that appeared with the improvement effect.

The compressive fracture of conventional fiber-reinforced concrete was caused not by the yield or drawing of fibers but by the fracture of a matrix with relatively low strength before fibers play a structural role. Accordingly, the compressive stress was considered to improve with the fiber volume fraction as sufficient adhesion occurred between the fiber and the high-strength cementitious matrix used in this study. Micro-cracks were caused in conventional concrete and fiber-reinforced concrete by the interface properties of the coarse aggregates and the cement paste. For SIFRCCs, a composite made of fine particles with no coarse aggregates, cracks at the interface could be reduced, and the adhesion performance of the filling slurry and steel fibers could be improved. Therefore, the effect of steel fibers, such as the fiber volume fraction, rather than the effect of the compressive stress caused by matrix fracture was considered to be reflected.

4.2. Elastic Modulus

The elastic modulus experiment results indicated that the elastic modulus showed insignificant differences with the fiber volume fraction; its value remained ~28 GPa. This was in contrast to the compressive stress experiment results. However, as shown in Figure 5, the strain at peak stress showed significant differences with increasing fiber volume fraction. These results implied that the energy absorption capacity improved as the fiber volume fraction increased.

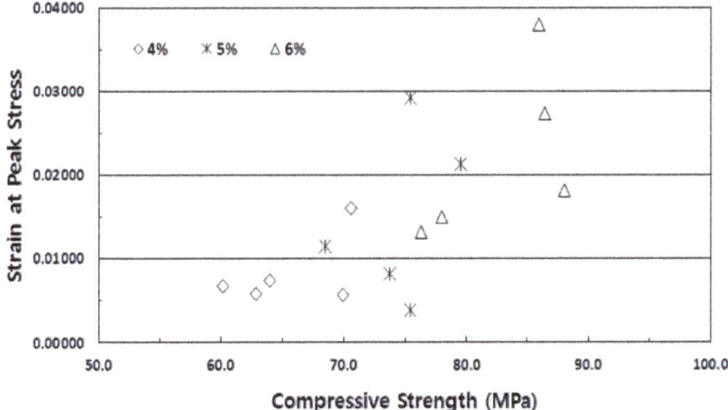

Figure 5. Strain at peak stress.

Figure 6 shows the elastic modulus estimation curve using the compressive stress as presented in KCI 2012 [30] and ACI 318 [31] and the compressive stress and elastic modulus with respect to the fiber volume fraction. The elastic modulus with respect to the fiber volume fraction measured in this study satisfied the lower limit of the elastic modulus estimation curve. This seemed to be because the elastic modulus prediction equation suggested in KCI 2012 [30] and ACI 318 [31] estimated the elastic modulus using only the compressive stress and unit weight.

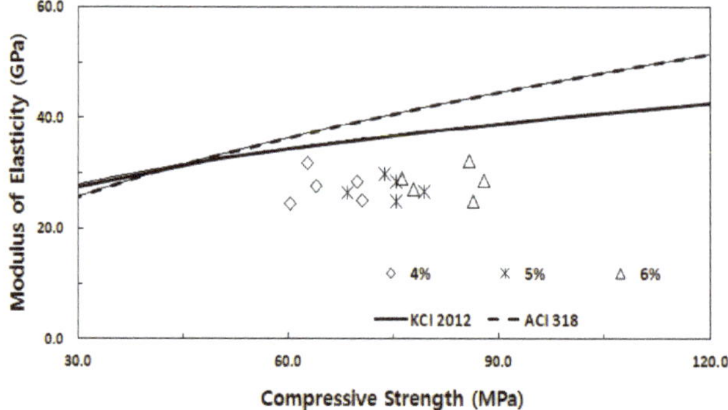

Figure 6. Correlations of compressive stress and elastic modulus between KCI 2012 [30] and ACI 318 [31].

4.3. Poisson's Ratio

Poisson's ratio of conventional concrete is known to be 0.16–0.20; that of fiber-reinforced high-strength concrete is expected to be higher. Figure 7 shows Poisson's ratio with respect to the fiber volume fraction of SIFRCCs. As with the elastic modulus experiment results, Poisson's ratio showed insignificant differences with the fiber volume fraction. The average Poisson's ratio was ~0.3; this was much higher than that of conventional concrete and was similar to that of steel. This was because the high-strength filling slurry matrix improved the interface adhesion between the matrix and the steel fibers. SIFRCCs showed homogeneous behavior owing to its high fiber volume fraction; this could suppress brittle fracture, a disadvantage of conventional concrete, and provide sufficient energy absorption capacity.

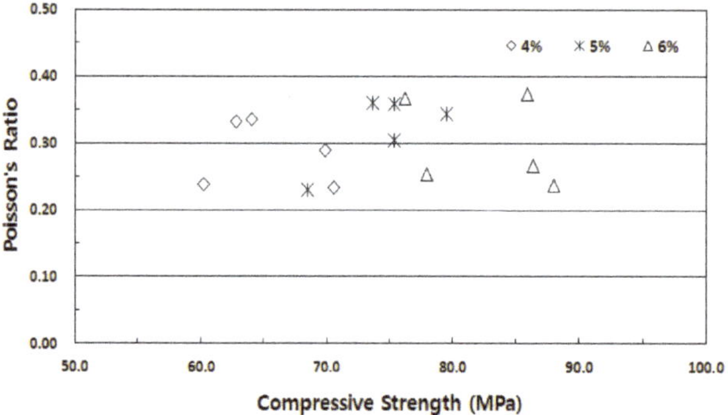

Figure 7. Correlation between compressive stress and Poisson's ratio with respect to fiber volume fraction.

4.4. Stress-Strain Relationship

The stress-strain experiment results of SIFRCCs with respect to the fiber volume fraction showed relatively high strain values for all variables compared to that of conventional concrete, with a strain of 0.008 or higher at peak stress. Figure 8 shows the results for the fiber volume fraction of 6%. In this case, the post-peak behavior exhibited a strain hardening behavior. Even after the elastic section in the stress-strain curve, it showed ductile behavior characteristics like those of metal. This showed that as in the experimental results obtained with high Poisson's ratio, the specimen exhibited sufficient energy absorption capacity even after the peak stress owing to its high fiber volume fraction. Figure 9 shows the results for the fiber volume fraction of 5%. As in the previous case, the post-peak behavior exhibited a strain hardening behavior in the plastic section of the stress-strain curve. Figure 10 shows the results for the fiber volume fraction of 4%. Unlike in the previous two cases, the post-peak behavior exhibited a strain hardening behavior to some degree in the plastic section of the stress-strain curve and a strain-softening behavior in the end. This was because the low fiber volume fraction of 4% produced only a small amount of fiber reinforcement that decreased the resistance to the expansion force in the vertical direction of the specimen axis. Figure 11 shows the results for the stress-strain curve with respect to fiber volume fraction.

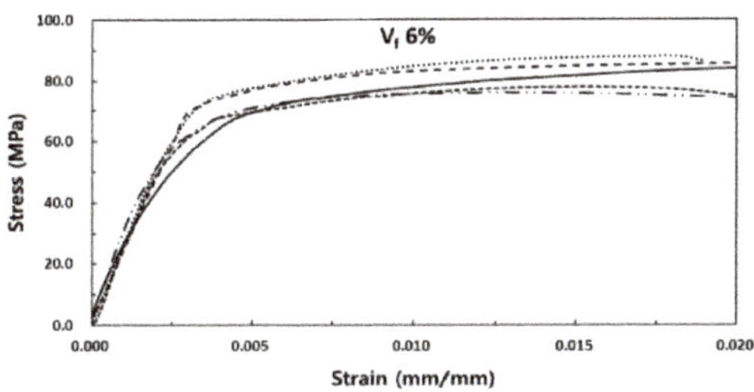

Figure 8. Stress-strain curve for the fiber volume fraction of 6%.

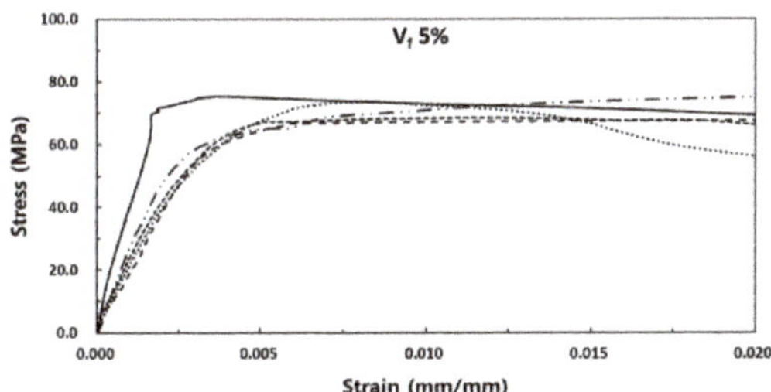

Figure 9. Stress-strain curve for the fiber volume fraction of 5%.

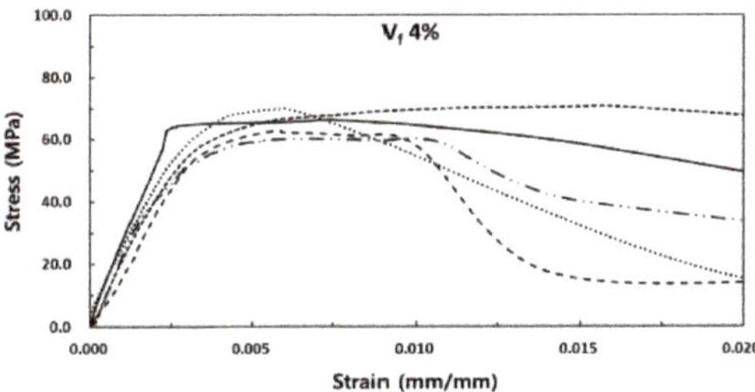

Figure 10. Stress-strain curve for the fiber volume fraction of 4%.

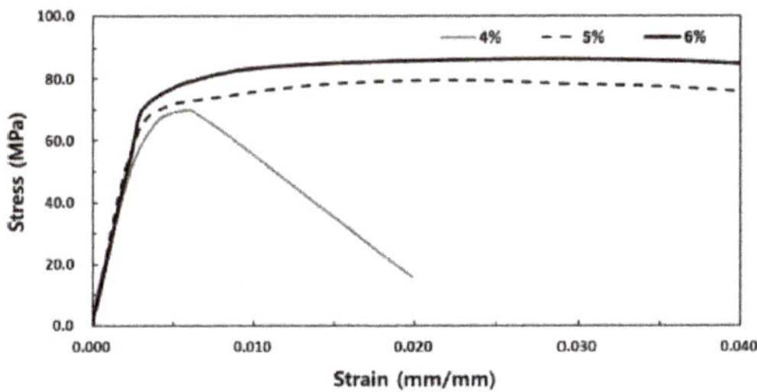

Figure 11. The stress-strain curve with respect to fiber volume fraction.

Table 5 lists the experimental results for maximum strength, elastic modulus, and strain at peak stress; these are generally used to define the stress-strain curve of the SIFRCCs. The relative error of compressive strength in the average value of the experimental results for each variable with respect to the fiber volume fraction was lower than 10%. The stress-strain experiment results revealed insignificant differences in the elastic modulus with the fiber reinforcement amount. This suggested that fiber reinforcement had no effect on the elastic modulus owing to the use of the same filling slurry matrix regardless of the fiber volume fraction.

Table 5. Experimental results for compressive behavior characteristics with respect to the fiber volume fraction.

Variables (V_f)	f_c (MPa)	E_c (MPa)	ε_{co} ($\times 10^{-6}$)	ν
4%	69.9	28,430	5.74	0.289
	70.6	25,098	16.08	0.233
	62.8	31,826	5.86	0.333
	64.0	27,667	7.40	0.336
	60.2	24,493	6.67	0.239
Average	**65.5**	**27,503**	**8.37**	**0.286**
5%	73.7	29,853	8.16	0.360
	68.5	26,548	11.52	0.231
	79.5	26,617	21.36	0.343
	75.4	24,804	3.84	0.305
	75.4	28,481	29.16	0.358
Average	**74.5**	**27,261**	**14.81**	**0.319**
6%	88.0	28,551	18.18	0.237
	78.0	26,967	15.02	0.253
	86.4	24,886	27.38	0.267
	85.9	32,182	38.02	0.373
	76.3	29,071	13.20	0.367
Average	**82.9**	**28,331**	**22.36**	**0.300**

However, the strain at peak stress also increased with the fiber volume fraction. In contrast to the above-mentioned compressive stress experiment results with respect to the fiber volume fraction, the compressive stress for the fiber volume fraction of 6% increased slightly by approximately 1.11 and 1.27 times compared to those for fiber volume fractions of 5% and 4%, respectively. However, the strain at peak stress for the fiber volume fraction of 6% increased significantly by 1.51 and 2.67 times compared to those for fiber volume fractions of 5% and 4%, respectively. The compressive fracture patterns obtained by performing the compressive behavior experiment, shown in Figure 12, indicated that the strain increased further as axial cracks in the specimen were restrained. This characteristic of the strain increase rate was considered to be caused by the restraining effect of the high fiber volume fraction, in which many steel fibers resisted the expansion force generated vertically to the specimen axis and thereby caused longitudinal split cracks. Moreover, the steel fiber reinforcement effect served to increase the size of the peak stress and the strain at peak stress in all compressive stress areas.

(a) 4% (b) 5% (c) 6%

Figure 12. Compressive fracture pattern with respect to fiber volume fraction.

5. Conclusions

A high-performance SIFRCCs that can prevent the fiber ball phenomenon, a disadvantage of conventional fiber-reinforced concrete, and maximize the fiber volume fraction was developed, and the compressive behavior characteristics with respect to the fiber volume fraction were analyzed. The following conclusions were derived from this study.

(1) The static compressive behavior characteristics with respect to the fiber volume fraction of SIFRCCs showed that the compressive stress increased in proportion to the increasing fiber volume fraction. The micro-cracks in conventional concrete and fiber-reinforced concrete were caused by the interface properties of coarse aggregates and cement paste. By contrast, SIFRCCs could reduce cracks at the interface because it was a composite made of fine particles with no coarse aggregates. Furthermore, the filling slurry and steel fibers improved the adhesion performance and reflected the effects of the fiber volume fraction and steel fibers instead of the compressive stress characteristics caused by matrix fracture. In addition, the high volume fraction of the steel fibers generated a restraining effect in the compressive stress test specimen, thereby affecting the increase in compressive stress.

(2) The elastic modulus experiment results showed that the elastic modulus did not increase with the increasing fiber volume fraction, unlike the compressive stress experiment results. However, for strain under compressive stress, the increased rate of the strain at peak stress showed a difference of up to 2.7 times depending on the fiber volume fraction. This means that with increasing fiber volume fraction, the post-peak behavior after peak stress showed a strain hardening behavior, implying that the energy absorption capacity improved at a higher fiber volume fraction.

(3) The characteristics of the strain increase rate were attributed to the restraining effect of the high fiber volume fraction in which many steel fibers resisted the expansion force generated vertically to the specimen axis that caused longitudinal split cracks. Furthermore, the reinforcing effect of steel fibers was considered to increase the peak stress and strain at peak stress in all compressive stress areas.

(4) Poisson's ratio experiment results showed insignificant differences in the elastic modulus with the fiber volume fraction. Similarly, Poisson's ratio showed insignificant differences with the fiber volume fraction. However, Poisson's ratio was ~0.3 for every variable; this was similar to that of steel. This was caused by the improved interface adhesion performance between the high-strength filling slurry matrix and the steel fibers. Therefore, the SIFRCCs with a high fiber volume fraction could suppress brittle fracture, a disadvantage of conventional concrete, and exhibit sufficient energy absorption capacity owing to its homogeneity.

Author Contributions: Conceptualization, S.K., C.P., and K.-K.Y.; methodology, S.K., C.P., and K.-K.Y.; writing—original draft preparation, S.K., S.H., C.P., and K.-K.Y. All authors have read and agreed to the published version of the manuscript.

Acknowledgments: This research was supported by a grant (19SCIP-B146646-02) from Construction Technology Research Project and was conducted under research project "Development of High-Performance Concrete Pavement Maintenance Technology to Extend Roadway Life (19TLRP-B146707-01)" funded by the Ministry of Land, Infrastructure, and Transport (MOLIT) and the Korea Agency for Infrastructure Technology Advancement (KAIA).

Conflicts of Interest: The authors declare no conflict of interest.

References

1. Kang, S.-T. Bond Strength of Steel Fiber Incorporated in Ultra High Performance Fiber-Reinforced Concrete. *J. Korea Concr. Inst.* **2013**, *25*, 547–554. [CrossRef]
2. Ku, D.-O.; Kim, S.-D.; Kim, H.-S.; Choi, K.-K. Flexural Performance Characteristics of Amorphous Steel Fiber-Reinforced Concrete. *J. Korea Concr. Inst.* **2014**, *26*, 483–489. [CrossRef]
3. Bindiganavile, V.; Banthia, N. Generating dynamic crack growth resistance curves for fiber-reinforced concrete. *Exp. Mech.* **2005**, *45*, 112–122. [CrossRef]
4. Kim, S.; Park, C. Flexural Behavior of High-Volume Steel Fiber Cementitious Composite Externally Reinforced with Basalt FRP Sheet. *J. Eng.* **2016**, *2016*, 2857270. [CrossRef]
5. Malvar, L.J.; Crawford, J.E.; Morrill, K.B. Use of Composites to Resist Blast. *J. Compos. Constr.* **2007**, *11*, 601–610. [CrossRef]
6. Otter, D.E.; Naaman, A.E. Strain rate effects on compressive properties of fiber reinforced concrete. In Proceedings of the International Symposium on Fibre Reinforced Concrete, Madras, India, 16–19 December 1987; Volume 2, pp. 225–235.
7. Jung, H.; Park, S.; Kim, S.; Park, C. Performance of SIFCON based HPFRCC under Field Blast Load. *Procedia Eng.* **2017**, *210*, 401–408. [CrossRef]
8. Choi, H.-K.; Bae, B.-I.; Choi, C.-S. Mechanical Characteristics of Ultra High Strength Concrete with Steel Fiber Under Uniaxial Compressive Stress. *J. Korea Concr. Inst.* **2015**, *27*, 521–530. [CrossRef]
9. Gopalaratnam, V.S.; Shah, S. Failure mechanism and fracture of fibre reinforced concrete. *J. Am. Concr. Inst.* **1987**, *105*, 1–26.
10. Propovics, S. A Numerical Approach to the Complete Stress-Strain Curve of Concrete. *Cement Concr. Res.* **1973**, *3*, 583–599. [CrossRef]
11. Sargin, M. *Stress-Strain Relationship for Concrete and the Analysis of Structural Concrete Sections*; University of Waterloo: Waterloo, ON, Canada, 1971; p. 167.
12. Carreira, D.J.; Chu, K.D. Stress-Strain Relationship for Plain Concrete in Compression. *ACI J.* **1985**, *82*, 797–804.
13. Collins, M.P.; Mitchell, D.; Macgregor, J.G. Structural Design Considerations for High-Strength Concrete. *Concr. Int. Des. Const.* **1993**, *15*, 27–34.
14. Wee, T.H.; Chin, M.S.; Mansur, M.A. Stress-Strain Relationship of High-Strength Concrete in Compression. *J. Mater. Civ. Eng.* **1996**, *8*, 70–76. [CrossRef]
15. Wang, P.T.; Shah, S.P.; Naaman, A.E. Stress-Strain Curves of Normal and Lightweight Concrete in Compression. *ACI J. Proc.* **1978**, *75*, 603–611.
16. Attard, M.M.; Setunge, S. Stress-strain relationship of confined and unconfined concrete. *ACI Mater. J.* **1996**, *93*, 432–442.
17. Graybeal, B.A. Compressive Behavior of Ultra-High-Performance Fiber-Reinforced Concrete. *ACI Mater. J.* **2007**, *104*, 146–152.
18. Gilani, A.M. Various Durability Aspect of Slurry Infiltrated Fiber Concrete. Ph.D. Thesis, Middle East Technical University, Ankara, Turkey, 7 September 2007.
19. Romualdi, J.P.; Batson, G.B. Mechanics of crack arrest in concrete. *J. Eng. Mech.* **1963**, *89*, 147–168.
20. Romualdi, J.P.; Mandel, J.A. Tensile strength of concrete affected by uniformly distributed and closely space short length of wire reinforcement. *ACI J.* **1964**, *61*, 657–672.
21. Snyder, M.L.; Landkard, D.R. Factors affecting the strength of steel fibrous concrete. *ACI J. Proc.* **1972**, *69*, 96–100.

22. American Society for Testing and Materials (ASTM). *Standard Specification for Portland Cement*; ASTM C150/C150M-19a; ASTM: West Conshohocken, PA, USA, 2019.
23. Kim, S.; Jung, H.; Kim, Y.; Park, C. Effect of steel fiber volume fraction and aspect ratio type on the mechanical properties of SIFCON-based HPFRCC. *Struct. Eng. Mech.* **2018**, *65*, 163–171.
24. Kim, S.W.; Cho, H.M.; Lee, H.Y.; Park, C.W. Flexural Performance Characteristics of High Performance Slurry Infiltrated Fiber Reinforced Cementitious Composite according to Fiber Volume Fraction. *J. Korea Inst. Struct. Maint. Insp.* **2015**, *19*, 109–115.
25. Kim, S.W.; Park, C.W.; Kim, S.W.; Cho, H.M.; Jeon, S.P.; Ju, M.K. Optimum Mix Proportions of In-fill Slurry for High Performance Steel Fiber Reinforced Cementitious Composite. *J. Korean Rec. Conct. Resources Inst.* **2014**, *2*, 196–201.
26. Kim, S.; Kim, Y.; Kim, S.W.; Kim, Y.J.; Park, C. Strengthening Effect of Externally Reinforced RC Structure with SIFRCC on Impact Load. In Proceedings of the ICPS5, Poznan, Poland, 19–23 August 2018; pp. 522–528.
27. American Society for Testing and Materials (ASTM). *Standard Test Method for Compressive Strength of Concrete Cylinders Cast in Place in Cylindrical Molds*; ASTM C873/C873M-15; ASTM: West Conshohocken, PA, USA, 2015.
28. American Society for Testing and Materials (ASTM). *Standard Test Method for Static Modulus of Elasticity and Poisson's Ratio of Concrete in Compression*; ASTM C469/C469M-14; ASTM: West Conshohocken, PA, USA, 2014.
29. Shah, S.P.; Rangan, B.V. Fiber Reinforced Concrete Properties. *Am. Concr. Inst. J. Proc.* **1971**, *68*, 126–137.
30. Korea Concrete Institute (KCI). *Concrete Design Code and Commentary*; Kimoondang Publishing Company: Seoul, Korea, 2012.
31. ACI Committee; International Organization for Standardization. *Building Code Requirements for Structural Concrete (ACI 318-11) and Commentary*; American Concrete Institute: Farmington Hills, MI, USA, 2008.

Article

Effective Bio-Slime Coating Technique for Concrete Surfaces under Sulfate Attack

Keun-Hyeok Yang [1], Hee-Seob Lim [2] and Seung-Jun Kwon [2,*]

[1] Department of Plant Architectural Engineering, Kyonggi University, Suwon 16227, Korea; yangkh@kqu.ac.kr
[2] Department of Civil Engineering, Hannam University, 70 Hannam-ro, Daedeok-gu, Daejeon 34430, Korea; heesubjm@naver.com
* Correspondence: jjuni98@hannam.ac.kr

Received: 27 February 2020; Accepted: 23 March 2020; Published: 26 March 2020

Abstract: The service life of concretes exposed to sulfate decreases as the concrete body expands due to the formation of gypsum and ettringite. Bacteria-based repair coating layers, which have been studied lately, are aerobic and very effective on the sulfate attack. In this study, bio-slime repair coating layers were fabricated using bacteria, and chloride diffusion experiments were performed. In addition, the service life of concrete under sulfate attack was evaluated using time-dependent diffusivity and a multi-layer technique. Chloride diffusivity was compared with sulfate diffusivity based on literature review, and the results were used to consider the reduction in the diffusion coefficient. In the analysis results, the service life of concrete was evaluated to be 38.5 years without bio-slime coating layer, but it was increased to 41.5–54.3 years using it. In addition, when the thickness of the bio-slime coating layer is 2.0 mm, the service life can be increased by 1.31–2.15 times if the sulfate diffusion coefficient of the layer is controlled at a level of $0.1 \sim 0.3 \times 10^{-12}$ m^2/s. Eco-friendly and aerobic bio-slime coating layers are expected to effectively resist sulfate under appropriate construction conditions.

Keywords: bio-slime; sulfate attack; chloride attack; service life; multi-layer diffusion; repair

---

1. Introduction

Concrete structures are used in various environments due to their economic efficiency and durability. Reinforced concrete (RC) structures are the combination of reinforcement, such as reinforcing steel and concrete. In recent years, studies have been concentrated on the corrosion of steel reinforcement due to salt damage and carbonation [1,2]. Most of these studies are focused on the corrosion of buried reinforcement or the corrosion initiation time. Concrete, however, can be degraded through the expansion of the concrete body or local cracks, as well as the corrosion of the internal steel reinforcement [3,4].

A representative deterioration phenomenon that affects the concrete body is the sulfate attack. In the deterioration of concrete by sulfate ions, they penetrate into concrete and react with calcium hydroxide, forming gypsum, or with monosulfates or C_3AH_6 (tricalcium aluminate hexahydrate—a C_3A hydrate), forming ettringite ($C_3A{\cdot}3CaSO_4{\cdot}32H_2O$), which causes the expansion of concrete. The generated ettringite causes cracks at the beginning and significantly affects safety through the expansion of the concrete body [5,6].

In general, sulfate control is conducted by controlling the penetration of sulfates and inhibiting the generation of gypsum and ettringite through the use of admixtures. Many studies have been conducted to improve resistance to deterioration caused by the penetration of sulfate ions, but few of them evaluated the service life through quantitative modeling. Studies on single and combined deterioration have been conducted, but they are focused on the evaluation of deterioration using experimental results. Until recently, the evaluation of deterioration or service life under sulfate attack was dominated by the evaluation of the strength reduction [7,8]. In addition, the material modeling levels of many studies were limited to a correlation with the amount of C_3A and the expansion volume [6,8].

Sulfate deterioration modeling is difficult because it is complex to simulate deterioration caused by the spalling of external concrete over time; it is also difficult to quantify the reaction of internal hydrates and sulfate ions. The Atkinson model used by ACI (American Concrete Institute) as well as the previously researched models only perform modeling of the deterioration depth and cannot consider the effects after cracking or surface spalling [9,10].

The deterioration of sewage facilities, which are lifelines, has become a problem, and ground subsidence caused by the deterioration of internal concrete and the leakage of joints has been reported [11,12]. For the internal erosion of concrete or pipe lines, bacteria-based self-healing materials can be excellent alternatives that can extend the service life of sewage pipes. Many studies have rapidly advanced aerobic bacteria-based concrete development technologies, and some technologies have reached the commercialization stage [13–16]. Of course, the residual viability of bacteria and the limitations of self-healing require consistent research, but bacteria-based self-healing materials can be used as excellent repair materials because they can complement such shortcomings as peeling-off, which is commonly found in organic repair materials [17,18]. The protective coating technology using fixed bacteria is expected to have a sustainable resistance to harsh environment that gradually degrades concrete performance, unlike conventional coating materials like epoxy that may cause delamination and peel-off [19].

The coating mortar with fixed bacteria was prepared using expanded vermiculite for a carrying bacteria. The glycocalyx was applied to bacteria in culture, which was used as a barrier to protect from the harmful ion attack outside [13].

In this study, bio-slime was prepared using bacteria to effectively control the penetration of sulfate ions, which are a major deterioration factor, and the service life under sulfate attack was evaluated considering the thickness of the coating layer. The time-dependent ion diffusion technique was used for the service life evaluation, and the results were compared with those of the existing deterioration depth model due to sulfate.

2. Preparation of Bio-Slime Coating Layers

Expanded vermiculite (EV) and super absorbent polymer (SAP) were used as immobilization materials to provide the living pores of bacteria and to create a neutral (less than pH 10) growth environment. The particle size of EV ranged from 0.25 to 0.36 mm while that of SAP ranged from 0.08 to 0.20 mm. Their densities were 0.25 and 0.70 g/cm^3, respectively. These materials had high moisture-retaining capacities and neutral pH. In particular, EV could effectively absorb cations (e.g., Ca^{2+} and Mg^{2+}) required for the growth of microorganisms (Rhodobacter capsulatus) because it had an excellent cation exchange capacity. As for bacteria, Rhodobacter capsulatus cultured at a concentration of 109 cel/mL was used, and the pH of the culture medium was slightly acid (pH = 6.8) to improve the highly alkaline environment.

Considering the durability improvement and eco-friendliness of the bacterial glycocalyx coating material, ordinary Portland cement (OPC, S company, Sejong, Korea), ground granulated blast furnace slag (GGBS) and fly ash (FA) were used as binders. In the case of aggregate, silica sands with particle sizes of 0.05–0.17 mm, 0.17–0.25 mm, and 0.25–0.70 mm were mixed at the same mass ratio and used. The porous materials for the immobilization of bacteria were used replacing 30% of the aggregate volume during mixing.

Table 1 shows the formulation of the mortar used as a coating material. Table 2 presents the mixing information of the bio-slime coating material used. Figure 1 shows the material preparation and the overview of the bio-slime coating layer. In order to promote sustainable and protective effect to the coating mortar, the glycocalyx was used as a coating materials. For the cultivation of bacteria, glycocalyx consists of polymer skin capsules and a slime layer. The composition ratio of glycocalyx depends on the characteristics of the bacterial strain and constituent of the cultivate media. In addition, the strength characteristics, sulfate resistance, and preparation of the bio-slime structure can be found in the existing literature [13].

Table 1. Mix proportions for mortar.

W/B	S/B	Unit Weight (kg/m^3)						
		Water	OPC	FA	GGBS	Silica Sand Size (mm)		
						0.05~0.17	0.17~0.25	0.25~0.7
35	2	135.9	158.5	90.6	203.8	196.3	196.3	196.3

Table 2. Mix proportions for bio-slime coating material.

Sample	Strain	Immobilization Material Type	Immobilization Material Substitution Ratio (%)
Non-bacteria	–	–	–
Non-immobilized bacteria	Rhodobacter capsulatus	–	–
Expanded vermiculite (EV) immobilized bacteria		EV	30
Super absorbent polymer (SAP) immobilized bacteria		SAP	30

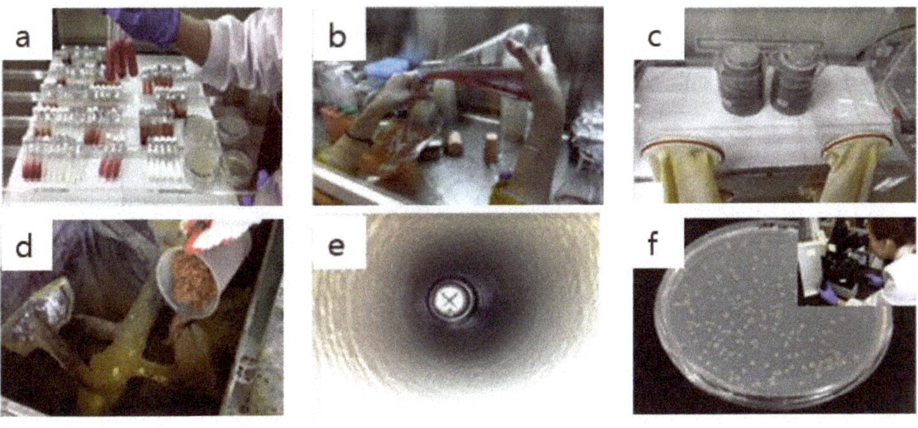

Figure 1. Procedures for producing bio-slime with bacteria and coating for sewage line: (**a**) bacteria lsolation and screening; (**b**) bacteria culture: 10^9 cell/mL. (pH = 6.8, anaerobic environment); (**c**) bacteria immobilization; (**d**) coating material formulation; (**e**) coating material construction by lining method; (**f**) confirm bacterial growth (recultured and microscopic).

3. Sulfate Penetration Analysis Considering Bio-Slime Coating Layer

3.1. Analysis without Coating Layer

The service life model for sulfate deterioration considered the model developed by Atkinson and Hearne [9,10]. This model was also developed as a software program, and it shows the deterioration depth over time considering the moisture inflow from the ground [20]. Equation (1) is the basic equation developed by Atkinson and Hearne. This model basically assumes that the formation of expandable ettringite inside concrete causes harmful expansion and cracking and assumes that a failure occurs while the deterioration side (X_{spall}) thickly peels off from the concrete surface when the deformation caused by the increased volume of ettringite exceeds the fracture energy of concrete.

$$R = \frac{Z_p}{t} = \frac{EB^2 c_0 D_i x \Phi_{Al_2O_3}}{0.10196\alpha\gamma(1-\nu)} \quad (1)$$

where, R is the deterioration rate of concrete by sulfate ions (m/s), Zp is the predicted sulfate penetration depth (m), t is the time (s), c_0 is the external sulfate concentration (mol/m^3), Di is the sulfate diffusion coefficient in concrete (m^2/s), E is the elastic modulus (kgf/m^2), α is the roughness coefficient of the area where performance deterioration occurs (1.0), B is the stress of 1 mol of sulfate that reacts in 1 m^3 of concrete (=1.8 × 10^{-6} m^3/mol), γ is the energy required for the concrete surface fracture (kgf/m), ν is Poisson's ratio, x is the cement content of the target structure (kg/m^3), and ΦAl_2O_3 is the aluminum oxide content of the target structure. As seen in Equation (1), the deterioration depth linearly increases with the increase in the diffusion coefficient under constant mix proportions. Accordingly, the service life also shows a linear reduction.

3.2. Analysis Considering Coating Layer

When the concrete surface is deteriorated or reinforced, diffusivity from the surface is increased or decreased. In this study, a repaired case, i.e. a case where the surface has a low diffusion coefficient, was assumed. As there is no modeling for the diffusion coefficient of sulfate ions, the diffusion theory of chloride ions was used. In a research that considers the diffusion theory of a multi-layer structure and the time-dependent diffusion coefficient, the flow of ions based on Fick's second law can be expressed as Equation (2) [21]. In the previous model [22], the time effect was not considered, so that the developed Equation (2) with time effect was used. The diffusion coefficient was controlled with time exponent m and specific period (t_c).

$$\left\{\begin{array}{l} \left\{\begin{array}{l} C_1(x,t) = C_S \sum\limits_{N=0}^{\infty} \alpha^n \left[erfc\left\{ \frac{2ne+x}{2\sqrt{\frac{D_1}{1-m}\left(\frac{t_0}{t}\right)^m t}} \right\} - \alpha \cdot erfc \frac{(2n+2)e-x}{2\sqrt{\frac{D_1}{1-m}\left(\frac{t_0}{t}\right)^m t}} \right] \\ C_2 = \frac{2kC_S}{k+1} \sum\limits_{n=0}^{\infty} \alpha^n erfc \left[\frac{(2n+1)e+k(x-e)}{2\sqrt{\frac{D_1}{1-m}\left(\frac{t_0}{t}\right)^m t}} \right] \\ (t < t_c) \end{array}\right\} \\ \left\{\begin{array}{l} C_1(x,t) = C_S \sum\limits_{N=0}^{\infty} \alpha^n \left[\begin{array}{l} erfc\left\{ \frac{2ne+x}{2\sqrt{\frac{D_2}{1-m}\left(\frac{t_0}{t_c}\right)^m \left[1-m+m\left(\frac{t_c}{t}\right)\right] t}} \right\} \\ -\alpha \cdot erfc \frac{(2n+2)e-x}{2\sqrt{\frac{D_2}{1-m}\left(\frac{t_0}{t_c}\right)^m \left[1-m+m\left(\frac{t_c}{t}\right)\right] t}} \end{array} \right] \\ C_2 = \frac{2kC_S}{k+1} \sum\limits_{n=0}^{\infty} \alpha^n erfc \left[\frac{(2n+1)e+k(x-e)}{2\sqrt{\frac{D_1}{1-m}\left(\frac{t_0}{t_c}\right)^m \left[1-m+m\left(\frac{t_c}{t}\right)\right] t}} \right] \\ (t \geq t_c) \end{array}\right\} \end{array}\right. \quad (2)$$

where, C_1 and C_2 are the chloride concentrations of the concrete surface and body (kg/m^3), respectively; D_1 and D_2 are the diffusion coefficients of the concrete surface and body (m^2/s), respectively; k is $(D_1/D_2)^{1/2}$; α is $(1-k)/(1+k)$; and e is the thickness of the surface reinforced through surface repair. The diffusion coefficient of the surface, D_1, was assumed to be smaller than the internal diffusion coefficient, D_2, to simulate the diffusion coefficient of the deteriorated concrete surface. In addition, m and t_c were assumed to be the time-dependent index of the diffusion coefficient and the time when the diffusion coefficient stabilizes (30 years), respectively.

4. Evaluation of the Service Life of a Concrete Structure Considering Bio-Slime

4.1. Derivation of the Bio-Slime Chloride Diffusion Coefficient

To evaluate the penetration of sulfate ions, the diffusion coefficient of chloride ions was indirectly derived and considered for analysis. It is very difficult to experimentally or analytically implement the diffusion coefficient of sulfate ions because it is difficult to directly implement the phase equilibrium and mobility of ions [3]. Previous studies showed that the ratio of sulfate to chloride ions is 1.06/2.06 under distilled water condition, indicating that sulfate ions are approximately 50% of chloride ions. They also experimentally derived the penetration depths of chloride ions and sulfate and found that the ratio of the square root of chloride ion diffusion coefficient and sulfate ion diffusion coefficient has a linear relationship with the penetration depth. The regression analysis of this relationship revealed that the sulfate ion diffusion coefficient is approximately 38.5% of the chloride ion diffusion coefficient, and this tendency is not significantly different from the 50% level, which is the diffusion coefficient ratio in aqueous solution [3].

In this study, the sulfate ion diffusion coefficient was assumed to be 40% of the chloride diffusion coefficient for analysis. The accelerated chloride diffusion coefficient was derived using the NTBULD 492 method, which is the non-steady-state diffusion coefficient. In the NTBUILD 492 method, 10% NaCl solution is applied for cathode, and 0.3 N NaOH solution is adapted for anode to accelerate penetration of chloride ion. Based on initial current and the criteria specified in the method, applied voltage and test period is determined. After the voltage is applied in each test period, silver nitrate solution (0.1 M, $AgNO_3$) is utilized as indicator. The diffusion coefficient from the test was calculated through Equations (3) and (4).

$$D_{rcpt} = \frac{RTL}{zF(U-2)} \times \frac{X_d - \alpha\sqrt{X_d}}{t} \tag{3}$$

$$\alpha = 2\sqrt{\frac{RT}{zFE}} \times erf^{-1} \times \left(1 - \frac{2C_d}{C_0}\right) \tag{4}$$

where D_{rcpt} and R are chloride diffusion coefficient (m^2/s) and universal gas constant (8.314 J/mol·K), respectively. T denotes absolute temperature (K) and L means thickness of specimen (m). z and F are ionic valence of 1.0 and Faraday constant (96,500 J/V·mol). U means applied potential (V), t denotes test duration time (s). The chloride concentrations of C_d and C_0 are the that at which the color changes and that in the cathode solution (mol/L), respectively. Figures 2 and 3 show the photo of the diffusion test and the results, respectively. A total of six slime coating types were considered, and excellent chloride diffusion coefficients of 0.71–3.39 × 10^{-12} m^2/s were derived.

Figure 2. Photo of accelerated chloride diffusion test.

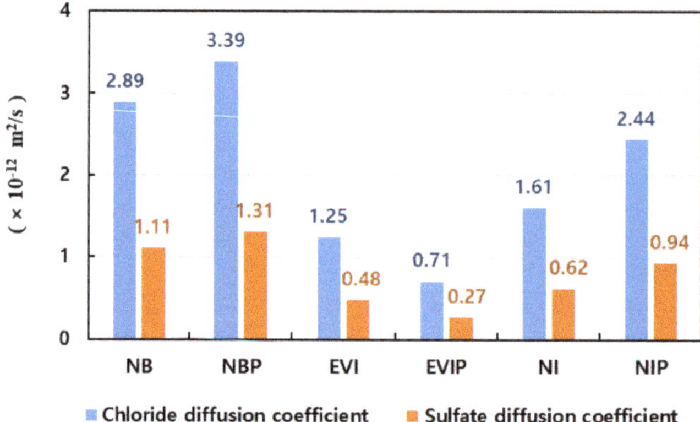

Figure 3. Results from chloride diffusion coefficient in various slime coating conditions. NB: Non-Bacteria; NBP: Non-Bacteria-Polymer; EVI: EV-Immob; EVIP: EV-Immob-Polymer; NI: Non-Immob; NIP: Non-Immob-Polymer.

4.2. Target Structure and Mix for Analysis

The target structure was RC box structure exposed to a high concentration of sulfate, and the mix proportions were assumed to have 30% of GGBFS replacement ratio and the design strength of 35 MPa, which was conventionally used for resisting chemical attack. Figure 4 shows the overview of the target structure. Table 3 show the analysis conditions for sulfate diffusion. In the analysis conditions, the diffusion coefficient of sulfate ions was assumed to be 40% of that of chloride ions, as discussed in Section 3.1. In addition, the sulfate concentration was assumed to be 5,000 ppm, which is considered a medium value in the 1500–10000 ppm range which is the third worst condition of the ACI 318 standard [23,24]. It is very difficult to set the critical sulfate concentration because there is no clear criterion. The condition at which expansion by ettringite begins during the penetration of Na_2SO_4 is usually known as approximately 0.12% [25], but 0.18%, which is 1.5 times higher, was considered under the assumption that the surface was continuously deteriorated.

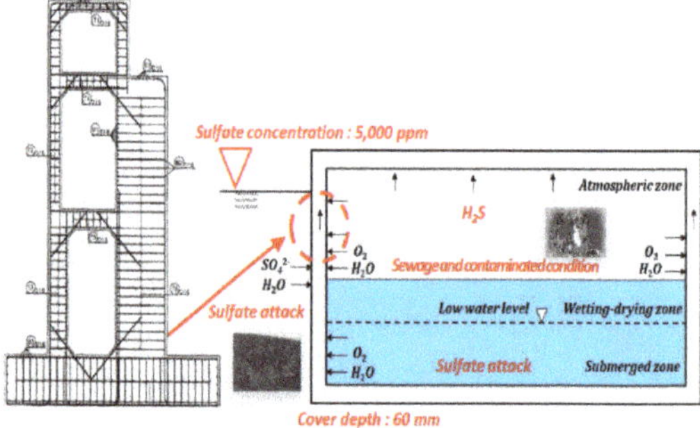

Figure 4. Schematic diagram for the RC structure exposed to sulfate attack.

Table 3. Analysis conditions for sulfate diffusion.

Item	Value
Exterior sulfate concentration (%)	5000 ppm
Diffusion coefficient in concrete	7.943×10^{-12} m²/s
Critical sulfate concentration	0.18% (1800 ppm)
Diffusion coefficient in slime-coating	0.60×10^{-12} m²/s
Thickness of slime coating layer	1.0~5.0 mm

In South Korea, the sulfate concentration (hydrogen sulfide) inside sewage pipes is less than 10 ppm, which is a very low level, and there is the influence of drainage, such as rainfall. Thus, very low deterioration depths have been reported [26].

4.3. Analysis of the Service Life under Different Conditions

(1) When the surface is not protected:

For the unprotected surface, analysis was conducted using two methods. When the Atkinson model was used, the input constants shown in Table 4 were applied. In addition, the service life according to the sulfate diffusion coefficient and that according to the external sulfate concentration are shown in Figure 5.

Table 4. Input constants for Atkinson model.

Input Variable	Unit	Input Value	Ground
c0	mol/m^3	52.067	5000 ppm sulfate concentration was converted into the molar weight (96.06 g/mol)
Di	m^2/s	0.7×10^{-12}	Mean value of typical sulfate diffusion coefficients ($(0-4) \times 10^{-12}$ m^2/s) based on the results of previous studies
E	Pa	$27{,}800 \times 10^6$	Typical elastic modulus of concrete
ν	–	0.177	Typical Poisson's ratio
x	kg/m^3	289.1	The amount of cement used in the mix design was used (slag substitution rate: 30%).
ΦAl_2O_3	mol/m^3	0.05	The mean value of cement manufacturers was used.

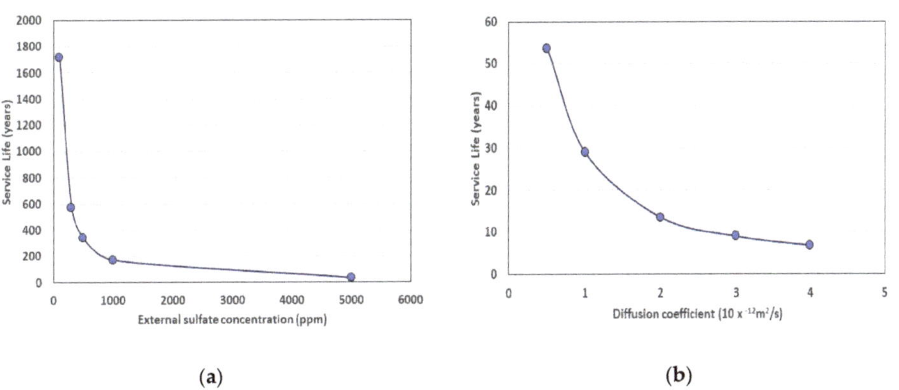

Figure 5. (**a**) Service life variation with exterior concentration; (**b**) service life with diffusion coefficient of sulfate ion.

When the cover thickness (60 mm) and external sulfate concentration (5000 ppm) of the sewage pipe, which was the target structure, were considered, its service life was evaluated to be 38.3 years when surface repair was not performed.

Figure 6 shows the sulfate behavior performed using the diffusion coefficient and Fick's second law. A software that applies the finite difference method was used. When 30% slag was used replacing cement, the diffusion coefficient (D_{ref}) was found to be 7.943×10^{-12} m^2/s and the time-dependent index (m) was 0.314 at the reference age. The service life for the 60 mm concrete cover thickness was evaluated to be 49.1 years.

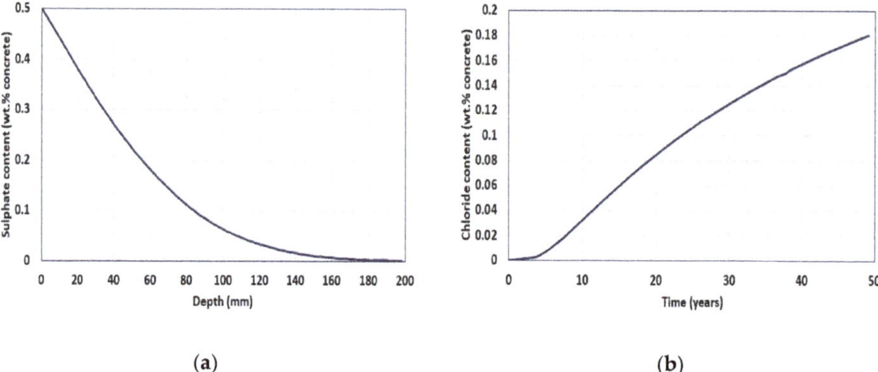

(**a**) (**b**)

Figure 6. (**a**) Sulfate ion profile with cover depth after 49.1 year; (**b**) increasing sulfate ion at steel location with time.

(2) When the surface is protected:

—Service life according to the slime thickness.

For the analysis of the surface protected by bio-slime, Equation (2) was used. When there was no surface repair, the service life was evaluated to be 38.3 years. As the thickness of the coating layer with a diffusion coefficient of 0.6×10^{-12} m^2/s increased to 1.0–5.0 mm, the service life increased from 38.3 to 54.3 years.

Figure 7 shows the service life behavior due to the coating layer change, and Figure 8 shows the sulfate diffusion behavior distribution according to the coating layer thickness after 50 years of sulfate exposure.

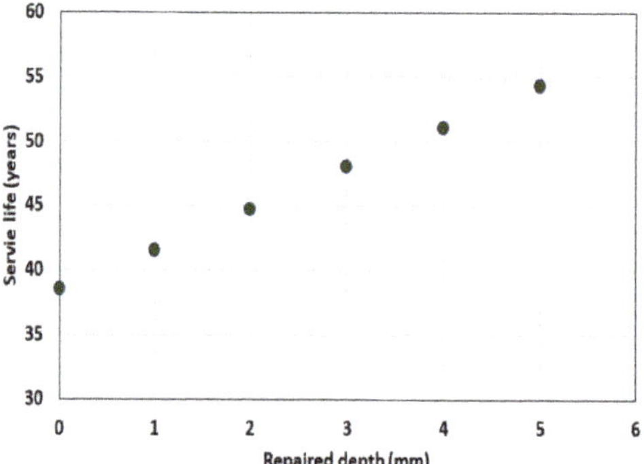

Figure 7. Service life variation with coating thickness.

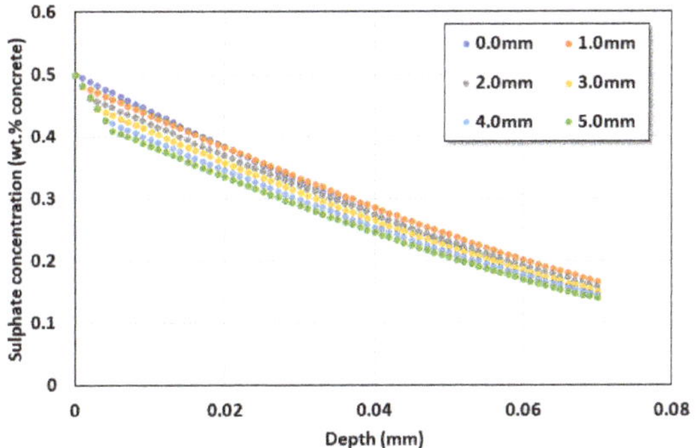

Figure 8. Sulfate ion profile after 50 years.

—Service life according to the external sulfate concentration.

After fixing the slime thickness at 2.0 mm and the diffusion coefficient of slime at 0.5×10^{-12} m^2/s, the service life was analyzed while the external sulfate ion concentration was increased from 0.3% to 0.5%. With the increase in the concentration of sulfate, the penetration of harmful ions to the inside was further accelerated. The sulfate concentration of the actual sewage pipe, however, was very low (100–200 ppm). As the sulfate concentration increased from 3000 to 7000 ppm, the service life significantly decreased from 138 to 29 years, as shown in Figure 9. In addition, Figure 10 shows the sulfate distribution behavior in each sulfate exposure environment after 30 years.

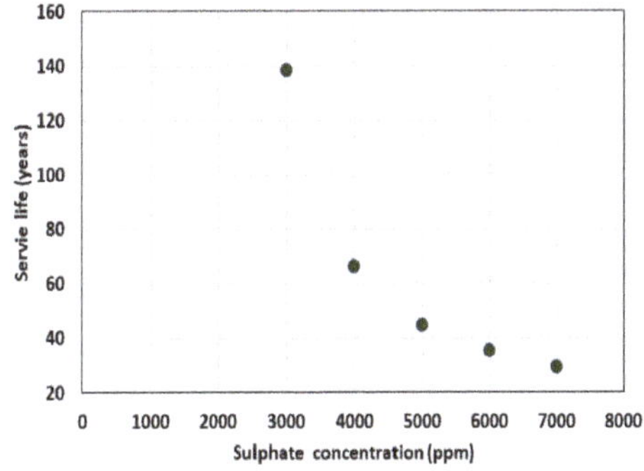

Figure 9. Service life variation with exterior conditions.

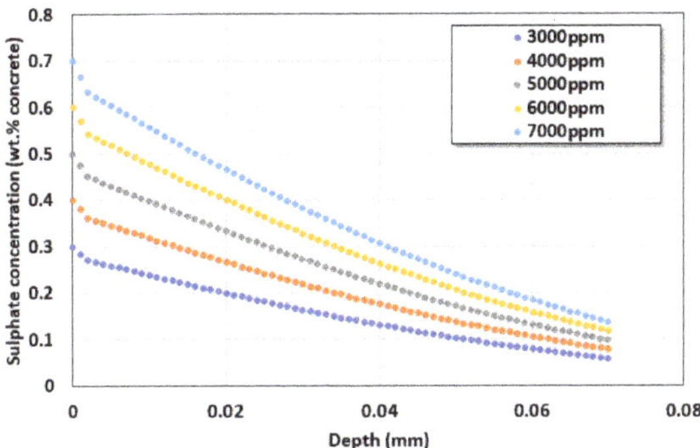

Figure 10. Sulfate ion profile after 30 years.

—Changes in service life due to changes in the diffusivity of the slime layer.

Actually, the sulfate diffusion coefficient of the repair slim layer has the sulfate diffusion characteristics of 0.27~1.31 × 10^{-12} m^2/s depending on the bioactive properties used. In this study, the service life was evaluated while the diffusion coefficient of the repair slime layer was increased from 0.1 × 10^{-12} to 1.2 × 10^{-12} m^2/s. Figures 11 and 12 show the service life according to the diffusion coefficient of the slime layer and the distribution of sulfate ions after 40 years.

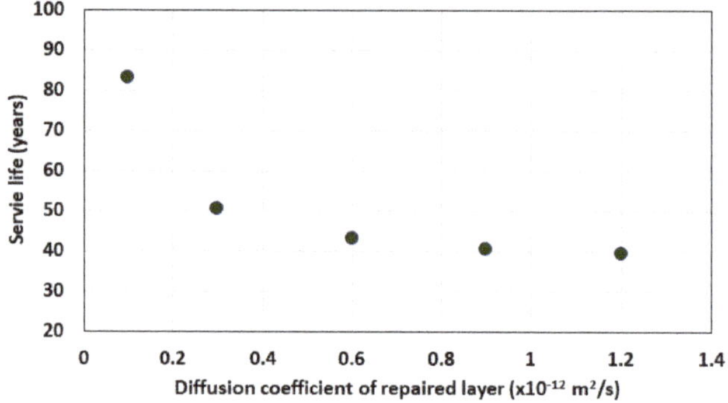

Figure 11. Service life variation with diffusion coefficient in slime coating.

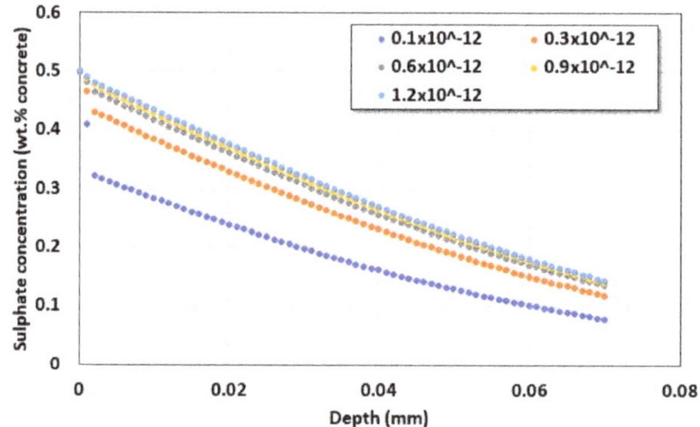

Figure 12. Sulfate ion profile after 40 years.

The analysis results showed that the service life was 83 years when the sulfate diffusion coefficient of the coating layer was 0.1 × 10^{-12} m^2/s, but it decreased to 50.5, 43, 40.5, and 39.5 years as the diffusion coefficient increased to 0.3, 0.6, 0.9, and 1.2 × 10^{-12} m^2/s, respectively.

5. Conclusions

In this study, the service life of a concrete sewage treatment structure was evaluated using the experimental values of the bio-slime coating layer and the diffusion model that considered multi-layers. The derived results are as follows:

(1) The literature survey revealed that the chloride and sulfate diffusion coefficients are proportional to the square root of the molar ratio. The experimental values and the previously proposed values indicated that the ratio of the sulfate diffusion coefficient to the chloride diffusion coefficient ranges from 0.38 to 0.50. In this study, the service life was evaluated using this relationship;

(2) When the Atkinson model was used for the sewage culvert box, its service life was evaluated to be 38.5 years under the conditions of a 60 mm cover thickness and a 5000 ppm sulfate concentration.

When the critical sulfate concentration was assumed to be 0.18%, the service life by the diffusion law was evaluated to be 49.1 years;

(3) The sulfate diffusion coefficient of the slim coating layer that considered the reduced diffusion ratio ranged from 0.27×10^{-12} to 1.31×10^{-12} m^2/s, resulting in the excellent diffusion reduction. When there was no coating layer, the service life was evaluated to be 38.5 years under the conditions that considered concrete properties (30% slag substitution, water to binder ratio of 0.4, and cover thickness of 60 mm). Simple bio-slime coating increased the service life to 41.5–54.3 years. In addition, although the thickness of the slime coating layer was 2.0 mm, the service life could be significantly increased to 50.5–83 years if the sulfate diffusion coefficient of the coating layer could be controlled between 0.1×10^{-12} and 0.3×10^{-12} m^2/s;

(4) The diffusivity of the bio-slime coating materials (EV-immob-polymer) derived by the experiment is approximately 0.3×10^{-12} m^2/s. As bio-slime coating materials are highly resistant to wetland and sulfate exposure conditions, they are expected to be very effective in extending the service life of the existing concrete structures by reducing deterioration.

Author Contributions: Conceptualization, K.-H.Y. and S.-J.K.; formal analysis, S.-J.K.; investigation, K.-Y.Y.; validation, H.-S.L.; writing—original draft preparation, H.-S.L.; writing—review and editing, S.-J.K. and H.-S.L.; funding acquisition, S.-J.K. All authors have read and agreed to the published version of the manuscript.

Funding: This research was supported by the Basic Science Research Program through the National Research Foundation of Korea (NRF) funded by the Ministry of Science, ICT & Future Planning (No. 2015R1A5A1037548) and also supported by a grant (17SCIP-B103706-03) from the Construction Technology Research Program funded by the Ministry of Land, Infrastructure and Transport of the Korean government.

Conflicts of Interest: The authors declare no conflict of interest.

References

1. Song, H.W.; Kwon, S.J. Permeability characteristics of carbonated concrete considering capillary pore structure. *Cem. Concr. Res.* **2007**, *37*, 909–915. [CrossRef]
2. Arya, C.; Buenfeld, N.R.; Newman, J.B. Factors influencing chloride binding in concrete. *Cem. Concr. Res.* **1990**, *20*, 291–300. [CrossRef]
3. Hilsdof, H.K.; Kroff, J. *Performance Criteria for Concrete Durability*; Rilem Report 12; Taylor & Francis: Milton Park, UK, 1995.
4. Kheder, G.F.; Al Rawi, R.S.; Al Dahi, J.K. Study of the behavior of volume change cracking in base-restraint concrete wall. *ACI Mater. J.* **1994**, *91*, 150–157.
5. Tulliani, J.M.; Montanaro, L.; Negro, A.; Collepardi, M. Sulfate attack of concrete building foundations induced by sewage waters. *Cem. Concr. Res.* **2002**, *32*, 843–849. [CrossRef]
6. Santhanam, M.; Cohen, M.D.; Olek, J. Modeling the effects of solution temperature and concentration during sulfate on cement mortars. *Cem. Concr. Res.* **2002**, *32*, 585–592. [CrossRef]
7. Zhang, W.Q.; Liu, L.X.; Dai, D.H. Experimental study of concrete corroding in brine and fresh water under dry-wet circulation. *J. Qinghai Univ.* **2006**, *4*, 25–29.
8. Zhang, M.; Yang, L.M.; Guo, J.J.; Liu, W.L.; Chen, H.L. Mechanical properties and service life prediction of modified concrete attacked by sulfate corrosion. *Adv. Civ. Eng.* **2018**, *2018*, 1–7. [CrossRef]
9. Atkinson, A.; Hearne, J.A. Mechanistic model for the durability of concrete barriers exposed to sulphate-bearing groundwaters. *Mrs Online Proc. Libr. Arch.* **1989**, *176*, 149–156. [CrossRef]
10. Lee, H.J.; Cho, M.S.; Lee, J.S.; Kim, D.G. Prediction model of life span degradation under sulfate attack regarding diffusion rate by amount of sulfate ions in seawater. *Int. J. Mater. Mech. Manuf.* **2013**, *1*, 251–255. [CrossRef]
11. Hadjmeliani, M. Degradation of sewage pipe caused Sinkhole: A real case study in a main road. In Proceedings of the AFM, Association Francaise de Mecanique, Lyon, France, 24–28 August 2015.
12. Nel, D.T.; Haarhoff, J. The failure probability of welded steel pipelines in dolomitic areas. *J. S. Afr. Inst. Civ. Eng.* **2011**, *53*, 9–21.
13. Yang, K.H.; Yoon, H.S.; Lee, S.S. Feasibility tests toward the development of protective biological coating mortars. *Constr. Build. Mater.* **2018**, *181*, 1–11. [CrossRef]

14. Siddique, R.; Chahal, N.K. Effect of ureolytic bacteria on concrete properties. *Constr. Build. Mater.* **2011**, *25*, 3791–3801. [CrossRef]
15. Zhong, W.; Yao, W. Influence of damage degree on self-healing of concrete. *Constr. Build. Mater.* **2008**, *22*, 1137–1142. [CrossRef]
16. Willems, A.; Gillis, M.; De Ley, J. Transfer of rhodocyclus gelatinosus to rubrivivax gelatinosus gen. nov., comb. nov., and phylogenetic relationships with Leptothrix, Sphaerotilus natans, Pseudomonas saccharophila, and Alcaligenes latus. *Int. J. Syst. Evol. Microbiol.* **1991**, *41*, 65–73. [CrossRef]
17. Kwon, S.J. Effect of mineral admixture on CO_2 emissions and absorption in relation to service life and varying CO_2 concentrations. *Int. J. Sustain. Build. Technol. Urban Dev.* **2016**, *7*, 165–173. [CrossRef]
18. Moon, H.Y.; Shin, D.G.; Choi, D.S. Evaluation of the durability of mortar and concrete applied with inorganic coating material and surface treatment system. *Constr. Build. Mater.* **2007**, *21*, 362–369. [CrossRef]
19. Tambe, S.P.; Jagtap, S.D.; Chaurasiya, A.K.; Joshi, K.K. Evaluation of microbial corrosion of epoxy coating by using sulphate reducing bacteria. *Prog. Org. Coat.* **2016**, *94*, 49–55. [CrossRef]
20. Fagerlund, G. Modeling the service life of concrete exposed to frost. In Proceedings of the International Conference on Ion and Mass Transport in Cement-Based Materials, American Ceramic Society, University of Toronto, Toronto, ON, Canada, 4–5 October 1999; pp. 195–217.
21. Lee, B.Y.; Ismail, M.A.; Kim, H.J.; Yoo, S.W.; Kwon, S.J. Numerical technique for chloride ingress with cover concrete property and time effect. *Comput. Concr.* **2017**, *20*, 185–196.
22. Andrade, C.; Diez, J.M.; Alonso, C. Mathematical modeling of a concrete surface "skin effect" on diffusion in chloride contaminated media. *Adv. Cem. Based Mater.* **1997**, *6*, 39–44. [CrossRef]
23. ACI Committee. *Building Code Requirements for Structural Concrete: (ACI 318-99) and Commentary (ACI 318R-99)*; American Concrete Institute: Farmington Hills, MI, USA, 1999.
24. ACI Committee. *ACI Committee 365: Service Life Prediction: State-of-the Art Report*; American Concrete Institute: Farmington Hills, MI, USA, 2000.
25. Mehta, P.K.; Monterio, P. *Concrete: Structure, Properties, and Methods*; Prentice-Hall International Series in Civil Engineering and Engineering Mechanics: Englewood Cliffs, NJ, USA, 1993.
26. Yungsan-Sumjin River Committee. *The Pollution Loadings Characterization by Drain Process Types in Sewer Systems*; Techinical Report; Honam University: Gwangju, Korea, 2007.

Article

Modeling and Design of SHPB to Characterize Brittle Materials under Compression for High Strain Rates

Tomasz Jankowiak [1], Alexis Rusinek [2,3] and George Z. Voyiadjis [4,*]

1 Institute of Structural Analysis, Poznan University of Technology, Piotrowo 5, 60-965 Poznań, Poland; tomasz.jankowiak@put.poznan.pl
2 Laboratory of Microstructure Studies and Mechanics of Materials, UMR-CNRS 7239, Lorraine University, 7 rue Félix Savart, BP 15082, 57073 Metz CEDEX 03, France; rusinek1@univ-lorraine.fr
3 Department of Mechanical Engineering, Chair of Excellence Universidad Carlos III de Madrid, Avda. de la Universidad 30, 28911 Leganés, Madrid, Spain
4 Computational Solid Mechanics Laboratory, Louisiana State University, Baton Rouge, LA 70803, USA
* Correspondence: voyiadjis@eng.lsu.edu

Received: 2 April 2020; Accepted: 6 May 2020; Published: 10 May 2020

Abstract: This paper presents an analytical prediction coupled with numerical simulations of a split Hopkinson pressure bar (SHPB) that could be used during further experiments to measure the dynamic compression strength of concrete. The current study combines experimental, modeling and numerical results, permitting an inverse method by which to validate measurements. An analytical prediction is conducted to determine the waves propagation present in SHPB using a one-dimensional theory and assuming a strain rate dependence of the material strength. This method can be used by designers of new SPHB experimental setups to predict compressive strength or strain rates reached during tests, or to check the consistencies of predicted results. Numerical simulation results obtained using LS-DYNA finite element software are also presented in this paper, and are used to compare the predictions with the analytical results. This work focuses on an SPHB setup that can accurately identify the strain rate sensitivities of concrete or brittle materials.

Keywords: concrete; dynamic compression; Split Hopkinson Pressure Bars (SPHB); brittle materials; simulation

1. Introduction

Protection of buildings and structures in emergency situations is a key issue that needs to be addressed [1]. Prevention in such situations is the most important element of a protection system, especially in cases of critical infrastructure facilities. However, despite precautions, accidents and disasters sometimes occur. These are often of an urgent nature and can result in constructions (or their parts) being subjected to fast dynamic loads that correspond to high energy transfers with short loading times, which can lead to the damage or failure of a structure. Such situations may be related to the occurrence of a sudden load (e.g., a shock wave) caused by the blast of an explosive material or the impact of a bullet or some other object flying at a high velocity [2]. Predicting the effects of such a sudden load is particularly difficult because there is 100% certainty of the scale of damage and failure (and any possible preservation of building integrity) only after it has already occurred. Of course, it is possible in some instances to conduct experimental research on a smaller scale, but this only applies to selected structural elements [3]. Another solution is the use of advanced computer simulations that can accurately determine the dynamic behavior of a building and its structure in the event of a sudden dynamic load. However, the precision of calculations in these cases depend on many factors, such as the methods used in the calculations, how precisely the geometry of the structure under consideration was modeled and the interaction of components and behaviors of the materials from which the analyzed object was built. In the case of building structures, brittle materials such as glass or concrete are also often used for construction. The latter is often used particularly for the construction of reinforced concrete structural elements. The behavior of concrete, especially in the case of dynamic loads when high deformation speeds occur, is very complex [4]. This aspect—namely, how to determine the sensitivity of concrete during dynamic compression and then correctly determine the parameters of a constitutive model that could be used in concrete simulations—will be considered in detail in this work.

A description of dynamic material behavior is generally difficult to assess in the case of brittle materials. Different experimental techniques must be coupled in order to cover such a wide range of strain rates, as shown in Table 1 and reported in [5,6]. For example, using metallic materials or polymers, the specimen must have a length of a few millimeters when using Kolsky bars [6–11]. However, in the case of concrete, the length must be equal to tens of millimeters (~50 mm) in order to obtain a representative macroscopic behavior, due to the microstructure and inclusions size. In this work, the aggregate size is between 0.0075 to 5 mm and corresponds to specific cases of clay, sand and gravel with a maximum diameter of 5 mm. For this reason, in brittle materials such as concrete, strain rates are relatively reduced due to the length of the specimen. Moreover, as the material behavior [12–14] in compression is different in comparison with tensile one [12,15,16], different experimental setups are necessary. In this work, a technique based on dynamic compression is described that allowed the material behavior of brittle materials to be defined for a large range of strain rates varying from 1 to 1000 (s^{-1}). The setup used was based on the Kolsky bar setup [7,8]. The equilibrium of the force impulses on both sides of the specimen during a test is crucial to use the elastic waves theory to determine the macroscopic behavior of the material sandwiched between the two elastic bars.

Table 1. Experimental techniques for specific strain rates.

Experimental Technique	Strain Rates in Metals (s^{-1})	Strain Rates in Brittle Materials (s^{-1})
Servo-hydraulic machines	10^{-6} to 10^{0}	
Specialized machines	10^{0} to 10^{2}	
Conventional Kolsky bar	10^{2} to 10^{4}	10^{1} to 10^{3}
Miniaturized Kolsky bar	10^{4} to 10^{5}	
Plate impact	10^{5} to 10^{7}	

A schematic description of a setup used for the brittle materials such as ceramic [17], glass or concrete is presented in Figure 1. Using this setup, it is possible to define dynamic material behaviors and strengths at high strain rates under compression [7,8]. This work presents an optimal configuration of the device by which to obtain the dynamic compressive strength of concrete at high strain rates close to 1000 (s^{-1}), and this comprised a projectile and a concrete specimen sandwiched between two long elastic bars termed the input and output bars (see Figure 1). During the test, the projectile (striker) impacts the input bar with an initial impact velocity V_0, inducing an incident elastic wave. The compressive incident wave (σ_I, ε_I) then propagates along the input bar with a velocity C_0. When the incident wave reaches the geometrical discontinuity between the input bar and the specimen, one part is reflected (σ_R, ε_R) and one part is transmitted (σ_T, ε_T), due to the difference of the mechanical impedance Z. Using the three wave measurements coupled to the theory of elastic waves [6,7,10,16], the average material behavior $\sigma\left(\varepsilon, \dot{\varepsilon}, T\right)$ and the dynamic compressive strength $f_{cd}\left(\dot{\varepsilon}_d\right)$ for brittle materials can be obtained (this will be described in the next section). It should be noted that the waves were measured in the middle of the bars using resistance gauges to avoid the problem of wave superposition.

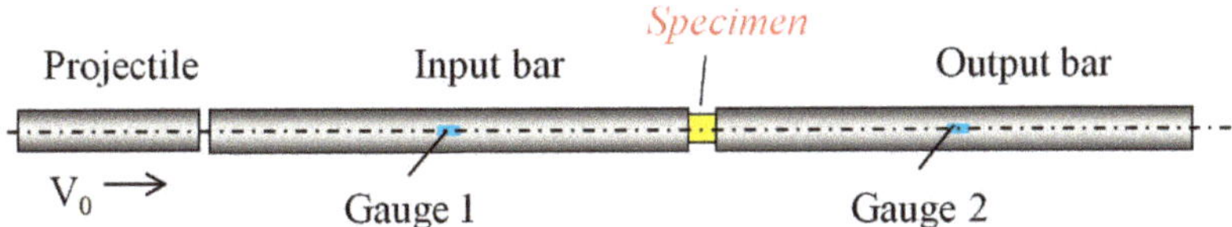

Figure 1. Schematic description of the split Hopkinson pressure bars (SHPBs) for dynamic compression.

A typical three-wave measurement is reported in Figure 2. The pulse shaping method is not used in order to avoid smoothing the elastic wave signal and obtain a representative measurement related to the split Hopkinson pressure bars (SHPBs) [18–21]. It should be noted that a pulse shaper is frequently used to test brittle materials [20,21], as it increases the rising time and causes the strain rate to be both more constant and lower (see Figure 2). A comparison of waves for two examples (with and without a copper shaper) is presented in Figure 2. The material parameters for the copper are taken from [22]. The dimensions of the cylindrical copper shaper are as follows: 20-mm diameter and 1-mm thickness.

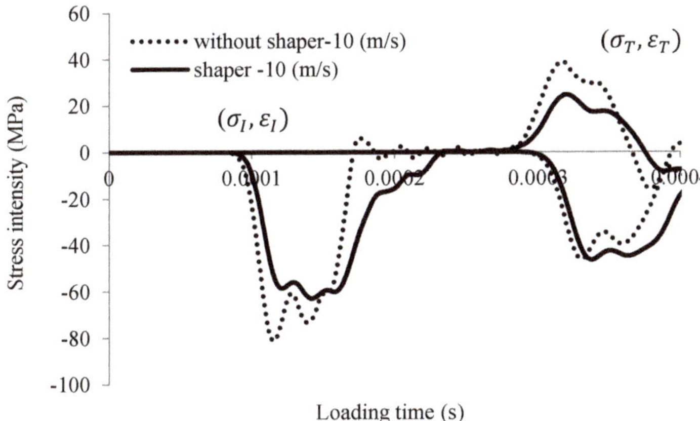

Figure 2. Elastic wave history signal using SHPB (numerical results with and without shaper) for an initial impact velocity of $V_0 = 10\ \text{m/s}$.

According to the CEB (Comité Euro-international du Béton) [23,24], the strength of concrete is highly strain-rate sensitive and can be defined using the compressive dynamic increase factor (CDIF) [23,24] as follows:

$$CDIF(\dot{\varepsilon}_d) = \frac{f_{cd}(\dot{\varepsilon}_d)}{f_{cs}} = \begin{cases} \left(\frac{\dot{\varepsilon}_d}{\dot{\varepsilon}_{cs}}\right)^{1.026\alpha} & if\ \dot{\varepsilon}_d \leq 30\ \text{s}^{-1} \\ \gamma\ \dot{\varepsilon}_d{}^{1/3} & if\ \dot{\varepsilon}_d > 30\ \text{s}^{-1} \end{cases}. \tag{1}$$

In Equation (1), f_{cd} is the dynamic compressive strength of concrete for a strain rate $\dot{\varepsilon}_d$ imposed on the material. The value $\dot{\varepsilon}_{cs} = 0.00003\text{s}^{-1}$ is defined as a reference strain rate [23,24], and corresponds to the value used to obtain the quasi-static compressive strength of concrete f_{cs}. The parameters γ and α are defined by the CEB with $\gamma = 10^{6.156\alpha - 0.49}$ and $\alpha = 1/(5 + 9f_{cs}/10)$. The values proposed by the CEB correspond to the experimental data reported by Bischoff and Perry (1991) [25]. The compressive dynamic increase factor (CDIF) is a function of strain rate (according to the CEB), as reported in Figure 3.

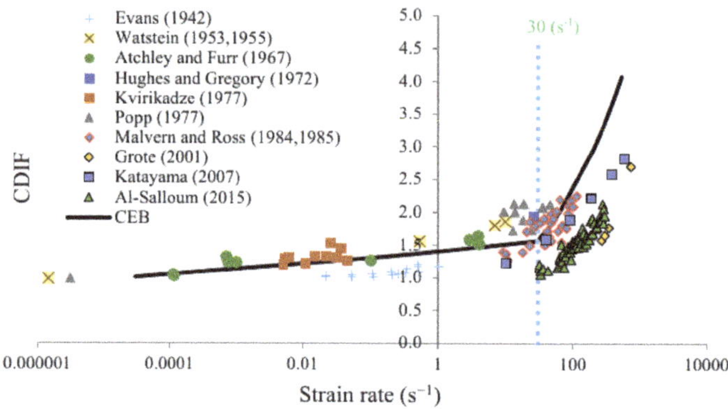

Figure 3. The compressive dynamic increase factor (CDIF) for concrete as a function of the strain rate.

Additionally, experimental data have also been reported to define the strain rate sensitivity of concrete under compression [25], and data expressing different trends other than those identified by the CEB can be seen in Figure 3 [24,26,27]. Despite these results, the authors assume in this paper that concrete under dynamic compression behaves in accordance with CEB recommendations, which does not limit the whole analysis, because this trend is only one example. This is particularly important when using the presented analysis to study the dynamic strength of modern high-strength concrete or other brittle-like materials. The effect of strain rate sensitivity on the dynamic strength of concrete is quite high during compression when compared with its effect in other materials [6,11,21,22]. As a result, it is important to consider it to properly estimate the dynamic behavior of structures subjected to extreme impulsive loading [23].

2. SHPB Technique for Concrete—Analytical Description

To understand the real experimental measurements for the dynamic compression of concrete (a highly strain-rate-sensitive material), a new analytical technique should be used in comparison with compression of the metals [22]. The new analytical description should take into account the strain rate sensitivity of the material during the splitting of the incident wave into reflected and transmitted waves. The procedure used to estimate mechanical properties and errors during experiments may be estimated by coupling predictions based on Equation (2) with a full 3D numerical model that considers a nonlinear process not considered in the simple elastic wave theory. Assuming no dispersions of the elastic waves due to the Pochhammer–Chree effect related to the bar geometry (by considering an instantaneous rising time, which is not the case during experiments), the average mechanical behavior (stress–strain curve) and the compressive strength are defined using Equations (2) coupled o the three waves described previously in Figure 2:

$$\dot{\varepsilon}(t) = \frac{2C_A}{L_C}\varepsilon_R(t) \tag{2-a}$$

$$\varepsilon(t) = \frac{2C_A}{L_C}\int_0^t \varepsilon_R(t)dt \tag{2-b}$$

$$\sigma(t) = E_A\left(\frac{r_A}{r_B}\right)^2\varepsilon_T(t) \tag{2-c}$$

All the necessary quantities including material parameters and geometry dimensions are defined explicitly in Tables 2 and 3. Analyzing the previous equations, the average stress in the specimen $\sigma(t)$ (Equation (2-c)) may be defined using the transmitted wave $\varepsilon_T(t)$. In addition, the reflected wave $\varepsilon_R(t)$ is used to calculate the strain rates $\dot{\varepsilon}(t)$ (Equation (2-a)) and strain $\varepsilon(t)$ (Equation (2-b)) induced to the concrete specimen with time. During the experiment, it was clear that the strain rate was non-constant and depended on the hardening and strength of the material. As the deformation of concrete or other brittle materials is smaller than in metals or polymers, a short projectile L_P can be used. It should be noted that the projectile length is proportional to the loading time of the specimen $t_{loading} = 2L_P/C_A$. To make clearer all quantities defined for the calculations, the following indexes are used: aluminum (A), concrete (C) and steel (S), as reported in Table 2. Moreover, for concrete testing, the material used to design the bars and the projectile is frequently made of an aluminum alloy due its low mechanical impedance Z (Z_A = 38.9 kg/s) (Equation (3-a)):

$$Z_A = AC_A\rho_A \tag{3-a}$$

$$Z_C = AC_C\rho_C \tag{3-b}$$

Table 2. Material parameters of the experimental setup from Figure 1.

Material	Value (Units)
Aluminum alloy	
Young's modulus, E_A	70,000 (MPa)
Density, ρ_A	2700 (kg/m^3)
Elastic wave speed, $C_A = \sqrt{E_A/\rho_A}$	5091.8 (m/s)
Concrete C30/37	
Young's modulus, E_C	26,357 (MPa)
Density, ρ_C	2450 (kg/m^3)
Elastic wave speed, $C_C = \sqrt{E_C/\rho_C}$	3280.0 (m/s)

Table 3. Geometry dimensions of the experimental setup from Figure 1.

Part	Value (Units)
Bars and projectile	
Input and output bar length, L_A	1 (m)
Length of the projectile, L_P	0.15 (m)
Radius of the projectile and bars, r_A	0.023 (m)
Specimen	
Length of the specimen, L_C	0.05 (m)
Radius of the specimen, r_C	0.02 (m)

The value obtained for aluminum is close to the one for the concrete specimen (Z_C = 22.7 kg/s) (Equation (3-b)). The value is obtained assuming a circular cross section $A = \pi r^2$ with a radius of r = 30 mm. For comparison, the mechanical impedance of a steel bar is equal to Z_S = 114 kg/s. Thus, the impedance of concrete is 1.71 times smaller than the impedance of an aluminum alloy and 5.04 times smaller than steel. Additionally, a subscript C means concrete, *A* refers to aluminum and *S* is steel. Therefore, the amplitude of the transmitted wave is reduced if steel bars are used to test the concrete material [19].

The incident stress σ_I and the elastic strain intensity ε_I can be determined if the initial impact velocity and the physical parameters of the bars are known [11,22]:

$$\sigma_I = \frac{\rho_A C_A V_0}{2} \text{ and } \varepsilon_I = \frac{\sigma_I}{E_A} \tag{4}$$

The strain and stress values depend on the material properties of the input bar and of the projectile and its initial impact velocity V_0. The intensity of the transmitted wave ε_T is calculated using Equations (1) and (5):

$$\sigma_T = f_{cd}\left(\dot{\varepsilon}_d\right)\left(\frac{r_c}{r_A}\right)^2 \text{ and } \varepsilon_T = \frac{\sigma_T}{E_A} \tag{5}$$

Finally, the reflected wave ε_R is calculated as the difference between the incident (Equation (4)) and transmitted (Equation (5)) waves, assuming force equilibrium:

$$\sigma_R = \sigma_I - \sigma_T \text{ and } \varepsilon_R = \frac{\sigma_R}{E_A} \tag{6}$$

The average strain rate and strain induced to the concrete specimen can be calculated as follows:

$$\dot{\varepsilon}_d = \frac{2C_A}{L_C}\varepsilon_R \text{ and } \varepsilon_d = \dot{\varepsilon}_d \Delta t \tag{7}$$

The set of Equations (1)–(7) were solved considering the geometry of the bars, the projectile, the specimen and the initial impact velocity of the projectile. The values f_{cd}, $\dot{\varepsilon}_d$, ε_d, $(\sigma_I, \varepsilon_I)$, $(\sigma_R, \varepsilon_R)$ and $(\sigma_T, \varepsilon_T)$ were calculated using the Newton–Raphson iterative algorithm.

Based on the previous analytical description, the material parameters and the geometry of the setup and specimen (Tables 2 and 3, respectively), it can be observed that the incident stress intensity increased with the initial impact velocity (Figure 4a). As reported before, at 30 m/s the stress level of the compressive wave was about 206 MPa (Figure 4a). This value must be smaller than the yield stress of the aluminum to avoid plastic deformation. The stress intensity in the specimen and its dynamic strength were calculated based on the transmitted wave amplitude using Equation (2-c). Equation (5) was used to predict the average value of the stress and strain if the transmitted wave in the middle of the output bar is known. Moreover, the reflected wave was used to calculate the strain rate and the strain in the specimen using Equation (2-a,b). To predict the average strain rate during the test, Equations (6) and (7) were used, and showed an increase in the strain rate tied to the initial impact velocity. As an example, for an initial impact velocity of V_0= 30 m/s, the stress intensity of the reflected wave was equal to 125 MPa, while the strain rate was close to 362 (s^{-1}) (Equations (6) and (7), respectively).

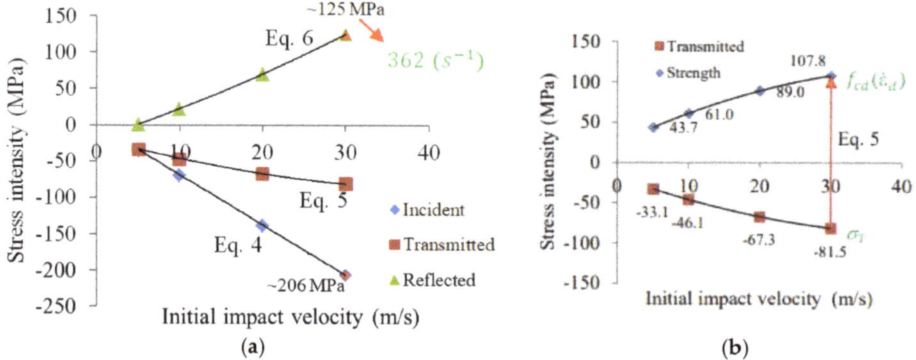

Figure 4. (**a**) Dependency of the incident, transmitted and reflected stress intensity for different initial impact velocities; (**b**) relation between the dynamic concrete strength and the intensity of the transmitted wave.

Both trends shown in Figure 4a are nonlinear (i.e., transmitted wave and reflected wave; Equations (5) and (6), respectively). Only the intensity of the incident wave is linear (Equation (4)). It should be added that the limit of this simplified analysis is the yield stress of the input bar material. It is also important to note that real maximum stress intensities of the incident wave were higher during short periods of time than the theoretical one calculated for the Pochhammer–Chree effect [6,11,22].

The following section concerns designing the experimental setup, and it addresses what to change in order to obtain the maximum possible strain rate from the experiments.

It is well known that increasing the bar radius and assuming a constant specimen radius (20 mm in this work) increases the strain rate, as shown in Figure 5. For example, assuming a bar radius of 23 mm (1.15 times larger than the concrete specimen radius), the maximum strain rate is about 531 (s^{-1}), while the maximum strain rate is about 714 (s^{-1}) for a radius of 43 mm (2.15 times larger than the concrete specimen radius) (Figure 5). However, this technique of changing the bar diameter is not the most appropriate way to increase the strain rate. Based on the previous results, Figure 5, it is observed that the rising strain rate is faster with the diameter bar increase allowing to reach a bigger value for the strain rate for an imposed impact velocity ($\sim 714\ s^{-1}|_{\varnothing=43\ mm}^{40\ m/s}$ and $\sim 531\ s^{-1}|_{\varnothing=23\ mm}^{40\ m/s}$). Moreover, it is possible to increase the strain rate by changing the initial length of the specimen. Based on the current configuration (SPHB with a bar diameter of 23 mm), the strain rate varies from 531 to 815 (s^{-1}) if the specimen length decreases from 50 to 30 mm (Figure 6). By coupling the changing of both variables (using a bar radius of 43 mm and a short specimen of 30 mm), it is possible to reach a maximum strain rate close to 1166 (s^{-1}). Concerning the analytical approach, the dynamic strength of concrete according to CEB was assumed to estimate the transmitted and reflected wave intensities. For a strain rate equal to 1166 (s^{-1}), the CDIF is equal to 5.3 (strength 159 MPa). However, the size cannot decrease continuously for concrete, since the material behavior must be representative and the aggregate sizes must be considered. Thus, the specimen must be designed considering the representative elementary volume (REV) [28,29].

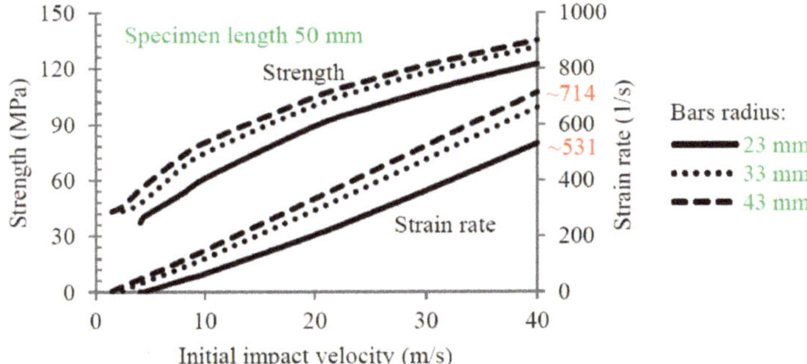

Figure 5. The effect of the initial impact velocity on the strength and strain rate for different radii of the transmitted bar (23, 33 and 43 mm).

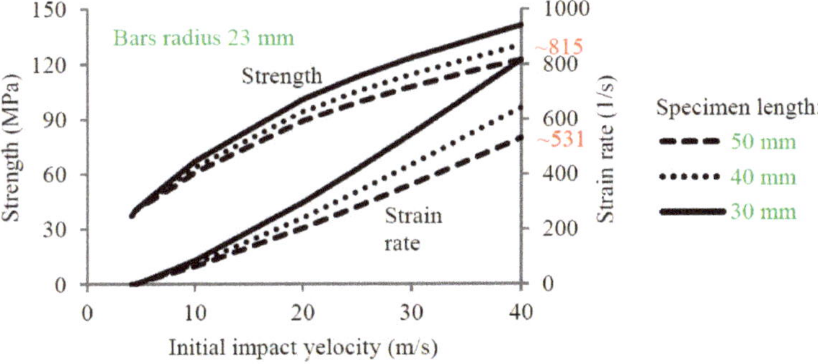

Figure 6. Effect of the initial impact velocity on the strength and strain rate for different lengths of the specimen (50, 40 and 30 mm).

The above calculations and analysis should be used to design the setup geometry and its configuration in order to obtain the correct and expected stress levels or strain rates during experiments. It can be useful to test specific concrete types such as ultra-high strength concrete (UHSC) with a static compressive strength close to 100 MPa. The proposed analytical approximation can be used with success to predict the effects of a strong strain rate sensitivity on the material behaviors of concrete or other brittle materials in SHPB testing.

3. Simulation of the SPHB Technique for Material Characterization of Concrete at High Strain Rates

The following numerical simulations present how it is possible to describe the dynamic compressive strength and strain rate of concrete during dynamic failure using the SHPB technique. Previously neglected effects (dispersion of the wave and loading time) were revealed in the results of a full three-dimensional model. The simulation corresponding to the process of elastic wave propagation and related to the SHPB technique was performed using an explicit integration scheme in LS-DYNA [30]. All parts of the SPHB setup were considered: projectile, input and output bars, as well as the specimen. Their geometry and dimensions are presented in Figure 1 and reported in Table 3. The model was discretized by hexagonal eight-node finite elements constant stress, with a total number of 50,000 elements (Figure 7). The finite element length in the projectile, bars and specimen was 0.006 m, while the fine mesh size was 0.003 m. The simulations using the above model considered all additional effects [22,31], including the punching effect and geometrical dispersion of the waves. The elastic properties are reported in Table 2. In addition, in order to define the behavior of the specimen, a continuous-damage surface cap model was used [32–34]. This advanced model for concrete considers strain rate sensitivity. Contact with a friction coefficient equal to zero was assumed between all parts of the SHPB. The results of the simulation were intended for comparison with the simplified theory (Section 1), which is why a no-friction condition was assumed. In general, the authors acknowledge that friction conditions are very important in many dynamic tests (e.g., with SPHB for metals [22]). As an example, using the configuration described in this paper and considering friction values equal to $\mu = 0.6$ and $\mu = 0$, a stress increase of approximately 7% and a strain rate decrease of approximately 5% could be observed.

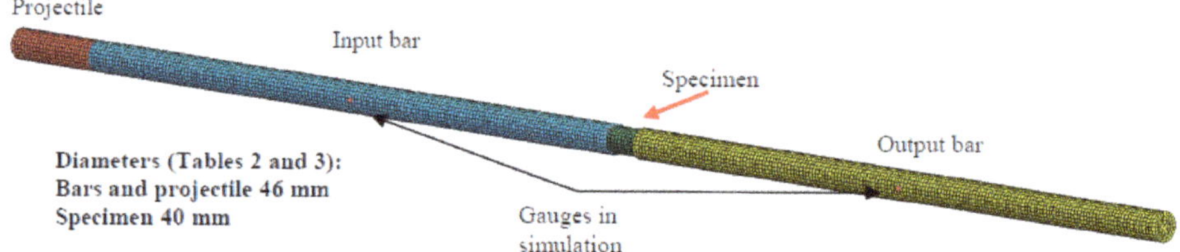

Figure 7. Discretization of the finite element model to simulate dynamic compression of concrete.

In addition, the continuous-damage surface cap model was used to simulate the concrete behavior and its properties at high strain rates. The parameters were obtained by calibrating a typical concrete; namely, C30 grade. A detailed description of the model is reported in [32–34]. In this paper, the assumptions of the model are presented and reported in Appendix A only considering the material parameters discussed in Section 3.

Based on the previous numerical solutions, experiments were mimicked and the same values were measured as the three elastic waves propagating along the two elastic bars. Before using the previous analysis (Equation (2-a–c)), the key point was to demonstrate that force equilibrium was reached during dynamic loading. In the following curves, 30 m/s was reported for impact velocity and 0.05 m was reported for length of the specimen in regards to the three waves after the shifting time to zero, Figure 8a. Based on this, a comparison can be done between the input and output forces. It can be observed that there was good agreement corresponding to force equilibrium (Figure 8b). It should be noted that force equilibrium was also reached also for other cases, including a shorter specimen with a length of 0.025 m.

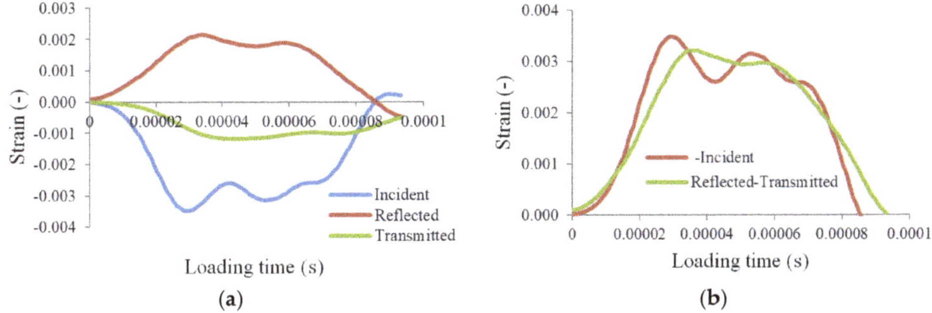

Figure 8. (**a**) Elastic waves measured during the dynamic compression of concrete with an impact v elocity 30 m/s (length of the specimen 0.05 m); (**b**) force equilibrium, input and output forces.

Once the force equilibrium has been reached and achieved a conservation of energy and quantity of movement, the theory of elastic waves may be used to described the material behavior of concrete under dynamic loading. Even if the transmitted wave seems large, the strain applied stays relatively small, as shown in Figure 9. The strain was less than 2%, which demonstrates the brittle behavior of concrete.

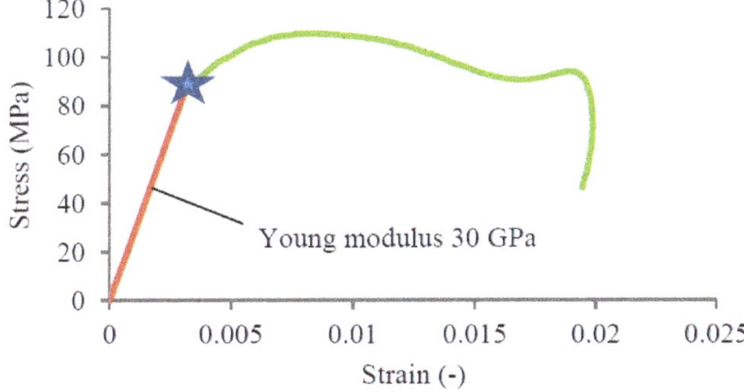

Figure 9. Material behavior description of concrete under dynamic compression based on the process of elastic wave propagation with an impact velocity of 30 m/s (strain rate 435 (s^{-1})).

The material behavior (Figure 9), which was defined by applying the elastic waves theory to the elastic waves propagating along the bars (Figure 8), was in agreement with previous experiments performed on concrete (e.g., [24,35,36]).

4. Parametric Study Concerning the Main Crucial Material Parameters

Concerning all assumptions described in Appendix A, the most important to consider in this analysis is the strain rate sensitivity of concrete under compression. Additionally, the damage mechanism and regularization process by the fracture energy will be discussed and presented in the subsequent sections.

4.1. Analysis of Strain Rate Sensitivity in Compression

The main parameters allowing the strain rate sensitivity of concrete under dynamic compression to be described are η_{0c} and N_c (see Appendix A). The usual values assumed for C30 concrete are 1.003×10^{-4} 1/s and 0.78, respectively [32,33]. Using these values and simulating material behaviors under dynamic compression, it can be observed that the predicted dynamic strength of concrete is below the CEB recommendation (Figure 10) [23,25]. Thus, the dynamic behavior of concrete has been underestimated in compression at high strain rates [25]. The numerical model has shown a trend during experiments [24,26,27] that is different from the recommendation of the CEB (see Figure 3). The main parameters used to model the behavior of C30 concrete under dynamic compression are presented in Table 4.

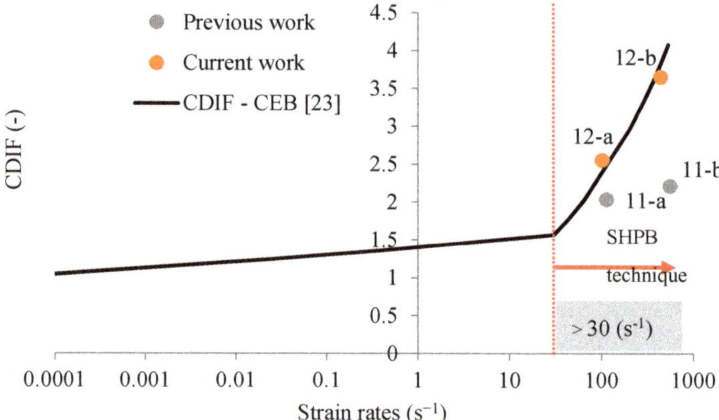

Figure 10. Comparison of the CEB recommendations with the strain rate sensitivity predicted for concrete by the continuous damage surface cap model.

Table 4. Main parameters used to predict the dynamic properties of concrete.

Parameter	Value (Units)
Articles [32,33]	
Fluidity in compression, η_{0c}	0.0001003 (s^{-1})
Power in compression, N_c	0.78 (-)
Current analysis	
Fluidity in compression, η_{0c}	0.00012 (s^{-1})
Power in compression, N_c	0.58 (-)

The real strain rate sensitivity presented in Figure 10 must be used to correctly simulate the material behavior in order to later define the response of a structure designed with concrete. A more appropriate approximation by which to estimate the real behavior of concrete under dynamic compression for a large range of strain rates using the SHPB technique can be obtained for $\eta_{0c} = 1.2 \times 10^{-4}$ 1/s and $N_c = 0.58$ (Figure 10). The CEB approximation function is identified based on experimental results reported in Figure 10. In addition, the strain rates for which the SHPB technique is valid are shown. Points 11-a, 11-b, 12-a and 12-b in Figure 10 are obtained based on the numerical results defined in Figure 11a,b and Figure 12a,b. The strain rate and the stress level correspond to the maximum value of the signal reached on time.

The numerical simulations of the dynamic compression of concrete were analyzed in detail for two initial impact velocities: 10 and 30 m/s. The results of the numerical simulations were compared with the CEB [23], which was the best fit for several experimental results.

To avoid plastic deformation of the bars and to have pure elastic wave propagation, the yield stress of a bar was assumed equal to 250 MPa (see Figure 4a). Simulations were performed for both sets of constitutive parameters presented in Table 4. The results of the previous parameters for both impact velocities are presented in Figure 11, and for the results of the current set of parameters are shown in Figure 12. The wave intensity from the simulations in term of stress in the middle of the input and output bars (lines "Input (Sim)" and "Output (Sim)", respectively) are compared to the analytical predictions described previously assuming an ideal rectangular elastic wave pulse (called

"Analytical"). For the plots in Figures 10 and 11, the stress history in the middle of the specimen (line "Specimen (Sim)") is also shown based on the simulations. The dashed lines (which describe the dynamic strength according to the CEB, based on the previous analysis) are also presented (line "Strength (Approx)"). At a lower initial impact velocity of 10 m/s, the strength of concrete agreed for both sets of parameters (Table 4 and Figure 11a,b). For higher initial impact velocities, only the new set of parameters described in this current work allowed results to be obtained that were in agreement with the CEB recommendations (Figure 12b).

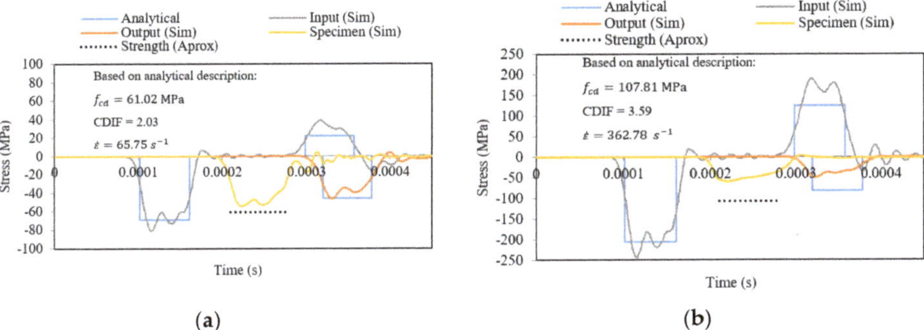

Figure 11. Waves in simulations of dynamic compression using SHPB for parameters from the previous work [32,33]; (**a**) 10 m/s, (**b**) 30 m/s.

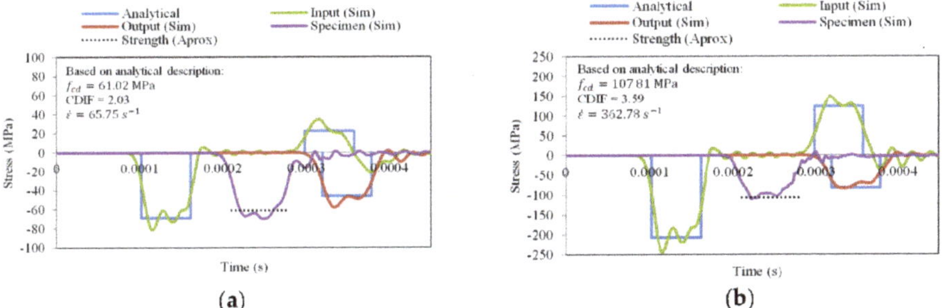

Figure 12. The waves in simulation of dynamic compression using SHPB for parameters from the current work; (**a**) 10 m/s, (**b**) 30 m/s.

The analysis showed that material model parameter identification should be conducted using the geometry and complexity of the whole SPHB experimental setup (including the projectile, bars and specimen) in cases of brittle materials with a high strain rate sensitivity.

4.2. Analysis of Mesh Size Sensitivity

The pathological influence of finite element discretization (mesh size) is an important problem during concrete simulation due to the softening effect (decreasing of the stress with strain) [37–39]. This problem has been investigated, and several results and studies have been reported in several articles and books. The two methods of regularization for this phenomenon have been well acknowledged (on the level of mathematics or numerical formulation). The first group of regularization methods has included strain rate-dependent models in which the stress state depends on the speed of deformation [40,41]. This method is mainly used under dynamic loadings. This group also includes the Cosserat and micropolar models, which are used for soil and granular media descriptions [42]. To model damage and failure in concrete, non-local models may be used [43,44], as well as higher-order gradient models [45–50]. Non-local models have introduced an averaging function that converts local variables into non-local ones according to a specific weight function. There was a possibility to regularize the problem at the level of numerical formulation by explicitly introducing the width of plastic zone deformation localization in the finite element [51]. In addition, automatic re-meshing of finite elements has been used based on a local error caused by large gradients of internal variables. Discontinuity of the displacement field inside the finite element has also been introduced [52], and this has led to regularizing the problem. All the aforementioned methods of regularization have been introduced into the material model with an internal characteristic length scale.

For the calculations presented in this article, the authors used a material model with a fracture damage energy coupled to visco-plasticity (see Equations (A1)–(A9) in Appendix A) to regularize the solutions [33], and the important material parameters G_{fc} and B (the fracture energy in uniaxial compression and the compression shape softening parameter, respectively) were considered. The values of these parameters for C30 are 6.838 MPa·mm and 100, respectively [32,33]. The simulations were done for two different finite element meshes. The coarse one was previously presented in Figure 7, and a fine one corresponded to the one-half element size. Finally, the size of the mesh in concrete was 6 or 3 mm (fine mesh), while it was 6 mm in the bars and projectile. The numerical results in term of elastic waves along the input and output bars are presented in Figure 13.

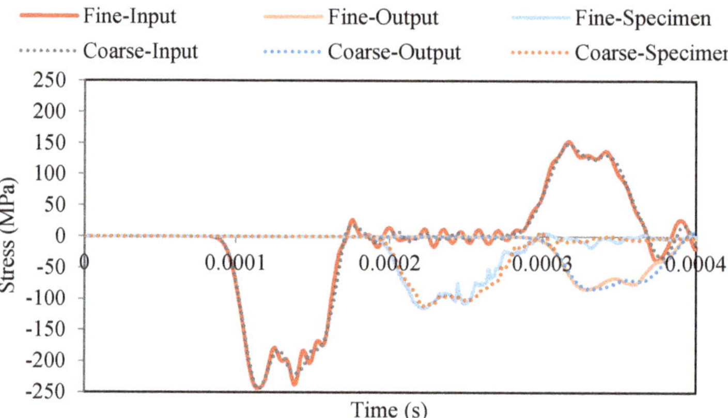

Figure 13. Mesh size sensitivity in a simulation of dynamic compression using SHPB for an initial impact velocity equal to 30 m/s.

The mesh size sensitivity of the numerical results is acceptable (small enough) to assume that the results were not mesh size dependent. Based on these curves the behavior (stress-strain and strain-rate curves) for both cases was also similar. The model combined with the analyzed material model behavior was correct, and a mesh size dependency was not observed.

4.3. Analysis of the Fracture Energy in Compression Sensitivity

The stress-strain curve may vary for different kinds of concrete. The slope of the curve in the unloading part is crucial. It is important to know whether the curve can be measured using the SPHB technique. The main parameter measured up to now has been strength. However, the addition of some components as aggregates or fiber reinforcements in the concrete can change other mechanical properties. A parametric study was conducted to prove that potential changes in the softening curve shape of a material can be defined and observed during dynamic testing. The analysis was conducted by considering the changes of the two material parameters G_{fc} and B (Table 5). During the simulations the values from Table 5 were used: Example 1 concerned the default values of both parameters; in Example 2, the value of the fracture energy in uniaxial compression G_{fc} was divided in two; in Example 3, the value was defined using a default fracture energy, but the compressive shape softening parameter B was 10 times smaller; and finally, in Example 4, both changes were used in the simulations.

Table 5. Main parameters used to predict the fracture energy sensitivity.

Examples	Parameter	Value (Units)
Example 1 (default)	Fracture energy in uniaxial compression, G_{fc}	6.838 (MPa·mm)
	Compressive shape softening parameter, B	100 (-)
Example 2	Fracture energy in uniaxial compression, G_{fc}	3.419 (MPa·mm)
	Compressive shape softening parameter, B	100 (-)
Example 3	Fracture energy in uniaxial compression, G_{fc}	6.838 (MPa·mm)
	Compressive shape softening parameter, B	10 (-)
Example 4	Fracture energy in uniaxial compression, G_{fc}	3.419 (MPa·mm)
	Compressive shape softening parameter, B	10 (-)

The results corresponding to all previously described examples are reported in Figure 14a,b. For better visibility, only the transmitted and reflected waves are presented in the figure. It is clear that by decreasing the fracture energy G_{fc} and the compressive shape softening parameter B, the material starts to become more brittle.

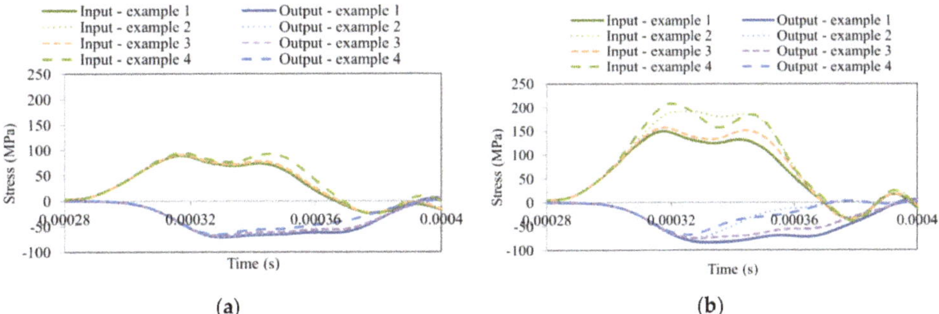

Figure 14. Parametric study concerning G_{fc} and B on the process of dynamic compression using SHPB for an initial impact velocity equal to (**a**) 20 m/s and (**b**) 30 m/s.

All effects are reported in Figure 14 in terms of elastic wave propagation. As can be observed, the material behavior definition based on the analysis of wave propagation changed depending on the parameters used (Table 5). The incident wave was not affected, since it is related to the initial velocity, the length of the projectile and the mechanical properties of the input bar. The analysis has shown that if the softening of the specimen changes (increasing: material becomes more brittle; decreasing: material becomes more ductile) then it is possible to see the effect on the elastic waves. The effect is visible on the transmitted and the reflected waves. The behavior in Example 4 is more brittle because the stress intensity is lower than in Example 1. This is observable especially for a higher impact velocity (30 m/s; Figure 14b) than for a lower impact velocity (20 m/s; Figure 14a). This is the main reason that the stress intensity of the reflected wave (strain rate of the test) was higher.

5. Conclusions

The dynamic behavior of concrete during impact or a blast is very often analyzed using numerical simulations. During these kinds of loadings, high strain rates are reached and observed in the material. To predict the material's behavior and dynamic strength, very precise tests and dynamic measurements are necessary, as has been discussed in this paper. Experimental results are then used to calibrate the concrete material model parameters. If the initial boundary value problem is used to simulate the dynamic behavior of the structure with the correct material model, its prediction will agree with the experimental observations. The following main concepts and results are presented in detail in this article:

- An analytical solution to predicting stress and strain wave intensities was presented. This could be used to simplify the design process of SHPB and check consistency in further experimental test results.
- The effect of the initial impact velocity of the projectile on the strength and strain rate reached in the specimen was determined for different bar diameters.
- A method to calibrate the material model for concrete including strain rate sensitivity was presented. A numerical simulation was used to find a correct value of the parameters that define the strain rate sensitivity. As discussed, the original parameters have very low values of dynamic strength for compressed concrete in comparison with the analytical solution.
- The presented analysis proved that the solution was not sensitive to mesh size. The important aspect is that possible changes in fracture energy during compression or the shape of the softening (descending) part of the curve can be identified using this experimental technique.
- This work assumes that the concrete specimen is in equilibrium during the simulations, and that the friction coefficient has limited influence on the final results.

For each case considered in this work, the previous dynamic experimental results recommended by the CEB as CDIF were compared with the numerical results. This work may be used to summarize the design process of SHPB for concrete to reach a certain strain rate. Extension of the analysis to other classes of concrete (concrete C30 was assumed) or glass is possible using the same procedure. These new materials (e.g., ultra-high performance concrete) can also be tested using the setup described herein. However, a limitation is imposed not to exceed the yield stress of the Hopkinson bars.

Author Contributions: Conceptualization, T.J. and A.R.; methodology, T.J.; software, T.J.; validation, T.J., A.R. and G.Z.V.; formal analysis, T.J.; investigation, T.J.; resources, T.J.; data curation, T.J.; writing—original draft preparation, T.J. and A.R.; writing—review and editing, T.J. and A.R.; visualization, T.J.; supervision, T.J. and A.R.; project administration, T.J.; funding acquisition, T.J. All authors have read and agreed to the published version of the manuscript.

Conflicts of Interest: The authors declare no conflict of interest.

Appendix A

The following explanation of the constitutive model for concrete is presented mainly based on the literature [32–34]. The authors present only the most important assumptions of the model discussed and analyzed in Section 3 of this paper.

The first assumption is that shear and the bulk moduli allow one to describe the elastic state of the material considered. The plasticity model depends on the first invariant of the stress tensor J_1, as well as on the second J_2'. For triaxial compression, the plastic potential is limited by the cap hardening function. The yield surface describes the shear failure limited to torsion and triaxial tension by the Rubin scaling function [32–34].

Using a visco-plastic formulation, it is possible to consider the strain rate sensitivity of a material. Thus, a visco-plastic stress tensor without damage σ_{ij}^{vp} is defined as follows:

$$\sigma_{ij}^{vp} = (1-\gamma)\sigma_{ij}^{T} + \gamma\sigma_{ij}^{p} \tag{A1}$$

where σ_{ij}^{T} is the trial elastic stress and σ_{ij}^{p} is the inviscid plastic stress tensor. The viscous variable γ is calculated using Equation (A2):

$$\gamma = \frac{\Delta t/\eta}{1+\Delta t/\eta} \tag{A2}$$

where η is the effective fluidity coefficient, which is calculated independently in compression (or in shear) and in tension using the following formulation:

$$\begin{aligned} \eta &= \eta_s + \left(\frac{J_1}{\sqrt{3J_2'}}\right)^{pwrc} (\eta_c - \eta_s) \quad \text{compression} \\ \eta &= \eta_s + \left(\frac{-J_1}{\sqrt{3J_2'}}\right)^{pwrt} (\eta_t - \eta_s) \quad \text{tension} \end{aligned} \tag{A3}$$

where η_t, η_c and η_s are calculated as follows:

$$\eta_t = \frac{\eta_{0t}}{\dot{\varepsilon}^{N_t}}, \eta_c = \frac{\eta_{0c}}{\dot{\varepsilon}^{N_c}} \text{ and } \eta_s = S_{rate}\eta_t \tag{A4}$$

where the material parameters η_{0t}, η_{0c}, N_t, N_c, S_{rate}, $pwrc$ and $pwrt$ describe the strain rate sensitivity of the material model, and the variable $\dot{\varepsilon}$ describes the effective strain rate in the material.

Using the continuous damage surface cap model [32–34], the scalar damage variable d is used to describe the stress tensor with damage σ_{ij}^d based on the visco-plastic stress tensor without damage σ_{ij}^{vp} as follows:

$$\sigma_{ij}^d = (1-d)\sigma_{ij}^{vp} \tag{A5}$$

Two independent damage mechanisms exist related to compression (or shear) and tension. Both are initiated by plasticity, since the initial damage surface coincides with the plastic shear function [33]. Therefore, two energetic criteria are used to describe the damage initiation (Equation (A6)) and its evolution (Equation (A7)).

The damages are initiated when the energy-type terms τ_c (in compression) and τ_t (in tension) exceed a damage threshold defined as r_{0c} or r_{0t}. The first invariant of the stress tensor J_1 is related to the compressive or tensile state, as follows:

$$\begin{aligned} \tau_c &= \sqrt{\tfrac{1}{2}\sigma_{ij}\varepsilon_{ij}} \quad \text{if} \quad \begin{cases} J_1 & \text{compression} \\ \tau_c \geq r_{0c} & \text{energy} \end{cases} \\ \tau_t &= \sqrt{E\varepsilon_{max}^2} \quad \text{if} \quad \begin{cases} J_1 & \text{tension} \\ \tau_t \geq r_{0t} & \text{energy} \end{cases} \end{aligned} \tag{A6}$$

The damage initiations are dependent on the elastic-plastic stress tensor σ_{ij} and strain tensor ε_{ij}, or to the maximum main strain ε_{max} for the case corresponding to tension. The damage thresholds are calculated using the internal procedure of LS-DYNA.

Two softening functions d are used independently related to compression (or in shear) and tension:

$$\begin{aligned} d(\tau_c) &= \tfrac{d_{max}}{B}\left[\frac{1+B}{1+Be^{-A(\tau_c - r_{0c})}} - 1\right] \quad \text{compression} \\ d(\tau_t) &= \tfrac{0.999}{D}\left[\frac{1+D}{1+De^{-C(\tau_t - r_{0t})}} - 1\right] \quad \text{tension} \end{aligned} \tag{A7}$$

where A and C are equal to the characteristic finite element length l_{ch}. Moreover, the variable A may be reduced as reported in Equation (A8). The internal parameter d_{max} is the maximum damage level:

$$A = A(d_{max} + 0.001)^{pmod} \tag{A8}$$

where parameters B, D and $pmod$ describe the shape of the softening functions in compression and in tension.

This model allows one to assure the uniqueness of the solution, which does not depend on mesh size and allows the problem to be regularized using material softening. The fracture damage energy G_f depends on the damage mechanism (state of stress):

$$\begin{aligned} G_f &= G_{fs} + \left(\frac{J_1}{\sqrt{3J_2'}}\right)^{pwrc}\left(G_{fc} - G_{fs}\right) \quad \text{compression} \\ G_f &= G_{fs} + \left(\frac{J_1}{\sqrt{3J_2'}}\right)^{pwrt}\left(G_{ft} - G_{fs}\right) \quad \text{tension} \end{aligned} \tag{A9}$$

where G_{fs}, G_{ft} and G_{fc} describe the fracture energy in shear, tension and compression, respectively; the coefficients *pwrc* and *pwrt* are the parameters used to describe the transition from shear to tension and from shear to compression [33], and have the same values as in Equation (A3). The fracture damage energy also depends also on the strain rate. All material parameters for concrete C30 have been previously reported (e.g., [32–34]). In Section 3, only the parameters strongly influencing the results are discussed.

References

1. Gebbeken, N.; Warnstedt, P.; Rüdiger, L. Blast protection in urban areas using protective plants. *Int. J. Prot. Struct.* **2017**, *9*, 226–247. [CrossRef]
2. Hao, Y.; Zhang, X. Aspecial issue on protective structures against blast and impact loading. *Int. J. Prot. Struct.* **2018**, *9*, 3. [CrossRef]
3. Sielicki, P.W.; Ślosarczyk, A.; Szulc, D. Concrete slab fragmentation after bullet impact: An experimental study. *Int. J. Prot. Struct.* **2019**, *10*, 380–389. [CrossRef]
4. Abdel-Kader, M. Numerical predictions of the behaviour of plain concrete targets subjected to impact. *Int. J. Prot. Struct.* **2018**, *9*, 313–346. [CrossRef]
5. Ramesh, K.T. High Rates and Impact Experiments. In *Springer Handbook of Experimental Solid Mechanics*; Sharpe, W.N.J., Ed.; Springer: Berlin/Heidelberg, Germany, 2008.
6. Field, J.E. Review of experimental techniques for high rate deformation and shock studies. *Int. J. Impact Eng.* **2004**, *30*, 725–775. [CrossRef]
7. Kolsky, H. An investigation of the mechanical properties of materials at very high rates of loading. *Proc. Phys. Soc. Sect. B* **1949**, *12*, 676–700. [CrossRef]
8. Hopkinson, B. A method of measuring the pressure in the deformation of high explosives or by the impact of bullets. *Philos. Trans. R. Soc. A* **1914**, *213*, 437–456.
9. Davies, R.M. A critical study of the Hopkinson pressure bar. *Philos. Trans. R. Soc. A* **1948**, *240*, 375–457.
10. Li, Q.M.; Meng, H. About the dynamic strength enhancement of concrete-like materials in a split Hopkinson pressure bar test. *Int. J. Solids Struct.* **2003**, *40*, 343–360. [CrossRef]
11. Jankowiak, T.; Rusinek, A.; Wood, P. Comments on paper: "Glass damage by impact spallation" by A. Nyoungue et al., Materials Science and Engineering A 407 (2005) 256–264. *Mater. Sci. Eng. A* **2013**, *564*, 206–212. [CrossRef]
12. Yang, F.; Ma, H.; Jing, L.; Zhao, L.; Wang, Z. Dynamic compressive and splitting tensile tests on mortar using split Hopkinson pressure bar technique. *Lat. Am. J. Solids Struct.* **2015**, *12*, 730–746. [CrossRef]
13. Piotrowska, E.; Forquin, P.; Malecot, Y. Experimental study of static and dynamic behavior of concrete under high confinement: Effect of coarse aggregate strength. *Mech. Mater.* **2016**, *92*, 164–174. [CrossRef]
14. Mu, Z.C.; Dancygier, A.N.; Zhang, W.; Yankelevsky, D.Z. Revisiting the dynamic compressive behavior of concrete-like materials. *Int. J. Impact Eng.* **2012**, *49*, 91–102. [CrossRef]
15. Kupfer, H.; Hilsdorf, H.K.; Rusch, H. Behavior of concrete under biaxial stresses. *J. Am. Concr. Inst.* **1969**, *66*, 656–666.
16. Harding, J.; Wood, E.D.; Campbell, J.D. Tensile testing of material at impact rates of strain. *J. Mech. Eng. Sci.* **1960**, *2*, 88–96. [CrossRef]
17. Subhash, G.; Ravichandran, G. Split-Hopkinson Pressure Bar Testing of ceramics. In *ASM Handbook, Volume 08—Mechanical Testing and Evaluation*; Kuhn, H., Medlin, D., Eds.; ASM International: Cleveland, OH, USA, 2000.

18. Xia, K.; Yao, W. Dynamic rock tests using split Hopkinson (Kolsky) bar system—A review. *J. Rock Mech. Geotech. Eng.* **2015**, *7*, 27–59. [CrossRef]
19. Baranowski, P.; Janiszewski, J.; Małachowski, J. Study on computational methods applied to modelling of pulse shaper in split—Hopkinson bar. *Arch. Mech.* **2014**, *66*, 429–452.
20. Frew, D.J.; Forrestal, M.J.; Chen, W. Pulse shaping techniques for testing brittle materials with a split Hopkinson pressure bar. *Exp. Mech.* **2002**, *42*, 93–106. [CrossRef]
21. Pająk, M.; Janiszewski, J.; Kruszka, L. Laboratory investigation on the influence of high compressive strain rates on the hybrid fibre reinforced self-compacting concrete. *Constr. Build. Mater.* **2019**, *227*, 116687. [CrossRef]
22. Jankowiak, T.; Rusinek, A.; Łodygowski, T. Validation of the Klepaczko-Malinowski model for friction correction and recommendations on Split Hopkinson Pressure Bar. *Finite Elem. Anal. Des.* **2011**, *47*, 1191–1208. [CrossRef]
23. Committee Euro-International du Beton (CEB). *Concrete Structures under Impact and Impulsive Loading in: CEB Bulletin d'information*; Committee Euro-International du Beton (CEB): Lausanne, France, 1988.
24. Al-Salloum, Y.; Almusallam, T.; Ibrahim, S.M.; Abbas, H.; Alsayed, S. Rate dependent behavior and modeling of concrete based on SHPB experiments. *Cem. Concr. Compos.* **2015**, *55*, 34–44. [CrossRef]
25. Bischoff, P.H.; Perry, S.H. Compressive behavior of concrete at high strain-rates. *Mater. Struct.* **1991**, *24*, 425–450. [CrossRef]
26. Grote, D.L.; Park, S.W.; Zhou, M. Dynamic behavior of concrete at high strain-rates and pressures: I. Experimental characterization. *Int. J. Impact Eng.* **2001**, *25*, 869–886. [CrossRef]
27. Katayama, M.; Itoh, M.; Tamura, S.; Beppu, M.; Ohno, T. Numerical analysis method for the RC and geological structures subjected to extreme loading by energetic materials. *Int. J. Impact Eng.* **2007**, *34*, 1546–1561. [CrossRef]
28. Drugan, W.J.; Willis, J.R. A micromechanics-based nonlocal constitutive equation and estimates of representative volume element size for elastic composites. *J. Mech. Phys. Solids* **1996**, *44*, 497–524. [CrossRef]
29. Gambin, B. *Wpływ Mikrostruktury na Własności Kompozytów Sprężystych, Piezoelektrycznych i Termosprężystych*; IPPT Reports on Fundamental Technological Research, 12; IPPT PAN: Warszawa, Poland, 2006; pp. 1–183.
30. Livermore Software Technology Corporation. *LS-DYNA Keyword User's Manual, (LSTC)*; Livermore Software Technology Corporation: Livermore, CA, USA, 2019.
31. Zhong, W.Z.; Rusinek, A.; Jankowiak, T.; Abed, F.; Bernier, R.; Sutter, G. Influence of interfacial friction and specimen configuration in Split Hopkinson Pressure Bar system. *Tribol. Int.* **2015**, *90*, 1–14. [CrossRef]
32. Jankowiak, T.; Łodygowski, T. Smoothed particle hydrodynamics versus finite element method for blast impact. *Bull. Pol. Acad. Sci. Tech. Sci.* **2013**, *61*, 111–121. [CrossRef]
33. U.S. Department of Transportation. *User's Manual for LS-DYNA Concrete Material Model 159, Public. Number FHWA-HRT-05-062*; U.S. Department of Transportation: Washington, DC, USA, 2007.
34. Jiang, H.; Zhao, J. Calibration of the continuous surface cap model for concrete. *Finite Elem. Anal. Des.* **2015**, *97*, 1–19. [CrossRef]
35. Guo, Y.B.; Gao, G.F.; Jing, L.; Shim, V.P.W. Response of high-strength concrete to dynamic compressive loading. *Int. J. Impact Eng.* **2017**, *108*, 114–135. [CrossRef]
36. Pajak, M.; Janiszewski, J.; Kruszka, L. Behavior of concrete reinforced with fibers from end-of-life tires under high compressive strain rates. *Eng. Trans.* **2019**, *67*, 119–131.
37. Voyiadjis, G.Z.; Taqieddin, Z.N.; Kattan, P.I. Theoretical Formulation of a Coupled Elastic-Plastic Anisotropic Damage Model for Concrete using the Strain Energy Equivalence Concept. *Int. J. Damage Mech.* **2009**, *18*, 603–638. [CrossRef]
38. Needleman, A. Material rate dependence and mesh sensitivity in localization problems. *Comput. Methods Appl. Mech. Eng.* **1988**, *67*, 69–85. [CrossRef]
39. De Borst, R.; Sluys, L.J.; Muhlhaus, H.B.; Pamin, J. Fundamental issues in finite element analyses of localization of deformation. *Eng. Comput.* **1993**, *2*, 99–121. [CrossRef]
40. Perzyna, P. The Thermodynamical Theory of Elasto-Yiscoplasticity (Review Paper). *Eng. Trans.* **2005**, *53*, 235–316.
41. Abed, F.H.; Voyiadjis, G.Z. A consistent modified Zerilli–Armstrong flow stress model for BCC and FCC metals for elevated. *Acta Mech.* **2005**, *175*, 1–18. [CrossRef]

42. Tejchman, J.; Gudehus, G. Shearing of a narrow granular strip with polar quantities. *Int. J. Numer. Anal. Methods Geomech.* **2001**, *25*, 1–18. [CrossRef]
43. Pijaudier-Cabot, G.; Bažant, Z.; Tabbara, M. Comparison of various models for strain-softening. *Eng. Comput.* **1988**, *5*, 141–150. [CrossRef]
44. Bobinski, J.; Tejchman, J. Modelling of size effects in concrete using elasto-plasticity with non-local softening. *Arch. Civ. Eng.* **2006**, *52*, 7–35.
45. De Borst, R.; Pamin, J.; Geers, M. On coupled gradient dependent plasticity and damage theories with a view to localization analysis. *Eur. J. Mech.—A/Solids* **1999**, *18*, 939–962. [CrossRef]
46. Pamin, J. Gradient plasticity and damage models: A short comparison. *Comput. Mater. Sci.* **2005**, *32*, 472–479. [CrossRef]
47. Voyiadjis, G.Z.; Song, Y. Strain Gradient Continuum Plasticity Theories: Theoretical, Numerical and Experimental Investigations. *Int. J. Plast.* **2019**, *55*. [CrossRef]
48. Voyiadjis, G.Z.; Song, Y. Effect of Passivation on Higher Order Gradient Plasticity Models for Non-proportional Loading: Energetic and Dissipative Gradient Components. *Philos. Mag. Struct. Prop. Condensed Matter* **2017**, *97*, 318–345. [CrossRef]
49. Al-Rub, R.K.A.; Voyiadjis, G.Z. Gradient-enhanced Coupled Plasticity-anisotropic Damage Model for Concrete Fracture: Computational Aspects and Applications. *Int. J. Damage Mech.* **2009**, *18*, 115–154. [CrossRef]
50. Cicekli, U.; Voyiadjis, G.Z.; Al-Rub, R.K.A. A Plasticity and Anisotropic Damage Model for Plain Concrete. *Int. J. Plast.* **2007**, *23*, 1874–1900. [CrossRef]
51. Pietruszczak, S.; Mróz, Z. Finite element analysis of deformation of strain softening materials. *Int. J. Numer. Methods Eng.* **1981**, *17*, 327–334. [CrossRef]
52. Melenk, J.; Babuska, I. The Partition of Unity Finite Element Method: Basic Theory and Applications. *Comput. Methods Appl. Mech. Eng.* **1996**, *39*, 289–314. [CrossRef]

Article

Effect of Pre-Wetted Zeolite Sands on the Autogenous Shrinkage and Strength of Ultra-High-Performance Concrete

Guang-Zhu Zhang and Xiao-Yong Wang *

Department of Architectural Engineering, Kangwon National University, Chuncheon-Si 24341, Korea; zhangks@kangwon.ac.kr

* Correspondence: wxbrave@kangwon.ac.kr; Tel.: +82-33-250-6229

Received: 13 April 2020; Accepted: 18 May 2020; Published: 20 May 2020

Abstract: In this study, the carrier effect of zeolite sands in reducing the autogenous shrinkage and optimizing the microstructure of ultra-high-performance concrete (UHPC) is studied. Pre-wetted calcined zeolite sand (CZ), calcined at 500 °C for 30 min, and natural zeolite sand (NZ), with 15 wt.% and 30 wt.% in UHPC, are used to partially replace standard sands. On that basis, a series of experiments are executed on the developed UHPC, including compressive strength, autogenous shrinkage, X-ray diffraction (XRD), and isothermal calorimetry experiments. With the increase of the zeolite sand content, the autogenous shrinkage of UHPC decreases gradually. Moreover, when the added CZ content is 30 wt.% (CZ30 specimen), it is effective in reducing autogenous shrinkage. Meanwhile, at the age of 28 days, the compressive strength of CZ30 is 97% of the control group. In summary, it is possible to effectively reduce the autogenous shrinkage of UHPC containing 30 wt.% CZ, without sacrificing its mechanical properties.

Keywords: calcined zeolite sand; ultra-high-performance concrete; pre-wetted; autogenous shrinkage; internal curing

---

1. Introduction

In recent years, due to the increasing demand for concrete, ordinary and high-performance concrete (HPC) may not meet actual engineering needs. Thus, ultra-high-performance concrete (UHPC) is of increasing interest to researchers. Compared with ordinary concrete, UHPC shows a more dense microstructure and excellent compressive strength, because it contains a large amount of cement-based materials, has a low water/binder (w/b) ratio, and uses a large number of superplasticizers [1,2]. In addition, the durability of ultra-high-performance concrete has been significantly improved, so UHPC could be used under more severe conditions. However, it must be considered that, due to the influence of a range of factors—such as a high cementitious material content and low w/b ratio—UHPC experiences autogenous shrinkage at an early age [3]. Compared with ordinary concrete, UHPC is more prone to cracking in the early stage [4]. The influence of early cracking, caused by restrained autogenous shrinkage, even limits the application of UHPC [5,6]. Therefore, solving the problem of the autogenous shrinkage of UHPC is urgently required.

Researchers have studied the following five methods to reduce the occurrence of the autogenous shrinkage of UHPC: (1) the control of the hydration reaction; (2) reduction of the internal restraint; (3) reduction of the surface tension of the pore solution; (4) formation of expansive products; and (5) replenishment of water through internal curing. Of these five methods, internal curing is considered to be the most effective and straightforward method [7]. Superabsorbent polymer (SAP) [8,9] and lightweight aggregates (LWA) [10,11] are considered to be two common internal curing materials. While SAP is very effective in reducing autogenous shrinkage, it also introduces additional pores into the concrete structure due to its water swelling effect, resulting in a reduction in the compressive strength of the concrete structure [12]. It was confirmed by Sun et al. [13] that the incorporation of SAP reduced the compressive strength, and the compressive strength decreased with the addition of SAP. In the study of Farzanian et al. [14], it was also concluded that the addition of SAP reduces the compressive strength of cement slurry in cement pastes with a high density of macrovoids. However, in other studies of concrete mixed with SAP, some researchers have come to the opposite conclusion. Bentz et al. [15] measured compressive strength development in experiments carried out for mortar mixes with w/b = 0.35, with and without SAP (0.4% relative to binder mass). After 28 days of curing, mortar with SAP showed higher compressive strength than the reference mortar; the values were 73 and 61 MPa, respectively. Similarly, Woyciechowski and Kalinowski [16] studied the influence of the dosing method of SAP on the effectiveness of the concrete. It was found that the 28-day compressive strength of concrete with an activated small particle size (150–850 μm) of SAP was higher than that of the control specimen. In summary, the effect of SAP on strength depends on the system being studied. Compared with SAP, LWA can reduce the autogenous shrinkage of concrete, but whether the compressive strength of the concrete structure is reduced depends on the type of LWA material [17]. Wang et al. [18] studied the effects of three different types of pre-wetted LWA (fly ash-clay ceramsite, shale ceramsite, and clay ceramsite) on the compressive strength and shrinkage of concrete and found that LWA will reduce the compressive strength of concrete. Liu et al. [10] also studied the effect of saturated coral aggregate (SCA) with UHPC and found that, although the autogenous shrinkage is reduced, the mechanical properties are also lost. The most widely used materials for LWA are different types of sand. However, the LWA material used in this paper is zeolite sand. Zeolite sand not only can effectively change the mechanical properties of concrete structures, but is also an environmentally friendly material that can be used for gas purification [19,20]. While ordinary natural zeolite sand can absorb a certain amount of water due to its fine pore structure, it has difficulty in completely releasing most of the absorbed water. The crystal structure of zeolite sands can be destroyed by heat treatment to significantly increase their water absorption capacity, while the zeolite sand porosity is changed by particle agglomeration [21]. According to the literature [22], increasing the water absorption of zeolite sand by calcination is very effective. Zhang et al. [23] used calcined zeolite particles with an average size of 0.18 mm as the internal curing agent of high-strength concrete, and it was confirmed that the calcined zeolite increased the internal relative humidity of the concrete and reduced the shrinkage. Zhang et al. [24] also applied pre-wetted calcined zeolite particles in a high-strength engineered cementitious composite, and more than 60% of autogenous and/or drying shrinkage at 28 days was reduced while the strength of the composite remained as high as the reference specimen. Some zeolite powders were also used to mitigate the autogenous shrinkage of concrete. Pezeshkian et al. [25] studied the effect of different percentages of silica fume replaced with natural zeolite powder (25%, 50%, 75%, and 100%) on the autogenous shrinkages and mechanical properties of UHPCs. It showed that with UHPC in which 50% in volume of natural zeolite was used as a substitute for silica fume, the 28- and 90-day compressive strengths were only slightly lower than that of reference specimen. Meanwhile, an increasing number of researchers have found that fine lightweight aggregates can react. Li et al. [26] analyzed the pore solution and found that the expanded shale and clay can reduce the alkalinity, as well as increase the aluminum content, in the pore solution. Suraneni et al. [27] found that finely ground lightweight aggregate is pozzolanic and participates in the hydration. In this paper, we also found that zeolite sand is not inert.

In practical engineering applications, most construction workers are highly interested in maintaining a similar compressive strength while reducing the autogenous shrinkage of concrete. The aim of this paper is to use pre-wetted zeolite sand to replace a portion of the standard sand in UHPC in order to significantly reduce the autogenous shrinkage of UHPC, without a significant loss of strength caused by adding the zeolite sand. The research methods implemented include water absorption, autogenous shrinkage, compressive strength, X-ray diffraction, and isothermal calorimetry tests.

The innovations of this paper are summarized as follows: Firstly, we find the use of 30 wt.% calcined zeolite sand can reduce the autogenous shrinkage of UHPC without reducing its compressive strength. Secondly, we find zeolite sand is not chemically inert, and the dissolution of fine zeolite sand particles changes the alkali ion concentration of the solution, accelerates the early binder hydration, and promotes the setting of the binder. Finally, the calcination can increase the water-absorbing capacity of zeolite sand and the internal curing effect of zeolite sand is enhanced after calcination.

2. Materials and Methods

2.1. Materials and Sample Preparation

In this paper, zeolite sands with minimum and maximum particle sizes of 1 mm and 3 mm, respectively, were used as internal curing agents of UHPC. To observe the influence of calcined zeolite sand on UHPC, natural zeolite sand particles were heated for 30 min in a muffle furnace at 500 °C. Figure 1 shows calcined zeolite sand after being calcined by a muffle furnace. The binder materials included Type I ordinary Portland cement (OPC) and silica fume (SF).

(a) (b)

Figure 1. Zeolite sands used to replace standard sand: (**a**) natural zeolite sand; (**b**) calcined zeolite sand.

The chemical compositions of cement, silica fume, and zeolite sand, measured by X-ray fluorescence (XRF), are shown in Table 1.

Table 1. Chemical compositions of cement, silica fume, and zeolite sand

Materials	SiO_2	Al_2O_3	Fe_2O_3	CaO	MgO	SO_3	ZnO	K_2O	P_2O_5	Loss
Cement (%)	21.65	5.57	2.45	62.68	2.60	2.34	0.11	1.08	0.10	0.46
Silica fume (%)	93.80	0.93	0.56	0.518	0.66	0.23	0.16	1.76	0.08	0.63
Zeolite sand (%)	65.35	13.58	1.61	1.95	1.35					15.8

The XRF fused cast bead method (ISO 12677:2003) was used in the sample preparation (XRF-Scientific, Tokyo, Japan). Samples of lithium tetraborate and lithium metaborate (each ~4 g) and a test sample (~0.8 g) were mixed for preparing the beads. The resistance furnace was heated to a fixed temperature of 1025 ± 25 °C [28].

Microscopic observation of the sheet samples was performed using a high-resolution field emission scanning electron microscope (S-4300, Hitachi, Tokyo, Japan) with an acceleration voltage of 1.5 kV and an emission current of 7000 nA. The zeolite sands were broken, and the inner sheet was selected for testing. Prior to microscopic observation, the surface of the samples was coated with platinum for 30 min using a Hitachi E-1010 ion sputterer (Tokyo, Japan) [28]. As shown in Figure 2, compared with natural zeolite sand, calcined zeolite sand has more pores and even a greater porosity distribution. A fine aggregate (standard sand), with a maximum particle size of 3.0 mm, was used. The SEM images of the natural zeolite sand and calcined zeolite sand are shown in Figure 2.

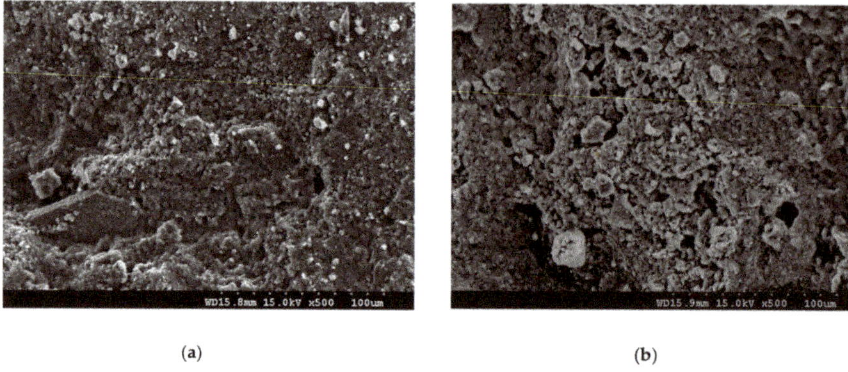

(a) (b)

Figure 2. SEM images: (**a**) natural zeolite sand; (**b**) calcined zeolite sand.

To achieve the objectives of this study, two sets of tests were conducted. In the first test group, the natural zeolite sand was used as the internal curing agent, and the additional amounts were 15 wt.% and 30 wt.% of standard sand, respectively. The UHPC specimens are marked as NZ15 and NZ30. In the second test group, calcined zeolite sand was used as an internal curing agent, and the additional amounts ware 15 wt.% and 30 wt.% of standard sand, respectively. The UHPC specimens are marked as CZ15 and CZ30. The water absorbed by the zeolite sands, before setting, has no effect on the porosity of the concrete and is therefore not considered to be part of the 'water/cement ratio' of a particular concrete mixture [29]. The mixture proportions of the UHPC specimens are shown in Table 2.

Table 2. Mix proportions of ultra-high-performance concretes (UHPCs)

Number	Binder / %		Sand/ Binder	Water/ Binder	Zeolite Sand/ Sand %	Superplasticizer/ Binder %
	Cement	Silica Fume				
Z0	85	15	1.5	0.2	0	3
NZ15	85	15	1.5	0.2	15	3
NZ30	85	15	1.5	0.2	30	3
CZ15	85	15	1.5	0.2	15	3
CZ30	85	15	1.5	0.2	30	3

This paper focuses on the internal curing effect of zeolite sand on strength and autogenous shrinkage. Since steel fiber is mainly used to improve the ductility of UHPC, and ductility is not the aim of this paper, we did not use fiber.

Figure 3 shows detailed information on the mixing procedure, adopted to produce self-curing UHPCs.

The natural zeolite sand and calcined zeolite sand, soaked in water for 48 h, were taken out. The soaked zeolite sands were unfolded with a paper towel, and a paper towel was placed on the surface of the zeolite sands to dry the surface of the zeolite sands [30]. The above steps were repeated until a color change in the paper towel could not be visually detected. At this point, it was assumed that the zeolite sands had reached a surface drying state [31]. All binders, standard sand, and zeolite sands are placed in a mortar mixer and stirred at a low speed for 30 s. Water, uniformly mixed with SP, was added to the solid material and stirred at a low speed for 150 s. The process was stopped for 30 s, and the powder that was not sufficiently contacted on the paddles and inner wall of the mortar mixer was scraped. The mixture was further stirred at a high speed for 180 s, and then the mortar was poured into a mold for shaping and curing.

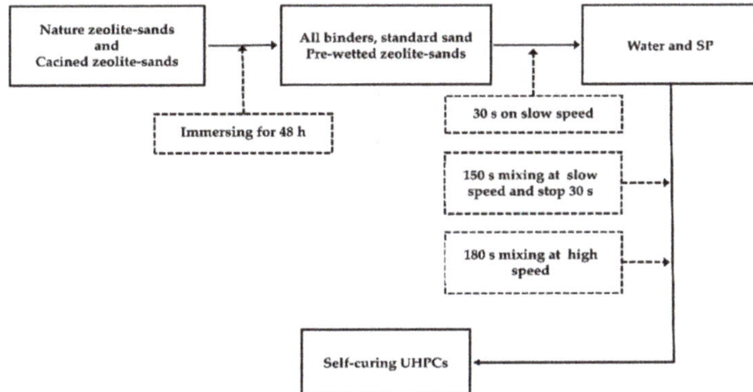

Figure 3. Detailed information on the mixing procedure adopted to produce self-curing UHPCs.

2.2. Test Methods

2.2.1. Water Absorption

Since the absorption of water by the zeolite sand specimens occurs over time, to determine the absorption of water changing with time, the volumetric flask test was used to determine the saturation of the zeolite sand specimens for 24 h. The specimens were placed in an oven at 105 °C for 24 h and cooled for another 24 h. The measuring method of water absorption was the same as that used in [32]. Of the sample, 100 g was put into a 250 mL volumetric flask. Deionized water was added to about 80% of the volumetric flask, which was then stirred manually for 2–3 min to remove bubbles between the zeolite sands. After stirring, and after the zeolite sand particles were submerged, the horizontal position of the liquid in the volumetric flask was observed, and extra water was added to make the water level in the volumetric flask reach its pre-marked standard horizontal position. As the deionized water was continuously absorbed by the zeolite sand particles, the level of the liquid decreased. Within the specified time intervals (approximately 10 min, 20 min, 30 min, 1 h, 3 h, 4 h, 5 h, 6 h, 24 h, 36 h, and 48 h), additional water was added to the flask to ensure that the liquid level reached the standard calibration. The total quality and actual measurement time were recorded. It should be noted that the flask needed to be stirred for about 30 s to remove the bubbles between the zeolite sand particles before each addition of water.

2.2.2. Autogenous Shrinkage

A corrugated tube method, based on ASTM C1698–09 [33], was used to measure the linear autogenous shrinkage of cement mortars. A corrugated tube with concrete was placed on the steel support, and one end of the corrugated tube was absorbed by a magnet. The other end was in contact with a linear variable differential transducer (LVDT) to measure the deformation throughout the experiment. The LVDT collected data through computation every 10 min. The whole setup was maintained in a constant temperature humidity chamber at 20 °C, until the test age of 7 days. It should be noted that in order to avoid a large quantity of entrapped air in the test specimens, mixtures were slowly poured into the corrugated mold with the vibrating table turned on.

2.2.3. Compressive Strength Test

The compressive strength of the designed UHPC was tested (Shin Gang, Seoul, Korea), based on the ASTM C39 [34]. The cement mortars were manufactured as a standard specimen, with a size of 50 mm × 50 mm × 50 mm, and the compressive strengths at 3, 7, and 28 d were measured. All specimens were sealed and cured at a temperature of 20 °C.

2.2.4. X-Ray Diffraction

X-ray powder diffraction (XRD) analysis was conducted on the raw materials (natural zeolite sand and calcined zeolite sand) and the samples (after 28 days of curing) after the hydration stoppage treatment. The solvent exchange method was used for hydration stoppage of the hydrated samples [35]. We ground the 5 g of hydrated samples into a powder. The powder was then immersed in 150 mL of isopropanol (CP grade) for 15 min. The suspension was vacuum filtered, rinsed once with isopropanol, and rinsed twice with ether (CP grade). The filtered solids were dried in a drying oven at 40 °C for 8 min. Finally, the dried powder was stored in a vacuum dryer. Testing of samples should be performed as soon as possible, preferably within two days. The raw materials and samples were ground until the sample surface was flat and smooth. The sample surface area and thickness was large enough to avoid beam overflow and sample transparency aberrations [36]. XRD was carried out using a Panalytical X'pert-pro MPD diffractometer (Panalytical, Almelo, The Netherlands), with CuKα radiation (λ = 1.5404 Å) [37]. The measurements were conducted in the 2θ range of 5°–70°, with a 2θ scan rate of 0.1°/min.

2.2.5. Isothermal Calorimetry

In previous studies, most of the specimens for isothermal calorimetry were pastes, without a fine aggregate. These studies ignored the effect of fine aggregates on the heat of hydration. Conversely, in this study, to clarify the effect of fine zeolite aggregates on the heat of hydration, a mortar specimen with a fine zeolite aggregate was used for isothermal calorimetry. The heat release rate and accumulated hydration heat of the adhesives were measured using a TAM-Air (TA Instruments, New Castle, DE, USA) for 72 h at 20 °C. Due to the low water binders (w/b = 0.2), all mortars were mixed by a mortar mixer and transferred to glass bottles [38].

3. Results and Discussions

3.1. Water Absorption

Figure 4 shows the water absorption after the natural zeolite sand and the calcined zeolite sand had been dried for 48 h. Both zeolite sands absorbed most of the water within the first 6 h.

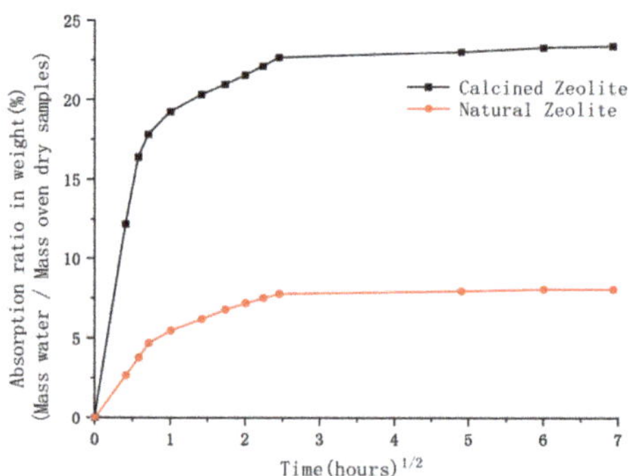

Figure 4. Square root of time vs. water absorption of zeolite sands, before and after calcining.

The water absorption of the natural zeolite sand reached 7.78% in the first 6 h, while the calcined zeolite sand reached 22.67%. The total water absorption increased with time, and the final water absorption of the natural zeolite sand reached 8.08%, while the calcined zeolite sand reached 23.43%. The water absorption of the calcined zeolite sand was higher than that of the natural zeolite sand. This is due to the initial physical bonding water present in the natural zeolite sand being evaporated during heating, resulting in more pores becoming empty. Similar results were obtained in [39]. The experimental results are in good agreement with the SEM images, shown in Figure 2.

3.2. Autogenous Shrinkage

All the mortars are placed in a corrugated tube, and the autogenous shrinkage of the mortars, during the 7-day period, was measured using the deformation measurement setup. The test data are shown in Figure 5.

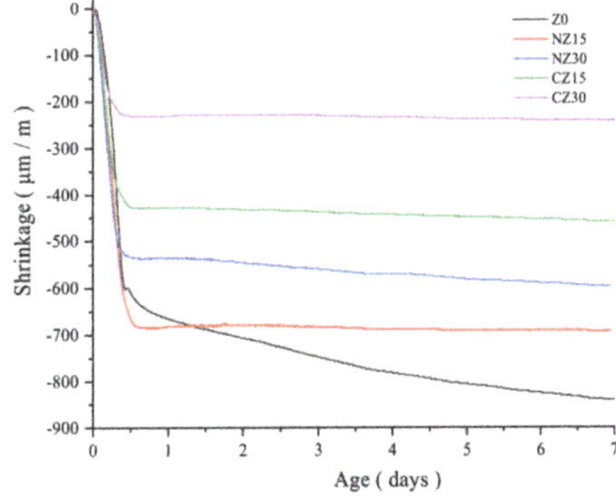

Figure 5. Development of the autogenous shrinkage of all mixtures for 7 days.

The control group is labeled as Z0, and the test pieces with the calcined zeolite sand, containing 15 wt.% and 30 wt.%, are labeled as CZ15 and CZ30, respectively. The test pieces with natural zeolite sand, containing 15 wt.% and 30 wt.%, are labeled as NZ15 and NZ30, respectively. It is observed in Figure 5 that the overall trend is that the shrinkage of the specimens decreases with the increased number of zeolite sand particles. Because zeolite sand absorbs additional water as a self-curing agent, parts of the self-curing agents are released with the binder hydration, and the released internal curing water fills the air space formed inside the UHPC due to chemical shrinkage, reducing the occurrence of autogenous shrinkage strain [40].

First, the shrinkage of Z0 increases rapidly within the first 12 h and slowly increases after 12 h. However, specimens containing calcined or natural zeolite sand remain unchanged after the inflection point. This is because the addition of zeolite sand is equivalent to adding a reservoir to the cement and, with the binder hydration, the internal curing water of zeolite sand is released and continues to participate in the hydration reaction, thus slowing down the occurrence of autogenous shrinkage. Similar results have been confirmed in previous studies [39,41].

Second, calcined zeolite sand is more effective in reducing the autogenous shrinkage of the specimens. The shrinkage of the specimens containing calcined zeolite sand (CZ15 = −458.293 μm/m, CZ30 = −239.947 μm/m) is lower than that of specimens containing natural zeolite sand (NZ15 = −692.952 μm/m, NZ30 = −595.542 μm/m). This means that the modification of zeolites is more helpful in improving the storage capacity of pores. While the specimens contain the same amount of zeolite sand, the calcined zeolite sand absorbed more internal curing water than the natural zeolite sand. The specimens containing the calcined zeolite sand released more internal curing water, and the released internal curing water increased the relative humidity inside the UHPC. Consequently, the shrinkage of the specimens with calcined zeolite sand is smaller than that of the specimens with natural zeolite sand.

3.3. X-Ray Diffraction (XRD)

XRD measurements were conducted on both the natural zeolite sand and calcined zeolite sand, and the experimental results are shown in Figure 6.

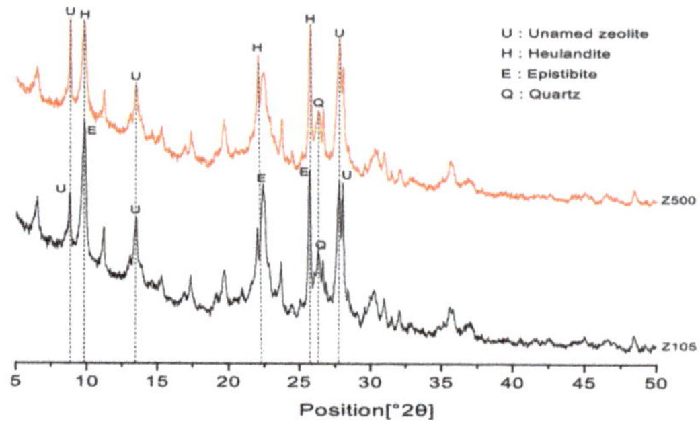

Figure 6. XRD patterns of natural zeolite sand and calcined zeolite sand.

Calcined zeolite sand belongs to the heulandite series, which includes a kind of thermally stable heulandite zeolite ($CaO{\cdot}Al_2O_3{\cdot}7SiO_2{\cdot}5H_2O$), synthesized by $CaO{\cdot}Al_2O_3{\cdot}7SiO_2$ in the temperature range of 250–360 °C. The calcined zeolite sand belongs to the epistilbite series ($CaO{\cdot}Al_2O_3{\cdot}5.5SiO_2{\cdot}5H_2O$), synthesized by the $CaO{\cdot}Al_2O_3{\cdot}5.5SiO_2$ composite [42]. Due to the change of temperature during the calcination process, the crystal shapes of the zeolites are changed, but there is no change in the chemical compositions. Previous studies have shown that calcining natural zeolite will lead to a collapse of crystal structures [43]. There is also an unnamed zeolite phase and mineral impurities, such as quartz.

The XRD patterns of mortars at the curing age of 28 days are presented in Figure 7.

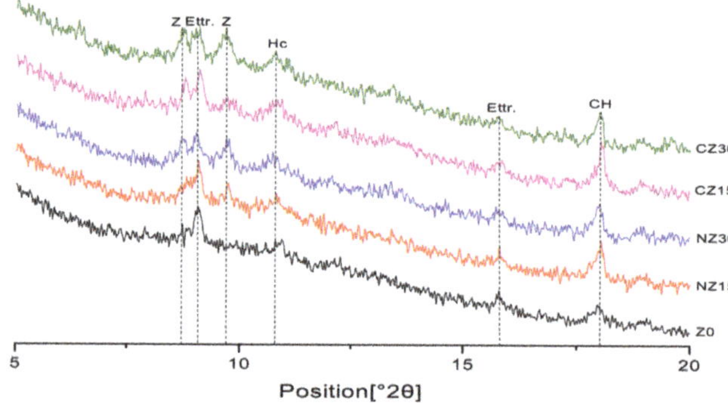

Figure 7. XRD patterns of UHPC casting for 28 days.

Since the diffraction peak of the sand would cause interference, a 2θ degree between 5 and 20° is selected. Compared with the control group, no new hydration products are generated, because no new peaks appeared. This is mainly because zeolite sands, as lightweight aggregates added to concretes, do not participate in a binder hydration reaction.

According to the peak of calcium hydroxide in the Figure 7, compared with the control group, Z0, the peak of calcium hydroxide increased significantly with the addition of zeolite sand.

This is because the specimens containing zeolite sands contain more internal curing water [44]. With the occurrence of the cement hydration reaction, the water in the reservoir is released to further promote cement hydration, and more hydration products are obtained. Since zeolite sands are placed into the system as lightweight aggregates, the zeolite sands themselves do not participate in the early cement hydration reactions. The $Ca(OH)_2$ peak of UHPC, containing calcined zeolite sand, is higher than that containing natural zeolite sand, mixed with the same content. This is because the calcined zeolite sand can absorb more internal curing water, which is later released to further participate in a hydration reaction. More hydration products are produced, leading to an increase in the peaks of the hydration products. The difference between the corresponding $Ca(OH)_2$ peaks in CZ15/30 and those in NZ15/30 are not obvious. XRD qualitative analysis method can determine the type and relative quantity of hydration productions, but it cannot determine the specific amount of hydration products. Therefore, the XRD qualitative analysis is not the best method of analyzing the degree of hydration. TGA can be used as a quantitative analysis method to measure the content of chemically bound water and calcium hydroxide. Compared with XRD qualitative analysis, TGA is a much better way to determine the degree of hydration, and a TGA experimental study will be conducted in the future.

3.4. Isothermal Calorimetry

All mortars were mixed by a mortar mixer and transferred to a TAM-Air. The heat release rate and accumulated hydration heat of the adhesives were measured using a TAM-Air for 72 h at 20 °C. According to the hydration heat release curve of mortars, the cement hydration could be divided into five periods: (1) initial (pre-induction) period; (2) induction period; (3) acceleration (post-induction) period; (4) deceleration period; and (5) diffusion period [45]. Because the mortar mixing takes place outside the glass bottles, it is difficult to fully capture the peak of the first period. In this study, Z0 represents a control group; CZ15 and CZ30 represent UHPCs doped with 15 wt.% and 30 wt.% of calcined zeolite sand, respectively; and NZ15 and NZ30 represent UHPC doped with 15 wt.% and 30 wt.% of natural zeolite sand, respectively.

As shown in Figure 8a, in the initial period, compared with the control group, Z0, the heat release rate of self-curing UHPC increased because of the addition of zeolite sand.

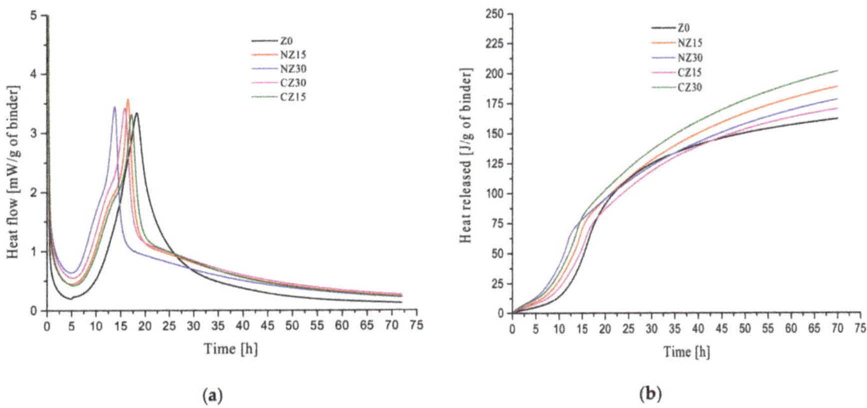

Figure 8. Isothermal calorimetry curves of UHPCs for 72 h: (**a**) heat flow curves; (**b**) cumulative heat curves.

For the UHPC containing calcined zeolite sand, as the content of calcined zeolite sand increased, the heat release rate of mortars accelerated. The heat release rate of mortars containing 30 wt.% of zeolite sand was the highest in the same group, which may be due to the addition of a large amount of zeolite sand, since soluble zeolite sand particles increase the content of alkali ions [1]. The same phenomenon is observed in mortars containing natural zeolite sand.

In the acceleration period, the time at which heat flow peaks of each hydration heat level appear are in the order of NZ30 > CZ30 > NZ15 > CZ15 > Z0. This is because some small zeolite sand particles are dissolved, and the alkali ion concentration is increased, which accelerates the binder hydration, causing the setting time to increase. During the deceleration period, it could be observed that the slope of the heat release rate curve of CZ15 was the same as that of CZ30 with respect to the hydration reaction. The slope of the heat release rate curve of NZ15 was the same as that of NZ30 with respect to the hydration reaction. Finally, in the diffusion period, the heat release rates of the mortars containing calcined zeolite sand and natural zeolite sand were higher than those of the control group, Z0.

Figure 8b describes the total heat release of the mortars within 72 h. It can be observed that the total heat released from self-curing UHPC is higher than that of the control group, Z0. The heat release of CZ30 is 260.84 J/g, which is the highest among all specimens. For the mortars mixed with natural zeolite sand, the heat release of NZ30 is higher than that of NZ15 at the initial stage of hydration, and the intersection point appears at about 20 h. After the intersection point, the heat release of NZ15 is higher than that of NZ30 at the later stage of hydration. At the age of 3 days, the heat release of NZ15 (188.71 J/g) is greater than that of NZ30 (178.81 J/g), which is caused by the heat release rate of NZ15 in the hardening deceleration stage being higher than that of NZ30.

3.5. Compressive Strength

The compressive strength, measured by cube specimens with dimensions of 50 mm × 50 mm × 50 mm, of UHPC at 3, 7, and 28 days, is shown in Figure 9.

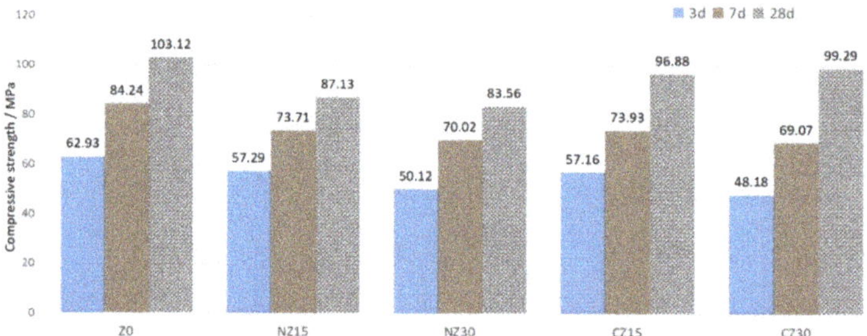

Figure 9. Compressive strength change of UHPCs at 3, 7, and 28 days.

At an age of 3 days, compared with the Z0, with the addition of zeolite sand, the compressive strength decreased. This is because zeolite sand is a porous material, and the addition of zeolite sand increases the porosity of concrete. Moreover, with the content of natural zeolite sand increasing from 15 wt.% to 30 wt.%, the compressive strength decreased because of the negative effect of porous zeolite sands. The compressive strengths of the NZ15 and NZ30 specimens are similar to those of CZ15 and CZ30 at early ages.

However, up to the age of 28 days, the compressive strengths of NZ15 (87.13 MPa) and NZ30 (83.56 MPa) are smaller (15.51% and 18.97%) than those of Z0 (103.12 MPa). The negative effect seems to have been compensated for. The compressive strength of CZ30 increases significantly, achieving 99.29 MPa, which is close to the compressive strength of Z0—about 103.12 MPa. The compressive strengths of CZ15 and CZ30 are 6% and 3.7%, which are lower than that of Z0. This may be due to the release of internal curing water in the zeolite sands, which fills the pores in the zeolite sands and improves the compressive strength of UHPC [39].

From 3 to 28 days, the incremental degree of strength for Z0 and CZ30 is different due to the effect of internal curing. The incremental degree of strength equals the ratio of the strength at 28 days to the strength at 3 days, and, regarding the increment degree of strength, Z0 increased by 1.64 (103.12/62.93 MPa) and CZ30 increased by 2.06 (99.29/48.18 MPa). The compressive strength of CZ30 increased more obviously. This is mainly because the addition of zeolite sand increased the porosity of the specimens. However, as the hydration reaction proceeds, the release of the internal curing water of the reservoir promotes the hydration of the cement, which significantly increases the compressive strength of the specimen at a later age. Therefore, the use of calcined zeolite sand as an autogenous curing agent has no obvious impairing effect on the compressive strength of UHPC.

3.6. Discussion

UHPC can be divided into two types, namely, precast UPHC and cast-in-place UHPC. For cast-in-place UHPC, autogenous shrinkage is an essential constraint because UHPC has a very low water-to-binder ratio. Construction companies are urgently seeking a solution to mitigate or minimize autogenous shrinkage without impairing the strength. In this study, we find the calcined zeolite sand is an effective internal curing material for UHPC. Compared with the control specimen, the autogenous shrinkage of concrete with 30% calcined zeolite sand is significantly lower while the 28-day compressive strength is similar.

On the other hand, the compressive strength of the UHPC designed in this paper did not reach 150 MPa (the general strength of UHPC). This may due to many factors affecting the strength of UHPC, such as: fibers [46], binder composition [47], particle size distribution, compactness of aggregate, and curing procedure [48]. The addition of fiber can improve the strength and ductility of concrete. The early-age high temperature curing may also contribute to the strength increment of concrete [49]. In addition, calcined-zeolite sand has some weak points; for example, the intrinsic porosity of the calcined zeolite sand may decrease the compactness of UHPC. These weak points should be further studied in future research [11,50].

4. Conclusions

Studying the effects of natural zeolite sand and calcined zeolite sand on the properties of UHPC, the following conclusions can be drawn from the present study:

1. Compared with natural zeolite sand, the water absorption of calcined zeolite sand is obviously increased. This is due to the initial physical bonding water present in the natural zeolite sand being evaporated during heating, resulting in more pores becoming empty.
2. The autogenous shrinkage of self-curing UHPC is significantly reduced due to the addition of zeolite sands. Among the resulting outcomes, the effect of adding calcined zeolite sand to UHPC is the most obvious. This is because calcined zeolite sand can absorb more internal curing water before mixing. With the occurrence of the hydration reaction, internal curing water is released to continue to participate in the hydration reaction, thereby reducing autogenous shrinkage.
3. For specimens with zeolite sand additions, XRD analysis showed that there were no new hydration-product peaks, indicating that zeolite sand particles do not participate in binder hydration. The peak of calcium hydroxide increased significantly with the addition of the zeolite sands. This is because the specimens containing zeolite sands contain more internal curing water. With the occurrence of the cement hydration reaction, the water in the reservoir is released to further promote cement hydration, and more hydration products are obtained.
4. It could be observed that the total heat released from self-curing UHPC is higher than that from the control group, Z0. In the initial period, the heat release rate of self-curing UHPC increased because of the addition of zeolite sand. As the content of calcined zeolite sand increased, the heat release rate of the mortars accelerated. In the acceleration period, the time at which heat flow peaks of each hydration heat level appeared is in the order of NZ30 > CZ30 > NZ15 > CZ15 > Z0. During the deceleration period, it could be observed that the slope of the heat release rate cures of CZ15 is the same as that of CZ30 with respect to the hydration reaction. The slope of the heat release rate cures of NZ15 is the same as that of NZ30 with respect to the hydration reaction.
5. At the age of 3 days, UHPC has a negative effect due to the addition of zeolite sands with pores, which reduces the compressive strength. However, at the age of 28 days, UHPC with calcined zeolite sand showed a good compressive strength, compared with the control group. This is due to the fact that calcined zeolite sand absorbs more internal curing water, and the internal curing water is released to participate in binder hydration and contributes to the development of compressive strength.

In summary, adding 30 wt.% of calcined zeolite sand as an internal curing agent has a positive effect in reducing the autogenic shrinkage strain of UHPC, without sacrificing compressive strength.

Author Contributions: Conceptualization, X.-Y.W.; Formal analysis, G.-Z.Z.; Funding acquisition, X.-Y.W.; Investigation, G.-Z.Z.; Methodology, X.-Y.W.; Writing—original draft, G.-Z.Z. and X.-Y.W.; Writing—review and editing, G.-Z.Z. and X.-Y.W. All authors have read and agreed to the published version of the manuscript.

Funding: This research is supported by the Basic Science Research Program through the National Research Foundation of Korea (NRF), funded by the Ministry of Science, ICT and Future Planning (no. 2015R1A5A1037548), and an NRF grant (no. NRF-2020R1A2C4002093).

Conflicts of Interest: The authors declared that they have no conflicts of interest to this work. We declare that we do not have any commercial or associative interest that represents a conflict of interest in connection with the work submitted.

References

1. Abdulkareem, O.M.; Fraj, A.B.; Bouasker, M.; Khelidj, A. Mixture design and early age investigations of more sustainable UHPC. *Constr. Build. Mater.* **2018**, *163*, 235–246. [CrossRef]
2. Li, P.P.; Yu, Q.L.; Brouwers, H.J.H. Effect of PCE-type superplasticizer on early-age behaviour of ultra-high performance concrete (UHPC). *Constr. Build. Mater.* **2017**, *153*, 740–750. [CrossRef]
3. Randl, N.; Steiner, T.; Ofner, S.; Baumgartner, E.; Mészöly, T. Development of UHPC mixtures from an ecological point of view. *Constr. Build. Mater.* **2014**, *67*, 373–378. [CrossRef]
4. Ghourchian, S.; Wyrzykowski, M.; Lura, P.; Shekarchi, M.; Ahmadi, B. An investigation on the use of zeolite aggregates for internal curing of concrete. *Constr. Build. Mater.* **2013**, *40*, 135–144. [CrossRef]
5. Buck, J.J.; McDowell, D.L.; Zhou, M. Effect of microstructure on load-carrying and energy-dissipation capacities of UHPC. *Cem. Concr. Res.* **2013**, *43*, 34–50. [CrossRef]
6. Yoo, D.-Y.; Min, K.-H.; Lee, J.-H.; Yoon, Y.-S. Shrinkage and cracking of restrained ultra-high-performance fiber-reinforced concrete slabs at early age. *Constr. Build. Mater.* **2014**, *73*, 357–365. [CrossRef]
7. Yang, L.; Shi, C.; Wu, Z. Mitigation techniques for autogenous shrinkage of ultra-high-performance concrete—A review. *Compos. Part B Eng.* **2019**, *178*, 107456. [CrossRef]
8. Justs, J.; Wyrzykowski, M.; Bajare, D.; Lura, P. Internal curing by superabsorbent polymers in ultra-high performance concrete. *Cem. Concr. Res.* **2015**, *76*, 82–90. [CrossRef]
9. Mo, J.; Ou, Z.; Zhao, X.; Liu, J.; Wang, Y. Influence of superabsorbent polymer on shrinkage properties of reactive powder concrete blended with granulated blast furnace slag. *Constr. Build. Mater.* **2017**, *146*, 283–296. [CrossRef]
10. Liu, J.; Ou, Z.; Mo, J.; Chen, Y.; Guo, T.; Deng, W. Effectiveness of Saturated Coral Aggregate and Shrinkage Reducing Admixture on the Autogenous Shrinkage of Ultrahigh Performance Concrete. *Adv. Mater. Sci. Eng.* **2017**, *2017*, 2703264. [CrossRef]
11. Meng, W.; Khayat, K. Effects of saturated lightweight sand content on key characteristics of ultra-high-performance concrete. *Cem. Concr. Res.* **2017**, *101*, 46–54. [CrossRef]
12. Liu, J.; Farzadnia, N.; Shi, C. Effects of superabsorbent polymer on interfacial transition zone and mechanical properties of ultra-high performance concrete. *Constr. Build. Mater.* **2020**, *231*, 117142. [CrossRef]
13. Sun, B.; Wu, H.; Song, W.; Li, Z.; Yu, J. Design methodology and mechanical properties of Superabsorbent Polymer (SAP) cement-based materials. *Constr. Build. Mater.* **2019**, *204*, 440–449. [CrossRef]
14. Farzanian, K.; Teixeiral, K.P.; Rocha, P.; Carneiro, L.D.S.; Ghahremaninezhad, A. The mechanical strength, degree of hydration, and electrical resistivity of cement pastes modified with superabsorbent polymers. *Constr. Build. Mater.* **2016**, *109*, 156–165. [CrossRef]
15. Bentz, D.P.; Geiker, M.; Jensen, O.M. On the mitigation of early age cracking. *Int. Semin. Self Desiccation III* **2002**, *3*, 195–204.
16. Woyciechowski, P.P.; Kalinowski, M. The Influence of Dosing Method and Material Characteristics of Superabsorbent Polymers (SAP) on the Effectiveness of the Concrete Internal Curing. *Materials* **2018**, *11*, 1600. [CrossRef]
17. Liu, K.; Yu, R.; Shui, Z.; Li, X.; Guo, C.; Yu, B.; Wu, S. Optimization of autogenous shrinkage and microstructure for Ultra-High Performance Concrete (UHPC) based on appropriate application of porous pumice. *Constr. Build. Mater.* **2019**, *214*, 369–381. [CrossRef]

18. Wang, X.F.; Fang, C.; Kuang, W.Q.; Lin, D.W.; Han, X.; Xing, F. Experimental investigation on the compressive strength and shrinkage of concrete with pre-wetted lightweight aggregates. *Constr. Build. Mater.* **2017**, *155*, 867–879. [CrossRef]
19. Ackley, M.W.; Rege, S.U.; Saxena, H. Application of natural zeolites in the purification and separation of gases. *Microporous Mesoporous Mater.* **2003**, *61*, 25–42. [CrossRef]
20. Klaysom, C.; Shahid, S. *Chapter 6—Zeolite-based mixed matrix membranes for hazardous gas removal. Advanced Nanomaterials for Membrane Synthesis and Its Applications*; Elsevier: Amsterdam, The Netherlands, 2019; pp. 127–157.
21. Ates, A.; Hardacre, C. The effect of various treatment conditions on natural zeolites: Ion exchange, acidic, thermal and steam treatments. *J. Colloid Interface Sci.* **2012**, *372*, 130–140. [CrossRef]
22. Inglezakis, V.J.; Zorpas, A.A. *Natural Zeolites Structure and Porosity*; Bentham Books: Bacau, Romania, 2012; p. 14, ISBN 9781608054466.
23. Zhang, J.; Wang, Q.; Ding, X.; Zheng, X. High-strength concrete mixture with calcined zeolite particles for shrinkage reduction. *Mag. Concr. Res.* **2019**, *71*, 690–699. [CrossRef]
24. Zhang, J.; Wang, Q.; Zhang, J. Shrinkage of internal cured high strength engineered cementitious composite with pre-wetted sand-like zeolite. *Constr. Build. Mater.* **2017**, *134*, 664–672. [CrossRef]
25. Pezeshkian, M.; Delnavaz, A.; Delnavaz, M. Effect of Natural Zeolite on Mechanical Properties and Autogenous Shrinkage of Ultrahigh-Performance Concrete. *J. Mater. Civ. Eng.* **2020**, *32*, 04020093. [CrossRef]
26. Li, C.; Thomas, M.D.A.; Ideker, J.H. A mechanistic study on mitigation of alkali-silica reaction by fine lightweight aggregates. *Cem. Concr. Res.* **2018**, *104*, 13–24. [CrossRef]
27. Suraneni, P.; Fu, T.; Azad, V.J.; Isgor, O.B.; Jason, W. Pozzolanicity of finely ground lightweight aggregates. *Cem. Concr. Compos.* **2018**, *88*, 115–120. [CrossRef]
28. Lin, R.-S.; Park, K.-B.; Wang, X.-Y.; Zhang, G.-Y. Increasing the early strength of high-volume Hwangtoh–cement systems using bassanite. *J. Build. Eng.* **2020**, *30*, 101317. [CrossRef]
29. Bentz, D.P.; Snyder, K.A. Protected paste volume in concrete: Extension to internal curing using saturated lightweight fine aggregate. *Cem. Concr. Res.* **1999**, *29*, 1863–1867. [CrossRef]
30. Holm, T.A.; Ooi, O.-S.; Bremner, T.W. Moisture Dynamics in Lightweight Aggregate and Concrete. In Proceedings of the 6th International Conference on the Durability of Concrete, Thessaloniki, Greece, 1–7 June 2003.
31. Castro, J.; Varga, I.D.L.; Weiss, J. Using Isothermal Calorimetry to Assess the Water Absorbed by Fine LWA during Mixing. *J. Mater. Civ. Eng.* **2012**, *24*, 996–1005. [CrossRef]
32. Castro, J.; Keiser, L.; Golias, M.; Weiss, J. Absorption and desorption properties of fine lightweight aggregate for application to internally cured concrete mixtures. *Cem. Concr. Compos.* **2011**, *33*, 1001–1008. [CrossRef]
33. ASTM C1698. *Standard Test Method for Autogenous Strain of Cement Paste and Mortar*; ASTM International: West Conshohocken, PA, USA, 2014.
34. ASTM International. *Standard Test Method for Compressive Strength of Cylindrical Concrete Specimens*; ASTM International: West Conshohocken, PA, USA, 2016.
35. Snellings, R.; Chwast, J.; Cizer, Ö.; De Belie, N.; Dhandapani, Y.; Durdzinski, P.; Elsen, J.; Haufe, J.; Hooton, D.; Patapy, C.; et al. Report of TC 238-SCM: Hydration stoppage methods for phase assemblage studies of blended cements—Results of a round robin test. *Mater. Struct.* **2018**, *51*, 111. [CrossRef]
36. Scrivener, K.; Snellings, R.; Lothenbach, B. *A Practical Guide to Microstructural Analysis of Cementitious Materials*; CRC Press: Boca Raton, FL, USA, 2016.
37. Lin, R.-S.; Wang, X.-Y.; Lee, H.-S.; Cho, H.-K. Hydration and Microstructure of Cement Pastes with Calcined Hwangtoh Clay. *Materials* **2019**, *12*, 458. [CrossRef] [PubMed]
38. Lin, R.-S.; Wang, X.-Y.; Zhang, G.-Y. Effects of Quartz Powder on the Microstructure and Key Properties of Cement Paste. *Sustainability* **2018**, *10*, 3369. [CrossRef]
39. Lv, Y.; Ye, G.; De Schutter, G. Investigation on the potential utilization of zeolite as an internal curing agent for autogenous shrinkage mitigation and the effect of modification. *Constr. Build. Mater.* **2019**, *198*, 669–676. [CrossRef]
40. Montanari, L.; Suraneni, P.; Weiss, W.J. Accounting for Water Stored in Superabsorbent Polymers in Increasing the Degree of Hydration and Reducing the Shrinkage of Internally Cured Cementitious Mixtures. *Adv. Civ. Eng. Mater.* **2017**, *6*, 583–599. [CrossRef]

41. Liu, J.; Shi, C.; Farzadnia, N.; Ma, X. Effects of pretreated fine lightweight aggregate on shrinkage and pore structure of ultra-high strength concrete. *Constr. Build. Mater.* **2019**, *204*, 276–287. [CrossRef]
42. Koizumi, M.; Roy, R. Zeolite Studies. I. Synthesis and Stability of the Calcium Zeolites. *J. Geol.* **1960**, *68*, 41–53. [CrossRef]
43. Seraj, S.; Ferron, R.D.; Juenger, M.C.G. Calcining natural zeolites to improve their effect on cementitious mixture workability. *Cem. Concr. Res.* **2016**, *85*, 102–110. [CrossRef]
44. Lura, P.; Jensen, O.M.; Breugel, K.V. Autogenous shrinkage in high-performance cement paste: An evaluationof basic mechanisms. *Cem. Concr. Res.* **2003**, *33*, 223–232. [CrossRef]
45. Yuan, B.; Yu, Q.L.; Brouwers, H.J.H. Reaction kinetics, reaction products and compressive strength of ternary activators activated slag designed by Taguchi method. *Mater. Des.* **2015**, *86*, 878–886. [CrossRef]
46. Abbas, S.; Soliman, A.M.; Nehdi, M.L. Exploring mechanical and durability properties of ultra-high performance concrete incorporating various steel fiber lengths and dosages. *Constr. Build. Mater.* **2015**, *75*, 429–441. [CrossRef]
47. Arora, A.; Aguayo, M.; Hansen, H.; Castro, C.; Federspiel, E.; Mobasher, B.; Neithalath, N. Microstructural packing- and rheology-based binder selection and characterization for Ultra-high Performance Concrete (UHPC). *Cem. Concr. Res.* **2018**, *103*, 179–190. [CrossRef]
48. Zhu, Y.; Zhang, Y.; Hussein, H.H.; Liu, J.; Chen, G. Experimental study and theoretical prediction on shrinkage-induced restrained stresses in UHPC-RC composites under normal curing and steam curing. *Cem. Concr. Compos.* **2020**, *110*, 103602. [CrossRef]
49. Garas, V.Y.; Kurtis, K.E.; Kahn, L.F. Creep of UHPC in tension and compression: Effect of thermal treatment. *Cem. Concr. Compos.* **2012**, *34*, 493–502. [CrossRef]
50. Valipour, M.; Khayat, K.H. Coupled effect of shrinkage-mitigating admixtures and saturated lightweight sand on shrinkage of UHPC for overlay applications. *Constr. Build. Mater.* **2018**, *184*, 320–329. [CrossRef]

Article

Reactive Powder Concrete Containing Basalt Fibers: Strength, Abrasion and Porosity

Stefania Grzeszczyk [1,*], Aneta Matuszek-Chmurowska [1], Eva Vejmelková [2] and Robert Černý [2]

[1] Department of Building Materials Engineering, Technical University of Opole, 45 061 Opole, Poland; a.matuszek-chmurowska@po.edu.pl

[2] Department of Materials Engineering and Chemistry, Czech Technical University in Prague, 166 29 Prague, Czech Republic; eva.vejmelkova@fsv.cvut.cz (E.V.); cernyr@fsv.cvut.cz (R.Č.)

* Correspondence: s.grzeszczyk@po.edu.pl; Tel.: +48-608-460-004

Received: 4 June 2020; Accepted: 29 June 2020; Published: 1 July 2020

Abstract: The paper presents the test results of basalt fiber impact on a compressive and flexural strength, resistance to abrasion and porosity of Reactive Powder Concrete (RPC). The reasons for testing were interesting mechanical properties of basalt fibers, the significant tensile strength and flexural strength, and in particular the resistance to high temperatures, as well as a relatively small number of RPC tests performed with those fibers and different opinions regarding the impact of those fibers on concrete strength. The composition of the concrete mix was optimized to obtain the highest packing density of particles in the composite, based on the optimum particle size distribution curve acc. to Funk. Admixture of basalt fibers was used in quantity 2, 3, 6, 8 and 10 kg/m^3, length 12 mm and diameter 18 µm. A low water-to-binder ratio, i.e., from 0.24, was obtained through application of a polycarboxylate-based superplasticizer. The introduction of up to 10 kg/m^3 of basalt fibers to RPC mix was proved to be possible, while keeping the same w/c ratio equal to 0.24, with a slight loss of workability of the concrete mix as the content of fibers increased. It was found that the increase of the fiber content in RPC to 10 kg/m^3, despite the w/c ratio was kept the same, caused reduction of the concrete compressive strength by 18.2%, 7.8% and 13.6%, after 2, 7, and 28 days respectively. Whereas, the flexural strength of RPC increased gradually (maximum by 15.9%), along with the fiber quantity increase up to 6 kg/m^3, and then it reduced (maximum by 17.7%), as the fiber content in the concrete was further increased. The reduction of RPC compressive strength, along with the increase in basalt fibers content, leads to the increase of the total porosity, as well as the change in pore volume distribution. The reduction of RPC abrasion resistance was demonstrated along with the increase of basalt fibers content, which was explained by the compressive strength reduction of that concrete. A linear relation between the RPC abrasion resistance and the compressive strength involves a high determination coefficient equal to 0.97.

Keywords: reactive powder concrete; strength; basalt fibers; abrasion; porosity

---

1. Introduction

Reactive Powder Concretes (RPC) are one of the most significant achievements in the field of the concrete technology in recent years [1,2]. They achieve considerable compressive strength values reaching even over 200 MPa in normal conditions, whereas with application of the hydrothermal treatment the strength values are much higher (from 300 to 500 MPa) [3–5]. Such high compressive strength values were achieved by elimination of coarse aggregate, which was replaced with fine ground quartz with particle size from 1 µm to 4 µm, and with sand with particle size from 200 µm to 400 µm. Apart from the increased quantity of cement (the cement content is usually 900 kg/m^3), the use of silica fumes, finely ground quartz powder, ground granulated blast furnace slag and fly ash at a low w/b ratio (from 0.20 to 0.23) allows to provide the maximum particle packing in the concrete. Whereas a larger quantity of steel fibers, in amount from 1.5 to 3.0%, assures achievement of the relatively high flexural and tensile strength of this concrete. Steel fibers play an important role in shaping of mechanical properties of this material [6,7]. Nowadays, the interest in modification of RPC properties grows through the application of various types of fibers, including basalt ones [8], and even of polypropylene [9]. Basalt fibers, due to their mechanical properties and resistance to high temperatures, are a subject of greater interest by concrete technologists. The tensile strength of basalt fibers is 700–1680 MPa, and the Young's modulus fluctuates within 70–90 MPa [10,11]. The advantages also include resistance to corrosion, both basic and acidic corrosion. Fibers are resistant to low and high temperatures, the range of working temperatures of application fluctuates from −260 °C to +750 °C [12].

To assess impact of the basalt fibers on the concrete properties, we have taken into account fibers content, size and the water-to-cement (w/c) ratio [13–16]. The influence of the experimental results of the basalt fibers concrete strength are ambiguous [14,17,18].

Kabay [15] tested the influence of basalt fibers in quantity 2% and 4% by mass, length 12 and 24 mm in high strength and normal strength concretes with w/c ratio equal to 0.45, and 0.60, respectively. He found that the addition of fibers caused the decrease in compressive strength and increase of the flexural strength and the fracture energy rose along with the increase in the fiber content of concrete.

Barabanshchikov and Gutskalov [13], while testing the impact of various quantity of basalt fibers (from 0.93% to 3.24% by volume) of various length 12.7 and 24.5 mm on properties of cement pastes, also found reduction of the compressive strength, whereas the increase of the flexural strength of specimens was the higher, the higher the fiber content was. It shall be noted, however, that the higher fiber content required the increase of w/c ratio from 0.31 to 0.40, which undoubtedly influenced the compressive strength. Furthermore, authors found that there was an optimum quantity of basalt fibers with the given length, which caused the highest increase of the flexural strength. In case of specimens tested, that quantity for fibers 12.7 mm long was 1.6% by volume.

Chaohua et al. [16] investigated the influence of basalt fibers, 12 and 22 mm long, in amount from 0.05% to 0.5% of the total volume of the concrete mix, with a relatively high w/c ratio equal to 0.6. They found a visible increase of flexural and tensile strength, but a slight increase of the compressive strength, and even its reduction at a later time. They stated that the content of fibers in amount of 0.3% in volume fraction was the most suitable in terms of the growth of the concrete mechanical parameters tested.

Tehmina Ayub et al. [14] observed a slight increase of the compressive strength of the High Performance Concrete with basalt fiber content 1% and 2% of volume fraction, and then the drop with a higher amount of fibers (3 vol.%). They noted the strength decrease occurred with addition of metakaolin and silica fumes. Authors observed the gradual increase of splitting tensile strength along with the increase of fiber content (from 1% to 3% by volume).

The increase of both compressive and flexural strength along with the increase of basalt fiber content from 0.25% to 1% by weight of cement, while keeping the same w/c ratio = 0.42, was found by authors of the paper [17]. The similar relation was observed by the authors of the paper [19] at the fiber content increase from 1% to 3% by weight of cement in sand basalt-fiber concrete, while keeping the w/c ratio within the range 0.32–0.35.

Not many papers on the RPC with addition of basalt fibers have been published so far. Such research was conducted by authors of the paper [8], who found that introduction of basalt fibers to RPC (w/c = 0.25) caused the increase of the compressive and flexural strength, when the content was from 1 kg/m^3 to 3 kg/m^3, whereas when the content of fibers increased to 5 kg/m^3, the gradual decrease of the compressive and flexural strength occurred.

Table 1 shows results of compressive and flexural strength tests of various types of concretes, with a different content of basalt fibers from 6 mm to 25 mm long, conducted by various authors [8,13–17,19]. The analysis of test results of flexural strength of concretes presented in Table 1 shows that, along with the fiber content increase in the concrete, the increase in the flexural strength also occurs to a various extent. [15–17]. The higher growth in the flexural strength (within the range from 54% to 64%) is achieved by samples with the w/c ratio 0.34–0.42 and the higher content of fibers, above 15 kg/m^3 [13,17]. In the case of RPC and hardened cement paste, which have a much higher content of fine particles, the flexural strength increases to a certain amount of basalt fibers in the material, and then it goes down with the further increase of the fiber content [8,13]. Application of the low w/c ratio (0.25) in RPC allows to introduce much smaller amount of fibers (up to 5 kg/m^3), than to the cement paste with the higher w/c (0.4).

Table 1. Impact of basalt fibers quantity on compressive and flexural strength according to various authors.

Authors	Type of Concrete	w/c	Length of Basalt Fibers (mm)	Fiber Content		Compressive Strength f_c (MPa) After 28 Days	Flexural Strength f_{cf} (MPa)	Comments
				By Volume (%)	By Weight (kg/m^3)		After 28 Days	
Chonghai D., Xinwei M. [8]	RPC	0.25	12		0	86	17.3	f_c–increase from 0 to 3 kg/m^3, above decrease
					2	101	19.8	f_{cf}–increase from 0 to 3 kg/m^3, above decrease
					3	115	20.4	
					4	106	19	
					5	102	20	
Barabanshchikov Y., Gutskalov I. [13]	Cement Paste	0.31	12.7	0	0	72.4	6.3	f_c–decrease from 0 to 94 kg/m^3
		0.34		0.93	26.9	47.8	10.9	f_{cf}–increase from 0 to 66.4 kg/m^3, above decrease
		0.36		1.61	47.6	41.5	15.9	
		0.38		2.29	66.4	37.2	16.8	
		0.4		3.24	94	36.2	14.9	
Ayub T. et al. [14]	HPC	0.4	25	0	0	71.9	-	f_c–increase from 0 to 60 kg/m^3, above decrease
				1	30 *	73.5	-	
				2	60 *	74.1	-	
				3	90 *	65.1	-	
Kabay N. [15]	HSC	0.45	12	0	0	69.3	7.7	f_c–decrease from 0 to 4 kg/m^3
				0.07	2	62.4	8	f_{cf}–increase from 0 to 4 kg/m^3
				0.14	4	56.9	8.5	
		0.6		0	0	49.4	6.3	
				0.07	2	44.8	6.7	
				0.14	4	46.7	7.1	
Jiang C. et al. [16]	FRC	0.6	12	0	0	45.1	7.6	f_c–increase from 0 to 3 kg/m^3, above decrease
				0.05	1.5 *	46.7	8.1	f_{cf}–increase from 0 to 3 kg/m^3, above decrease
				0.1	3.0 *	47.2	8.2	
				0.3	9.0 *	46.3	8.4	
				0.5	15.0 *	45	8.3	
Ahmed T. et al. [17]	FRC	0.42	6–10	0	0	34.6	3.7	f_c–increase from 0 to 30 kg/m^3
				0.25	7.5 *	37.5	5.8	f_{cf}–increase from 0 to 30 kg/m^3
				0.5	15.0 *	42.7	7.4	
				0.75	22.5 *	50.4	9.5	
				1	30.0 *	64.1	10.2	
Morozov N.M. et al. [19]	Sand FRC	0.32		0%	0	92	-	f_c–increase from 0 to 16.5 kg/m^3
				2% **	11	113	-	
				3% **	16.5	121	-	

* A conversion factor of amount basalt fibers was used: 0.1% by volume = 3 kg/m^3 according to [13]. ** By weight of cement.

In case of the compressive strength, the increase can be also observed [17,19], as well as reduction of the strength along with the basalt fiber content in the concrete [13–15], and also the increase of the strength, and then its drop when a specific amount of fibers in the concrete is exceeded [8,16]. Reduction of the concrete compressive strength is observed both in case the fiber content is lower: 4.0 kg/m^3 by 18% [15], and in case the fiber content is higher: 94 kg/m^3 by 50% [13]. In general, the increase of fiber content makes their proper distribution in the concrete mix more difficult and may lead to higher concrete porosity, resulting in the reduction of the compressive strength [16,20]. Furthermore, the higher content of basalt fibers in the mix, to provide its workability, requires the increase of the w/c ratio, which in consequence also leads to reduction of the concrete compressive strength.

There are many factors mentioned above, which have the impact on the strength of concretes with basalt fibers, but the impact of the concrete composition cannot be ignored, including the influence of mineral admixtures to seal the cement matrix and the matrix contact with fibers.

There are multiple factors that affect the strength of concrete, and with addition of fibers, it is difficult to determine the impact on each of them individually. Based on research conducted by various authors, general trends in the change in strength can be indicated under the influence of basalt fibers introduced to the concrete. Although, most researchers use basalt fibers with an optimal length of 12 mm due to the strength of concrete [15], this parameter should also be taken into accoung when the strength of concrete with fibers are analysed.

Few papers referring to tests of RPC with basalt fibers have resulted in research to be undertaken, in order to determine the optimum quantity of basalt fibers in RPC, in terms of the compressive and flexural strength. Testing included the impact of basalt fibers on the RPC strength, resistance to abrasion and porosity.

Taking the basalt fibers resistance to high temperatures into account, the objective of tests performed was to determine composition of the RPC with the highest compressive and flexural strength, which will be subject to high temperatures, as in case of tests of RPC with steel fibers, conducted by the authors of this paper [7].

2. Materials and Methods

To prepare the Reactive Powder Concrete (RPC), the following ingredients were used: Portland cement CEM I 52,5 R (Cement Plant, Rejowiec, Poland) with specific surface area 410 m^2/kg, silica fumes, quartz powder 0/0.2 mm, quartz sand 0/0.4 mm and basalt fibers 12 mm long and diameter 18 µm. Polycarboxylate superplasticizer was added in quantity of 2.5 wt.% of cement. The chemical composition of the cement, silica fumes, quartz powder and quartz sand is given in Table 2, and Figures 1 and 2 shows the particle size distribution of the cement, silica fumes, quartz powder and quartz sand determined by means of a laser size analyzer Mastersizer 3000 (Malvern, UK). Waste silica fume containing more than 62% particle > 40 µm.

Table 2. Chemical composition of RPC ingredients (% by mass).

Ingredient	SiO_2	Fe_2O_3	Al_2O_3	CaO	MgO	SO_3	Na_2O	K_2O
Cement	21.83	2.00	4.38	65.68	0.93	3.29	0.29	0.32
Silica fume	93.3	2.3	1.5	1.0	-	-	-	1.6
Quartz powder	99.0	0.05	0.29	< 0.1	< 0.1	-	0.2	< 0.1
Quartz sand	98.6	0.03	0.75	-	-	-	-	-

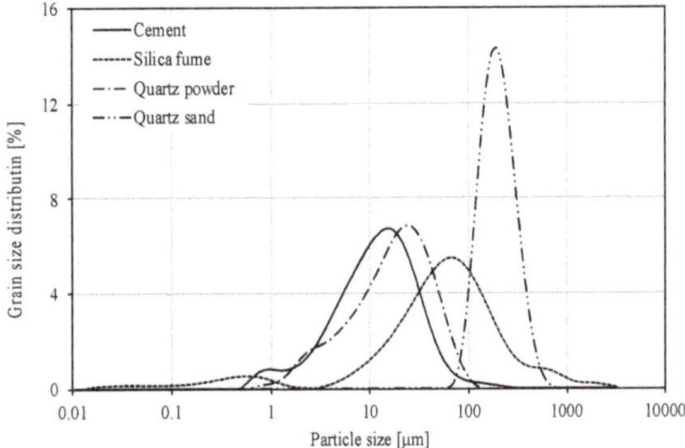

Figure 1. Particle size distribution of cement, silica fume, quartz powder and quartz sand.

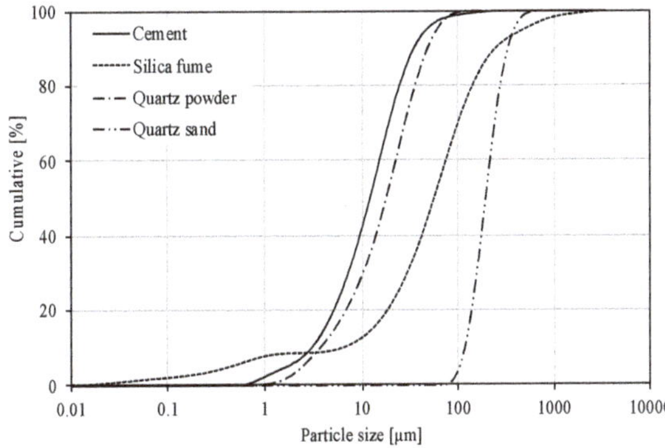

Figure 2. Cumulative particle size distribution of cement, silica fume, quartz powder and quartz sand.

Optimization of the concrete mix composition that assured the maximum packing density of particles, was performed based on Funk and Dinger's curve [21] (Figure 3) following formula,

$$y_i = \left(\frac{d_i^n - d_{Min}^n}{d_{Max}^n - d_{Min}^n} \right) \times 100\%, \quad (1)$$

where y_i—cumulative percentage of i-fraction content, d_i—i-fraction diameter [μm], d_{Max}—maximum particle diameter [μm], d_{Min}—minimum particle diameter [μm], n—constant value equal to 0.37.

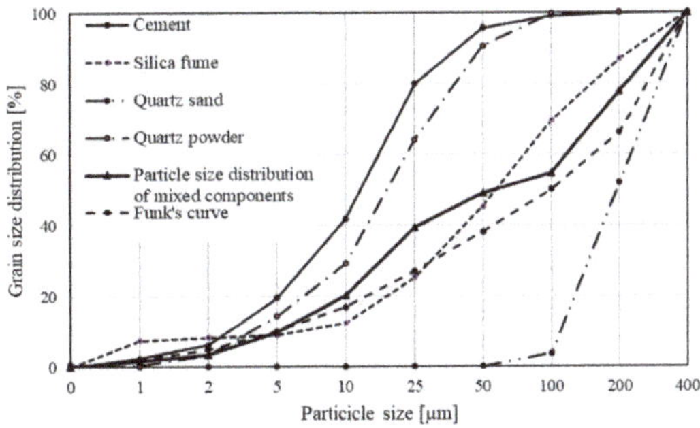

Figure 3. The particle size distribution of RPC mix and it's components.

For that purpose, also other computational models are used [22,23].

Composition of RPC mixes is given in Table 3. The content of basalt fibers in the mix were 2; 3; 6; 8 and 10 kg/m^3, at the same w/c ratio amounting to 0.24.

Table 3. Mix proportion of RPC [kg/m^3].

Cement	Silica Fume	Quartz Powder	Quartz Sand	Superplasticizer	Water	w/c
876	175	105	902	22	210	0.24

The particle size distribution of the concrete was determined by means of the particle size laser analyzer Mastersizer 3000 within the range 0.01–3500 μm.

X-ray diffraction (XRD) tests were conducted by means of Philips X'PertSystem diffractometer (Amsterdam, The Netherlands). The measurement was conducted within a range from 5° to 60° 2θ. CuKα radiation was used.

Tests of RPC consistence were performed in line with PN-EN 1015-3 standard (Methods of test for mortar for masonry. Part 3: Determination of consistence of fresh mortar (by flow table)) [24]. Consistence was determined based on measurement of the concrete flow diameter.

The tests of compressive strength and flexural strength of RPC specimens were performed in line with PN-EN 1015-11 standard (Methods of test for mortar for masonry. Part 11: Determination of flexural and compressive strength of hardened mortar) [25]. Tests were performed on specimens 40 mm × 40 mm × 160 mm each time. The flexural strength was calculated according to the following formula,

$$f = 1.5\frac{Fl}{bd^2}, \tag{2}$$

where f—flexural strength [MPa], F—maximum load [N], l—distance between axes of cylinders [mm], b—width of specimen [mm], d—height of specimen [mm].

The compressive strength was calculated according to the following formula:

$$f_c = \frac{F}{A_c}, \tag{3}$$

where: fc—compressive strength [MPa], F—maximum load [N] and A_c—section area of specimen [mm^2].

Resistance to abrasion of RPC specimens was tested in accordance with PN-EN 1338 standard (Concrete paving blocks. Requirements and test methods) [26]. Measurement of abrasion was performed on Böhme's disc (EMEL, Poland). Specimens were subject to abrasion load equal to 294 ± 3 N during 16 cycles. The abrasion resistance was calculated according to the following formula,

$$A = \Delta V = \frac{\Delta m}{\rho_R}\Delta l \times 5, \tag{4}$$

where Δm—specimen mass loss (wastage) after 16 cycles [g], ρ_R—specimen density [g/mm^3].

Infrared Spectra (IR) of basalt fiber specimen in the form of potassium bromide pellets (5 mg/500 mg KBr) recorded on Thermo Nicolet spectrophotometer (ThermoFisher Scientific, Waltham, MA, USA), Nexus model.

The microstructure of RPC was tested by scanning electron microscopy (SEM). Analyses were performed by means of the scanning microscope NOVA NANO SEM 200 (FEI Europe B.V., Eindhoven, The Netherlands). Observations were conducted with magnification from 200 to 10,000. Specimens for testing were prepared by spraying a layer of gold on the surface in high vacuum conditions. Energy Dispersive X-Ray Spectra (EDS) were obtained for selected points.

Testing of RPC porosity were performed with the use of a mercury porosimeter PoreMaster 60 (Quantachrome Instruments, Boynton Beach, FL, USA), within a pressure range from 1 to 400 MPa.

The results of the tests were presented in a form of differential curves and volumetric distribution curves of pores with different size.

3. Results and Discussion

3.1. Test of Basalt Fibers

Identification tests of basalt fibers were conducted by means of XRD test, infrared spectroscopy and scanning electron microscopy.

An XRD pattern of basalt fibers specimen is presented in Figure 4. A reflex of metallic iron occurred in XRD pattern next to a significant background rise associated with amorphous glass. It is very likely that, during the melting of basalt dust, iron crystallized out without binding with the glass. Notably, basalt contains quite a lot of magnetite, its reduction and crystallization in a form of metallic iron cannot be excluded.

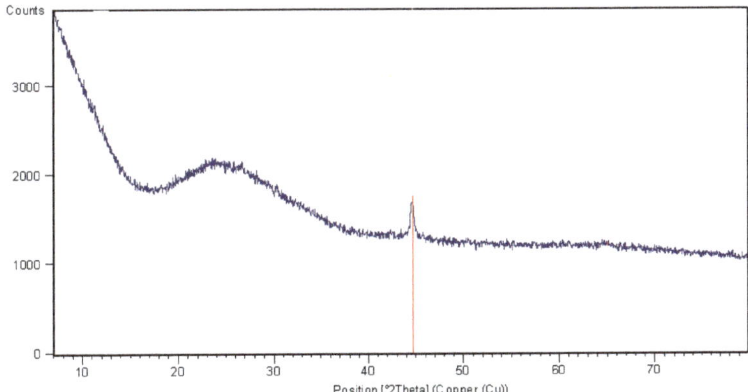

Figure 4. XRD pattern of basalt fibers.

Figure 5a shows SEM image of basalt fibers. As you can see, roughness occurs here and there on the smooth surface of fibers. EDS spectra (Figure 5b,c) made in two different points on basalt fibers (points 1 and 2)—Figure 5a) demonstrated the presence of carbon. A much lower intensity of a line belonging to the carbon is observed on the EDS spectrum of the smooth surface of the fiber (Figure 5c) than on the rough surface (Figure 5b). The presence of carbon on basalt fibers surface may indicate that fibers are coated with the organic substance, more quantity of it in some places.

IR spectroscopy tests were performed to identify the organic substance. The IR spectra of basalt fibers is presented in Figure 6, whereas Figure 7 shows a fragment of the spectra characteristic for vibration of aliphatic groups of CH_2 and CH_3.

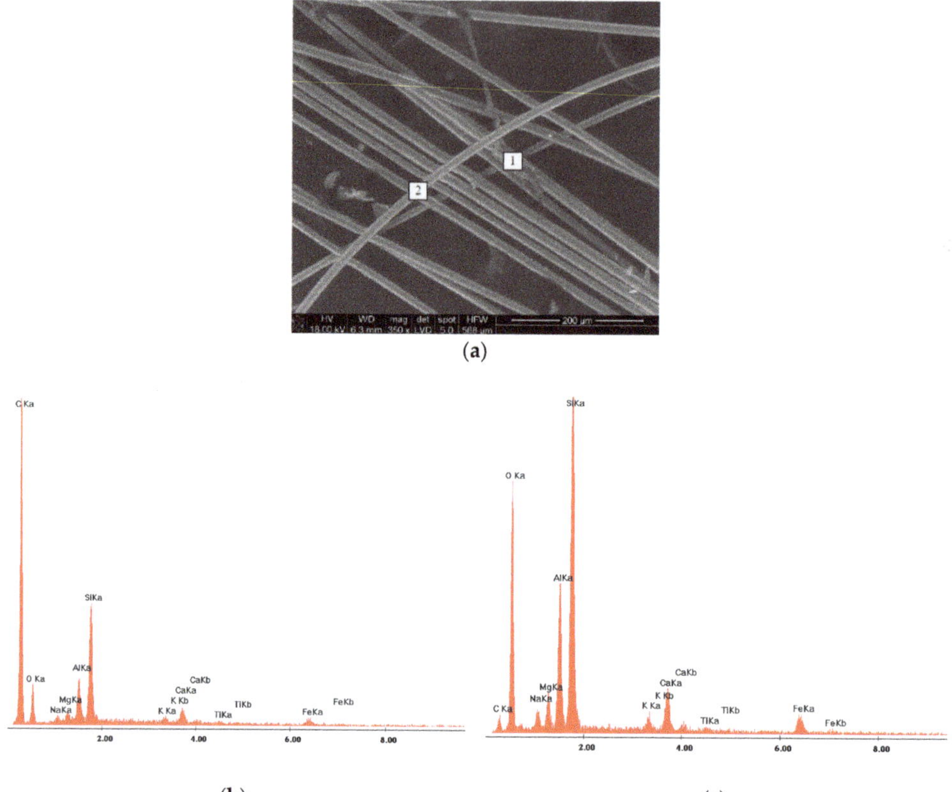

(**a**)

(**b**) (**c**)

Figure 5. (**a**) SEM of basalt fibers; (**b**) EDS analysis in point 1; (**c**) EDS analysis in point 2.

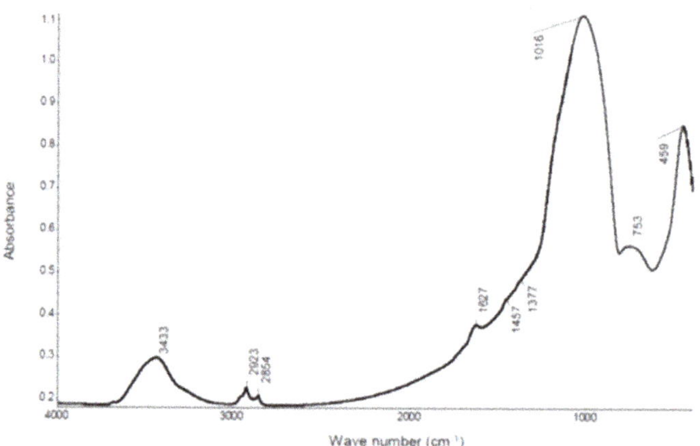

Figure 6. IR spectra of basalt fibers in range from 4000 to 500 cm^{-1}.

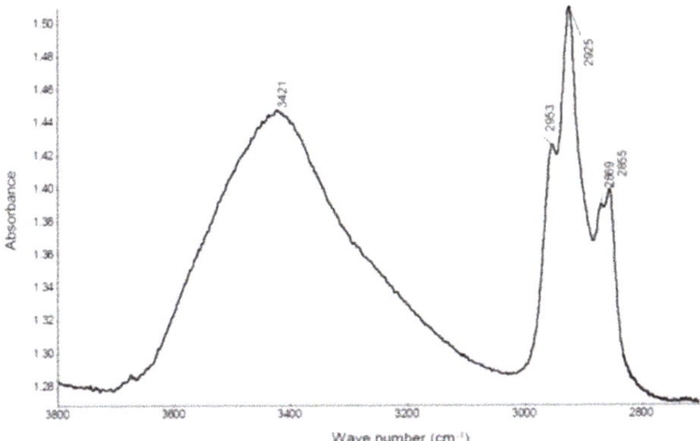

Figure 7. IR spectra of basalt fibers in range from 3800 to 2700 cm^{-1}.

Analysis of IR spectra demonstrated the presence of bands belonging to CH_2 and CH_3 groups in the spectrum: 2953 cm^{-1}—$_{STR/AS}CH_3$, 2925 cm^{-1}—$_{STR/AS}CH_2$, 2869 cm^{-1}—$_{STR/S}CH_3$, 2855 cm^{-1}—$_{STR/S}CH_2$. These bands are well visible in Figure 7. Furthermore, in the IR spectrum (Figure 6) there are also trace bands coming from deformation (bending) vibration of CH_3, CH_2 groups (1457 cm^{-1}) and scissor vibration of CH_3 groups (1377 cm^{-1}). The results of the above testing indicate that the surface of basalt fibers contains a hydrocarbon modifier. It is also supported by results of the SEM testing. EDS spectra made in two different points: 1 (Figure 5b) and 2 (Figure 5c) marked in the microscope image of basalt fibers (Figure 5a) showed that the line belonging to carbon is much more intense in point 1 than on the fiber smooth surface (point 2). Point 2 clearly shows presence of the modifier on the fiber surface.

The lack of bands coming from oxygen functional groups in the spectra: $-\overset{||}{C}-$, $-C-O-C-$, $-\overset{||}{C}-O-$, which may come from the substance that modifies the fibers surface, allows to suspect that the hydrocarbon (oil) is the modifier itself or siloxane ethers (silicones) e.g., silicone oil.

The IR spectra shows an intensive, broad band belonging to siloxane groups Si–O–Si $_{STR/AS}$ (1016 cm^{-1}). Unfortunately, based on the presence of this band, it cannot be stated definitively whether the silicone oil is the modifier, as basalt fibers also contain Si–O–Si groups. The authors of the paper [10] write about modification of fiber surface with silicon oil, in order to increase the adhesion to the polymer matrix. The presence of oil on the surface of fibers, found in testing (IR and SEM), undoubtedly facilitates dispersion of fibers in the RPC mix, but it may have an adverse effect on adhesion of basalt fibers to the cement matrix, which may contribute to reduction of the concrete compressive strength.

3.2. Consistence of Concrete Mix

Results of RPC mix consistence tests after 15 min with a flow table test method [24], are presented in Table 4 and Figure 8.

Table 4. Flow diameter of concrete mix [mm].

RPC	C0	C2	C3	C6	C8	C10
Fibers content [kg/m^3]	0	2	3	6	8	10
w/c	0.24	0.24	0.24	0.24	0.24	0.24
Flow diameter	240	240	230	220	200	180

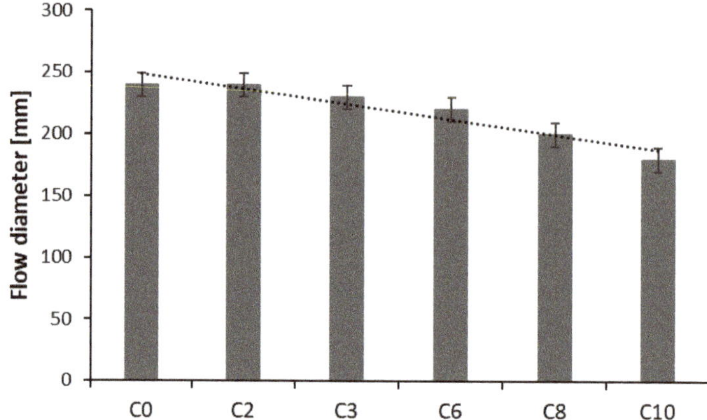

Figure 8. Impact of basalt fibers quantity on RPC flow diameter.

As data in Table 4 show, the increase of the basalt fibers content from 2 kg/m^3 to 10 kg/m^3 in RPC mix, while keeping the same w/c = 0.24, causes reduction of the flow diameter from 240 mm to 180 mm, which indicates a drop in concrete mix flowability. It results from a commonly known fact of a larger water demand of concrete mixes with the addition of fibers, in order to obtain the required workability.

3.3. Compressive and Flexural Strength

The results of compressive strength tests of Reactive Powder Concretes are shown in Table 5 and Figure 9, and results of flexural strength tests in Table 6 and Figure 10.

Table 5. Compressive strength of RPC containing basalt fibers [MPa].

RPC		C0	C2	C3	C6	C8	C10
Fiber content [kg/m^3]		0	2	3	6	8	10
w/c		0.24	0.24	0.24	0.24	0.24	0.24
Compressive strength after days	2	105.0	100.4	88.7	88.3	86.7	85.9
	7	114.9	116.2	112.2	108.9	106.2	105.9
	28	149.6	143.6	135.6	132.3	131.5	129.3

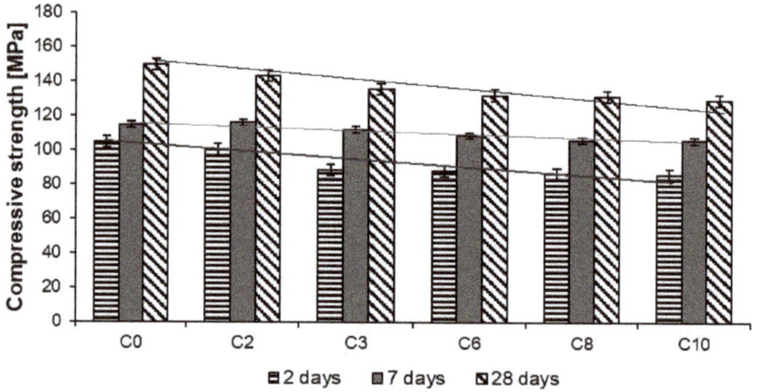

Figure 9. Compressive strength of RPC.

Table 6. Flexural strength of RPC containing basalt fibers [MPa].

RPC		C0	C2	C3	C6	C8	C10
Fiber content [kg/m^3]		0	2	3	6	8	10
w/c		0.24	0.24	0.24	0.24	0.24	0.24
Flexural strength after days	2	15.8	15.6	16.2	16.5	13.0	13.1
	7	18.9	20.6	20.5	21.9	17.1	17.5
	28	21.9	23.4	23.2	24.6	19.2	19.4

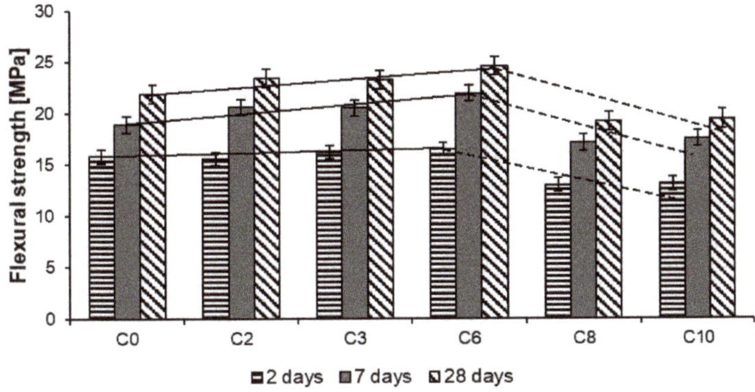

Figure 10. Flexural strength of RPC.

Based on data given in Table 5 and in Figure 9 it appears that addition of basalt fibers, in quantity from 2 kg/m^3 to 10 kg/m^3, while keeping the same w/c ratio, causes a reduction in the compressive strength, which is the higher in value, the higher the fiber content is. The lowest compressive strength values were obtained by concrete with the highest content of fibers (10 kg/m^3). The elimination of w/c coefficient influence the compressive strength of RPCs with different amounts of basalt fibers, which indicates that the main factor determining the strength is the amount of fibers. Increasing the amount of fibers in the concrete mix makes their dispersion difficult, which in consequence, does not allow a homogeneous microstructure of the material to be obtained. This can cause an increase in porosity, resulting in a decrease in the compressive strength of the concrete.

The analysis of RPC flexural strength test results, while keeping the same w/c ratio, showed that as the content of basalt fibers increased from 2 kg/m^3 to 6 kg/m^3, increases the RPC flexural strength. Whereas, when the quantity of fibers was higher (8 kg/m^3 and 10 kg/m^3), a reduction of the concrete flexural strength was observed, but the difference in RPC strength with 8 kg/m^3 and 10 kg/m^3 content is small (Table 6). To sum up, it can be stated that there is a certain amount of basalt fibers added to RPC, at which it reaches the highest flexural strength. For the RPC tested this value is 6 kg/m^3. A similar phenomenon was observed by authors of the paper [8], except that the quantity of basalt fibers at which they found the highest flexural strength was 3 kg/m^3. To explain that phenomenon, it would be advisable to perform testing of basalt fibers distribution in the RPC. However, proposing a model of basalt fiber distribution in the RPC composite space, based on statistical grounds, analogically to steel fibers, is hindered. The methods applied for testing of steel fibers distribution in concretes, e.g., electromagnetic induction and ultrasonic wave propagation [27], will not meet their task in case of basalt fibers.

3.4. Porosity

A few factors have the impact on RPC compressive strength reduction along with the increase of the fiber content [15,18]. One of them is the concrete porosity. According to [16], admixion of the basalt fibers increases the concrete porosity. In case of RPC tested it was found that introduction of basalt fibers caused the increase of the total porosity along with the growth of fibers content, as well as a change of pores volume distribution (Table 7). A reduction in the total porosity in time was also observed.

Table 7. Total porosity and volumetric percentage of pores of different sizes in RPC.

Type of Concrete	Total Porosity [%] After Days		Percentage of Pores [%]				
			< 20 nm	20–200 nm	200–2000 nm	2000–20,000 nm	<20,000 nm
C0	2	10.9	39.8	55.5	2.5	0.5	2.0
	7	8.2	64.9	25.1	1.5	0.2	8.2
	28	4.4	77.1	8.0	3.6	2.5	8.2
C2	2	10.6	73.9	12.6	0.8	0.6	10.8
	7	8.5	64.9	12.2	1.5	0.4	20.4
	28	4.0	62.5	9.1	14.6	1.8	11.9
C6	2	21.8	52.1	38.7	1.4	0.8	6.7
	7	10.4	39.2	52.5	2.3	0.8	4.8
	28	15.7	69.4	16.6	2.4	0.8	10.5
C10	2	21.2	46.5	45.3	2.4	0.6	5.4
	7	25.8	22.9	68.6	3.4	1.0	3.8
	28	15.5	74.1	11.9	2.7	1.7	8.0

The results of porosity testing of RPC with and without basalt fibers in quantity of: 2 kg/m^3, 6 kg/m^3 and 10 kg/m^3 are presented in Figures 11–14 and Table 7.

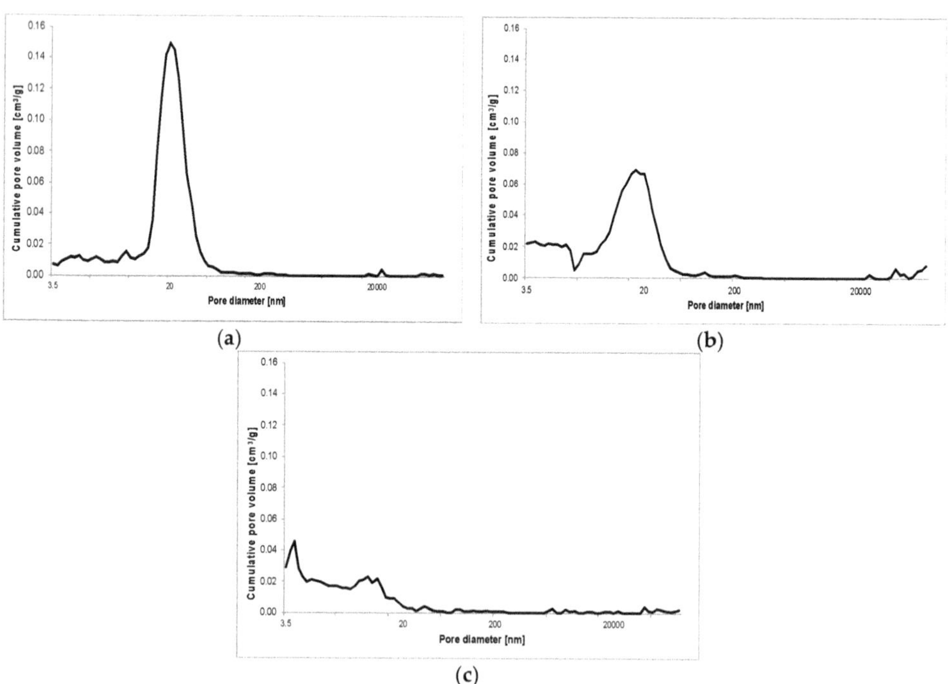

Figure 11. Pore volume distribution vs. pore diameter—RPC without basalt fibers (C0) after 2 (**a**), 7 (**b**) and 28 (**c**) days.

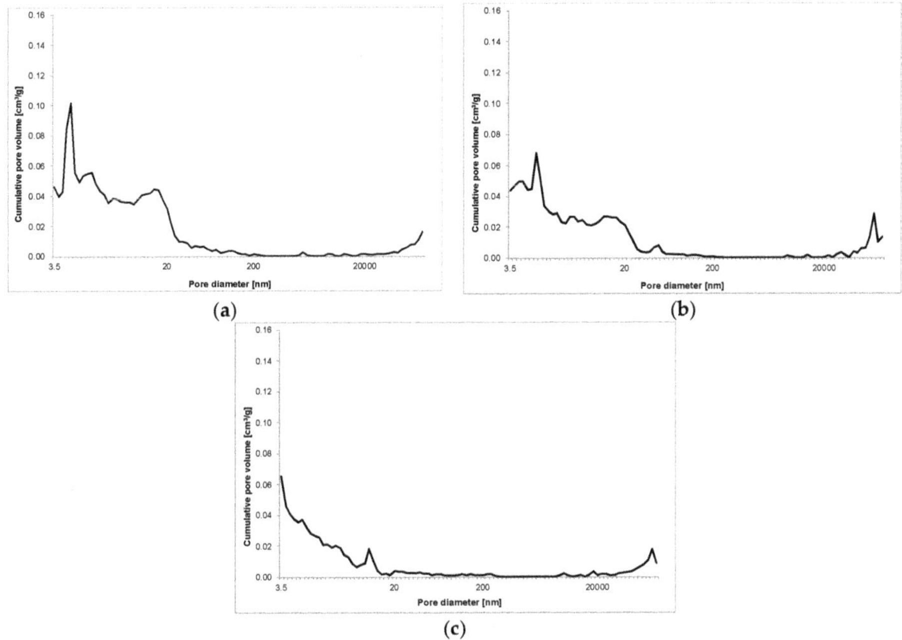

Figure 12. Pore volume distribution vs. pore diameter—RPC with basalt fibers (C2) after 2 (**a**), 7 (**b**) and 28 (**c**) days.

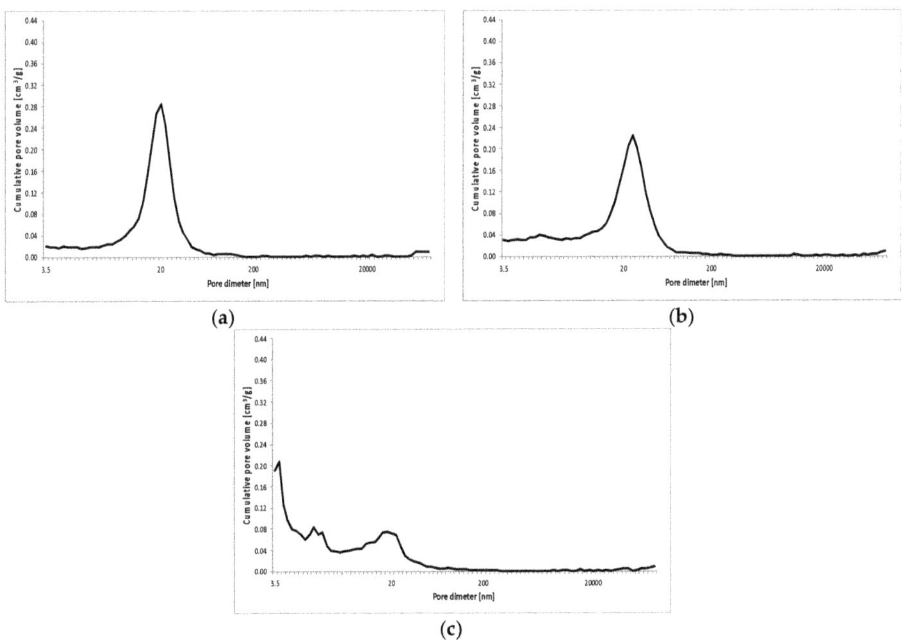

Figure 13. Pore volume distribution versus pore diameter—RPC without basalt fibers (C6) after 2 (**a**), 7 (**b**) and 28 (**c**) days.

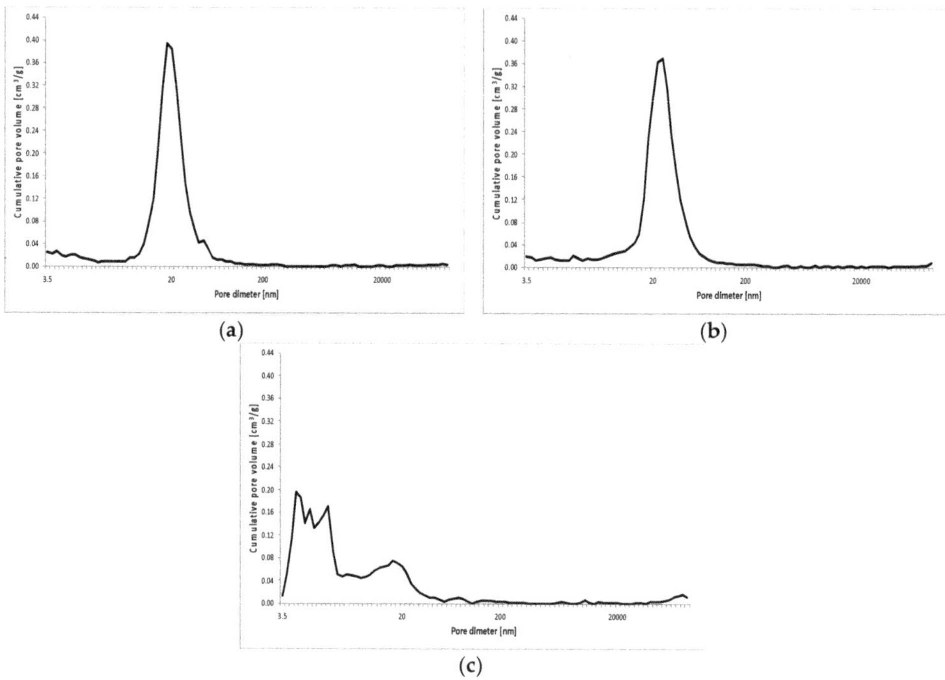

Figure 14. Pore volume distribution vs. pore diameter—RPC with basalt fibers (C10) after 2 (**a**), 7 (**b**) and 28 (**c**) days.

As shown in Figure 11, Figure 12, Table 7, the total porosity of RPC without basalt fibers and a small amount of fibers (2 kg/m^3) is comparable after 2, 7 and 28 days of curing of specimens. Whereas, the distribution of pore volume is different. RPC with basalt fibers contains definitely more pores with a larger diameter above 20,000 nm, but fewer smaller pores with the diameter from 200 nm to 2000 nm compared to the concrete without fibers, while keeping almost the same total porosity in both concretes. After 2 days of curing, the concrete with fibers showed nearly five times higher amount of pores with diameter above 20,000 nm and about five times lower amount of pores within the range 20–200 nm, compared to the content of those pores in the concrete without fibers. In the later period (7 and 28 days) those differences were smaller. In addition, a reduction in total porosity in time (from 2 to 28 days) can be clearly seen.

The addition of basalt fibers to RPC in quantity of 6 kg/m^3 and 10 kg/m^3, causes a definite growth of RPC total porosity, even by over 10 vol.% along with the increase of the fibers content, but similar to the lower fiber content, the total porosity is reduced over time. A reduction in pore content < 20 nm can be observed, as well as a definite higher amount (by several times) of pores within a range 20–200 nm, compared to the content of these pores in the specimen with 2 kg/m^3 of basalt fibers. After 7 days, the quantity of these pores is four or even five times higher for RPC specimens with fiber content 6 kg/m^3 and 10 kg/m^3, respectively. Whereas after 28 days, a visible growth of pores < 20 nm is observed with simultaneous reduction of pores within 20–200 nm. The content of pores > 20,000 nm is generally lower than 10 vol.% for RPC with basalt fiber content 6 kg/m^3 and 10 kg/m^3 in the time tested, after 2, 7 and 28 days.

In general, the content of pores within the ranges 200–2000 nm and 2000–20,000 nm, both in RPC without fibers, as well as with the various content of basalt fibers, is relatively small. In the first case it generally does not exceed 3.0 vol.% and in the second one 2.0 vol.%.

The change in pore volume distribution in RPC, along with an increase in the basalt fibers content, most probably results from the impediment of fiber dispersion in the concrete mix with a very low w/c ratio, and may have the impact on the concrete strength. Cement pastes compressive strength reduction is related to the growth of pore volume with diameters larger than 20 nm [28,29]. The reduced content of pores < 20 nm with simultaneous increase of pores in the range 20–200 nm, with the basalt fibers content above 2 kg/m^3 in RPC, as found in this paper, may contribute to the reduced compressive strength of RPC with the higher fibers content.

Furthermore, the growth of RPC total porosity found along with the increase of basalt fibers content, while keeping the constant water-cement ratio, is one of significant factors having an impact on the reduction of the concrete compressive strength. An increase in porosity, along with the increase in basalt fibers in the concrete, as well as the effect of that phenomenon on its strength, was observed by the authors of the paper [16]. Whereas, authors of the paper [15,16] emphasize that introduction of the larger quantity of basalt fibers to the concrete mix, generally requires the increase of the w/c ratio to provide concrete workability, and in consequence it affects the reduction of the concrete compressive strength.

3.5. Abrasion

Results of abrasion tests of RPC without fibers and containing 2, 6 and 10 kg/m^3 of basalt fibers are presented in Table 8 and Figures 15 and 16.

Table 8. Abrasion and strength of RPC with basalt fibers.

RPC	C0	C2	C6	C10
Fiber content [kg/m^3]	0	2	6	10
Abrasion [mm^3/50 cm^2]	15,000	16,000	20,000	22,000
Compressive strength after 28 days [MPa]	149.6	143.6	132.3	129.3
Flexural strength after 28 days [MPa]	21.9	23.4	24.6	19.4

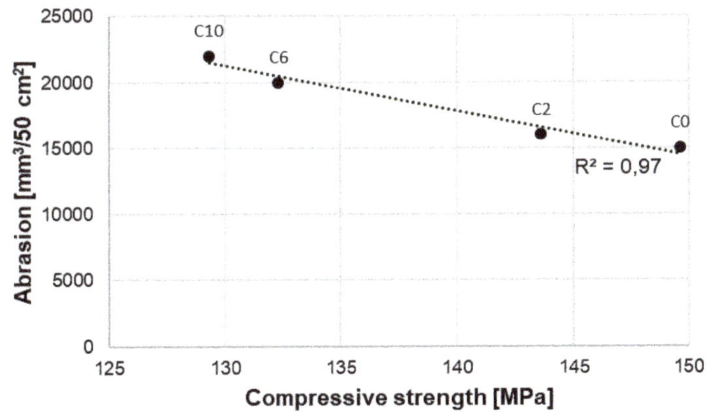

Figure 15. Abrasion—compressive strength relation.

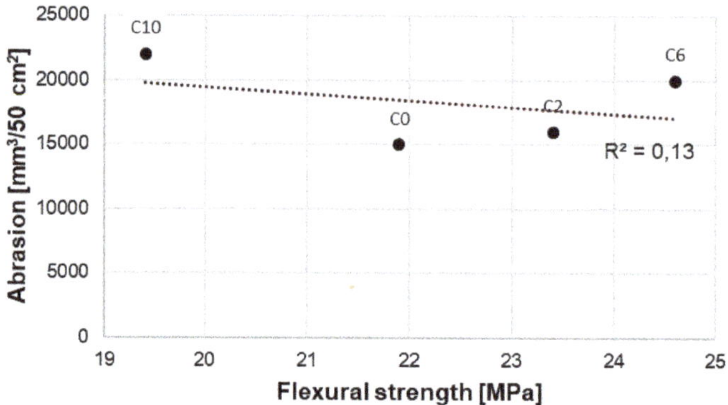

Figure 16. Abrasion—flexural strength relation.

The analysis of abrasion test results of RPC with addition of basalt fibers showed that the resistance to abrasion was associated with the concrete compressive strength (Figure 15). Increased compressive strength of a concrete causes enhanced resistance to abrasion. The highest resistance to abrasion was reached by RPC with basalt fiber content of 2 kg/m^3, which showed the highest compressive strength. Whereas, as the content of basalt fibers in RPC grows, the resistance to abrasion is reduced, which is associated with reduction of the compressive strength of this concrete. No linear relation was found between the abrasion resistance of RPC with basalt fibers and the flexural strength (Figure 16). The above confirms the value of R^2 determination coefficients, which is high ($R^2 = 0.97$) for abrasion dependence on the compressive strength (Figure 15), and low ($R^2 = 0.13$) for abrasion dependence on the flexural strength (Figure 16). It means that no linear relationship between the flexural strength and abrasion occurs, which was found also by means of Pearson correlation coefficient equal to −0.32 (weak correlation). Whereas, for data presented in Figure 15, Pearson correlation coefficient is equal to −0.96.

Different results of testing were obtained by authors of the paper [15], who found that abrasion of the RPC containing basalt fibers depends on the flexural strength rather than on the compressive strength.

The microstructure of RPC with basalt fibers in amount of 2 kg/m^3 showed the highest compressive strength after 28 days of curing is presented in Figures 17–19.

The observation of RPC microstructure under the scanning microscope showed the presence a high content of C-S-H phase bonded to quartz grains (Figure 17) and basalt fibers, which in general, are arranged parallel to each other in the cement matrix (Figures 18 and 19). The high content of C-S-H, analogous to RPC with steel fibers [30], is mainly caused by a large quantity of cement and addition of silica fumes and quartz powder, known for their high reactivity to calcium ions.

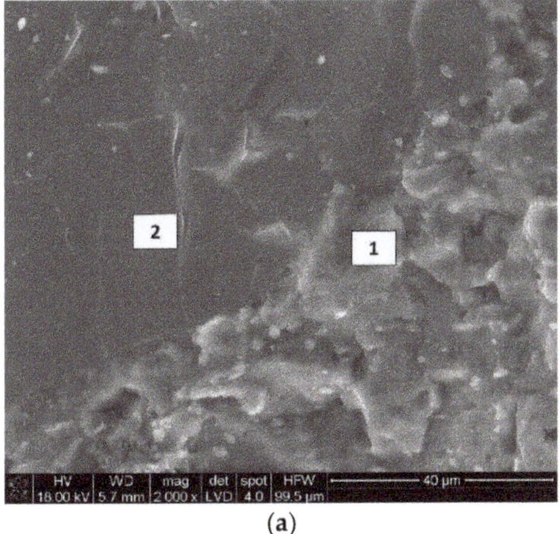

(**a**)

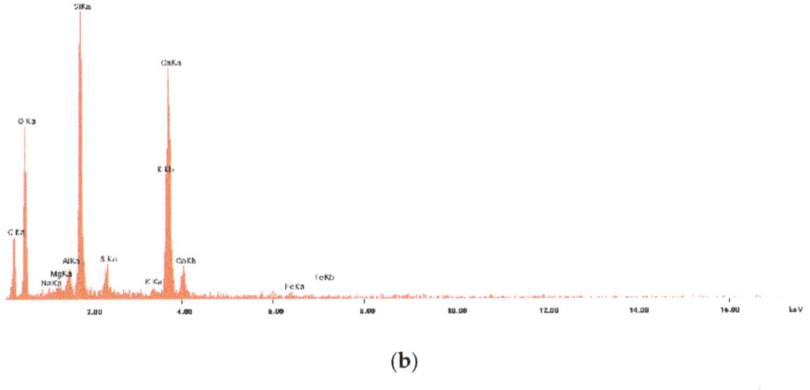

(**b**)

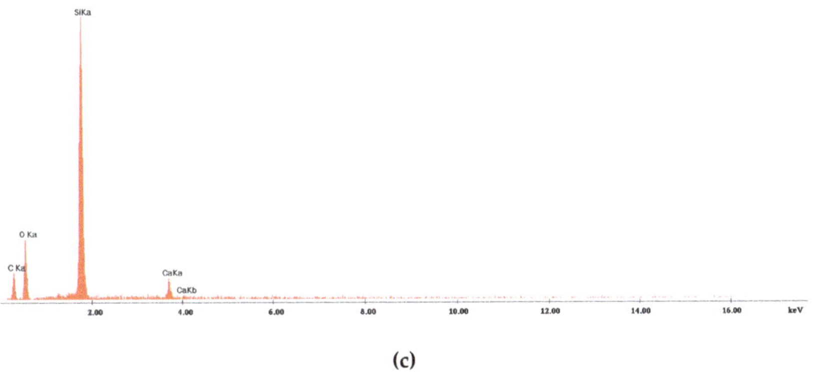

(**c**)

Figure 17. (**a**) RPC microstructure. Very good adhesion of C-S-H (1) on the quartz grain (2) is visible; (**b**) EDS analysis in point 1; (**c**) EDS analysis in point 2.

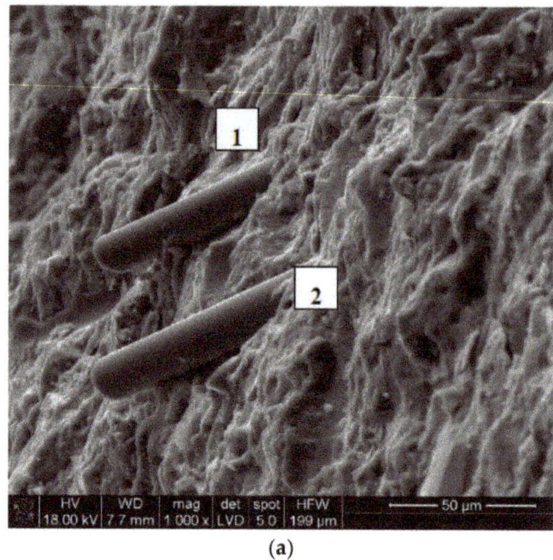

(**a**)

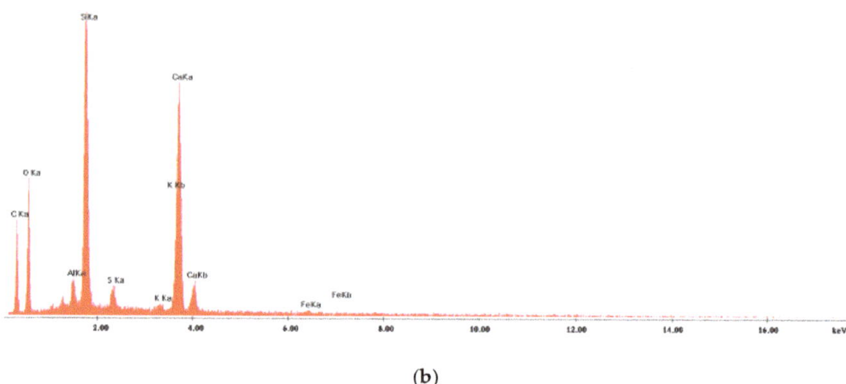

(**b**)

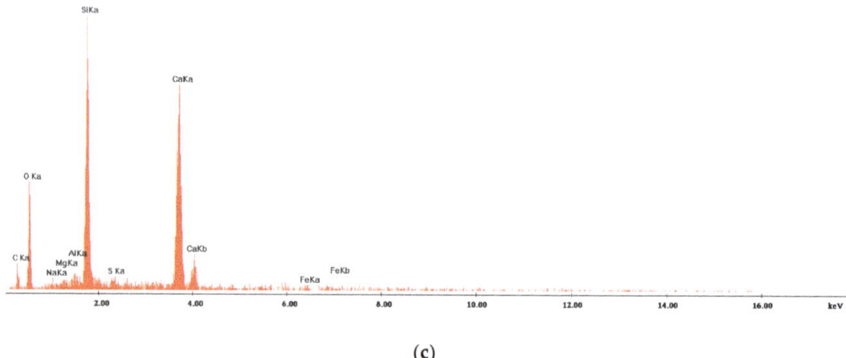

(**c**)

Figure 18. (**a**) RPC microstructure. Visible basalt fiber and C-S-H phase well adhering to the fiber (1, 2); (**b**) EDS analysis in point 1; (**c**) EDS analysis in point 2.

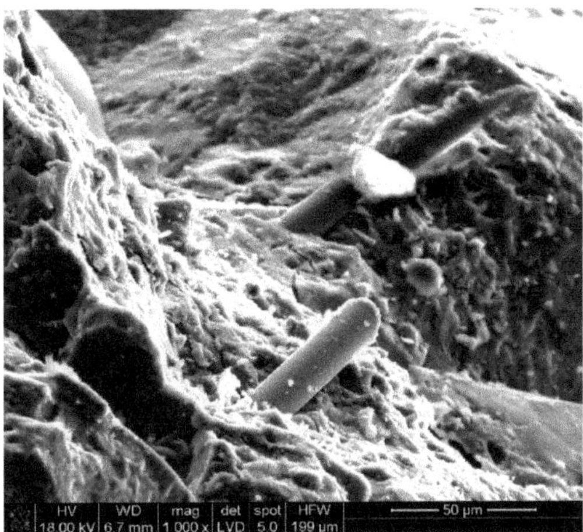

Figure 19. RPC microstructure. Visible parallel arrangement of basalt fibers.

4. Conclusions

The paper impact of basalt fiber on the properties of the RPC mix and hardened concrete.

The results obtained show that the addition of basalt fibers to the RPC leads to a decrease in the workability of concrete mix. It results from a commonly known fact of a larger water demand of concrete mixes with addition of fibers in order to get the required workability.

While keeping the same w/c ratio equal to 0.24, the compressive strength of RPC containing basalt fibers reduces along with the increase of the fiber content in the concrete. The highest drop of the compressive strength from 15.5% to 18.2% with the fiber content increase from 2 kg/m^3 to 10 kg/m^3 is observed after two days. A similar relation is observed after 7 and 28 days of specimen curing, however, the reduction of the compressive strength along with the fiber content increase occurs in a lesser extent over time and it is: 7.8%, and 13.6%, respectively for specimens containing the highest quantity of basalt fibers (10 kg/m^3). A reduction of RPC compressive strength is caused by the presence of a larger quantity of pores along with the increase of the fiber content in this concrete. Whereas, the clear reduction of the total porosity of specimens over time (from 2 to 28 days), due to the increase of cement hydration products formed, weakens the impact of fiber quantity on reduction of RPC compressive strength.

Introduction of basalt fibers to RPC in quantity up to 6 kg/m^3 causes a slight gradual growth in the flexural strength (maximum by 15.9%). Whereas, with the higher content of fibers (8 and 10 kg/m^3), reduction of the flexural strength occurs (maximum by 17.7%), but the differences in flexural strength of RPC specimen with the content 8 kg/m^3 and 10 kg/m^3 are minimal and they are from 0.1 MPa to 0.4 MPa. The increase of the flexural strength along with the increase of the fiber content up to 6 kg/m^3 goes up over time and after 2, 7 and 28 days it amounts to: 4.4%, 15.9%, and 12.3%, respectively. The higher content of basalt fibers in the RPC mix at the lower w/c ratio used (0.24), deteriorates their dispersion, which in consequence, may cause a reduction of the flexural strength.

The reduced resistance to abrasion resistance of RPC along with the increase of basalt fibers content is caused by the reduction of the concrete compressive strength. A strong relationship was found between the RPC resistance to abrasion and its compressive strength. The determination coefficient R^2 for this relationship is 0.97. Whereas, there was no significant relationship between the resistance to abrasion and the flexural strength ($R^2 = 0.13$). The compressive strength is generally recognized as the most important factor that affects the concrete resistance to abrasion. The results abrasion resistance tests of RPC with basalt fibers also confirmed this relationship.

The introduction of basalt fibers to RPC causes the increase of the total porosity, as well as the change in distribution of pores as the content of basalt fibers grows. At the fiber content of 2 kg/m^3, the content of larger pores > 20,000 nm grows, and the content of 20–200 nm pores goes down. Whereas at the higher content of fibers (6 kg/m^3 and 10 kg/m^3), reduction of pores below 20 nm occurs, as well as the increase of pores within a range 200–2000 nm, but the content of pores above 20,000 nm goes down. The above changes in the total quantity of pores and their distribution correspond well with the reduction of RPC compressive strength, as the content of basalt fibers in that concrete increases.

Author Contributions: Conceptualization, S.G. and A.M.-C.; methodology, S.G.; A.M.-C., E.V. and R.Č.; investigation, A.M.-C.; data curation, S.G., A.M.-C.; writing—original draft preparation, S.G. and A.M.-C.; writing—review and editing, S.G., A.M.-C., E.V. and R.Č.; supervision, S.G. and R.Č. All authors have read and agreed to the published version of the manuscript.

Funding: The research was supported by the bilateral Polish-Czech scientist exchange program for years 2020-2021, co-financed by the Polish National Agency for Academic Exchange (NAWA) (Project No. PPN/BCZ/2019/1/00029/U/00001) and by the Ministry of Education, Youth and Sports of the Czech Republic (Project No. 8J20PL031). The Czech Science Foundation Project No. 20-00653S is acknowledged as well.

Conflicts of Interest: The authors declare no conflict of interest.

References

1. Richard, P.; Cheyrezy, M. Composition of reactive powder concretes. *Cem. Concr. Res.* **1995**, *25*, 1501–1511. [CrossRef]
2. Aitcin, P.-C. Cements of yesterday and today, concrete of tomorrow. *Cem. Concr. Res.* **2000**, *30*, 1349–1359. [CrossRef]
3. Blais, P.Y.; Couture, M. Prestressed Pedestrian Bridge–World's First Reactive Powder Concrete Structure. *PCI J.* **1999**, *44*, 60–71. [CrossRef]
4. Matte, V.; Moranville, M. Durability of reactive powder composites: Influence of silica fume on the leaching properties of very low water/binder pastes. *Cem. Concr.Comp.* **1999**, *21*, 1–9. [CrossRef]
5. Staquet, S.; Espion, E. Influence of cement and silica fume type on compressive strength of reactive powder concrete. In Proceedings of the 6th Int. Symp. UHS/HPC, Leipzig, Germany, 22–24 June 2002; pp. 1421–1436.
6. Glinicki, M.A. Concrete with structural fibers. In Proceedings of the XXV Workshops—The Work of Construction Designer, Szczyrk, Poland, 10–13 March 2010; pp. 279–308. (In Polish).
7. Scheinherrová, L.; Vejmelková, E.; Keppert, M.; Bezdička, P.; Doleželová, M.; Krejsová, J.; Grzeszczyk, S.; Matuszek-Chmurowska, A.; Černý, R. Effect of Cu-Zn coated steel fibers on high temperature resistance of reactive powder concrete. *Cem. Concr. Res.* **2019**, *117*, 45–57.
8. Chonghai, D.; Xinwei, M. Experimental research on mechanical properties of basalt fiber reinforced reactive powder concrete. *Adv. Mat. Res.* **2014**, *893*, 610–613.
9. Hager, I.; Zdeb, T.; Krzemień, K. The impact of the amount of polypropylene fibers on spalling behaviour and residua mechanical properties of reactive powder concrete. *MATEC Conf.* **2013**, *6*, 1–8.
10. Fiore, V.; Scalici, T.; Di Bella, G.; Valenza, A. A review on basalt fiber and its composites. *Comp. B* **2015**, *74*, 74–94.
11. Singua, K. A short review on basalt fiber. *Int. J. Tex. Sci.* **2012**, *1*, 18–19.
12. Karwowska, J.; Łapko, A. The usefulness of modern fiber-reinforced concrete in building structures. *Civ. Envir. Eng.* **2011**, *2*, 41–46.
13. Barabanshchkov, Y.; Gutskalov, I. Strength and deformability of fiber reinforced cement paste on the basis of basalt fiber. *Adv. Civ. Eng.* **2016**, 1–5. [CrossRef]
14. Ayub, T.; Shafiq, N.; Nuruddin, M.F. Mechanical properties of high-performance concrete reinforced with basalt fibers. *Pro. Eng.* **2014**, *77*, 131–139. [CrossRef]
15. Kabay, N. Abrasion resistance and fracture energy of concretes with basalt fiber. *Con. Build. Mat.* **2014**, *50*, 95–101. [CrossRef]
16. Jiang, C.; Fan, K.; Wu, F.; Chen, D. Experimental study on the mechanical properties and microstructure of chopped basalt fiber reinforced concrete. *Mat. Des.* **2014**, *58*, 187–193. [CrossRef]
17. Ahmed, T.; Alam, A.; Chufal, M.S. Experimental study on mechanical properties of basalt fiber reinforced concrete. *Int. J. Sci. Res.* **2013**, *4*, 468–472.

18. Li, J.J.; Zhao, Z.M. Study on mechanical properties of basalt fiber reinforced concrete. In Proceedings of the 5th Int. Conf. EMCPE, Zhengzhou, China, 11–12 April 2016; pp. 583–587.
19. Morozov, N.M.; Borovskich, I.V.; Khozin, V.G. Sand basalt-fiber concrete. *World Appl. Sci. J.* **2013**, *25*, 832–838.
20. Atiş, C.D. Abrasion-porosity-strength model for fly ash concrete. *J. Mater. Civ. Eng.* **2013**, *15*, 408–410. [CrossRef]
21. Funk, J.; Dinger, D. *Predictive Process Control of Crowded Particulate Suspensions—Applied to Ceramic Manufacturing*; Kluver Academic Publishers: Dordrecht, The Netherlands, 1994.
22. Larrard, F.; Sedran, T. Optimalization of ultra-high-performance concrete by use of a packing model. *Cem. Concr. Res.* **1994**, *24*, 997–1009. [CrossRef]
23. Zdeb, T. Ultra-high performance concrete–properties and technology. *Bull. Pol. Ac.: Tech.* **2013**, *61*, 183–193. [CrossRef]
24. PN-EN 1015-3. *Methods of Test for Mortar for Masonry—Part 3: Determination of Consistence of Fresh Mortar (by Flow Table)*; BSI Group Polska Sp. z o.o.: Warsaw, Poland, 2000.
25. PN-EN 1015-11. *Methods of Test for Mortar for Masonry—Part 11: Determination of Flexural and Compressive Strength of Hardened Mortar*; BSI Group Polska Sp. z o.o.: Warsaw, Poland, 2001.
26. PN-EN 1338. *Concrete Paving Blocks. Requirements and Test Methods*; BSI Group Polska Sp. z o.o.: Warsaw, Poland, 2005.
27. Głodkowska, W. Waste Sand Fiber Composite—Models of Description of Properties Application. *Annu. Set Environ. Prot.* **2018**, *20*, 3. (In Polish)
28. Odler, I.; Rößler, M. Investigations on the relationship between porosity, structure and strength of hydrated Portland cement pastes. II. Effect of pore structure and of degree of hydration. *Cem. Concr. Res.* **1985**, *15*, 401–410. [CrossRef]
29. Kumar, R.; Bhattachrjee, B. Porosity, pore size distribution and in situ strength of concrete. *Cem. Concr. Res.* **2003**, *33*, 155–164. [CrossRef]
30. Grzeszczyk, S.; Matuszek-Chmurowska, A.; Černý, R.; Vejmelková, E. Microstructure of reactive powder concrete. *Cem. Lime Concr.* **2018**, *1*, 1–15.

Article

Characterization of Porous Cementitious Materials Using Microscopic Image Processing and X-ray CT Analysis

Jinyoung Yoon [1], Hyunjun Kim [2], Sung-Han Sim [2] and Sukhoon Pyo [3,*]

[1] Department of Civil and Environmental Engineering, The Pennsylvania State University, State College, PA 16801, USA; jpy5278@psu.edu

[2] School of Civil, Architectural Engineering and Landscape Architecture, Sungkyunkwan University, Suwon 16419, Korea; hyunjun@skku.edu (H.K.); ssim@skku.edu (S.-H.S.)

[3] School of Urban and Environmental Engineering, Ulsan National Institute of Science and Technology (UNIST), Ulsan 44919, Korea

* Correspondence: shpyo@unist.ac.kr

Received: 4 June 2020; Accepted: 10 July 2020; Published: 12 July 2020

Abstract: The use of lightweight concrete has continuously increased because it has a primary benefit of reducing dead load in a concrete infrastructure. Various properties of lightweight concrete, such as compressive strength, elastic modulus, sound absorption performance, and thermal insulation, are highly related to its pore characteristics. Consequently, the identification of the characteristics of its pores is an important task. This study performs a comparative analysis for characterizing the pores in cementitious materials using three different testing methods: a water absorption test, microscopic image processing, and X-ray computed tomography (X-ray CT) analysis. For all 12 porous cementitious materials, conventional water absorption test was conducted to obtain their water permeable porosities. Using the microscopic image processing method, various characteristics of pores were identified in terms of the 2D pore ratio (i.e., ratio of pore area to total surface area), the pore size, and the number of pores in the cross-sectional area. The 3D tomographic image-based X-ray CT analysis was conducted for the selected samples to show the 3D pore ratio (i.e., ratio of pore volume to total volume), the pore size, the spatial distribution of pores along the height direction of specimen, and open and closed pores. Based on the experimental results, the relationships of oven-dried density with these porosities were identified. Research findings revealed that the complementary use of these testing methods is beneficial for analyzing the characteristics of pores in cementitious materials.

Keywords: microscopic image processing; X-ray CT analysis; porous cementitious materials; 3D tomographic image

---

1. Introduction

In recent years, lightweight cementitious materials have extensively been applied in a concrete infrastructure due to their primary benefit of reducing dead load in structures. Many types of lightweight cementitious materials—such as lightweight aggregate concrete (LWAC), pervious concrete, and aerated concrete—have been developed for various purposes. LWAC typically comprises cement, lightweight aggregate (LWA), water, and mineral and chemical admixtures. In general, the ranges of density and compressive strength of LWAC are 1460–1910 kg/m^3 and 36.5–60.0 MPa, respectively [1–4]. Because of its low density and moderate strength level for structural purposes, LWAC has successfully been applied for bridge components [5,6]. Pervious concrete, also called porous concrete, has a macro-porous structure using gap-graded coarse aggregates without fine aggregates in the mixture [7,8]. In pervious concrete, the coarse aggregate particles are typically coated in a thick layer of cementitious paste, resulting in a formation of highly interconnected macro-pore network [9]. To enhance the pore ratio of pervious concrete, natural and artificial fibers can be used despite the reduction of strength and freeze-thaw resistance [10]. Pervious concrete has widely been used for pavements, vegetation concrete beds, and noise absorbing concrete [11,12]. Its ranges of porosity and compressive strength are typically 15–25% and 22–39 MPa, respectively [12]. Aerated concrete has a large number of uniformly distributed small pores that can be generated by adding metallic powder (e.g., aluminum and zinc) and/or a foaming agent (e.g., glue resin and saponin) [13–15]. Because of its good heat conservation and low density (300–1800 kg/m^3), aerated concrete can be used as a high-efficiency heat-insulating material in roofs, walls, and floors—as a form of block and panel—despite its relatively low mechanical properties [16–18]. As various properties of lightweight cementitious materials, such as density, compressive strength, elastic modulus, water drainage, heat conservation, and noise absorption, are highly related to their pore characteristics, various methods for analyzing the pore structures have been introduced [19–24].

Even though many testing methods have been utilized for analyzing the pore structures of cementitious materials, a suitable testing method should be carefully selected by considering the properties of testing specimens and the limitations of techniques. The most convenient and widely used method is the water absorption test, which measures the difference between the saturated and oven-dried masses of a testing specimen [25,26]. This test provides the total volume of water permeable pores by filling internal pores with water. Similarly, a gas pycnometer can determine the porosity and density of a testing sample using nearly ideal gas (e.g., helium or nitrogen) [27,28]. Although this method limits the size of a testing specimen, due to its small chamber size, the volume of the solid phase can be measured more accurately by detecting the pressure change caused by the gas displacement than with the water absorption test [28]. However, neither the water saturation nor the gas pycnometer method can provide pore characteristics, such as pore size and their spatial distribution. The mercury intrusion porosimetry (MIP) method can evaluate various pore properties (porosity and pore size) by measuring the volume of intruded non-wetting liquid (e.g., mercury) at different pressures [29,30]. However, the MIP method also has some drawbacks regarding the accessibility issue of mercury in the pore throat, resulting in a true pore diameter measuring error [31,32].

As a way to make up for the weakness of the aforementioned methods, the image processing and X-ray computed tomography (X-ray CT) can be applied for analyzing the characteristics of pores based on two-dimensional (2D) and three-dimensional (3D) visual images of specimens, respectively. Using a digital camera and/or a microscope, the image processing technique was conducted to investigate the size of pores, its spatial distributions, and 2D pore ratios [33,34]. This image processing approach generally consists of a collection of high-resolution images, a conversion of an RGB image into a grayscale image, a determination of a threshold, and image binarization [35]. With the use of a high-quality binary image, the detection and quantification of pores can be carried out well. However, the difference between the porosity and 2D pore ratio obtained from the image processing should be considered. Similarly, X-ray computed tomography (X-ray CT) is also available for pore structure analysis based on a 3D tomographic image of a sample [36,37]. Because X-ray CT analysis scans the entire testing specimen and virtually reconstructs actual pore structures, this method can provide information about the internal pore microstructure, 3D pore ratio, pore size, and their spatial distribution [38,39]. Nevertheless, this test is rather expensive for the analysis and requires time-consuming post processing due to a large scanning data that needs to be analyzed. Despite their drawbacks, these testing methods are beneficial for investigating the characteristics of the pore structure based on the images. However, as a comparative analysis of different methods—conventional method, image processing, and X-ray CT—has not been fully studied yet, further research is still needed to select a suitable testing method for highly porous cementitious materials.

In this study, a comparative analysis that uses three different methods—conventional water absorption test, microscopic image processing and X-ray CT analysis—was conducted for investigating highly porous structures of cementitious materials. All 12 porous cementitious material series were prepared using different amounts of pore generation materials of aluminum powder and natural fibers. To analyze the pore structures, a conventional water absorption test was first conducted in order to obtain their water permeable porosities. In addition, their bulk specific gravities were measured. The microscopic image processing method, which uses a local thresholding algorithm, was adopted to characterize the pores in terms of the 2D pore ratio, the pore size, and the number of pores in a cross-section. X-ray CT analysis was also used to provide information about the pore structures—the 3D pore ratio, the pore size, the number of pores, and their spatial distribution—on the basis of a 3D tomographic image.

2. Materials and Mix Proportions

To fabricate highly porous cementitious materials, aluminum powder (ECKART, Hartenstein, Germany) and natural fibers (Soo Industry, Gyeongju, Republic of Korea) were added to the mixtures. The water-to-binder ratio (w/b) of cementitious materials was fixed at 0.3. The binders used in the cementitious materials were CEM I 42.5 R ordinary Portland cement (OPC) (Ssangyong Cement, Seoul, Republic of Korea) and silica fume (Elkem, Oslo, Norway). The specific gravity and Blaine fineness of the OPC were 3160 kg/m^3 and 330 m^2/kg, respectively. Silica fume had a specific gravity of 2270 kg/m^3 and replaced 10 wt% of the OPC in order to control viscosity, as well as prevent segregation. The length of the kenaf natural fibers was approximately 1 mm. These fibers consisted of cellulose of 45–57 wt%, hemicellulose of 22 wt%, and lignin of 8–13 wt%, having a tensile strength and Young's modulus of 930 MPa and 53 GPa, respectively. Its specific gravity was 1800 kg/m^3, which was determined using a gas pycnometer (Micromeritics AccuPyc 1330, Norcross, GA, USA). The amount of natural fibers used in the cementitious materials was controlled up to 5.0 wt% of mixtures. Aluminum powder was a flake type with a 99.7% purity. Its specific gravity and covering capacity were 2700 kg/m^3 and 1100–1450 m^2/kg, respectively. The size of the aluminum powder particles ranged from 12 to 86 μm, with a mean size of 34 μm, as measured using laser diffraction (Sympatec Helos, Clausthal-Zellerfeld, Germany). The amount of aluminum powder was controlled up to 0.1 wt% of binders. A polycarboxylate-based superplasticizer (MasterGlenium SKY 8808, Ludwigshafen, Germany) was added in order to enhance the workability, as well as the homogeneity of mixtures. The dosage of superplasticizer was fixed at 2.0 wt% of binders. As provided in Table 1, a total of 12 mixtures were fabricated following the mixing process. OPC, silica fume, and aluminum powder (if any) were first dry mixed. Then, superplasticizer in liquid form was added to the mixing water. Natural fibers (if any) were added to the superplasticizer-concentrated water. Subsequently, all constituents were mixed in a planetary mixer for 10 min. The fresh mixtures were then cast into a 50 mm cube mold in order to create the specimens to be used for analyzing pore structure and density. The samples were cured in a water chamber for 28 days at a temperature of 23 °C.

Table 1. Mix proportions of porous cementitious materials.

Label	Mix Proportion (g)						
	w/b[1]	Water	Cement	Silica Fume	NF[2]	AP[3]	SP[4]
PCM1-1	0.3	660	2000	200	0	0	44.0
PCM1-2						1.1	
PCM1-3						2.2	
PCM2-1					28.6	0	
PCM2-2						1.1	
PCM2-3						2.2	
PCM3-1					85.8	0	
PCM3-2						1.1	
PCM3-3						2.2	
PCM4-1					143.0	0	
PCM4-2						1.1	
PCM4-3						2.2	

w/b[1]: water-to-binder ratio,NF[2]: Natural fibers, AP[3]: Aluminum powder, SP[4]: Superplasticizer.

3. Test Methods for Analyzing the Pore Structures

3.1. Water Absorption Test

A water absorption test was conducted to determine the water permeable porosity of cementitious materials following ASTM (American Society for Testing and Materials) C 642. First, the oven-dried mass (M_{oven}) of the testing specimen was measured, which was previously dried in an oven at a temperature of 110 ± 5 °C for not less than 24 h. Afterward, the saturated surface-dried mass (M_{SSD}) was determined by measuring the mass of the specimen that was immersed in water for 7 days to saturate porous structure of samples. Subsequently, the water absorption capacity (P_{wat}) was calculated as [25]

$$P_{wat}[\%] = (M_{SSD} - M_{oven})/M_{oven} \times 100 \quad (1)$$

The water absorption capacity stands for the water permeable porosity of the testing sample. Considering very low density of testing specimens, their dry bulk densities were measured following ASTM C 1693 [40]. The dry bulk density of the specimen was equal to the oven-dried mass divided by the volume of the sample (i.e., 125,000 mm^3 in this study). Both the water permeable porosity and the dry bulk density were determined by the average value of 6 specimens in the form of 50 mm cubes.

3.2. Microscopic Image Processing

The aims of microscopic image processing are to distinguish the cementitious matrix and pores and to provide information about the pore size, the pore ratio, and the number of pores in a tested cross-section. This method was performed in three parts: acquisition of microscopic image, image binarization, and characterization of pores (see Figure 1). The cross-sections of specimens were prepared using a cutter (Allied High Tech Products Inc PowerCut 10x, Compton, CA, USA) at the height of approximately 25 mm from the bottom. The cross-sectional images of these samples were collected using an optical microscope (ZEISS Axio Zoom V16 and ZEISS PlanApo Z 0.5x, Oberkochen, Germany). It should be noted that no observable cracks were identified in microscopic images, which indicates that the cutting process did not affect the characterization of pore in cementitious materials. The size of the observed area via the optical microscope was a 28.64 × 35.70 mm^2 rectangular region with a 12.97 µm/pixel resolution. Through the use of a lateral light source, a shadow was created in the pore, which enabled image processing to classify both the solid phase and the pores well. Subsequently, the image obtained was converted into grayscale, with a brightness range from 0 to 255. During the binarization process, the brightness of each pixel in an image was compared to a threshold. As provided in Equation (2), the brightness of pixels that was less than the threshold was designated as zero, while those having higher values became one—resulting in a binary image (B_{img}).

$$B_{img}(i,j) = 0 \text{ if } M_{img}(i,j) < t, \text{otherwise } B_{img}(i,j) = 1, \text{ where } 1 \le i \le a \text{ and } 1 \le j \le b, \quad (2)$$

where $M_{img}(i,j)$ is the pixel intensity of grayscale image, t is the brightness threshold, and a and b are the number of pixels along the width and height. Lastly, using the binary image, the 2D pore ratio—the ratio of the number of pixels for pores to the total number of pixels in the image—and pore size—equivalent circular area diameter of irregularly shaped pores—were numerically calculated. Because these values are influenced by the quality of binary image, a suitable threshold value was carefully determined in order to obtain a reliable classification of pores and cementitious matrix.

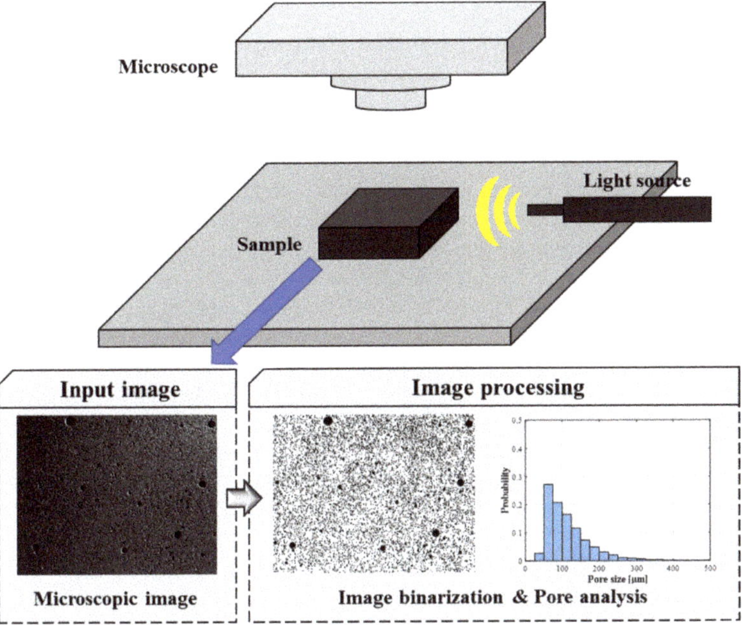

Figure 1. Schematic diagram for the microstructural image processing method.

The threshold value can be determined using the global and local thresholding methods. The global method (e.g., the Otsu method) uses one threshold value for the binarization of the entire image [41,42]. Although this method is computationally simple and quick, the threshold value depends on the operator or histogram of pixel intensity. Consequently, a poor binarization result can be obtained if the image had a noisy or complex background [42]. On the other hand, the local thresholding method uses the brightness of detected pixels in a neighborhood around a pixel (i.e., window) for the calculation of the unique local threshold value. Although this local method demands more computational cost, the local characteristics of pixels in varying background images can be identified well, including illuminated or degraded images [43,44].

Because non-uniform illumination microscopic images—caused by the lateral light source (see Figure 1)—were used in this study, a local thresholding method was selected. More specifically, Sauvola's image binarization algorithm—used for document analysis and concrete crack identification—was adapted, as provided in Equation (3) [44,45].

$$T_{Sauvola} = m \cdot (1 - k \times (1 - s/R)) \tag{3}$$

where $T_{Sauvola}$ is threshold value, k is the sensitivity, R is the normalized standard deviation, and m and s are the average and standard deviation of the brightness value of a pixel in a selected local area (window), respectively. In this study, the size of the window and the sensitivity used for the computation were user-defined values. The sensitivity value was fixed at 0.5, which showed a good result in the previous study [44]. Although this sensitivity caused several false detections in PCM1-1 and PCM1-2, due to their rough surfaces, the errors were very small and acceptable because of their lower porosity. The optimal window size was determined to be in the range of 10–300 pixels when comparing changes of the total number of detected pixels for the pores. This value could reach the plateau value, if the classification of the cementitious matrix and the pores were performed well.

3.3. X-ray CT Analysis

X-ray CT has strong advantages for analyzing the internal structure of cementitious materials, providing information about internal defects (i.e., pores), as well as visualized 3D anatomical images [46–48]. 3D images from X-ray CT can be obtained through two steps: image acquisition and reconstruction. First, a microfocus X-ray beam passes through a sample and 2D X-ray projection images are acquired. The sample that is mounted on the stage is rotated in the 180°–360° range and additional 2D projections are obtained, typically 500–1000 images. Afterward, these projection images are converted into a complete 3D tomographic image of the sample, through a reconstruction process using a back-projection algorithm. Further analysis of the 3D image provides the volumetric information of the solid phase and internal defects on the basis of 3D volumetric elements (voxels). In the tomographic image, a denser phase has a high X-ray attenuation and is displayed as a bright color, whereas a low-density phase—consisting of air or gas—would appear as quite dark due to little X-ray attenuation. Typically, no special pre-processing is required for sample preparation. In a porosity analysis using a software of VG Studio and myVGL 3.0 (Volume Graphics, Heidelberg, Germany), the defect detection algorithm compared the potential defect with its local neighborhood. If the appearance of a potential defect was very similar to the surrounding structure, its probability was reduced, which started from zero. Based on preliminary tests, probability threshold of 0.5 was selected to classify pores and cementitious matrix. The value for minimum pore size was no less than 8 voxels edge lengths (2 × 2 × 2 voxels).

In this study, X-ray CT (Nikon Metrology XT H 320, Tokyo, Japan) was used for characterizing the porous internal structures of cementitious materials as shown in Figure 2. A 50 mm cube specimen was mounted on the 360° rotational stage and the 3D image was acquired at a 230 kV accelerating voltage and a 300 μA current. A total of 1140 pictures were obtained from each X-ray CT analysis, taking 354 ms for each image projection. These 2D images were converted into 3D tomographic images consisting of voxels with an edge length of 48 μm. A total of four cementitious samples were examined through the X-ray CT test. Their porosity and pore size results were compared to the water absorption capacity and microscopic image processing-based analysis. Furthermore, the volume of open and closed pores in cementitious materials can be analyzed. The open and closed pores stand for a cavity or channel with access to an external surface and a pore not connected to the surface, respectively. Because of these advantages of X-ray CT, this test could be applied for various purposes, such as internal structure analysis of asphalt concrete [49] and permeability of oil-well cement [50].

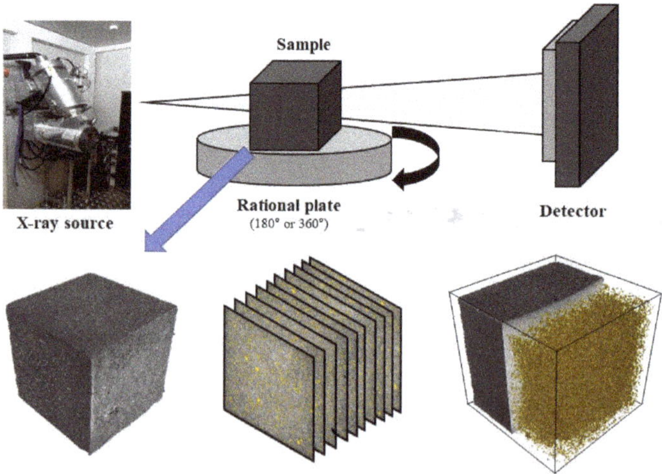

Figure 2. Schematic diagram of the X-ray CT analysis.

4. Experimental Results and Discussion

4.1. Density and Water Absorption Capacity

It is known that porous cementitious materials generally have a low density with a high-water absorption capacity [37], which was also identified in this study. As shown in Figure 3, the specimens—which incorporated aluminum powder and natural fibers—showed a decrease in the oven-dried density and an increase in the water absorption capacity. In the PCM1 series, which contained 0–0.1 wt% of aluminum powder, the densities were drastically decreased from 1859 kg/m^3 to 944 kg/m^3. The PCM2-1, PCM3-1, and PCM4-1 samples, which incorporated 1.0–5.0 wt% of natural fibers, also showed a decrease in densities from 1797 kg/m^3 to 944 kg/m^3. The synergistic effect of these two materials was also observed. For example, the PCM4-3 sample, which contained 0.1 wt% of aluminum powder and 5.0 wt% of natural fibers, showed the lowest density of 792 kg/m^3. The decrease of the oven-dried density for cementitious materials was highly related to the reaction of aluminum powder and the dispersion of natural fibers. More specifically, the chemical reaction of aluminum powder with calcium hydroxide and water generated air bubbles (hydrogen gas) in the fresh state, resulting in the formation of porous structures [51,52]. In case of natural fibers, the dispersion of fibers during the mixing process can create entrapped air pores in the matrix phase [10,37]. Therefore, the use of these pore generation materials was effective for making highly porous cementitious materials that have low densities.

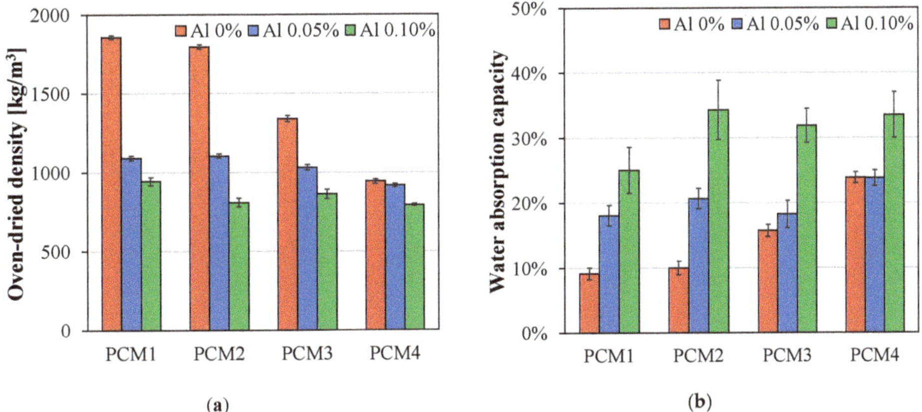

Figure 3. (**a**) Oven-dried density and (**b**) water absorption capacity of the tested porous cementitious materials.

It was also revealed that the water absorption capacities of the samples were inversely proportional to their densities, as shown in Figure 3b. In the PCM1 series, which incorporated 0–0.1 wt% of aluminum powder, the water permeable porosity significantly increased from 9.2% to 25.1%. The PCM2-1, PCM3-1, and PCM4-1 samples, which contained 1.0–5.0 wt% of natural fibers, also showed an increase in the porosity from 10.0% to 23.9%. The combination of aluminum powder and natural fibers significantly increased the porosity of the samples. Based on this experimental work, the oven-dried density and water absorption capacity of samples were used as references for a comparison with the results obtained from microscopic image processing and X-ray CT analysis.

4.2. Microscopic Image Processing Analysis

Following Section 3.2, microscopic image processing was conducted in order to characterize the porous structures of cementitious materials in terms of their 2D pore ratio, the pore size and the number of pores. Because of the non-uniform illumination characteristics of microscopic images, the local thresholding method (Sauvola's method) was applied for the image binarization. In this section, the optimal window size for the local method was identified to obtain a high-quality binary image. Using this binary image, the characteristics of pores were numerically calculated.

4.2.1. Image Binarization Using the Local Thresholding Method

In the analysis of microscopic image processing, a suitable user-defined parameter—window size—for the local method was first determined to obtain high quality binary image. In Figures 4–6, the results obtained from representative samples—the less porous PCM1-1 and the highly porous PCM3-3—are provided in order to show the effects of different window sizes on image binarization. First, the total number of pixels for pores in the cross-section was counted using different window sizes of 10–300 pixels (see Figure 4). As the size of the window increased, the number of pixels for pores also rapidly increased, showing a plateau convergence at a 50-pixel window for PCM1-1 and PCM3-3. This result indicated that the classification of pores can be achieved well using a window size at this convergence value. Therefore, the window sizes of 50, 100, and 200 pixels were selected to evaluate the quality of binary images. As shown in Figure 5, the binary images for the PCM1-1 sample did not show a big difference, regardless of the window size. On the contrary, the large window size of 200 × 200 pixels showed good binarization performance for the highly porous PCM3-3 sample (see Figure 6). Small windows were not enough to cover several large-sized pores in the PCM3-3 sample, resulting in a false binarization for the inside of some large pores. Hence, the 200 × 200 window size was selected for Sauvola's algorithm in microscopic image processing.

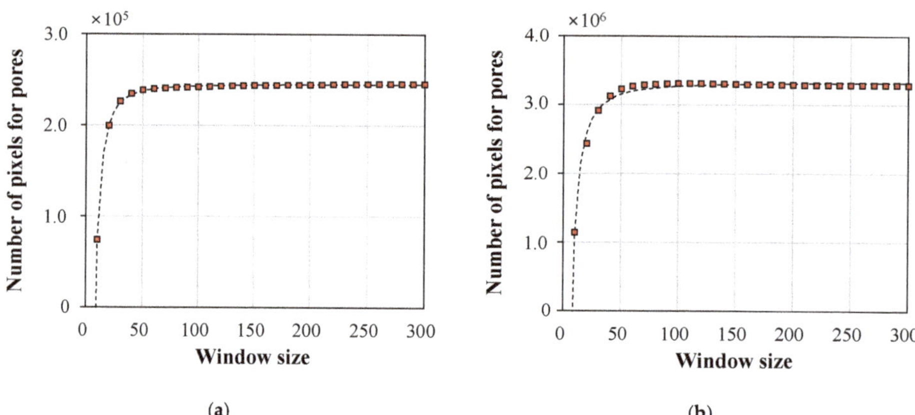

Figure 4. The relationship of the number of pixels designated as pores with the window size: (**a**) PCM1-1 and (**b**) PCM3-3.

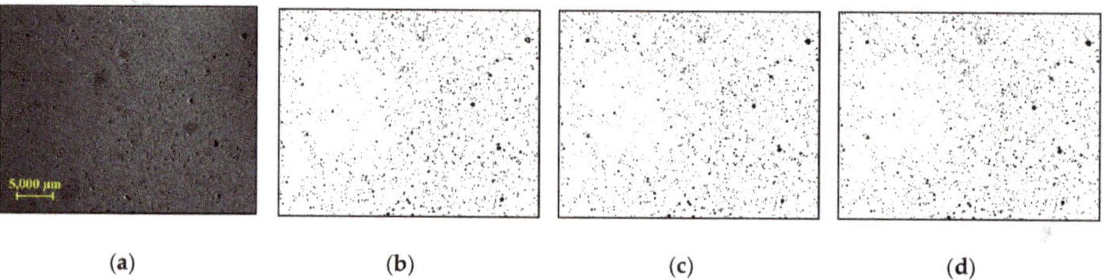

Figure 5. PCM1-1: (**a**) microscopic image; (**b–d**) binary images using 50 × 50, 100 × 100, and 200 × 200 windows, respectively.

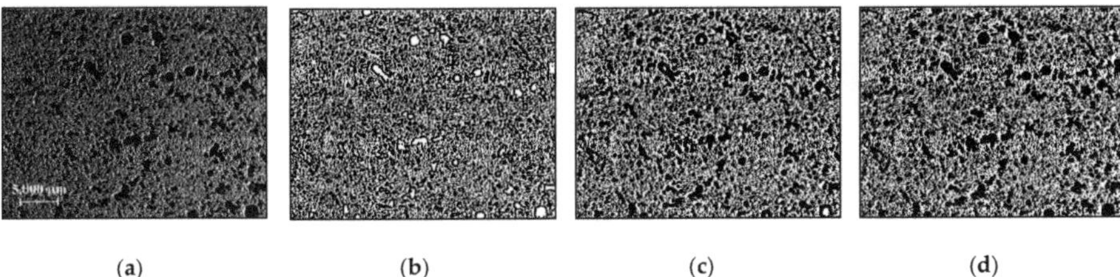

(a) (b) (c) (d)

Figure 6. PCM3-3: (**a**) microscopic image; (**b–d**) binary images using 50 × 50, 100 × 100, and 200 × 200 windows, respectively.

4.2.2. Characteristics of Pores

Using binary images, the characteristics of pores can be identified in terms of the 2D pore ratio, the pore size, and the number of pores, as provided in Table 2. In the PCM1 series, the 2D pore ratio increased from 4.0% to 43.9% with aluminum powder of 0–0.1 wt%. The PCM2-1, PCM3-1, and PCM4-1 samples, containing natural fibers of 1.0–5.0 wt%, showed a 2D pore ratio of 6.7–26.5%. Here, PCM3-3 showed the highest 2D pore ratio of 54.2%, using the combination of aluminum powder of 0.1 wt% and natural fibers of 3.0 wt%.

Table 2. Microscopic image processing-based pore analysis using the local method.

Measurement	Label	Aluminum Powder		
		0%	0.05%	0.10%
2D Pore Ratio	PCM1	4.0%	35.6%	43.9%
	PCM2 (NF 1%)	6.7%	36.9%	49.6%
	PCM3 (NF 3%)	19.3%	34.7%	54.2%
	PCM4 (NF 5%)	26.5%	48.9%	51.1%
Mean Pore Size (μm)	PCM1	103.2	167.3	200.9
	PCM2 (NF 1%)	91.7	161.0	242.1
	PCM3 (NF 3%)	101.7	147.4	212.9
	PCM4 (NF 5%)	141.2	173.3	215.0
Total Number of Pores	PCM1	3568	10,317	8397
	PCM2 (NF 1%)	8234	11,669	4914
	PCM3 (NF 3%)	18,421	13,384	5282
	PCM4 (NF 5%)	10,513	7984	6023

In the analysis of the mean pore size, it was revealed that the mean pore size was increased with a dosage of 0–0.1 wt% aluminum powder. The mean pore size of the PCM1 series ranged from 103.2 μm for PCM1-1 to 200.9 μm for PCM1-3 because of pore generation resulting from the aluminum powder reaction. In addition, the increase in the mean pore size was observed for PCM2-1, PCM3-1, and PCM4-1 as the amount of natural fibers increased. The relationship between the 2D pore ratio and the mean pore diameter was linearly proportional for all 12 samples, as shown in Figure 7.

The number of pores did not proportionally increase as the amount of aluminum powder and natural fibers increased. This is because the number of pores for a given observed area was decreased as the size of pores increased for highly porous cementitious materials. Meanwhile, PCM1-1, PCM2-1, and PCM3-1 samples having low 2D pore ratio and relatively small mean pore size showed an increase of the number of pores as the amount of natural fiber increased.

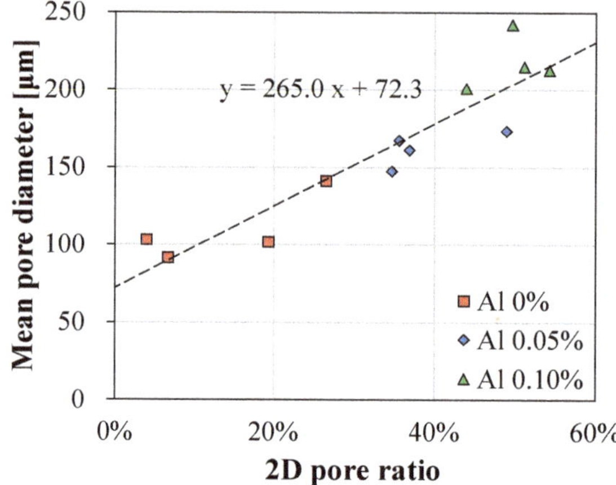

Figure 7. The relationship of the 2D pore ratio with the mean pore diameter obtained using the local thresholding image processing method.

4.3. X-ray CT Analysis

X-ray CT analysis provided cross-sectional and 3D tomographic images of the porous cementitious materials, as well as information about the 3D pore ratio, the pore size, and the number of pores. In order to show the effects of aluminum powder and natural fibers, four samples—PCM1-1, PCM 1-3, PCM3-1, and PCM3-3—were selected for the X-ray CT analysis. Figure 8 shows the cross-sectional images of PCM1-1 and PCM3-3. As discussed in Section 4.2, the PCM1-1 control sample had a less porous structure in comparison to the PCM3-3, which incorporated aluminum powder of 0.1 wt% and natural fiber of 3.0 wt%. In Figure 8, the colors of open and closed pores are designated as black and yellow, respectively.

(a) (b)

Figure 8. Cross-sectional images of (**a**) PCM1-1 and (**b**) PCM3-3 from the X-ray CT analysis.

Based on the volumes of the solid phase and the closed pores, the volume of the open pores, as well as the porosity of the samples, can be calculated. It was assumed that the porous concrete specimen consisted of the cement matrix phase, as well as open and closed pores. Then, the volume of the open pores was calculated by subtracting the volumes of the solid phase and the closed pores from the sample volume. The volume of the 50 mm cube specimen was assumed to be 125,000 mm^3. The volumes of the solid phase and the open and closed pores were provided in Table 3. The total 3D pore ratios of the PCM1-1, PCM1-3, PCM3-1, and PCM3-3 samples were 8.4%, 21.3%, 9.3%, and 23.3%, respectively. The difference between the water permeable porosity, the 2D pore ratio, and the 3D pore ratio is discussed in Section 4.4. Their mean pore sizes were 347.2 μm, 302.2 μm, 335.7 μm, and 360.4 μm, respectively, where the high values were caused by the limited voxel resolution of X-ray CT. The total numbers of closed pores for PCM1-1, PCM1-3, PCM3-1, and PCM3-3 were 194,425, 212,650, 133,471, and 272,721, respectively.

Table 3. X-ray CT-based analysis for the porous structures of PCM1-1, PCM1-3, PCM3-1, and PCM3-3.

Label	Volume (mm^3)			3D Pore Ratio (Open/Closed)	Mean Pore Size (μm)
	Solid Phase	Closed Pore	Open Pore		
PCM1-1	114,451	5616	4933	8.4% (3.9%/4.5%)	3472
PCM1-3	98,395	21,443	5162	21.3% (4.1%/17.2%)	3022
PCM3-1	113,337	7325	4339	9.3% (3.5%/5.9%)	3357
PCM3-3	95,860	18,312	9528	23.3% (8.7%/14.6%)	3604

X-ray CT analysis can be applied for evaluating the distribution of the pore ratio based on the image processing technique [53]. Here, the cross-sectional images obtained from X-ray CT were collected along the height direction at 1 mm intervals. As shown in Figure 9, the 2D pore ratios of PCM1-1, PCM1-3, PCM3-1, and PCM3-3 were 1.5–3.3%, 4.7–6.2%, 28.9–32.4%, and 29.5–37.8%, respectively. This result revealed that the PCM1-1, PCM1-3, and PCM3-1 samples had homogeneous distributions of the pore ratio along the height direction. In case of the PCM3-3 sample, which incorporated both aluminum powder and natural fibers, the pore ratio was inconsistent along the height. This is because some natural fibers agglomerated on the top surface, resulting in the formation of a fiber ball, as well as in the reduction of the pore ratio in the matrix [54].

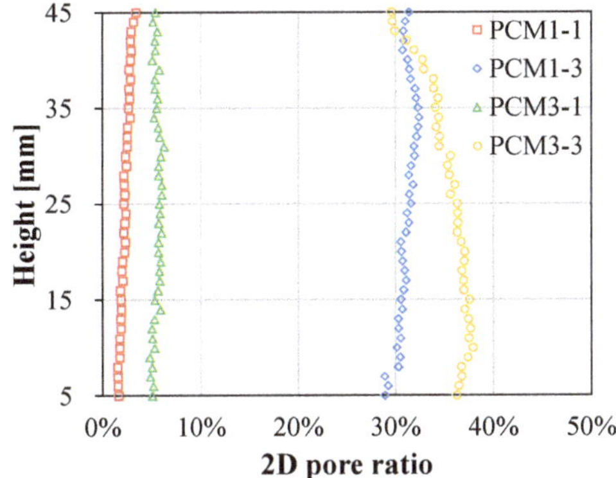

Figure 9. Distribution of the 2D pore ratios obtained through X-ray CT analysis along the height direction.

The difference between the average 2D pore ratio and 3D pore ratio obtained from the X-ray CT were identified. The average 2D pore ratios were 2.2%, 31.0%, 5.4%, and 35.2% for PCM1-1, PCM1-3, PCM3-1, and PCM3-3, respectively. These results were different from their 3D pore ratios of 8.4%, 21.3%, 9.3%, and 23.3%, respectively. This indicated that the less porous sample had a relatively low 2D pore ratio, while the highly porous one had a high 2D pore ratio. This was caused by the different calculation for area of circle and volume of sphere. A detailed discussion is provided in Section 4.4.

As provided in this experimental result, X-ray CT analysis was beneficial for analyzing various characteristics of pores, such as open and closed pores, pore distribution in 3D space, and 3D tomographic images. Furthermore, the homogeneous pore distribution can be evaluated using the cross-sectional image processing. These results are highly related to the unique properties of porous cementitious materials, such as heat insulation, sound absorption, and water drainage. Consequently, more diverse applications of the X-ray CT method are expected in the field of construction materials.

4.4. Comparative Analysis

In this section, the characteristics of the porous structures obtained from the water absorption test, microscopic image processing, and X-ray CT analysis were compared. First, the relationships of the oven-dried densities and porosities determined by these different testing methods were investigated, as shown in Figure 10. The regression curves for the water permeable porosity and the 3D pore ratio showed similar trends. This is because both methods determined the porosities of the specimens by considering all 3D pores in them. Furthermore, testing four specimens for X-ray CT analysis—covering a wide range of porosities and densities—was adequate to estimate the trend for all 12 samples. It should be noted that the difference of the 3D pore ratio and the water permeable porosity for the highly porous samples might be attributed by the classification errors of the cementitious matrix and the pores due to the limited voxel resolution of X-ray CT.

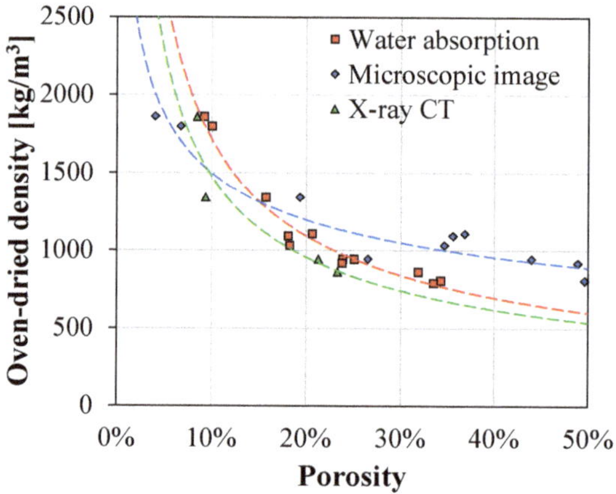

Figure 10. The relationship of the oven-dried density with the porosities obtained using the water absorption test, microscopic image processing, and X-ray CT analysis.

On the other hand, the results from microscopic image processing showed a relatively low porosity for the high-density sample and a high porosity for the low density one. The low 2D pore ratio might be attributed to small-sized pores, which were quite bright and designated as matrix due to shallow depths. The high 2D pore ratio obtained from microscopic image processing was caused by different calculations. Let us assume that a pore with a 1 mm radius was located inside a cube that is 2 mm in length. The 3D pore ratio would be 0.52 and the 2D pore ratio, at the center, would be 0.79. Because the cross-sections of the highly porous samples were almost fully filled with pores, the 2D pore ratio was always higher than the water permeable porosity and the 3D pore ratio obtained using X-ray CT. Furthermore, closed pores could represent a portion of the pores in a high-resolution microscopic image—and they cannot be saturated by water. For these reasons, there was a difference between the results obtained from the 3D-based testing methods—water absorption test and X-ray CT—and the 2D-based image processing. Hence, considering the different characteristics of each testing method, their complementary uses are recommended for analyzing the porous structures of cementitious materials well.

5. Conclusions

This experimental study conducts a comparative analysis for characterizing the pores in cementitious materials by adopting microscopic image processing and X-ray CT. All 12 porous cementitious samples were fabricated using various dosages of aluminum powder and natural fibers. The porous structures were evaluated in terms of the pore size, the number of pores, the spatial distribution of pores along the height, and open and closed pores. The key observations and findings of this study are summarized as follows:

1. In microscopic image processing, the local thresholding method was adopted by considering non-uniform illumination images caused by a lateral light source. As a preliminary study, user-defined parameters of window size and sensitivity were carefully selected as 200 × 200 and 0.5, respectively. Consequently, microscopic image processing was successfully performed and various characteristics of pores were provided using high quality binary images. Furthermore, the linear relationship between the 2D pore ratio and the mean pore diameter was identified.
2. X-ray CT analysis was conducted for the representative samples with a wide range of porosities. This 3D tomographic image-based analysis provided various unique characteristics of pores, such as open and closed pores and the distribution of pores in the 3D space. However, a high-resolution 3D tomographic image is required in order to obtain a more accurate analysis on the porous structures.
3. To compare the properties of porosity obtained using different testing methods, the relationship of porosity with oven-dried density was investigated. The regression curves obtained for the water permeable porosity and the 3D pore ratio using X-ray CT showed similar trends. On the other hand, the results obtained using microscopic image processing provided a low 2D pore ratio for dense materials and a high 2D pore ratio for porous materials due to both the calculations used and the portion of closed pores that the samples contained.

As technology advances, it is expected that a high-resolution 3D tomographic image from X-ray CT would provide a more accurate analysis for the characteristics of porous cementitious materials. Furthermore, it might also be beneficial for investigating the relationships of porous cementitious materials with their unique properties, such as heat conservation, noise absorption, and water drainage.

Author Contributions: Conceptualization, J.Y., H.K.; methodology, J.Y., H.K.; software, J.Y., H.K.; writing—original draft preparation, J.Y.; writing—review and editing, H.K., S.-H.S., S.P.; visualization, J.Y.; supervision, S.-H.S., S.P.; project administration, S.P.; funding acquisition, S.P. All authors have read and agreed to the published version of the manuscript.

Funding: This research was funded by the National Research Foundation of Korea (NRF) grant number NRF-2019R1F1A1060906 funded by the Korea government (MSIT). The opinions expressed in this paper are those of the authors and do not necessarily reflect the views of the sponsors.

Conflicts of Interest: The authors declare no conflict of interest.

References

1. Bogas, J.A.; Gomes, A.; Pereira, M.F.C. Self-compacting lightweight concrete produced with expanded clay aggregate. *Constr. Build. Mater.* **2012**, *35*, 1013–1022. [CrossRef]
2. Rossignolo, J.A.; Agnesini, M.V.C.; Morais, J.A. Properties of high-performance LWAC for precast structures with Brazilian lightweight aggregates. *Cement Concrete Compos.* **2003**, *25*, 77–82.
3. Andrade, L.B.; Rocha, J.C.; Cheriaf, M. Influence of coal bottom ash as fine aggregate on fresh properties of concrete. *Constr. Build. Mater.* **2009**, *23*, 609–614. [CrossRef]
4. Choi, B.S.; Yoon, H.S.; Moon, H.K.; Yang, K.H. Environmental impact assessment of lightweight aggregate concrete according to replacement ratio of artificial lightweight fine aggregates. *J. Korea Concr. Inst.* **2018**, *30*, 297–304.
5. Chai, Y.H. Service Performance of Long-Span Lightweight Aggregate Concrete Box-Girder Bridges. *J. Perform. Constr. Facil.* **2016**, *30*, 04014196. [CrossRef]

6. Qian, Z.; Chen, L.; Jiang, C.; Luo, S. Performance evaluation of a lightweight epoxy asphalt mixture for bascule bridge pavements. *Constr. Build. Mater.* **2011**, *25*, 3117–3122. [CrossRef]
7. Neithalath, N.; Sumanasooriya, M.S.; Deo, O. Characterizing pore volume, sizes, and connectivity in pervious concretes for permeability prediction. *Mat. Charact.* **2010**, *61*, 802–813. [CrossRef]
8. Zaetang, Y.; Wongsa, A.; Sata, V.; Chindaprasirt, P. Use of lightweight aggregates in pervious concrete. *Constr. Build. Mater.* **2013**, *48*, 585–591. [CrossRef]
9. Chen, Y.; Wang, K.; Wang, X.; Zhou, W. Strength, fracture and fatigue of pervious concrete. *Constr. Build. Mater.* **2013**, *42*, 97–104. [CrossRef]
10. Kim, H.-H.; Kim, C.-S.; Jeon, J.-H.; Park, C.-G. Effects on the Physical and Mechanical Properties of Porous Concrete for Plant Growth of Blast Furnace Slag, Natural Jute Fiber, and Styrene Butadiene Latex Using a Dry Mixing Manufacturing Process. *Materials* **2016**, *9*, 84. [CrossRef]
11. Kant Sahdeo, S.; Ransinchung, G.D.; Rahul, K.L.; Debbarma, S. Effect of mix proportion on the structural and functional properties of pervious concrete paving mixtures. *Constr. Build. Mater.* **2020**, *255*, 119260. [CrossRef]
12. Chindaprasirt, P.; Hatanaka, S.; Chareerat, T.; Mishima, N.; Yuasa, Y. Cement paste characteristics and porous concrete properties. *Constr. Build. Mater.* **2008**, *22*, 894–901. [CrossRef]
13. Narayanan, N.; Ramamurthy, K. Structure and properties of aerated concrete: A review. *Cem. Concr. Compos.* **2000**, *22*, 321–329. [CrossRef]
14. Benazzouk, A.; Douzane, O.; Mezreb, K.; Quéneudec, M. Physico-mechanical properties of aerated cement composites containing shredded rubber waste. *Cem. Concr. Compos.* **2006**, *28*, 650–657. [CrossRef]
15. Yoon, H.S.; Yang, K.H. Optimum mixture proportioning of autoclave lightweight aerated concrete considering required foaming rate and compressive strength. *J. Korea Concr. Inst.* **2019**, *31*, 123–130. [CrossRef]
16. He, X.; Zheng, Z.; Yang, J.; Su, Y.; Wang, T.; Strnadel, B. Feasibility of incorporating autoclaved aerated concrete waste for cement replacement in sustainable building materials. *J. Cleaner Prod.* **2020**, *250*, 119455. [CrossRef]
17. Song, Y.; Li, B.; Yang, E.H.; Liu, Y.; Ding, T. Feasibility study on utilization of municipal solid waste incineration bottom ash as aerating agent for the production of autoclaved aerated concrete. *Cem. Concr. Compos.* **2015**, *56*, 51–58. [CrossRef]
18. Liu, Y.; Leong, B.S.; Hu, Z.T.; Yang, E.H. Autoclaved aerated concrete incorporating waste aluminum dust as foaming agent. *Constr. Build. Mater.* **2017**, *148*, 140–147. [CrossRef]
19. Stroeven, P.; Hu, J.; Koleva, D.A. Concrete porosimetry: Aspects of feasibility, reliability and economy. *Cem. Concr. Compos.* **2010**, *32*, 291–299. [CrossRef]
20. Elsharief, A.; Cohen, M.D.; Olek, J. Influence of lightweight aggregate on the microstructure and durability of mortar. *Cem. Concr. Res.* **2005**, *35*, 1368–1376. [CrossRef]
21. Chung, S.Y.; Abd Elrahman, M.; Kim, J.S.; Han, T.S.; Stephan, D.; Sikora, P. Comparison of lightweight aggregate and foamed concrete with the same density level using image-based characterizations. *Constr. Build. Mater.* **2019**, *211*, 988–999. [CrossRef]
22. Ćosić, K.; Korat, L.; Ducman, V.; Netinger, I. Influence of aggregate type and size on properties of pervious concrete. *Constr. Build. Mater.* **2015**, *78*, 69–76. [CrossRef]
23. Yoon, J.Y.; Kim, J.H. Mechanical properties of preplaced lightweight aggregates concrete. *Constr. Build. Mater.* **2019**, *216*, 440–449. [CrossRef]
24. Yoon, J.Y.; Lee, J.Y.; Kim, J.H. Use of raw-state bottom ash for aggregates in construction materials. *J. Mater. Cycles Waste Manag.* **2019**, 1–12. [CrossRef]
25. ASTM C642-13, Standard Test Method for Density, Absorption, and Voids in Hardened Concrete 2013. Available online: https://www.astm.org/Standards/C642 (accessed on 6 July 2020).
26. Cnudde, V.; Cwirzen, A.; Masschaele, B.; Jacobs, P.J.S. Porosity and microstructure characterization of building stones and concretes. *Eng. Geol.* **2009**, *103*, 76–83. [CrossRef]
27. Semel, F.J.; Lados, D.A. Porosity analysis of PM materials by helium pycnometry. *Powder Metall.* **2006**, *49*, 173–182. [CrossRef]
28. Yang, X.; Sun, Z.; Shui, L.; Ji, Y. Characterization of the absolute volume change of cement pastes in early-age hydration process based on helium pycnometry. *Constr. Build. Mater.* **2017**, *142*, 490–498. [CrossRef]

29. Chen, S.J.; Li, W.G.; Ruan, C.K.; Sagoe-Crentsil, K.; Duan, W.H. Pore shape analysis using centrifuge driven metal intrusion: Indication on porosimetry equations, hydration and packing. *Constr. Build. Mater.* **2017**, *154*, 95–104. [CrossRef]
30. Shah, V.; Bishnoi, S. Analysis of Pore Structure Characteristics of Carbonated Low-Clinker Cements. *Transp. Porous Media.* **2018**, *124*, 861–881. [CrossRef]
31. Zhou, J.; Ye, G.; van Breugel, K. Characterization of pore structure in cement-based materials using pressurization-depressurization cycling mercury intrusion porosimetry (PDC-MIP). *Cem. Concr. Res.* **2010**, *40*, 1120–1128. [CrossRef]
32. Diamond, S. Mercury porosimetry: An inappropriate method for the measurement of pore size distributions in cement-based materials. *Cem. Concr. Res.* **2000**, *30*, 1517–1525. [CrossRef]
33. Chung, S.Y.; Sikora, P.; Rucinska, T.; Stephan, D.; Abd Elrahman, M. Comparison of the pore size distributions of concretes with different air-entraining admixture dosages using 2D and 3D imaging approaches. *Mater. Charact.* **2020**, *162*, 110182. [CrossRef]
34. Martin, W.D.; Putman, B.J.; Kaye, N.B. Using image analysis to measure the porosity distribution of a porous pavement. *Constr. Build. Mater.* **2013**, *48*, 210–217. [CrossRef]
35. Choi, J.; Lee, Y.; Kim, Y.Y.; Lee, B.Y. Image-processing technique to detect carbonation regions of concrete sprayed with a phenolphthalein solution. *Constr. Build. Mater.* **2017**, *154*, 451–461. [CrossRef]
36. Zhang, J.; Ma, G.; Ming, R.; Cui, X.; Li, L.; Xu, H. Numerical study on seepage flow in pervious concrete based on 3D CT imaging. *Constr. Build. Mater.* **2018**, *161*, 468–478. [CrossRef]
37. Yoon, J.; Kim, H.; Koh, T.; Pyo, S. Microstructural characteristics of sound absorbable porous cement-based materials by incorporating natural fibers and aluminum powder. *Constr. Build. Mater.* **2020**, *243*, 118167. [CrossRef]
38. Yu, F.; Sun, D.; Hu, M.; Wang, J. Study on the pores characteristics and permeability simulation of pervious concrete based on 2D/3D CT images. *Constr. Build. Mater.* **2019**, *200*, 687–702. [CrossRef]
39. Zhou, H.; Li, H.; Abdelhady, A.; Liang, X.; Wang, H.; Yang, B. Experimental investigation on the effect of pore characteristics on clogging risk of pervious concrete based on CT scanning. *Constr. Build. Mater.* **2019**, *212*, 130–139. [CrossRef]
40. ASTM C1693-11, Standard Specification for Autoclaved Aerated Concrete 2017. Available online: https://www.astm.org/Standards/C1693.htm (accessed on 6 July 2020).
41. Wong, H.S.; Head, M.K.; Buenfeld, N.R. Pore segmentation of cement-based materials from backscattered electron images. *Cem. Concr. Res.* **2006**, *36*, 1083–1090. [CrossRef]
42. Otsu, N. Threshold selection method from gray-level histograms. *IEEE Trans. Syst. Man Cybern.* **1979**, *9*, 62–66. [CrossRef]
43. Gatos, B.; Pratikakis, I.; Perantonis, S.J. Adaptive degraded document image binarization. *Pattern Recognit.* **2006**, *39*, 317–327. [CrossRef]
44. Sauvola, J.; Pietikäinen, M. Adaptive document image binarization. *Pattern Recognit.* **2000**, *33*, 225–236. [CrossRef]
45. Kim, H.; Ahn, E.; Cho, S.; Shin, M.; Sim, S.H. Comparative analysis of image binarization methods for crack identification in concrete structures. *Cem. Concr. Res.* **2017**, *99*, 53–61. [CrossRef]
46. Manahiloh, K.N.; Muhunthan, B.; Kayhanian, M.; Gebremariam, S.Y. X-ray Computed Tomography and Nondestructive Evaluation of Clogging in Porous Concrete Field Samples. *J. Mater. Civ. Eng.* **2012**, *24*, 1103–1109. [CrossRef]
47. Bernardes, E.E.; Mantilla Carrasco, E.V.; Vasconcelos, W.L.; de Magalhães, A.G. X-ray microtomography (μ-CT) to analyze the pore structure of a Portland cement composite based on the selection of different regions of interest. *Constr. Build. Mater.* **2015**, *95*, 703–709. [CrossRef]
48. Hong, S.; Liu, P.; Zhang, J.; Xing, F.; Dong, B. Visual & quantitative identification of cracking in mortar subjected to loads using X-ray computed tomography method. *Cem. Concr. Compos.* **2019**, *100*, 15–24.
49. Liu, T.; Zhang, X.N.; Li, Z.; Chen, Z.Q. Research on the homogeneity of asphalt pavement quality using X-ray computed tomography (CT) and fractal theory. *Constr. Build. Mater.* **2014**, *68*, 587–598. [CrossRef]
50. Golubev, I.; Karpova, Y. Quality improvement of oil-contaminated wastewater, meant for injection into formation, using two-stage treatment technology. *J. Ecol. Eng.* **2017**, *18*, 48–52. [CrossRef]
51. Holt, E.; Raivio, P. Use of gasification residues in aerated autoclaved concrete. *Cem. Concr. Res.* **2005**, *35*, 796–802. [CrossRef]

52. Wang, F.; Liu, Z.; Hu, S. Early age volume change of cement asphalt mortar in the presence of aluminum powder. *Mater. Struct.* **2010**, *43*, 493–498. [CrossRef]
53. Zhao, Y.; Wang, X.; Jiang, J.; Zhou, L. Characterization of interconnectivity, size distribution and uniformity of air voids in porous asphalt concrete using X-ray CT scanning images. *Constr. Build. Mater.* **2019**, *213*, 182–193. [CrossRef]
54. Chakraborty, S.; Kundu, S.P.; Roy, A.; Basak, R.K.; Adhikari, B.; Majumder, S.B. Improvement of the mechanical properties of jute fibre reinforced cement mortar: A statistical approach. *Constr. Build. Mater.* **2013**, *38*, 776–784. [CrossRef]

 materials

Article

CO_2 Curing Efficiency for Cement Paste and Mortars Produced by a Low Water-to-Cement Ratio

Seong Ho Han, Yubin Jun, Tae Yong Shin and Jae Hong Kim *

Department of Civil and Environmental Engineering, Korea Advanced Institute of Science and Technology, Daejeon 34141, Korea; ha8890@kaist.ac.kr (S.H.H.); ssjun97@gmail.com (Y.J.); tyshin@kaist.ac.kr (T.Y.S.)

* Correspondence: jae.kim@kaist.ac.kr

Received: 3 July 2020; Accepted: 28 August 2020; Published: 2 September 2020

Abstract: Curing by CO_2 is a way to utilize CO_2 to reduce greenhouse gas emissions. Placing early-age cement paste in a CO_2 chamber or pressure vessel accelerates its strength development. Cement carbonation is attributed to the quickened strength development, and CO_2 uptake can be quantitatively evaluated by measuring CO_2 gas pressure loss in the pressure vessel. A decrease in CO_2 gas pressure is observed with all cement pastes and mortar samples regardless of the mix proportion and the casting method; one method involves compacting a low water-to-cement ratio mix, and the other method comprises a normal mix consolidated in a mold. The efficiency of the CO_2 curing is superior when a 20% concentration of CO_2 gas is supplied at a relative humidity of 75%. CO_2 uptake in specimens with the same CO_2 curing condition is different for each specimen size. As the specimen scale is larger, the depth of carbonation is smaller. Incorporating colloidal silica enhances the carbonation as well as the hydration of cement, which results in contributing to the increase in the 28-day strength.

Keywords: CO_2 curing; size effect; colloidal silica; cement-based material; casting method

---

1. Introduction

Greenhouse gas emissions in the industrial sector are of serious concern. In the construction industry, a large amount of CO_2 is emitted during the production of cement by the calcination process. Various studies have been conducted to reduce the amount of CO_2 emitted in the manufacturing process or to utilize emitted or captured CO_2 for sustainable development [1,2]. CO_2 curing for cement-based materials has been demonstrated as one possible way of utilizing CO_2 [3–7].

The carbonation of calcium silicates such as C_3S, β-C_2S, and γ-C_2S in Portland cement generally occurs more quickly than their hydration. Therefore, the CO_2 curing of early-age concrete facilitates faster development of its strength [8]. The carbonation of anhydrous alite (C_3S) and belite (C_2S) is expressed as [9]:

$$3C_3S + (3-x)CO_2 + nH \rightarrow C_xSH_n + (3-x)CaCO_3 \quad (1)$$

and

$$2\beta{-}C_2S + (2-x)CO_2 + nH \rightarrow C_xSH_n + (2-x)CaCO_3 \quad (2)$$

where C_xSH_n refers to the calcium silicate hydrates of $(CaO)_x(SiO_2)(H_2O)_n$, simply denoted by C-S-H. The carbonation products are the calcium silicate hydrates and calcium carbonate ($CaCO_3$). In addition, the calcium hydroxide, a product by the calcium silicate hydration, is also carbonated:

$$Ca(OH)_2 + CO_2 \rightarrow CaCO_3 + H_2O \quad (3)$$

The calcium carbonate, produced at an early age, precipitates in pores of the cement paste. Consequently, cement-based materials obtain pore refinement, leading to enhanced durability and strength [10].

The degree of carbonation was usually estimated by the mass-curve or mass-gain method [11–13]. Equation (4) calculates CO_2 uptake by measuring the increase in the mass of intact samples. The increase in the mass corresponds to the mass of reacted CO_2. The mass-curve method [11] then evaluates

$$CO_2 \text{ uptake } (\%) = \frac{\text{Mass of specimen subjected to } CO_2}{\text{Initial mass of specimen}} \times 100 \quad (4)$$

The mass-curve method needs caution when monitoring the increase in the mass of samples in a chamber. The water presented in a sample partially vaporizes due to the heat production from the carbonation, and excess water vapor condenses on the chamber wall. The mass-gain method compensates the water amount lost by vaporization to reduce experimental error [12–14]. Equation (5) then expresses a calculation of CO_2 uptake:

$$CO_2 \text{ uptake } (\%) = \frac{(\text{The mass increase of a cast sample by } CO_2 \text{ curing}) + (\text{water loss})}{\text{Mass of raw materials for a sample}} \times 100 \quad (5)$$

Another way to consider the CO_2 uptake evaluation is by measuring the decrease in CO_2 gas pressure in a sealed reactor. The decrease in the CO_2 pressure monitored for the whole carbonation process can be converted into the CO_2 mole consumed for the carbonation. Such a pressure monitoring method is less prone to error. This paper, therefore, proposes the pressure monitoring method.

The efficiency of CO_2 curing is related to water in the pore system of a sample. Water invariably allows for the mixing and subsequent workability of cement-based materials. However, if there is much free water in the sample, water also fills the pore of a sample fully, resulting in hindering CO_2 gas from entering the sample interior. CO_2 diffusion into a sample is limited if its pores are fully saturated. On the other hand, CO_2 in gaseous form does not react, and so its dissolution in liquid water precedes for the carbonation. Sufficient water, that is, more than the reacting amount accounted for in Equations (1) and (2), is required for the CO_2 dissolution. Previous studies [15,16] suggested the use of a relatively low water-to-cement ratio (W/C less than 0.25) in CO_2 curing, and made samples by compaction molding. The cement compacts produced by the low water-to-cement ratio had a large amount of air-filled pores, which resulted in a higher CO_2 diffusion and CO_2 uptake.

Controlling the pore system, by the use of nano-sized particles, also affects the efficiency of CO_2 curing. The nucleation effect of nano-sized limestone powder on CO_2 curing was reported [17,18]. In addition to the nano-sized limestone powder, this study further investigates the effect of incorporating colloidal silica. The nano-sized silica particles reportedly nucleate the hydration of cementitious materials in accompany with their pozzolanic reaction [19]. As a result, it reduces the degree of chloride ion penetration [20,21] even though the increase in compressive strength is not substantial [22]. Lastly, the optimization for CO_2 curing conditions, together with the effect of specimen size, is also investigated for the purpose of controlling the pore system.

2. Experimental Details

2.1. Materials

Ordinary Type I Portland cement and ISO standard sand (ISO 679) [23] were used to produce samples in this study. Table 1 shows the chemical composition of the cement determined by X-ray fluorescence spectrometry. The specific density of the sand was 2620 kg/m^3. Its grain size ranged from 0.08 to 1.60 mm. The colloidal silica (commercial grade) used in this study mainly contained particles from 10 to 20 nm, and the SiO_2 content in the aqueous solution was 40%. The pH of the colloidal silica ranged from 9.5 to 10.5.

Table 1. Chemical composition of ordinary Portland cement (wt.%).

CaO	SiO_2	Al_2O_3	Fe_2O_3	SO_3	MgO	Na_2O	K_2O
67.29	17.18	4.13	4.17	3.16	1.94	0.22	1.23

2.2. Sample Preparation

Table 2 lists the mix proportions of the samples. A planetary mixer was used for a total of 5 min mixing. The mixes were then fabricated by two methods: (1) Compacting or (2) conventional consolidating-in-mold procedure following ASTM C109 [24]. The compacting method was applied to the samples with a relatively low W/C: Paste (W/C = 0.15) and Mortar (W/C = 0.35), which considered the condition of brick production in practice. Each mix was filled in a confined mold, and then it was compacted by 5 kN compression for 30 min. The dimensions of the paste and mortar compacts were as a 40-mm cube. In contrast, the dimensions for the samples, Paste (W/C = 0.4) and Mortar (W/C = 0.5), produced by the conventional procedure [24], were various as 25-, 40-, and 50-mm cubes so as to analyze the specimen size effect on CO_2 curing. The sealed curing for the premature sample in a mold proceeded for 24 h at approximately 25 °C.

Table 2. Mix proportion of samples.

Label	Fabrication Method	Mix Proportion (g)			
		Water	Cement	Sand	Colloidal Silica
Paste (W/C = 0.15)	Compacting †	150	1000	0	-
Mortar (W/C = 0.35)		157.5	450	1350	-
Mortar (W/C = 0.35) with colloidal silica		157.5	450	1350	18
Paste (W/C = 0.4)	Consolidating-in-mold ‡	400	1000	0	-
Mortar (W/C = 0.5)		225	450	1350	-
Mortar (W/C = 0.5) with colloidal silica		225	450	1350	18

† Each stiff sample in a confined mold was compressed by 5 kN (within 30 min). ‡ The mortar flow of each sample was within 150 to 250 mm.

An additional two mortar samples incorporating colloidal silica were produced to analyze the effect of colloidal silica. Their mix proportions and fabrication methods were the same as the samples in Table 2, with 4% colloidal silica per cement mass.

2.3. CO_2 Curing and Successive Hydration

Table 3 details the schedule of the casting and curing conditions. We considered two conditions for the CO_2 curing after demolding. The first CO_2 curing condition was at 20% CO_2 concentration, where the relative humidity (RH) was 75% ± 5% and the temperature was 25 °C under ambient pressure. Each sample was placed in a controlled chamber, and the 20% concentration CO_2 curing continued for 28 days.

The other was 3 bar pressure CO_2 curing. The samples in a pressure vessel, vacuum-sealed beforehand, were subjected to 99.9% purified CO_2 gas for 3 h. The initial CO_2 gas pressure was set above 340 kPa, but not to exceed 380 kPa (approximately 350 kPa). Each sample in Table 2 was subjected to 3 bar CO_2 curing for 3 h, and then successive hydration followed: 21 h moisture curing for Paste (W/C = 0.15) and Mortar (W/C = 0.35), and 28 days water curing for Paste (W/C = 0.4) and Mortar (W/C = 0.5). The moisture curing was conducted under 85% ± 5% RH and a temperature of approximately 25 °C. The water curing was conducted at approximately 23 °C.

The control samples were produced by the moisture curing for the Paste (W/C = 0.15) and Mortar (W/C = 0.35). The control samples for Paste (W/C = 0.4) and Mortar (W/C = 0.5) took the water curing. The conditions for the controlled curing were the same as the conditions for the successive hydration after the 3 bar CO_2 curing.

Table 3. Curing condition and sequence for samples.

Label	Fabrication Time	Initial Curing for 24 h		Curing Until 28 Days
		3 h	21 h	
Paste (W/C = 0.15) Mortar (W/C = 0.35)	0.5 h	Moisture curing (Control)		
		3 bar CO_2 curing	Moisture curing	
		20% concentration CO_2 curing		
Paste (W/C = 0.4) Mortar (W/C = 0.5)	24 h	Water curing (Control)		
		3 bar CO_2 curing	Water curing	
		20% concentration CO_2 curing		

2.4. Measurements

Figure 1 shows a pressure vessel designed to have a single inlet valve and a single outlet valve. The internal temperature and pressure were monitored during the 3 bar CO_2 curing. Pressure loss caused by cement carbonation was measured using a pressure digital gauge (PDR1000; Pressure Development of Korea Co., Daejeon, Korea). Its measurement range was from −100 kPa to 400 kPa, and its precision was 0.1%. The sampling rate for the pressure measurement was 1 record per second.

Figure 1. Pressure vessel.

The compressive strengths of the samples were measured at the age of 1, 3, 7, 14, and 28 days in accordance with ASTM C109 [24]. The standard test method suggests a loading rate within 0.9 to 1.8 kN/s. In this study, the loading rate for the paste and mortar samples were assigned as 1.0 kN/s and 1.5 kN/s, respectively. The strength measurements took an average of the results of three replicated samples.

Each broken specimen following the strength measurement was used to evaluate the depth of carbonation. The depth of carbonation was determined using 1% phenolphthalein indicator in the broken specimen. The sprayed phenolphthalein solution on the specimen remains colorless when the specimen was carbonated, resulting in pH < 9.

3. Results

3.1. CO_2 Uptake by Pressure Monitoring

Figure 2a shows the pressure loss of each pressure vessel during the 3 bar CO_2 curing of Mortar (W/C = 0.35) and Mortar (W/C = 0.5) as an example. The values were calibrated by considering a CO_2 leakage of 0.696 kPa/h in the pressure vessel. The CO_2 pressure decreased over time, where the initial pressure was approximately 350 kPa as injected. Taking the slope at each point, the carbonation rate in a unit of kPa/h was evaluated as shown in Figure 2a. CO_2 in the pressure vessel was also dissolved in the water phase of a sample, but the dissolution in water phase went to equilibrium quickly. The rate of the pressure decrease, disregarding the initial pressure records right after the initiation of the monitoring, therefore directly indicated the CO_2 amount being carbonated.

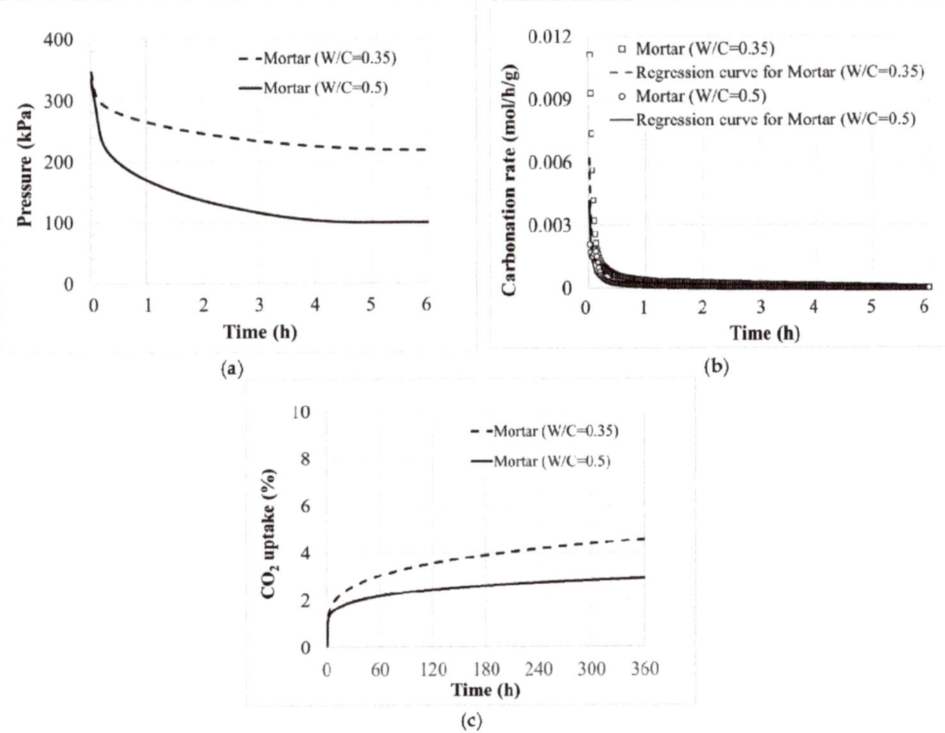

Figure 2. Pressure loss and CO_2 uptake of Mortar (W/C = 0.35) and Mortar (W/C = 0.5). (**a**) Decrease of CO_2 pressure, (**b**) Carbonation rate, (**c**) CO_2 uptake.

The consumed CO_2 was then calculated with the ideal gas equation, $PV = nRT$, where R is the gas constant of 8.314 J/mol/K. The volume of the pressure vessel was $V = 0.004$ m^3. The temperature slightly changed over time, but it was averaged at $T = 24$ °C or 301 K. The consumed CO_2, $n = PV/RT$ per unit time, represented the carbonation rate in a unit of mol/h. Normalizing the carbonation rate with the cement mass required for producing the samples in the pressure vessel yielded its value per cement mass, as shown in Figure 2b. Integrating the carbonation rate for the time of the CO_2 curing gave the CO_2 uptake of each sample, where the percentage was calculated with the molecular weight of CO_2 (44.01 g/mol) as shown in Figure 2c.

In Figure 2a, it can be seen that Mortar (W/C = 0.5) had a fast loss of pressure. However, as shown in Figure 2c, the actual amount of carbonation should be divided by the cement mass compared to the total mass. The result indicates that the CO_2 uptake of the Mortar (W/C = 0.35) is larger than the CO_2 uptake of Mortar (W/C = 0.5). As a result, it could be confirmed that the Mortar (W/C = 0.35) compact showed quicker carbonation, and with that, CO_2 uptake, compared with the Mortar (W/C = 0.5) sample.

Table 4 compares the CO_2 uptake per cement mass, where its value at 3 h is representatively reported. First of all, as expected, the Paste (W/C = 0.15) and Mortar (W/C = 0.35) compacts proportioned by a low W/C ratio had a higher CO_2 uptake compared with the Paste (W/C = 0.4) and Mortar (W/C = 0.5) samples. The compacts made of low W/C had a high air-filled porosity, resulting in easy CO_2 diffusion inside. Second, there was a size effect on the CO_2 uptake of the paste samples. The 40-mm cube specimens showed a higher CO_2 uptake than the 25-mm cubes of the replicated samples. Lastly, incorporating colloidal silica increased the CO_2 uptake of the mortar samples. This will be discussed later in detail.

Table 4. Results of CO_2 uptake.

Sample	CO_2 Uptake at 3 h (%)	
	40-mm Cube	25-mm Cube
Paste compact (W/C = 0.15)	17.54	15.80
Paste sample (W/C = 0.4)	1.43	0.80
Mortar compact (W/C = 0.35)	9.51	–
Mortar sample (W/C = 0.5)	2.60	–
Mortar compact (W/C = 0.35) with colloidal silica	14.85	–
Mortar sample (W/C = 0.5) with colloidal silica	3.11	–

3.2. Compressive Strength

Figure 3 compares the 1-day strengths of the 40-mm cube samples. The CO_2 curing was influential in the early-age strength development of the paste samples. The strengths of Paste compact (W/C = 0.15) and Paste sample (W/C = 0.4) subjected to CO_2 curing were higher than the control sample (moisture curing for 24 h). The 20% CO_2 curing for 24 h was much more effective in the Paste samples (W/C = 0.4), while the 3 bar CO_2 curing was better in the Paste compacts (W/C = 0.15). The effectiveness of the CO_2 curing condition bifurcated with the mortar samples. The 20% CO_2 curing resulted in a higher strength regardless of the casting method, whereas the 3 bar CO_2 curing failed.

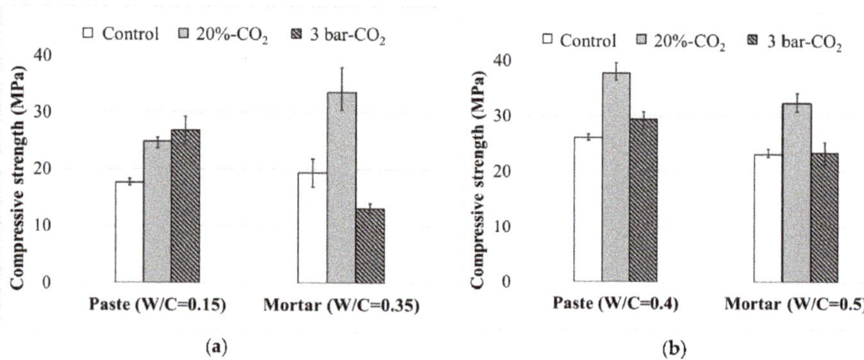

Figure 3. Strength of 1-day samples fabricated by (**a**) compacting and (**b**) consolidating-in-mold method.

Figure 4 shows the strength development of the 40-mm cube samples of Mortar (W/C = 0.5), where each trend was fitted in a hyperbolic equation. Following the trend of the early-age strength, as illustrated in Figure 3, the 20% CO_2 curing provided a higher strength gain than the other curing conditions. Incorporating colloidal silica intensified the effect of the 20% CO_2 curing, which resulted in higher 28-day strength. However, the effect of the 3 bar CO_2 curing was negligible, as shown in Figure 4, and even negative for Mortar (W/C = 0.5) incorporating colloidal silica.

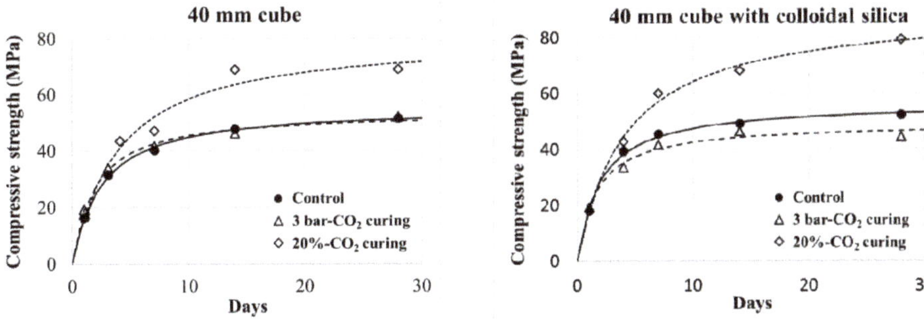

Figure 4. Strength development of Mortar (W/C = 0.5).

4. Discussions

4.1. Optimization for CO_2 Curing Condition

Both the initial pressure level and the duration in which the samples remained in the pressure vessel affect the efficiency of the CO_2 curing. A high pressure of CO_2 reportedly accelerates the carbonate reaction at an early age [25,26]. In this study, the initial pressure was therefore controlled for all samples: 3 bar, strictly inbetween 340 to 380 kPa, as stated in the previous section. For the effect of the duration, a preliminary test was conducted to optimize the CO_2 curing condition using the pressure vessel. The duration of the CO_2 curing then took the period in which the carbonation rate slowed to a crawl. As shown in Figure 2, the 50-mm cube Mortar (W/C = 0.35) and Mortar (W/C = 0.5) samples were cured for more than 6 h. The carbonation rate of Mortar (W/C = 0.35) is 1.71×10^{-4} mol/h/g at 3 h, and 0.98×10^{-4} mol/h/g at 6 h. The carbonation rate of Mortar (W/C = 0.5) is 5.40×10^{-5} mol/h/g at 3 h, and 3.02×10^{-5} mol/h/g at 6 h. The carbonation rate decreased from 3 h to 6 h was within 5% compared to the carbonation rate decreased during the initial 1 h. The carbonation under the initial pressure of 350 kPa was almost accomplished within 3 h.

The time of demolding is also critical for the effectiveness of CO_2 curing on the sample produced by the conventional consolidating-in-mold procedure [27]. The time of sealed curing in a mold affects the air-filled pore system [16]. Cement hydration is expected to consume water and at the same time, also produce solid hydrates in the pores. The former increases the volume of air-filled pores, but the latter adversely decreases the total amount of pores. Another preliminary test was conducted for this case. The Paste (W/C = 0.4) and Mortar (W/C = 0.5) were sealed in 40-mm cube molds for 6, 12, 18, and 21 h, and then they were subjected to 3 bar CO_2 curing after their demolding. Table 5 lists their CO_2 uptakes. Note that the Mortar (W/C = 0.5) sample was broken when it was demolded at 6 h. Paste (W/C = 0.4) and Mortar (W/C = 0.5) had the highest CO_2 uptake with the demolding time of 12 h and 18 h, respectively. After that, the CO_2 uptake monotonically decreased with the demolding time. The air-filled pores were expected to decrease stably with the cement hydration. The time of 21 h for the demolding was therefore taken for the period when the air-filled pores showed stable change.

Table 5. CO_2 uptake depending on sealed time.

Sample	Demolding Time (h)	CO_2 Uptake (%)
Paste (W/C = 0.4)	6	0.76
	12	1.31
	18	0.89
	21	0.74
Mortar (W/C = 0.5)	6	–
	12	3.06
	18	3.77
	21	2.74

4.2. Effect of Specimen Size

The size effect on the strength of cement-based materials is inherent, and CO_2 curing affects the size effect of sample because of inconsistent CO_2 diffusion. Figure 5 shows the carbonation depth of Mortar (W/C = 0.5) subjected to 20% CO_2 curing for 28 days, where the area of carbonation can be clearly compared. The 25-mm cube specimen was fully carbonated, but its 40-mm and 50-mm cubes were not fully carbonated, displaying a crimson color inside (pH > 9).

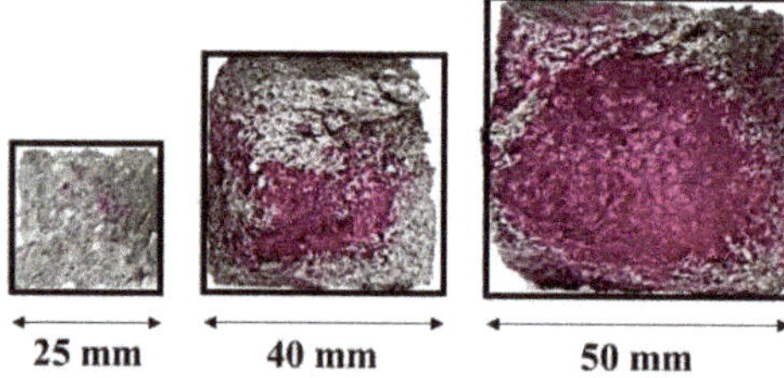

Figure 5. Carbonation depth of Mortar (W/C = 0.5) cured in 20% concentration CO_2 for 28 days.

The size effect law [28,29] helps us to understand the measurement of the compressive strengths of concrete. A large concrete cylinder provides a lower strength than that of a small cylinder which is geometrically similar to the large one. The tendency could be fitted with a size-effect equation [30]. Applying it to the current measurement generates an equation predicting the strength of a D-sized cube, $f_{cu}(D)$, based on that of a 25-mm cube:

$$f_{cu}(D) = \frac{f_{cu}(25)}{\left[1+\left(\frac{D}{\lambda_0 d_a}\right)\right]^{1/2}}\mathrm{B} + \alpha f_{cu}(25) \tag{6}$$

where α, B, λ_0, and d_a can be considered empirical constants. Each parameter, notably λ_0 and d_a, has a physical meaning; however, here it is important that the parameters are constant. The variation of each strength is then explained by a linear relation of:

$$\frac{\partial f_{cu}(D)}{\partial f_{cu}(25)} = \frac{1}{\left[1+\left(\frac{D}{\lambda_0 d_a}\right)\right]^{1/2}}\mathrm{B} + \alpha \tag{7}$$

Using Equation (7) allows us to consider the strength development of each sample. For example, Δf_{cu} (25) is calculated by the difference in the 25-mm cube strength at a certain age compared with 28 days (the reference age). Figure 6 comparatively shows the strength variation (development) of Mortar (W/C = 0.5). The linear trend lines, whose slopes correspond with the constant in Equation (6), do not change according to CO_2 curing when the 25- and 40-mm cube strengths are compared: $\Delta f_{cu}(25)$ and $\Delta f_{cu}(40)$ in the left figure. However, the linear trend lines of the samples subjected to the 20% CO_2

curing go off on the trend with the 50-mm cube strength ($\Delta f_{cu}(25)$ and $\Delta f_{cu}(50)$ in the right figure). The resultant nonlinear trend indicates that the size-effect parameters in Equation (6) need corrections or its reformulation. Partial carbonation on the edge of the 50-mm cube specimen, as shown in Figure 5, breaks the assumption of a geometrically similar specimen, which results in the nonlinear trend. In order to fit into the size effect law, the degree of carbonation in specimens with different size should be similar under the same CO_2 curing condition. However, in this study, as the specimen size is larger, the depth of carbonation is smaller. This result may show that the size effect law of the specimen in CO_2 curing does not fit.

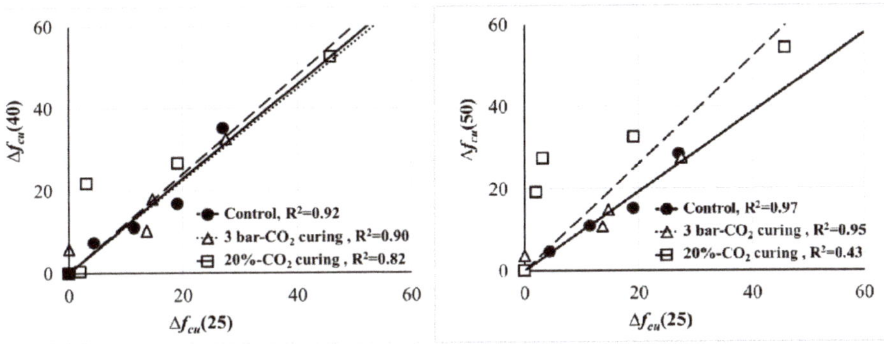

Figure 6. Comparison of the size effect on the strength development of mortar (W/C = 0.5).

4.3. *Effect of Colloidal Silica on CO_2 Curing*

Colloidal silica reportedly contributes to a low diffusivity in a hardened cement paste and a low degree of carbonation because it has the filling/nucleating effect [20,31]. However, the CO_2 curing produced an opposite effect. Incorporating the colloidal silica in the mortar samples subjected to the 3 bar CO_2 curing increased the CO_2 uptake approximately 56% for Mortar (W/C = 0.35) and 19% for Mortar (W/C = 0.5), as documented in Table 4. Incorporating non-reactive nanoparticles reportedly result in a slight increase in the rate of cement hydration because they provide supplementary nucleation sites [32–34]. Colloidal silica provides nucleation sites for the carbonation as well as the hydration of cement. Colloidal silica is more effective in the high-rate carbonation during which cement hydration is dormant. As a result, carbonation by 3 bar CO_2 curing (at the age of 3 h) is accelerated.

Conversely, as shown in Figure 4, the strength of the mortar incorporating colloidal silica was not enhanced, but even weakened by the 3 bar CO_2 curing. The carbonation products, mostly calcite, are not helpful for improving strength. Calcites are crystallized, and they do not provide a binding force among aggregates. Their inclusion in a paste matrix even cuts a binding link of the main hydrates (C–S–H). The strength of the mortar is consequently less developed.

When the samples incorporating colloidal silica were continuously subjected to 20% CO_2 curing, their strengths increased up to 15% at 28 days. Colloidal silica certainly accelerates carbonation with 20% CO_2 curing. Here, as opposed to the short-period 3 bar CO_2 curing, carbonation continues concurrently with cement hydration in 20% CO_2 curing. Calcite crystals produced by the carbonation put C–S–H on themselves at nano-scale, and the C–S–H layer is consequently reinforced by the distributed calcite [35–37]. The continuous 20% CO_2 curing consequently improves the strength of the mortar sample.

5. Conclusions

Curing by CO_2 can accelerate and improve the strength of cement-based materials via cement carbonation. In this study, 3 bar CO_2 curing was applied to premature cement paste and mortar for 3 h, and then successive conventional curing followed for cement hydration. As a result, the strength of a paste compact (W/C = 0.15) increased a lot, providing a high CO_2 uptake. That of a Paste (W/C = 0.4) consolidated in a mold also displayed a meaningful increase. However, despite cement carbonation, 3 bar CO_2 curing resulted in an adverse effect in terms of the strength of a mortar compact (W/C = 0.35), while that of a mortar (W/C = 0.5) consolidated in a mold was unchanged. In contrast, continuous 20% CO_2 curing increased the strengths of all the cement paste and mortar samples. Partial carbonation inside a specimen affects the size effect on the strength of the cement mortar. Incorporating colloidal silica provides more nucleation sites for cement carbonation, with the result that the effect of 20% CO_2 curing is slightly improved.

Author Contributions: Conceptualization, S.H.H.; Methodology, J.H.K; Validation, J.H.K.; Formal Analysis, S.H.H.; Investigation, S.H.H. and T.Y.S.; Data Curation, Y.J.; Writing—Original Draft Preparation, S.H.H; Writing—Review & Editing, Y.J. and J.H.K.; Visualization, S.H.H.; Supervision, J.H.K.; Funding Acquisition, J.H.K. All authors have read and agreed to the published version of the manuscript.

Funding: This work was supported by a Korea Institute of Energy Technology Evaluation and Planning (KETEP) grant funded by the Korean government (MOTIE) (No. 20188550000580).

Conflicts of Interest: The authors declare no conflict of interest.

References

1. Gartner, E.; Hirao, H. A review of alternative approaches to the reduction of CO_2 emissions associated with the manufacture of the binder phase in concrete. *Cem. Concr. Res.* **2015**, *78*, 126–142. [CrossRef]
2. Huang, C.-H.; Tan, C.-S. A Review: CO_2 Utilization. *Aerosol Air Qual. Res.* **2014**, *14*, 480–499. [CrossRef]
3. Klemm, W.A.; Berger, R.L. Accelerated curing of cementitious systems by carbon dioxide. Part I. Portland cement. *Cem. Concr. Res.* **1972**, *2*, 567–576. [CrossRef]
4. Watanabe, K.; Yokozeki, K.; Ashizawa, R.; Sakata, N.; Morioka, M.; Sakai, E.; Daimon, M. High durability cementitious material with mineral admixtures and carbonation curing. *Waste Manag.* **2006**, *26*, 752–757. [CrossRef]
5. Monkman, S.; Shao, Y. Carbonation curing of slag-cement concrete for binding CO_2 and improving performance. *J. Mater. Civ. Eng.* **2010**, *22*, 296–304. [CrossRef]
6. Rostami, V.; Shao, Y.; Boyd, A.J.; He, Z. Microstructure of cement paste subject to early carbonation curing. *Cem. Concr. Res.* **2012**, *42*, 186–193. [CrossRef]
7. El-Hassan, H.; Shao, Y. Early carbonation curing of concrete masonry units with Portland limestone cement. *Cem. Concr. Compos.* **2015**, *62*, 168–177. [CrossRef]
8. Young, J.F.; Berger, R.L.; Breese, J. Accelerated Curing of Compacted Calcium Silicate Mortars on Exposure to CO2. *J. Am. Ceram. Soc.* **1974**, *57*, 394–397. [CrossRef]
9. Berger, R.L.; Young, J.F.; Leung, K. Acceleration of Hydration of Calcium Silicates by Carbon Dioxide Treatment. *Nat. Phys. Sci.* **1972**, *240*, 16–18. [CrossRef]
10. Shaikh, F.U.A.; Supit, S.W.M. Mechanical and durability properties of high volume fly ash (HVFA) concrete containing calcium carbonate ($CaCO_3$) nanoparticles. *Constr. Build. Mater.* **2014**, *70*, 309–321. [CrossRef]
11. Mahoutian, M.; Ghouleh, Z.; Shao, Y. Carbon dioxide activated ladle slag binder. *Constr. Build. Mater.* **2014**, *66*, 214–221. [CrossRef]
12. Monkman, S.; Shao, Y. Assessing the carbonation behavior of cementitious materials. *J. Mater. Civ. Eng.* **2006**, *18*, 768–776. [CrossRef]
13. Shi, C.; He, F.; Wu, Y. Effect of pre-conditioning on CO_2 curing of lightweight concrete blocks mixtures. *Constr. Build. Mater.* **2012**, *26*, 257–267. [CrossRef]
14. Matsushita, F.; Aono, Y.; Shibata, S. Carbonation degree of autoclaved aerated concrete. *Cem. Concr. Res.* **2000**, *30*, 1741–1745. [CrossRef]
15. Shi, C.; Wu, Y. Studies on some factors affecting CO_2 curing of lightweight concrete products. *Resour. Conserv. Recycl.* **2008**, *52*, 1087–1092. [CrossRef]
16. Zhang, D.; Li, V.C.; Ellis, B.R. Optimal Pre-hydration Age for CO_2 Sequestration through Portland Cement Carbonation. *ACS Sustain. Chem. Eng.* **2018**, *6*, 15976–15981. [CrossRef]
17. Tu, Z.; Guo, M.Z.; Poon, C.S.; Shi, C. Effects of limestone powder on $CaCO_3$ precipitation in CO_2 cured cement pastes. *Cem. Concr. Compos.* **2016**, *72*, 9–16. [CrossRef]

18. Shao, Y.; Rostami, V.; He, Z.; Boyd, A.A. Accelerated carbonation of portland limestone cement. *J. Mater. Civ. Eng.* **2014**, *26*, 117–124. [CrossRef]
19. Aly, M.; Hashmi, M.S.J.; Olabi, A.G.; Messeiry, M.; Abadir, E.F.; Hussain, A.I. Effect of colloidal nano-silica on the mechanical and physical behaviour of waste-glass cement mortar. *Mater. Des.* **2012**, *33*, 127–135. [CrossRef]
20. Zhang, M.H.; Li, H. Pore structure and chloride permeability of concrete containing nano-particles for pavement. *Constr. Build. Mater.* **2011**, *25*, 608–616. [CrossRef]
21. Givi, A.N.; Rashid, S.A.; Aziz, F.N.A.; Salleh, M.A.M. Investigations on the development of the permeability properties of binary blended concrete with nano-SiO_2 particles. *J. Compos. Mater.* **2011**, *45*, 1931–1938. [CrossRef]
22. Jo, B.W.; Kim, C.H.; Lim, J.H. Characteristics of cement mortar with nano-SiO_2 particles. *ACI Mater. J.* **2007**, *104*, 404–407. [CrossRef]
23. ISO. *ISO 679:2009 Cement—Test Methods—Determination of Strength*; ISO: Geneva, Switzerland, 2009.
24. ASTM C109. *Standard Test Method for Compressive Strength of Hydraulic Cement Mortars (Using 2-in. or [50-mm] Cube Specimens)*; ASTM International: West Conshohocken, PA, USA, 2013. [CrossRef]
25. Shi, C.; Liu, M.; He, P.; Ou, Z. Factors affecting kinetics of CO_2 curing of concrete. *J. Sustain. Cem. Mater.* **2012**, *1*, 24–33. [CrossRef]
26. Kashef-Haghighi, S.; Shao, Y.; Ghoshal, S. Mathematical modeling of CO_2 uptake by concrete during accelerated carbonation curing. *Cem. Concr. Res.* **2015**, *67*, 1–10. [CrossRef]
27. El-Hassan, H.; Shao, Y.; Ghouleh, Z. Effect of initial curing on carbonation of lightweight concrete masonry units. *ACI Mater. J.* **2013**, *110*, 441–450. [CrossRef]
28. Bažant, Z.P. Size effect in blunt fracture: Concrete, rock, metal. *J. Eng. Mech.* **1984**, *110*, 518–535. [CrossRef]
29. Kim, J.-K. Size effect in concrete specimens with dissimilar initial cracks. *Mag. Concr. Res.* **1990**, *42*, 233–238. [CrossRef]
30. Yi, S.-T.; Yang, E.-I.; Choi, J.-C. Effect of specimen sizes, specimen shapes, and placement directions on compressive strength of concrete. *Nucl. Eng. Des.* **2006**, *236*, 115–127. [CrossRef]
31. Lim, S.; Mondal, P. Effects of incorporating nanosilica on carbonation of cement paste. *J. Mater. Sci.* **2015**, *50*, 3531–3540. [CrossRef]
32. Péra, J.; Husson, S.; Guilhot, B. Influence of finely ground limestone on cement hydration. *Cem. Concr. Compos.* **1999**, *21*, 99–105. [CrossRef]
33. Sato, T.; Diallo, F. Seeding effect of nano-$CaCO_3$ on the hydration of tricalcium silicate. *Transp. Res. Rec.* **2010**, *2141*, 61–67. [CrossRef]
34. Jayapalan, A.R.; Lee, B.Y.; Fredrich, S.M.; Kurtis, K.E. Influence of additions of anatase TiO_2 nanoparticles on early-age properties of cement-based materials. *Transp. Res. Rec.* **2010**, 41–46. [CrossRef]
35. Farahi, E.; Purnell, P.; Short, N.R. Supercritical carbonation of calcareous composites: Influence of curing. *Cem. Concr. Compos.* **2013**, *43*, 48–53. [CrossRef]
36. Šavija, B.; Luković, M. Carbonation of cement paste: Understanding, challenges, and opportunities. *Constr. Build. Mater.* **2016**, *117*, 285–301. [CrossRef]
37. Farahi, E.; Purnell, P.; Short, N.R. Supercritical carbonation of calcareous composites: Influence of mix design. *Cem. Concr. Compos.* **2013**, *43*, 12–19. [CrossRef]

Article

Performance Comparison between Densified and Undensified Silica Fume in Ultra-High Performance Fiber-Reinforced Concrete

Sung-Hoon Kang [1], Sung-Gul Hong [1,*] and Juhyuk Moon [2,3,*]

1 Department of Architecture and Architectural Engineering, Seoul National University, 1 Gwanak-ro, Gwanak-gu, Seoul 08826, Korea; medesis@snu.ac.kr

2 Department of Civil and Environmental Engineering, Seoul National University, 1 Gwanak-ro, Gwanak-gu, Seoul 08826, Korea

3 Institute of Construction and Environmental Engineering, Seoul National University, 1 Gwanak-ro, Gwanak-gu, Seoul 08826, Korea

* Correspondence: sglhong@snu.ac.kr (S.-G.H.); juhyukmoon@snu.ac.kr (J.M.); Tel.: +82-2-880-8360 (S.-G.H); +82-2-880-1524 (J.M.)

Received: 5 August 2020; Accepted: 31 August 2020; Published: 3 September 2020

Abstract: Silica fume (SF) is a key ingredient in the production of ultra-high performance fiber-reinforced concrete (UHPFRC). The use of undensified SF may have an advantage in the dispersion efficiency inside cement-based materials, but it also carries a practical burden such as high material costs and fine dust generation in the workplace. This study reports that a high strength of 200 MPa can be achieved by using densified SF in UHPFRC with Portland limestone cement. Additionally, it was experimentally confirmed that there was no difference between densified and undensified SFs in terms of workability, compressive and flexural tensile strengths, and hydration reaction of the concrete, regardless of heat treatment, because of a unique mix proportion as well as mixing method for dispersing agglomerated SF particles. It was experimentally validated that the densified SF can be used for both precast and field casting UHPFRCs with economic and practical benefits and without negative effects on the material performance of the UHPFRC.

Keywords: ultra-high performance fiber-reinforced concrete; densified silica fume; agglomeration; pozzolanic reaction; densification

1. Introduction

Ultra-high performance fiber-reinforced concrete (UHPFRC) is a construction material that has excellent mechanical properties, durability, and flowability. Its development was attributed to the achievement of optimal packing density by the use of ultrafine particles such as silica fume (SF) and quartz powder, the prevention of brittle failure by the inclusion of the high volume of short fibers, and advances in the technology of chemical admixtures [1,2]. The demand for this commercial material has increased in earnest since the beginning of this century, mainly related to innovative and sustainable structures such as building facades, infrastructures, and protection or explosion-proof facilities [3,4]. In these circumstances, practical factors such as price competitiveness, sustainability, accessibility of raw materials, and worker safety are becoming increasingly important concerns. With regard to the sustainability of this material, one major challenge was to reduce its enormous content of Portland cement (up to 1000 kg/m^3) or unnecessary unhydrated cement [5,6]. Therefore, the incorporation of various supplementary cementitious materials (SCMs) such as limestone powder, calcined clay, fly ash, ground granulated blast furnace slag, steel slag, and metakaolin has been attempted in order to replace a portion of the cement [7–13]. Considering the practical aspect in particularly, cement replacement by limestone powder has been regarded as one of the most effective solutions [14–19].

To enhance UHPFRC's market competitiveness, it is also important to reasonably determine the type and content of silica fume (SF), considering economic efficiency and usability. SF is one of the most widely used ultrafine particles today, but, at the same time, it is also one of the most expensive construction materials due to its limited supply [20]. Therefore, inevitably, the more the material is incorporated, the higher the material cost of concrete [21]. After pioneering studies on the optimization of packing density by the use of ultrafine particles [1,2,22], it has been generally agreed that the optimal ratio of SF to the weight of cement is 25% in UHPFRC [6,23]. However, this ratio is at least two to three times higher than other types of concrete [20], and this gap is even more overwhelming when compared based on the content per unit volume of concrete [24].

Although SF makes a significant contribution to UHPFRC's material price, the use of alternative materials must be approached cautiously because SF is one of the most important raw materials affecting almost all material properties of UHPFRC. The ball bearing effect, which acts as a lubricant by the combination of SF and superplasticizer, and the micro filling effect that fills the space between fine particles or refines the capillary pore structure contribute decisively to exhibiting the unique characteristics and superior performance of UHPFRC [25]. The yield stress of fresh UHPFRC is reduced by the ball bearing effect, which guarantees its superior self-compacting ability [23]. The micro filling effect makes cement paste extremely compact and increases the interfacial bond between aggregates and the paste [26,27]. SF is also responsible for making the fiber matrix interface area; as an example, the interfacial bond strength within the steel fiber matrix is significantly affected by the content of the ultrafine particles [28]. SF particles that can be adsorbed on a cement particle up to millions also affect the chemical reaction of UHPFRC [29]. By providing a site for nucleation of calcium silicate hydrate (C-S-H), it accelerates cement hydration and directly participates in secondary hydration (i.e., a pozzolanic reaction) [30,31]. In particular, the latter not only reduces the content of portlandite, which negatively affects the strength and durability of the cement paste, but it also forms additional C-S-H, which improves these properties [32]. For these reasons, although it has been previously reported that SCMs composed of amorphous silica can replace the chemical roles of SF [33], it has been suggested that, with these materials, it is difficult to completely replace all the complicated roles of the SF in UHPFRC, especially in terms of the physical effects carried out by spherical particles with submicron size [9,24,25]. This further supports the extreme difficulty in finding materials that can replace SF in UHPFRC.

As another alternative, the use of densified SF can be considered, and this can be a practical compromise as it has been the type of product commonly used in the concrete industry [29,34]. When collected in a silo as a produced form, SF originally has a density of 125 to 150 kg/m^3 [35]. However, an additional process for densification of ultrafine particles proceeds due to the improvement of convenience in handling and transportation (also related to material cost) and workers' health problems caused by fine dust (also related to labor cost) [36]. When air is blown into the silo, the particles roll and agglomerate in tens of μm to several mm, so that its density increases in the range of 480 to 720 kg/m^3 [37]. Although undensified SF with a density of less than 350 kg/m^3 is also used in limited amounts for special applications such as those involving pre-bagged products and grouts [38], it is not a widely used form for concrete production due to its high price, caused by low demand and the inconvenience involved with transportation and storage, as well as dust generation in the workplace [37]. However, despite disadvantages in usability and price competitiveness, undensified products have been preferred as suitable SFs in the production of UHPFRC because they can guarantee better dispersion than a densified product. Indeed, improved homogeneity compared to other types of concrete was one of the basic principles in the development of UHPFRC, and it has been considered that homogeneous dispersion of ultrafine particles is a prerequisite prior to particle size and specific surface area (SSA) of ultrafine particles [1]. For this reason, despite the practically great advantages, the feasibility of the densified product and its effect on the material properties of UHPFRC have been barely investigated.

When studying concrete with SF, its agglomeration tendency should be taken into account since SF particles do not exist in the form of independent nanoparticles but almost always as lumps [34,39]. This situation has been commonly reported in concrete containing coarse aggregate despite partial crushing of the lumps during the mixing process [40,41]. The agglomeration phenomenon is no exception in the SF manufactured as a collected state [42]. In other words, undensified SF can also not be free from agglomeration issues, so it is difficult to guarantee its complete dispersion within UHPFRC [43]. Moreover, due to the nature of the agglomeration, it is also considered very challenging to accurately measure the particle size or size distribution of SF [41]. Although ultrasonic dispersion or sonification before measurement is somewhat effective [44], it did not show completely satisfactory results because it could not affect the strong electrical attraction of the whole particles [41,45–47]. In general, laser diffraction (LD) and dynamic light scattering (DLS) are used for particle size analysis of powder materials, with advantages such as short analysis time and good reproducibility [48,49]. These techniques are suitable for particles with sizes between 0.5–3500 μm and 0.5–3 to 5 μm, respectively [49,50]. In this context, it has been suggested that it is reasonable to exclude the results of >1 μm range that are clearly larger than the size of separated SF particles while including the ultrasonic dispersion process [34,35]. Similarly, DLS techniques can be considered to characterize silica nanoparticles as they can naturally exclude large fractions of agglomerates depending on the measurement range of the device [25,32,51].

The main purpose of this study is to investigate the feasibility of using densified SF in order to improve the usability and price competitiveness of UHPFRC.

2. Materials and Methods

Three types of SF products with different manufacturers, manufacturing methods, and ages were prepared and then their effects on the material properties of the concrete were experimentally examined. First of all, to characterize SF products, chemical analysis by X-ray fluorescence (XRF), morphological analysis by high-resolution field emission scanning electron microscopes (FE-SEM), individual particle size calculation by the image processing technique, particle size distribution measurement by DLS, and SSA measurement by the BET method were carried out. Additionally, the performance of UHPFRC with different SF products was evaluated by measuring slump spread and mechanical properties such as compressive and flexural tensile strengths. Finally, X-ray diffraction (XRD) and thermogravimetric (TG) analyses were performed to compare the hydration mechanism due to the different form of SF on the properties of the concrete.

2.1. Preparation of UHPFRC Samples

For the experiment, UHPFRC was manufactured using Portland limestone cement, SF, quartz powder (SiO_2 > 97 wt %), quartz sand (SiO_2 > 90 wt %), water, polycarboxylate (PCE)-based superplasticizer, and steel fiber (Φ0.2 mm × 13 mm, tensile strength > 2500 MPa), as in previous studies [16,17]. The mineralogical composition of the used cement is monoclinic alite (46.79 wt %), triclinic alite (6.94 wt %), monoclinic belite (2.19 wt %), orthorhombic belite (1.38 wt %), anhydrite (0.61 wt %), aluminate (0.88 wt %), gypsum (1.09 wt %), dolomite (12.36 wt %), calcite (21.38 wt %), and amorphous content (6.38 wt %) measured by quantitative XRD analysis. The results of particle size analysis of the raw materials by LD are shown in Figure 1. Considering the measurement range of the equipment used (Mastersizer 3000, Malvern Panalytical, UK), the size distributions of cement, quartz powder, and quartz sand were measured. The size of quartz powder was in the range of 1 to 20 μm; moreover, its sizes were located within the range of the particle size of cement (0.4 to 100 μm), confirming that the quartz powder can properly perform its primary role in UHPFRC, i.e., filling the space between cement particles [1].

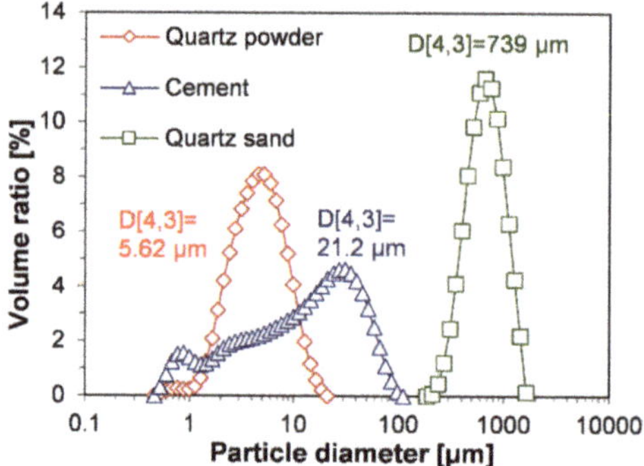

Figure 1. Particle size distribution of raw materials by laser diffraction.

Three different SF products (labeled as SF1_U, SF2_U, and SF2_D) were prepared as experimental variables. SF1_U is a high quality, undensified SF product available worldwide. It is named Microsilica-Grade 940U (Elkem, Oslo, Norway) and its bulk density is 200 to 350 kg/m^3 according to the manufacturer's specifications. Around three years had passed since the purchase date of the product, but when estimated based on the date of manufacture, a much longer period should have elapsed considering import, distribution, and storage by domestic dealers. SF2_U and SF2_D are undensified and densified SF products manufactured in the same factory (POSCO, Pohang, Korea) and have bulk densities of ~180 and ~500 kg/m^3, respectively. These domestic products, which do not require shipping or long-term transport and storage, had not been in our possession more than one year from the date of manufacture. As shown in Table 1, analysis results of chemical composition by a wavelength dispersive XRF spectrometer (S4 PIONEER, Bruker, Germany) verify that all products met the purity requirement of SF (>96 wt %) suitable for UHPFRC [23]. One reason that purity is important is that the inclusion of impurities, especially unburned carbon, has serious adverse effects on the hydration, mechanical properties, and workability of UHPFRC [24]. The results of XRF analysis also confirm that all SF products have the ability to function as highly reactive pozzolans and that their chemical compositions can be excluded from experimental parameters.

Table 1. Chemical composition of silica fume samples.

Sample	SiO_2	K_2O	Al_2O_3	MgO	CaO	Na_2O	SO_3	Fe_2O_3	Others [1]	LOI [2]	Total
SF1_U	96.00	0.83	0.72	0.40	0.27	0.26	0.19	0.10	0.20	1.00	99.98
SF2_U	96.88	0.36	-	0.17	0.16	-	0.46	0.53	0.09	1.36	100.00
SF2_D	96.29	0.40	0.31	0.22	0.13	-	0.40	0.71	0.13	1.40	99.99

[1] Components with less than 0.1% of content (Cl, P_2O_5, ZnO, MnO, CuO, Ru, Pd); [2] Loss on ignition.

UHPFRC samples were prepared based on the mix proportions shown in Table 2; this was done with a 5-L planetary mixer. SF was blended with quartz sand at low speed (140 rpm) for 5 min to break up the lumped, ultrafine particles by collision with hard grains. This is possible because the collisions between particles enable dispersive movement, which causes collapse and dispersion of the agglomerated particles [52]. The remaining powders (cement and quartz powder) were then blended for another 5 min after being placed into the mixer. After 10 min of the dry blending process, water and superplasticizer were poured into the mixer and low speed mixing was maintained until the mixtures were agglomerated. As shown in Figure 2, the agglomeration of the mixture proceeded in two stages, during which local agglomeration of the relatively small grains was followed by the formation of one large agglomerate due to the agglomeration of the small grains.

Table 2. Mix proportions of UHPFRC (by weight of cement).

Cement	Silica Fume	Quartz Powder	Quartz Sand	Water	Superplasticizer [1]	Steel Fiber [2]
1	0.25	0.35	1.1	0.25	0.0012	2%

[1] Solid contents; [2] By volume of UHPFRC.

Figure 2. Manufacturing process of UHPFRC: (**a**) formation of small grains by local agglomeration; (**b**) agglomeration of small grains into large lumps; (**c**) dispersion after solid-suspension transition.

When mixing is maintained for a certain period of time after the liquid materials are added, a fluid bond is formed between the particles so that the materials in the mixer begin to agglomerate, as shown in Figure 2a. The interparticle force between particles increases due to the surface tension of the water and the capillary pressure inside the bond, which requires the mixer to increase its power to further disperse the particles; eventually, dispersion efficiency increases [53]. When all the ingredients in the mixer were agglomerated, the power of the mixer reached its maximum, as shown in Figure 2b. Immediately after this, a transition occurred between the solid aggregate and the dispersed suspension; at this time, the particles inside the aggregate were evenly distributed and the space between the particles was filled with liquid. Once this transition occurred, the capillary force disappeared so that the power consumption of the mixer was drastically reduced. Consequently, the mixture began to have self-compacting abilities, as shown in Figure 2c. After the steel fibers were added into the mixture, which had been changed to the suspension state, they were mixed at a high speed (285 rpm) for 2 min as the last step of the manufacturing process.

The freshly prepared UHPFRC was filled in the prepared mold. The concrete was poured slowly at one end of the mold when casting prismatic specimens to exclude the effect of fiber orientation. All specimens were cured in a sealed state at 20 °C on the first day. The next day, heat treatment was applied. In other words, the specimens were cured at 90 °C for 2 days after demolding, as the standard heat treatment method. From the third day to the experimental day, specimens were cured in a chamber set at 20 °C and 60% relative humidity (RH). In addition, the samples without heat treatment were also prepared. In this case, the specimens were exposed to air drying conditions (20 °C and RH 60%) from the next day after casting.

2.2. Experimental Methods

For morphological analysis of raw materials, microscopic observation was performed using FE-SEM equipment (JSM-7800F Prime, JEOL Ltd., Tokyo, Japan). Powder samples were fixed on a carbon tape bonded holder and the upper surface of the holder was coated with carbon to prevent the charging effect. The holder was then inserted into the microscope and images of the particles were captured at a magnification of up to ×100,000. Additionally, the SEM images of SF were used to obtain its size distribution based on individual spherical particles. Since the results can significantly vary depending on the number of particles or images, the image processing technique was performed according to the method specified in ISO 13322-1 [54].

As another method for obtaining the particle size distribution of SF, an analysis using DLS was performed. To improve dispersion efficiency, the sample was sonicated for 5 min in an ultrasonic bath (53 kHz and 200 W) and then the size distribution was measured at a set temperature (25 °C) by the device (Zetasizer Nano ZSP, Malvern Panalytical, UK). Moreover, SSA of SF samples was determined by the BET method, for which the nitrogen adsorption desorption isotherms were recorded using a surface area analyzer (Autosorb IQ-MP/XR, Quantachrome Instrument, Boynton Beach, FL, USA).

To measure the material properties of UHPFRC, a series of experiments was conducted on workability, compressive strength, and flexural tensile strength. Workability of the fresh concrete was evaluated by the use of a flow table test without shock [55]. Compressive strength tests were performed according to ASTM C109 [56]; on the 28th day, cube specimens with one side of 50 mm were loaded by a universal testing machine and the average of the three specimens was determined as the strength value. On the same day, flexural tensile strength was measured by the three-point bending test method specified in ISO 679 [57]. For this, line loads were applied to three prismatic specimens (40 mm × 40 mm × 160 mm) by the machine.

XRD and TG analysis were performed to investigate the effect of SF on the hydration reaction of UHPFRC. To improve the precision of the analysis results, paste samples (excluding the materials that do not contribute to the chemical reaction, such as fine aggregate and steel fibers) were additionally prepared on the same day as the UHPFRC samples. On the 28th day, the hydrated paste was crushed and ground into a powder. Thereafter, the hydration reaction of the samples was stopped using isopropanol and diethyl-ether, according to the solvent exchange method [58,59]. After the pretreatment, the powders were placed on sample holders and then mounted on an X ray diffractometer (SmartLab SE, Rigaku, Tokyo, Japan). XRD patterns were collected by Cu·Kα_1 radiation (λ = 1.5406 Å) under the established conditions, such as voltage: 40 kV; current: 40 mA; step size: 0.02°, and scanning speed: 1°/min. The patterns collected between 5° and 70° were analyzed by SmartLab Studio II software (Rigaku, Tokyo, Japan), equipped with an NIST inorganic crystal structure database and crystallography open database.

TG analysis was performed to quantitatively estimate the consumption of portlandite due to the pozzolanic reaction. Around 50 mg of powder samples was weighed in an alumina made sample holder which was placed on an analytical instrument (SDT 650, TA Instruments, New Castle, DE, USA). Under an environment where nitrogen gas was introduced at the rate of 100 mL/min, the samples were heated from 30 to 1000 °C at the rate of 10 °C/min. The weight loss with increasing temperature was recorded and the derivative thermogravimetric (DTG) was presented in the graph along with the weight loss to clearly identify the sudden weight change in a specific temperature range.

3. Results and Discussion

3.1. Observation of Agglomerated SF Particles

Figure 3 shows SEM images of cement and quartz powders at various magnifications. As depicted below, these μm sized particles do not aggregate together but exist separately. The cement is composed of particles with a wide range of sizes from ~1 to ~50 μm (Figure 3a–c). In the case of quartz powder, the large particles were around 20 to 30 μm and the small ones were around several μm (Figure 3d–f). Compared to the cement, quartz powder showed a narrow particle size range because the crushed rock was sieved within a specific size range during its manufacture. The observation of particle size by SEM tended to be consistent with the result of LD-based analysis, as shown in Figure 1. This implies that particle size analysis for completely separated powders can ensure good reliability.

Figure 3. SEM images of cement (**a**–**c**) and quartz powder (**d**–**f**) with various resolutions.

Prior to SEM observation of the SF particles, visual inspection was performed as shown in Figure 4. Although the sample SF1_U is an undensified product, many lumps large enough to match the size of the coarse aggregates were observed. This was probably due to the agglomeration during a long period of shipment, transportation, and storage. Moreover, during this period, a vertical load was applied by the other products stacked on top of it. On the other hand, such large lumps were not found in the sample SF2_U, which should have been less affected by the factors related to aggregation after manufacturing. As expected, the sample SF2_D consisted of globular lumps due to the densification process during manufacturing. The size of the lumps was up to several mm, and, unlike irregular ones in the sample SF1_U, they were almost perfectly spherical. This means that, once lumps are formed during the densification process, they are difficult to agglomerate thereafter, and thus this type of product has advantages for quality control.

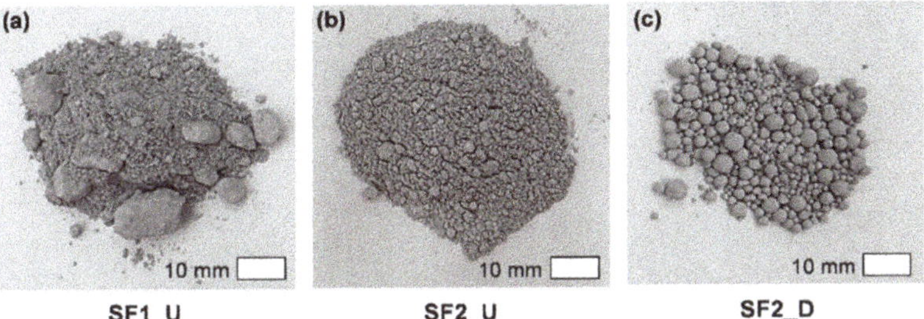

Figure 4. Visual observation of SF samples as a state used for manufacture of UHPFRC: (**a**) SF1_U, (**b**) SF2_U, and (**c**) SF2_D.

The true nature of these lumps can be viewed in detail by the use of SEM images (Figure 5). Figure 5a shows dust-like lumps that were observed. As mentioned in the introduction, the spherical nanoparticles never existed independently. The size of the lumps was various, and large ones of several tens of μm were also seen. The magnified images are presented in Figure 5b,c, in which myriad nanoparticles were agglomerated together to form the lumps. It has been reported that such lumps can consist of up to several tens of millions of nanoparticles [41]. As shown in Figure 5d–f, where individual nanoparticles are observed, it was confirmed that their size is tens to hundreds of nm based on separated spheres. In particular, some particles are not separate spheres but rather they appear to melt and stick together (yellow arrows in Figure 5d,e). Another interesting discovery was that of broken particles (white arrows in Figure 5f), indicating that the SF particles also have the shape of a hollow sphere, like other spherical SCMs such as fly ash.

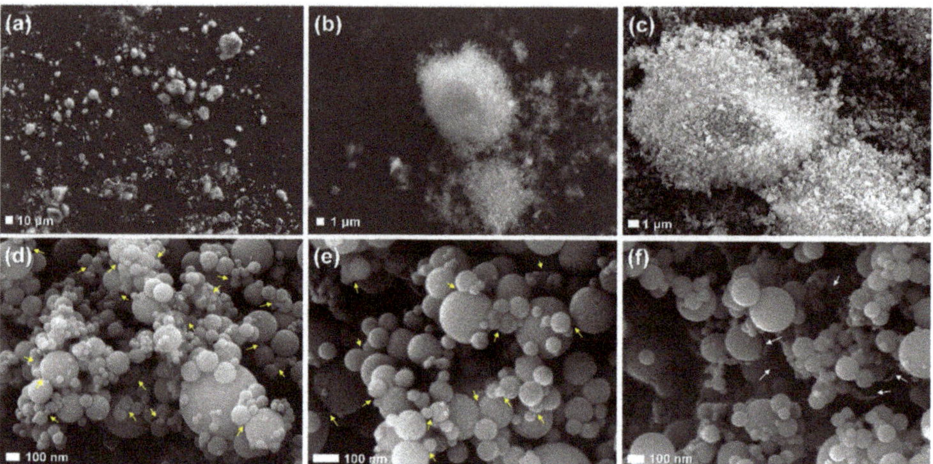

Figure 5. SEM images of SF1_U samples with various resolutions: agglomerated lumps (**a**–**c**) and spherical nanoparticles (**d**–**f**) inside lumps.

It is generally known that SF particles are mostly (>95%) composed of nanoparticles [27]. However, it should be noted that a significant part of the SF particles has a form of linked spheres, regardless of the densification process. Spherical nanoparticles that are connected as a result of sintering or fusion have been previously observed [39]. As also found in this study (Figure 5d,e), consequently, general forms of SF particles are aggregates of several spheres rather than independent spheres [37]. This type of connection is completely different from the agglomeration due to the densification process because it is a very strong and irreversible bond. When high purity quartz is converted to silicon at a high temperature (~2000 °C), SiO_2 vapor is generated; the vapor oxidizes and condenses at a low temperature while forming spherical particles with a size of 100 to 200 nm, but when they come into contact with each other in the molten state, primary aggregation occurs [60]. Due to this, dozens to hundreds of spheres are connected, and, typically, the size of these clusters has been reported to be 500 to 800 nm [41]. Such clusters formed by primary aggregation cannot be separated unless they are broken. For this reason, there has been confusion about the effective size of SF in all cases: immediately after manufacture, at the point of use, and after being incorporated into concrete [35].

Figure 6 shows the particle size distribution of SF samples by DLS and image processing techniques. By comparing the results of SF2_U (DLS) and SF2_D (DLS), the effectiveness of the technique (DLS with ultrasonic treatment)—which excludes the effect of the densification process—can be verified. As evident from visual observation (Figure 4), there was a significant difference in the degree of aggregation between the two samples, but, nevertheless, there was almost no difference in particle size distribution, as shown in Figure 6. In both samples, the main peak was formed between 70 nm and 1 μm. However, in the size of the peak formed between 1 and 10 μm, the densified sample was larger than the undensified sample due to the contribution of large, aggregated particles. For this reason, SF2_D showed a higher value than SF2_U in the average particle size, as shown in Table 3. Meanwhile, by comparing the two yellow and blue curves, differences can be observed between the two products classified by the manufacturer. Regarding the proportion of the particles with sizes of >400 nm, the SF1_U sample had a higher value than SF2_U and SF2_D, which also resulted in a difference in average particle size (Table 3). This is presumably due to an increase in the agglomeration of small particles during the manufacturing process or thereafter. Moreover, the agglomeration phenomenon that had become stronger over a long period of time could have lowered the dispersion efficiency by the ultrasonic treatment.

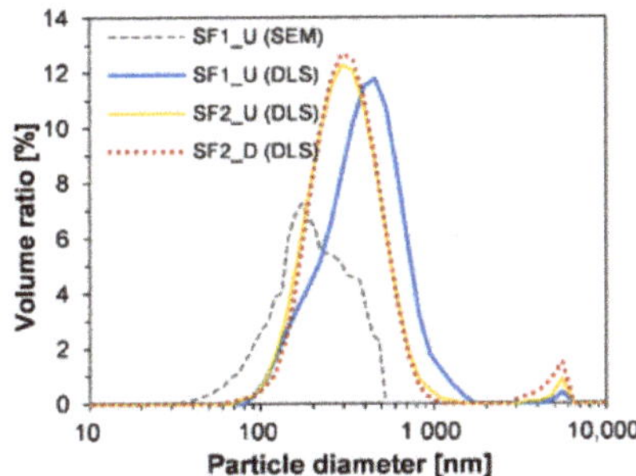

Figure 6. Particle size distribution of SF samples by DLS and SEM-based image processing techniques.

Table 3. Average particle size and specific surface area of SF samples.

Sample Name	SF1_U	SF2_U	SF2_D
Average particle size by DLS (nm)	440	300	311
Specific surface area by BET method (m^2/g)	23.8	25.3	22.9

Outcomes of the difference between the image processing and DLS techniques can be confirmed by comparing the two results of SF1_U (SEM) and SF1_U (DLS) (see gray dotted line and blue solid line in Figure 6). As expected, compared to DLS, the image processing technique formed the particle size distribution in a small size range, since it did not reflect the agglomeration phenomenon at all. Both methods have their own limitations. In particular, DLS can overestimate the particle size of SF because it measures hydrodynamic size (not a static or dry particle size) [61]. Nevertheless, this technique can be considered more suitable than the SEM-based image processing technique to measure the size distribution of SF particles. This is because effective particle sizes have more significant meanings than those that are independent due to their agglomerating nature [22]. Furthermore, the use of a high-resolution microscope is time consuming and expensive, and only a limited number of particles included in the 2D image are considered in the results. Indeed, results from previous studies have suggested that the actual size observed by SEM is unable to reflect the realistic particle size distribution because SF is always present in a clustered or agglomerated state [25,41].

The SSAs of the samples are shown in Table 3. The measured value was 22.9 to 25.3 m^2/g and the deviation between the minimum and maximum was around 10%. This is consistent with the results of the previous study in that SSA of SF was in the order of 20 m^2/g, despite diversity in type or manufacturer [41]. Moreover, there was no significant difference among the three samples of SSA, which is one of the most important parameters for the chemical effect of SCM on cement-based materials [32,45]. Unlike other natural or industrial byproduct-based SCMs, one of the advantages of SF is very low quality variation in terms of chemical composition, particle size, and SSA, which can vary by factory and manufacturer [62]. In this aspect, the measurement results in this study showed this benefit well.

3.2. Material Properties of UHPFRC with Various Types of Silica Fume

The measured flow diameters were 270, 260, and 255 mm for the fresh UHPFRC samples containing SF1_U, SF2_U, and SF2_D, respectively. According to the international standard on UHPFRC, their workability class is Cv, meaning viscous UHPFRC with self-compacting ability [63]. When comparing the two UHPFRC samples containing SF2_U and SF2_D, there was almost no difference, although the flow diameter was slightly higher than when undensified SF was used. If other conditions are the same, as the SF is more evenly dispersed in UHPFRC, the yield stress is further reduced due to the ball bearing effect [23], thereby increasing the self-compacting ability [52]. This implies the possibility that there was no significant difference in the degree of dispersion of SF particles inside the two fresh concrete samples. Meanwhile, between the two samples containing undensified SFs, when SF2_U was used, the flow diameter was 10 mm lower than the result of SF1_U. Regarding the decrease in flowability, both the dispersion efficiency of SF and the difference in SSA can be complicated because, as the SSA increases, water demand and friction between the particles increase [48]. Nonetheless, overall, there was no significant difference in workability among the samples, suggesting that there was no significant difference in the SSA and the dispersion efficiency of SF in the concrete samples.

The experimental results of compressive strength and flexural tensile strength are presented in Figure 7a. Above all, every sample exhibited a very high compressive strength of around 200 MPa. Compared to the sample [UHPFRC], SF2_U showed the highest strength. [UHPFRC] SF1_U was ~10 MPa lower but [UHPFRC] SF2_D was ~5 MPa lower. This is an important result for the feasibility of using densified SF in UHPFRC because it confirms the negative effects that can be caused by the densification process of SF, which are negligibly small in the mechanical properties of UHPFRC. In addition, the concrete's excellent strength can be ensured by using this commercially optimized product. Based on the results of [UHPFRC] SF2_U, the difference in concrete strength with the same type of product from another company ([UHPFRC] SF1_U) was more pronounced than with the other type of product from the same company ([UHPFRC] SF2_D). Various material properties affect SF concrete performance, including chemical composition, SSA, particle size or size distribution, the amount of silanol groups (Si-OH) on the particle surface, and post production age [64]. This suggests the diversity of properties related to SF1_U and SF2_U which affect the performance of UHPFRC. On the other hand, in terms of the difference in such properties of SF2_U and SF2_D, factors other than the densification process can be ignored. For this reason, the feasibility of using densified SF in UHPFRC can be proposed. Results from the flexural tensile strength test further support this suggestion. When compared to 45.1 MPa of the sample [UHPFRC] SF2_U, the strength of sample [UHPFRC] SF2_D was only 1.1 MPa (or 2.4%) lower. However, the strength of the sample [UHPFRC] SF1_U was 8.1 MPa (18%) lower than this. The flexural strength results were consistent with those of compressive strength, as confirmed by the clear linear correlation in Figure 7b. In summary, the parameters such as the densification process and the manufacturer or storage period affected the compressive strength and the flexural tensile strength with the same tendency. More importantly, the former parameter had little effect on both compressive and tensile properties of UHPFRC.

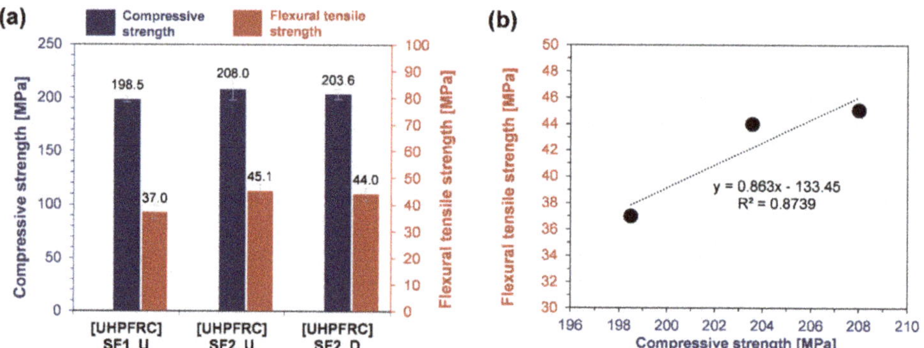

Figure 7. Results of strength tests: (**a**) compressive and flexural tensile strengths; (**b**) the relationship between the two strengths.

Direct measurement of the dispersion degree of SF in cement-based materials is extremely challenging [34]. Thus, indirectly, mechanical properties can be considered as an indicator for evaluating this. To date, the benefits expected from the use of undensified SF (such as improvement in dispersion efficiency in concrete and thereby improvement in mechanical properties) have been reported. This can be explained by enhancements in the role of SF such as the micropore filling effect, pozzolanic reaction, and the provision of nucleation sites. However, comparisons between different types of SF products have been mainly investigated in cement composites other than UHPFRC, such as the concrete and mortar types that contain coarse aggregates or have relatively high *w/c* [48,65]. Moreover, results from previous studies confirmed that a large amount of SF lumps (>10 μm) existed in such cement composites [34,39]. In practical situations, the only way to disperse SF products inside concrete or mortar is with mechanical crushing by a mixer, which is also one of the most effective methods [66]. When mixing concrete, crushing and shearing actions are transferred to the lumps of SF particles by a mixer and the dispersion efficiency is highly dependent on the mixture composition and mixing method. However, it has been consistently reported that densified SF cannot be satisfactorily dispersed in the cement composites, which is completely different from UHPFRC in terms of the mixing method and procedure [66–69].

When UHPFRC is manufactured, SF particles should be effectively dispersed by a suitable mixing method and a superplasticizer; otherwise, fundamental principles such as the optimization of mixture composition or packing density are meaningless and, in the end, the excellent performance of the concrete cannot be guaranteed [23]. The manufacturing process of this concrete includes one unique state in which all the ingredients in the mixer are agglomerated like flour dough (Figure 2b). At this point, when the mixer's power consumption reaches its maximum, a strong shearing action is applied to the agglomerate so that the components, including the SF particles, can be effectively dispersed. Moreover, in the presence of steel fibers, additional dispersion is possible in the subsequent high-speed mixing process. Although the perfect dispersion of individual SF particles might be impossible [39], the lumps formed by the densification process are weakly connected and thus can be easily crushed by agitation. Therefore, UHPFRC's unique formulation and manufacturing process can be effective in removing reversible aggregation by the densification process. As a result, there may not be a significant difference between densified and undensified products in dispersion efficiency in this concrete.

Along with compressive strength and durability, excellent crack resistance performance is one of the main features of UHPFRC, which can contribute to the construction of innovative and sustainable concrete structures [70]. This is the reason that the strength under flexural or tensile loading is an important property of this concrete [71]. Regarding this, one notable result was that, unlike the compressive strength, which increased due to the promoted pozzolanic reaction, there was no noticeable change in flexural tensile strength despite the increase in temperature (60 to 90 °C) and duration (up to 4 days) during heat treatment [72]. This is because the properties of steel fibers (aspect ratio, shape, surface treatment, etc.) and the distribution state of the fibers inside the concrete (volume ratio, direction, degree of dispersion, etc.) have a decisive effect on flexural or tensile properties [73–76], but the difference between these factors could be neglected in this study. Apart from such promotion of the pozzolanic reaction and the change in fiber parameters, the effect of the mixture composition on tensile strength has also been reported. In the study by Chan and Chu, the interfacial bond strength of the fibers increased proportionally with the contents of undensified SF (between 0 and 30 wt % by cement), which was attributed to the improvement in friction and resistance force of the fibers as the interface became denser [28]. In addition to these previous results, those from this study revealed that the type of SF also affects the tensile properties of UHPFRC, and more importantly, there is no significant reduction in the strength despite the use of densified SF.

3.3. Hydration Reaction of Heat-Treated UHPFRC with Various Types of Silica Fume

Figure 8a,b show the results of XRD and TG analysis of heat-treated paste samples at 28 days. Typical phases identified in the XRD pattern of this concrete are quartz, calcite, cement clinkers, ettringite, and portlandite, but the last two phases were not detected in this study. The absence of ettringite at 28 days confirms that it was decomposed at an elevated temperature (>70 °C), and delayed ettringite formation did not occur during the subsequent 25 days of curing [77]. In particular, the cause of the latter is the extremely compact microstructure and low *w/c* of UHPFRC; the effect of heat treatment makes this condition even more dramatic because elevated temperature curing (or heat treatment) significantly promotes both cement hydration and pozzolanic reaction. These further fill capillary pores while consuming a limited amount of water [78]. However, despite the heat treatment, unhydrated cement clinkers still existed (as shown in Figure 8a). This is the result of an insufficient amount of water to fully hydrate the cement and the lack of space for further forming hydration products [79,80]. Additionally, these conditions do not provide water and space for the reformation of ettringite [81], and, because of its excellent watertightness, external water cannot penetrate into the concrete, even under water curing conditions [80]. The weight loss and DTG curves shown in Figure 8b clearly confirm that ettringite was not formed after decomposition. This is because sudden weight loss at around 105 °C is associated with dehydration of ettringite, and a sharp peak on the DTG curve due to this loss could not be observed [58]. Therefore, the gentle peak formed between 40 and 300 °C can only be attributed to the dehydration of C-S-H.

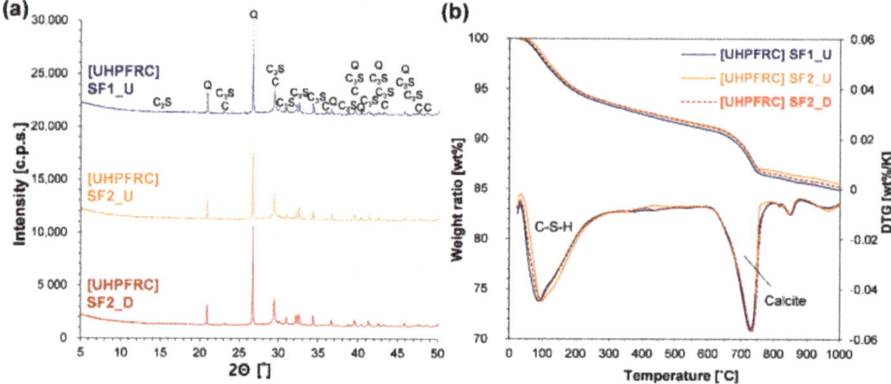

Figure 8. Analysis of crystal phases and hydration products of heat-treated paste samples: (**a**) XRD patterns at 28 days (Q: quartz, C: calcite, C_3S: tricalcium silicate); (**b**) TG and DTG curves at 28 days.

Another notable result from XRD and TG analysis is the absence of portlandite. The presence of portlandite was clearly confirmed in the previous study [72], in which all conditions other than the cement (i.e., raw materials, mixing method, and curing program) were the same as with this study. Additionally, in other studies, although the pozzolanic reaction was significantly promoted by the heat treatment at 90 °C, portlandite was not completely consumed [31,82,83]. In the study by Korpa et al., which includes in-depth discussions of the phase development of UHPFRC conditions based on XRD and TG analyses, the presence of portlandite and hence the ongoing progress of the pozzolanic reaction after 28 days was reported [84]. In UHPFRC, this crystal has been reported to be completely consumed by the pozzolanic reaction that was promoted by the heat treatment at >200 °C [85]. Unlike previous studies mentioned, the cement used in this study contained a significant amount of limestone powder, which was likely to contribute to the complete consumption of portlandite. The cause can be explained as follows: substitution of limestone powder instead of cement at a given amount of water increases the effective *w/c* (0.3 in this study). Additionally, since portlandite is produced by the primary cement hydration, a decrease in the content of primary hydration products due to the cement dilution effect can deplete portlandite even when the amount of accessible water is increased. The previous result also supports this, in that the portlandite content in UHPFRC decreased proportionally with the increase in the content of limestone powder [17].

However, even when a significant portion (up to 74%) of cement was replaced with limestone powder, portlandite could not be completely consumed without heat treatment [15,17]. This is because the reactivity of SF at room temperature is significantly low, and thus, even if a small amount of portlandite is formed, the role of SF related to consumption of this crystal is bound to be limited. The activation energy of SF required to participate in the pozzolanic reaction has been reported to be approximately 80 kJ/mol [86]. In this regard, heat treatment can greatly accelerate the reactivity of SF. It has also been reported that the solubility of amorphous silica increases proportionally with the temperature of water or solution [87]. Pfeifer et al. noted that the reaction degree of SF in UHPFRC was around 5% under 28 days of the ambient curing condition (20 °C), but the degree rapidly increased to around 45% (around nine times) when the standard heat treatment was applied to the concrete [31]. They also reported that an increase in effective *w/c* significantly improves the proportion of SF participating in the chemical reaction (around three times when *w/c* increases from 0.2 to 0.4). For this reason, even under the conditions in which a small amount of portlandite is formed due to the use of limestone powder, complete consumption of portlandite in UHPFRC would not be possible without heat treatment.

In addition to reducing the content of the cement, the use of limestone powder in UHPFRC has several advantages; autogenous shrinkage is alleviated by the increase in effective *w/c* [17,18] and the initial hydration is accelerated due to the provision of nucleation site for the formation of C-S-H [16]. Furthermore, the complete consumption of portlandite found in this study can provide additional advantages regarding the use of Portland limestone cement in UHPFRC. The pozzolanic reaction by SF or the consumption of portlandite has been known to persist for several years [88], and even after heat treatment, the compressive strength of UHPFRC increases continuously for 6 to 8 years due to this chemical reaction [89]. Indeed, an inversely linear relationship between the compressive strength and portlandite content has been reported in UHPFRC. This can be explained by the formation of additional C-S-H and refinement of the pore structure and removal of portlandite which has a morphologically undesirable effect on the strength of concrete and hardness of the interfacial transition zone [72]. Moreover, in terms of durability, this crystal can cause an expansion reaction with ions penetrated from the outside, thereby leading to cracking. Consequently, since the pozzolanic reaction has a decisive influence on the long-term strength and durability of UHPFRC [24], the complete consumption of portlandite shown herein suggests that the maximum mechanical properties and durability of this concrete can be guaranteed within 28 days. The stabilization of long-term properties is certainly an advantage when considering only practical usability because it can provide a reliable design strength for informing the work of practitioners and structural engineers [23].

In addition to the reactivity of SF, the chemical composition, particle size distribution, SSA, and degree of dispersion are all important factors in the chemical reaction of concrete [37,90]. In this regard, the results in Figure 8, in which no portlandite was detected in any of the samples, confirm that there might be no difference between densified and undensified SF products in the hydration and pozzolanic reaction of UHPFRC. This also implies that, between them, there was no difference in the degree of dispersion. This can be more evident if the same result is confirmed under the situation in which portlandite is not fully consumed.

3.4. Feasibility of Using Densified Silica Fume in Field Casting UHPFRC

An important application type of UHPFRC is field casting concrete (e.g., overlaying of concrete decks or slabs, jacketing of beam or columns, and filing material for precast concrete segments) [4,91–94]. In this case, heat treatment is not practically applied. The use of densified SF can be more essential for this type. This is because, in general, powder materials are continuously input and mixed into a mixer at an outdoor construction site; in this environment, the use of undensified SF can further deteriorate the limitations such as workplace dust generation, transportation, and the storage of raw materials. Thus, to fully examine the practical feasibility of densified SF, it is also necessary to conduct an investigation of UHPFRC cured without heat treatment. Moreover, when heat treatment is not applied, the mechanical properties of UHPFRC can change more sensitively depending on the type of SF, because, as mentioned earlier, the physical role (i.e., micro filling effect) rather than the chemical role (i.e., the pozzolanic reaction) can significantly contribute to the properties. In this case, a decrease in concrete performance due to a decrease in the dispersion efficiency of SF can be more pronounced.

Figure 9 shows the results of TG analysis of the paste samples cured under ambient conditions for 28 days. To exclude factors other than the densification process, paste samples containing SF products from the same company were compared. Analyses of the results verify that the type of SF classified by this process does not affect the hydration reaction of UHPFRC, regardless of heat treatment. However, unlike results from the heat-treated samples (Figure 7b), a sharp peak was formed around 105 °C. This clearly confirms the presence of ettringite. The integration result of this sharp peak area was 5.76% and 5.75% in the SF2_U and SF2_D samples, respectively, confirming that there is no difference in the main hydration products (C-S-H and ettringite). This also indicates the possibility that there was no difference in the chemical reaction related to the formation of portlandite. Moreover, the result that no protrusion or peak was formed between 150 and 200 °C on the DTG curve shows that it is extremely difficult to form a new phase (hemi or mono carboaluminate) by direct reaction of calcite in the UHPFRC condition, regardless of the heat treatment. This is consistent with the results of previous studies [15–17].

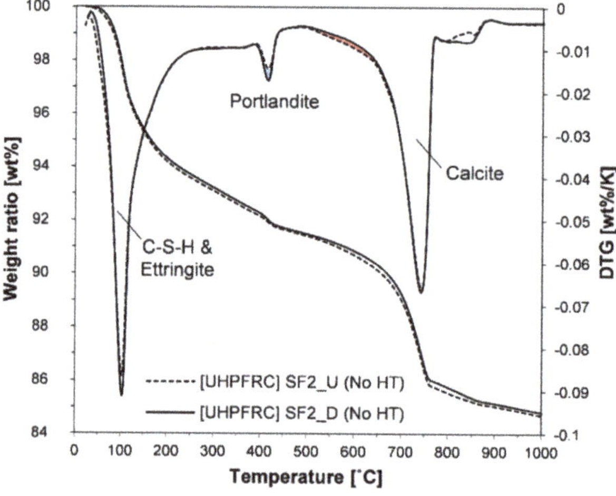

Figure 9. TG and DTG curves of ambient-cured paste samples at 28 days.

When observing the peak at 400 °C on the DTG curve, there was a difference in the sizes of portlandite peaks between the two samples. The quantitative analysis results are presented in Table 4. The contents in the table were determined according to the tangential method along with the normalization method based on the weight at 550 °C [58]. The portlandite content of the sample with densified SF was 1.1 wt %, which was 0.21 wt % higher than the other sample of 0.89 wt %. However, based on the weight ratio of calcium oxide, most of the differences in the portlandite content are attributed to the formation of calcite (difference in CaO, $_{\text{Portlandite}}$ = 0.16; difference in CaO, $_{\text{calcite}}$ = 0.12). This also confirms that there was almost no difference between the two samples in the degree of pozzolanic reaction. The contribution of carbonation to the differences in portlandite content can also be visually shown in the blue and red shaded areas on the DTG curve. Since there was no difference between the two SF samples in the chemical reaction, the difference in the physical role (i.e., micro filling effect) caused by the dispersion efficiency can be considered as another potential factor affecting the mechanical properties. However, there was also no difference between the two samples in the compressive strength of UHPFRC at 7 and 28 days (Table 4). This demonstrates that the densified

SF does not have any negative effect on either the physical or chemical roles of SF in the concrete, compared to the undensified SF.

Table 4. Portlandite and calcite contents and compressive strength of ambient-cured UHPFRC.

Sample	Quantitative Analysis by TG (wt %)				Compressive Strength (MPa)	
	Portlandite	CaO, Portlandite	Calcite	CaO, Calcite	7 Days	28 Days
[UHPFRC] SF2_U (No HT)	0.89	0.67	10.12	5.67	117.01 ± 1.38	154.01 ± 5.12
[UHPFRC] SF2_D (No HT)	1.10	0.83	9.92	5.55	117.17 ± 2.55	154.56 ± 4.02

SF products are only manufactured in limited regions of the world. Therefore, the products are inevitably transported and stored for a long time in a stacked state. Compared to undensified products, densified products are much freer from further agglomeration and quality changes after being manufactured [20]. The agglomeration by the densification process is reversible so that agglomerated SF particles can be effectively dispersed by an optimized mixing method, which is practically the only way this can be achieved. Along with this, the dispersion efficiency also depends on the chemical admixture used (performance or content), mixture composition, and water-to-powder ratio [37]. In particular, when a mixture of high volume powders is lumped and stiff like flour dough and a high mixing energy is applied to the mixture by a suitable mixer (e.g., twin shaft, planetary, or intensive mixer, etc.), the SF agglomerate can be effectively broken; eventually, their dispersion efficiency can be greatly increased [29,60]. In addition, the inclusion of PCE-based superplasticizer under these conditions and the application of high-speed mixing makes the homogeneous dispersion of powders including SF more effective. The composition of UHPFRC is characterized by a very low water-to-powder ratio and a very high content of PCE-based superplasticizer; this causes the dry materials to clump together in the mixing process. This is a clearly unique feature of UHPFRC that differs from other types of cement-based materials, and thus it can provide conditions for efficiently dispersing densified SF (equivalent to undensified products in terms of dispersion efficiency). Indeed, as all the results of this study consistently indicated, there was no notable difference between densified and undensified products in terms of the material properties of UHPFRC. Therefore, densified SF can be used to manufacture UHPFRC. In other words, although using this commercially optimal type of product, a compressive strength of >200 MPa can be achieved within 28 days, without disadvantages in workability and tensile properties.

4. Conclusions

This study was undertaken with the hypothesis that the composition and mixing method of UHPFRC are both unique and thus this condition is effective for dispersing the reversibly aggregated particles in densified SF. The UHPFRC is characterized by very high content in powders, PCE-based superplasticizer, and short steel fibers. Furthermore, a high-speed mixing process is included under all these conditions. In particular, since SF affects all important chemical and physical properties of the concrete (such as microstructure, hydration reaction, self-compacting ability, tensile and compressive strengths, and durability), the difference in dispersion degree should greatly affect the performance of the concrete. Experimental results on the comparison of undensified and densified SF demonstrated the validity of our hypothesis:

- Visual inspection and SEM image analysis confirmed that SF is composed of spherical nanoparticles, but, regardless of the type of SF product, they existed in the form of agglomerated lumps and the sizes of large ones reached several millimeters. The particle size analysis based on SEM images formed the size distribution in a smaller range compared to the results obtained by the DLS technique. The difference between the two techniques was attributed to the link of nanoparticles at a high temperature, the densification process or the agglomeration of nanoparticles thereafter, or the difference in dispersion efficiency during the ultrasonic treatment.

- The material properties of UHPFRC with densified and undensified SF were compared (their conditions other than the densification process were the same). Experimental results showed that there was no significant difference in workability, compressive strength, or flexural tensile strength between the two samples. Analysis of the hydration reaction based on XRD and TG also showed that there was almost no difference between the two samples in the formation or consumption of the main hydration products.
- When the samples were heat treated at 90 °C, portlandite was not identified because the chemical reaction related to the formation of this crystal was accelerated. This means that the pozzolanic reaction, which decisively affects the long-term strength and durability of the concrete, can be terminated significantly early due to the influence of limestone powder contained instead of Portland cement. Therefore, even when densified SF is used under standard heat treatment conditions, UHPFRC's very high ultimate compressive strength (>200 MPa) can be ensured before 28 days.
- The results that the densified and undensified SFs did not differ in the hydration reaction and mechanical properties were also valid under air-dried curing conditions, without heat treatment. Thus, it was concluded that densified SF can be used for both precast and field casting UHPFRCs.

Author Contributions: Conceptualization, S.-H.K. and S.-G.H.; methodology, S.-H.K. and S.-G.H.; investigation, S.-H.K.; writing—original draft preparation, S.-H.K.; writing—review and editing, J.M.; visualization, S.-H.K.; supervision, J.M. All authors have read and agreed to the published version of the manuscript.

Funding: This research was funded by the Korea Agency for Infrastructure Technology Advancement (KAIA), grant number 20NANO-B156177-01, and the Korea Agency for Infrastructure Technology Advancement (KAIA) was funded by the Ministry of Land, Infrastructure and Transport.

Acknowledgments: The Institute of Engineering Research at Seoul National University provided research facilities for this work.

Conflicts of Interest: The authors declare no conflict of interest.

References

1. Richard, P.; Cheyrezy, M. Composition of reactive powder concretes. *Cem. Concr. Res.* **1995**, *25*, 1501–1511. [CrossRef]
2. De Larrard, F.; Sedran, T. Optimization of ultra-high-performance concrete by the use of a packing model. *Cem. Concr. Res.* **1994**, *24*, 997–1009. [CrossRef]
3. Brühwiler, E.; Denarié, E. Rehabilitation of concrete structures using ultra-high performance fibre reinforced concrete. In Proceedings of the Second International Symposium on Ultra High Performance Concrete, Kassel, Germany, 5–7 March 2008; pp. 895–902.
4. Bastien-Masse, M.; Brühwiler, E. Experimental investigation on punching resistance of R-UHPFRC–RC composite slabs. *Mater. Struct.* **2016**, *49*, 1573–1590. [CrossRef]
5. Schmidt, M.; Fehling, E. Ultra-high-performance concrete: Research, development and application in Europe. *ACI Spec. Publ.* **2005**, *228*, 51–78. [CrossRef]
6. Wille, A.E.N.K.; Gustavo, J.P.-M. Ultra-high performance concrete with compressive strength exceeding 150 MPa (22 ksi): A simpler way. *ACI Mater. J.* **2011**, *108*. [CrossRef]
7. Fontana, P.; Lehmann, C.; Müller, U.; Meng, B. Reactivity of mineral additions in autoclaved UHPC. In Proceedings of the International RILEM Conference on Material Science, Aachen, Germany, 6–8 September 2010; pp. 69–77.
8. Huang, W.; Kazemi-Kamyab, H.; Sun, W.; Scrivener, K. Effect of replacement of silica fume with calcined clay on the hydration and microstructural development of eco-UHPFRC. *Mater. Des.* **2017**, *121*, 36–46. [CrossRef]

9. Hafiz, M.A.; Skibsted, J.; Denarié, E. Influence of low curing temperatures on the tensile response of low clinker strain hardening UHPFRC under full restraint. *Cem. Concr. Res.* **2020**, *128*, 105940. [CrossRef]
10. Ibrahim, M.A.; Farhat, M.; Issa, M.A.; Hasse, J.A. Effect of material constituents on mechanical and fracture mechanics properties of ultra-high-performance concrete. *ACI Mater. J.* **2017**, *114*, 453–465. [CrossRef]
11. Zhang, X.; Zhao, S.; Liu, Z.; Wang, F. Utilization of steel slag in ultra-high performance concrete with enhanced eco-friendliness. *Constr. Build. Mater.* **2019**, *214*, 28–36. [CrossRef]
12. Norhasri, M.S.M.; Hamidah, M.S.; Fadzil, A.M. Inclusion of nano metaclayed as additive in ultra high performance concrete (UHPC). *Constr. Build. Mater.* **2019**, *201*, 590–598. [CrossRef]
13. Mo, Z.; Wang, R.; Gao, X. Hydration and mechanical properties of UHPC matrix containing limestone and different levels of metakaolin. *Constr. Build. Mater.* **2020**, *256*, 119454. [CrossRef]
14. Yu, R.; Spiesz, P.; Brouwers, H.J.H. Development of an eco-friendly Ultra-High Performance Concrete (UHPC) with efficient cement and mineral admixtures uses. *Cem. Concr. Compos.* **2015**, *55*, 383–394. [CrossRef]
15. Huang, W.; Kazemi-Kamyab, H.; Sun, W.; Scrivener, K. Effect of cement substitution by limestone on the hydration and microstructural development of ultra-high performance concrete (UHPC). *Cem. Concr. Compos.* **2017**, *77*, 86–101. [CrossRef]
16. Kang, S.-H.; Jeong, Y.; Tan, K.H.; Moon, J. The use of limestone to replace physical filler of quartz powder in UHPFRC. *Cem. Concr. Compos.* **2018**, *94*, 238–247. [CrossRef]
17. Kang, S.-H.; Jeong, Y.; Tan, K.H.; Moon, J. High-volume use of limestone in ultra-high performance fiber-reinforced concrete for reducing cement content and autogenous shrinkage. *Constr. Build. Mater.* **2019**, *213*, 292–305. [CrossRef]
18. Li, P.P.; Brouwers, H.J.H.; Chen, W.; Yu, Q.L. Optimization and characterization of high-volume limestone powder in sustainable ultra-high performance concrete. *Constr. Build. Mater.* **2020**, *242*, 118112. [CrossRef]
19. Li, P.P.; Yu, Q.L.; Brouwers, H.J.H.; Chen, W. Conceptual design and performance evaluation of two-stage ultra-low binder ultra-high performance concrete. *Cem. Concr. Res.* **2019**, *125*, 105858. [CrossRef]
20. ACI Committee 234. *Guide for the Use of Silica Fume in Concrete*; American Concrete Institute: Farmington Hills, MI, USA, 2000; p. 51.
21. Ahmad, S.; Mohaisen, K.O.; Adekunle, S.K.; Al-Dulaijan, S.U.; Maslehuddin, M. Influence of admixing natural pozzolan as partial replacement of cement and microsilica in UHPC mixtures. *Constr. Build. Mater.* **2019**, *198*, 437–444. [CrossRef]
22. De Larrard, F. Ultrafine particles for the making of very high strength concretes. *Cem. Concr. Res.* **1989**, *19*, 161–172. [CrossRef]
23. Fehling, E.; Schmidt, M.; Walraven, J.; Leutbecher, T.; Fröhlich, S. *Ultra-High Performance Concrete UHPC: Fundamentals, Design, Examples*; John Wiley & Sons: Hoboken, NJ, USA, 2014.
24. Kang, S.-H.; Hong, S.-G.; Moon, J. The use of rice husk ash as reactive filler in ultra-high performance concrete. *Cem. Concr. Res.* **2019**, *115*, 389–400. [CrossRef]
25. Oertel, T.; Hutter, F.; Tänzer, R.; Helbig, U.; Sextl, G. Primary particle size and agglomerate size effects of amorphous silica in ultra-high performance concrete. *Cem. Concr. Compos.* **2013**, *37*, 61–67. [CrossRef]
26. Loukili, A.; Khelidj, A.; Richard, P. Hydration kinetics, change of relative humidity, and autogenous shrinkage of ultra-high-strength concrete. *Cem. Concr. Res.* **1999**, *29*, 577–584. [CrossRef]
27. Siddique, R.; Khan, M.I. *Supplementary Cementing Materials*; Springer: Berlin/Heidelberg, Germany, 2011. [CrossRef]
28. Chan, Y.-W.; Chu, S.-H. Effect of silica fume on steel fiber bond characteristics in reactive powder concrete. *Cem. Concr. Res.* **2004**, *34*, 1167–1172. [CrossRef]
29. Holland, T.C. *Silica Fume User's Manual*; Federal Highway Administration: Washington, DC, USA, 2005; p. 193.
30. Lothenbach, B.; Scrivener, K.; Hooton, R.D. Supplementary cementitious materials. *Cem. Concr. Res.* **2011**, *41*, 1244–1256. [CrossRef]
31. Pfeifer, C.; Moeser, B.; Weber, C.; Stark, J. Investigations of the pozzolanic reaction of silica fume in Ultra-high performance concrete (UHPC). In Proceedings of the International RILEM Conference on Material Science-MATSCI, Aachen, Germany, 6–8 September 2010; pp. 287–298.
32. Oertel, T.; Hutter, F.; Helbig, U.; Sextl, G. Amorphous silica in ultra-high performance concrete: First hour of hydration. *Cem. Concr. Res.* **2014**, *58*, 131–142. [CrossRef]

33. Van Tuan, N.; Ye, G.; van Breugel, K.; Copuroglu, O. Hydration and microstructure of ultra high performance concrete incorporating rice husk ash. *Cem. Concr. Res.* **2011**, *41*, 1104–1111. [CrossRef]
34. St John, D.A. *The Dispersion of Silica Fume*; Industrial Research Ltd.: Lower Hutt, New Zealand, 1994; p. 29.
35. Scrivener, K.; Young, J.F. *Mechanisms of Chemical Degradation of Cement-Based Systems*; E&FN Spon: London, UK, 1997.
36. Pedro, D.; de Brito, J.; Evangelista, L. Evaluation of high-performance concrete with recycled aggregates: Use of densified silica fume as cement replacement. *Constr. Build. Mater.* **2017**, *147*, 803–814. [CrossRef]
37. Bapat, J.D. *Mineral Admixtures in Cement and Concrete*; CRC Press: New York, NY, USA, 2012.
38. Adil, G.; Kevern, J.T.; Mann, D. Influence of silica fume on mechanical and durability of pervious concrete. *Constr. Build. Mater.* **2020**, *247*, 118453. [CrossRef]
39. St John, D.A.; McLeod, L.C.; Milestone, N.B. *An Investigation of the Mixing and Properties of DSP Mortars Made from New Zealand Cements and Aggregates*; Industrial Research Ltd.: Lower Hutt, New Zealand, 1993; p. 69.
40. Diamond, S.; Sahu, S.; Thaulow, N. Reaction products of densified silica fume agglomerates in concrete. *Cem. Concr. Res.* **2004**, *34*, 1625–1632. [CrossRef]
41. Diamond, S.; Sahu, S. Densified silica fume: Particle sizes and dispersion in concrete. *Mater. Struct.* **2006**, *39*, 849–859. [CrossRef]
42. Kolderup, H. Particle size distribution of fumes formed by ferrosilicon production. *J. Air Pollut. Control. Assoc.* **1977**, *27*, 127–130. [CrossRef]
43. Möser, B.; Pfeifer, C. Microstructure and durability of ultra-high performance concrete. In Proceedings of the Second International Symposium on Ultra High Performance Concrete, Kassel, Germany, 5–7 March 2008; pp. 417–424.
44. Wang, X.; Huang, J.; Dai, S.; Ma, B.; Tan, H.; Jiang, Q. Effect of silica fume particle dispersion and distribution on the performance of cementitious materials: A theoretical analysis of optimal sonication treatment time. *Constr. Build. Mater.* **2019**, *212*, 549–560. [CrossRef]
45. Arvaniti, E.C.; Juenger, M.C.G.; Bernal, S.A.; Duchesne, J.; Courard, L.; Leroy, S.; Provis, J.L.; Klemm, A.; De Belie, N. Determination of particle size, surface area, and shape of supplementary cementitious materials by different techniques. *Mater. Struct.* **2015**, *48*, 3687–3701. [CrossRef]
46. Rodríguez, E.D.; Soriano, L.; Payá, J.; Borrachero, M.V.; Monzó, J.M. Increase of the reactivity of densified silica fume by sonication treatment. *Ultrason. Sonochem.* **2012**, *19*, 1099–1107. [CrossRef] [PubMed]
47. Ma, R.; Guo, L.; Sun, W.; Rong, Z. Well-dispersed silica fume by surface modification and the control of cement hydration. *Adv. Civ. Eng.* **2018**, *2018*, 6184105. [CrossRef]
48. Bianchi, G.Q. Application of Nano-Silica in Concrete. Ph.D. Thesis, Eindhoven University of Technology, Eindhoven, The Netherlands, 2014.
49. Foerter-Barth, U.; Teipel, U. Characterization of particles by means of laser light diffraction and dynamic light scattering. In *Developments in Mineral Processing*; Massacci, P., Ed.; Elsevier: Amsterdam, The Netherlands, 2000; Volume 13, pp. C1-1–C1-8.
50. Malm, A.V.; Corbett, J.C.W. Improved dynamic light scattering using an adaptive and statistically driven time resolved treatment of correlation data. *Sci. Rep.* **2019**, *9*, 13519. [CrossRef]
51. Land, G.; Stephan, D. The influence of nano-silica on the hydration of ordinary Portland cement. *J. Mater. Sci.* **2012**, *47*, 1011–1017. [CrossRef]
52. Lowke, D.; Schiessl, P. Effect of mixing energy on fresh properties of SCC. In Proceedings of the 4th International RILEM Symposium on Self-Compacting Concrete, Chicago, IL, USA, 30 October–2 November 2005; p. 6.
53. Dils, J.; De Schutter, G.; Boel, V. Influence of mixing procedure and mixer type on fresh and hardened properties of concrete: A review. *Mater. Struct.* **2012**, *45*, 1673–1683. [CrossRef]
54. *ISO 13322-1:2014. Particle Size Analysis—Image Analysis Methods—Part 1: Static image analysis methods*; International Organization for Standardization: Geneva, Switzerland, 2014; p. 24.
55. *ASTM C230/C230M-14. Standard Specification for Flow Table for Use in Tests of Hydraulic Cement*; ASTM International: West Conshohocken, PA, USA, 2014. [CrossRef]
56. *ASTM C109/C109M-16a. Standard Test Method for Compressive Strength of Hydraulic Cement Mortars (Using 2-in. or [50-mm] Cube Specimens)*; ASTM International: West Conshohocken, PA, USA, 2016. [CrossRef]
57. *ISO 679. Cement–Test Methods–Determination of Strength*; International Organization for Standardization: Geneva, Switzerland, 2009; p. 29.

58. Scrivener, K.; Snellings, R.; Lothenbach, B. *A Practical Guide to Microstructural Analysis of Cementitious Materials*; CRC Press: Boca Raton, FL, USA, 2016.
59. Snellings, R.; Chwast, J.; Cizer, Ö.; De Belie, N.; Dhandapani, Y.; Durdzinski, P.; Elsen, J.; Haufe, J.; Hooton, D.; Patapy, C.; et al. Report of TC 238-SCM: Hydration stoppage methods for phase assemblage studies of blended cements—Results of a round robin test. *Mater. Struct.* **2018**, *51*, 111. [CrossRef]
60. Hewlett, P. *Lea's Chemistry of Cement and Concrete*, 4th ed.; Butterworth-Heinemann: Oxford, UK, 2010.
61. Kaasalainen, M.; Aseyev, V.; von Haartman, E.; Karaman, D.Ş.; Mäkilä, E.; Tenhu, H.; Rosenholm, J.; Salonen, J. Size, stability, and porosity of mesoporous nanoparticles characterized with light scattering. *Nanoscale Res. Lett.* **2017**, *12*, 74. [CrossRef]
62. Snellings, R.; Mertens, G.; Elsen, J. Supplementary Cementitious Materials. *Rev. Mineral. Geochem.* **2012**, *74*, 211–278. [CrossRef]
63. *NF P18-470. Ultra-High Performance Fibre-Reinforced Concrete—Specifications, Performance, Production and Conformity*; Association Française de Normalisation (AFNOR): Saint-Denis, France, 2016; p. 94.
64. Sobolev, K.; Flores, I.; Hermosillo, R.; Torres-Martínez, L.M. Nanomaterials and nanotechnology for high-performance cement composites. In Proceedings of the ACI Session on Nanotechnology of Concrete: Recent Developments and Future Perspectives, Denver, CO, USA, 7 November 2006; pp. 91–118.
65. Deshini, A.; Ioannides, A.M. Undispersed agglomerates and the strength of microsilica concrete. *Int. J. Pavement Eng.* **2012**, *13*, 226–234. [CrossRef]
66. Lagerblad, B.; Utkin, P. *Silica Granulates in Concrete: Dispersion and Durability Aspects*; Cement och Betong Institutet: Stockholm, Sweeden, 1993; p. 44.
67. Marusin, S.L.; Shotwell, L.B. Alkali-silica reaction in concrete caused by densified silica fume lumps: A case study. *Cem. Concr. Aggreg.* **2000**, *22*, 90–94. [CrossRef]
68. Shayan, A.; Quick, G.W.; Lancucki, C.J. Morphological, mineralogical and chemical features of steam-cured concretes containing densified silica fume and various alkali levels. *Adv. Cem. Res.* **1993**, *5*, 151–162. [CrossRef]
69. Baweja, T.C.D.; Bucea, L. *Investigation of Dispersion Levels of Silica Fume in Pastes, Mortars, and Concrete*; American Concrete Institute: Farmington Hills, MI, USA, 2003.
70. Cao, Y.; Yu, Q.L.; Brouwers, H.J.H.; Chen, W. Predicting the rate effects on hooked-end fiber pullout performance from Ultra-High Performance Concrete (UHPC). *Cem. Concr. Res.* **2019**, *120*, 164–175. [CrossRef]
71. Hafiz, M.A.; Denarié, E. Tensile response of UHPFRC under very low strain rates and low temperatures. *Cem. Concr. Res.* **2020**, *133*, 106067. [CrossRef]
72. Kang, S.-H.; Lee, J.-H.; Hong, S.-G.; Moon, J. Microstructural investigation of heat-treated ultra-high performance concrete for optimum production. *Materials* **2017**, *10*, 1106. [CrossRef]
73. Zhou, B.; Uchida, Y. Influence of flowability, casting time and formwork geometry on fiber orientation and mechanical properties of UHPFRC. *Cem. Concr. Res.* **2017**, *95*, 164–177. [CrossRef]
74. Abrishambaf, A.; Pimentel, M.; Nunes, S. Influence of fibre orientation on the tensile behaviour of ultra-high performance fibre reinforced cementitious composites. *Cem. Concr. Res.* **2017**, *97*, 28–40. [CrossRef]
75. Huang, H.; Gao, X.; Li, Y.; Su, A. SPH simulation and experimental investigation of fiber orientation in UHPC beams with different placements. *Constr. Build. Mater.* **2020**, *233*, 117372. [CrossRef]
76. Larsen, I.L.; Thorstensen, R.T. The influence of steel fibres on compressive and tensile strength of ultra high performance concrete: A review. *Constr. Build. Mater.* **2020**, *256*, 119459. [CrossRef]
77. Taylor, H.; Famy, C.; Scrivener, K. Delayed ettringite formation. *Cem. Concr. Res.* **2001**, *31*, 683–693. [CrossRef]
78. Heinz, D.; Urbonas, L.; Gerlicher, T. Effect of heat treatment method on the properties of UHPC. In Proceedings of the 3rd International Symposium on Ultra High Performance Concrete and Nanotechnology for High Performance Construction Materials, Kassel, Germany, 7–9 March 2012; pp. 283–290.
79. Jensen, O.M.; Hansen, P.F. Water-entrained cement-based materials: I. Principles and theoretical background. *Cem. Concr. Res.* **2001**, *31*, 647–654. [CrossRef]
80. Justs, J.; Wyrzykowski, M.; Bajare, D.; Lura, P. Internal curing by superabsorbent polymers in ultra-high performance concrete. *Cem. Concr. Res.* **2015**, *76*, 82–90. [CrossRef]
81. Heinz, D.; Ludwig, H.-M. Heat treatment and the risk of DEF delayed ettringite formation in UHPC. In Proceedings of the 1st International Symposium on Ultra-High Performance Concrete, Kassel, Germany, 13–15 September 2004; pp. 717–730.

82. Selleng, C.; Meng, B.; Fontana, P. Phase composition and strength of thermally treated UHPC. In Proceedings of the 4th International Symposium on Ultra-High Performance Concrete and High Performance Materials, Kassel, Germany, 9–11 March 2016; pp. 7–8.
83. Selleng, C.; Fontana, P.; Meng, B. Possibilities for improving the properties of UHPC by means of thermal treatment. In Proceedings of the AFGC-ACI-fib-RILEM PRO 106: Ultra-High Performance Fibre-Reinforced Concrete (UHPFRC 2017), Montpellier, France, 2–4 October 2017; pp. 83–92.
84. Korpa, A.; Kowald, T.; Trettin, R. Phase development in normal and ultra high performance cementitious systems by quantitative X-ray analysis and thermoanalytical methods. *Cem. Concr. Res.* **2009**, *39*, 69–76. [CrossRef]
85. Cheyrezy, M.; Maret, V.; Frouin, L. Microstructural analysis of RPC (Reactive Powder Concrete). *Cem. Concr. Res.* **1995**, *25*, 1491–1500. [CrossRef]
86. Jensen, O.M.; Hansen, P.F. Influence of temperature on autogenous deformation and relative humidity change in hardening cement paste. *Cem. Concr. Res.* **1999**, *29*, 567–575. [CrossRef]
87. Alexander, G.B.; Heston, W.M.; Iler, R.K. The solubility of amorphous silica in water. *J. Phys. Chem.* **1954**, *58*, 453–455. [CrossRef]
88. Zhang, M.-H.; Gjørv, O.E. Effect of silica fume on cement hydration in low porosity cement pastes. *Cem. Concr. Res.* **1991**, *21*, 800–808. [CrossRef]
89. Schachinger, I.; Hilbig, H.; Stengel, T. Effect of curing temperature at an early age on the long-term strength development of UHPC. In Proceedings of the Second International Symposium on Ultra High Performance Concrete, Kassel, Germany, 5–7 March 2008; pp. 205–212.
90. Chung, D.D.L. Review: Improving cement-based materials by using silica fume. *J. Mater. Sci.* **2002**, *37*, 673–682. [CrossRef]
91. Talayeh, N.; Eugen, B. Experimental investigation on reinforced ultra-high-performance fiber-reinforced concrete composite beams subjected to combined bending and shear. *ACI Struct. J.* **2013**, *110*, 251–261. [CrossRef]
92. Lampropoulos, A.P.; Paschalis, S.A.; Tsioulou, O.T.; Dritsos, S.E. Strengthening of reinforced concrete beams using ultra high performance fibre reinforced concrete (UHPFRC). *Eng. Struct.* **2016**, *106*, 370–384. [CrossRef]
93. Ford, E.L.; Hoover, C.G.; Mobasher, B.; Neithalath, N. Relating the nano-mechanical response and qualitative chemical maps of multi-component ultra-high performance cementitious binders. *Constr. Build. Mater.* **2020**, *260*, 119959. [CrossRef]
94. Ali Dadvar, S.; Mostofinejad, D.; Bahmani, H. Strengthening of RC columns by ultra-high performance fiber reinforced concrete (UHPFRC) jacketing. *Constr. Build. Mater.* **2020**, *235*, 117485. [CrossRef]

Article

Fresh and Hardened Properties of Portland Cement-Slag Concrete Activated Using the By-Product of the Liquid Crystal Display Manufacturing Process

Sung Choi [1] and Sukhoon Pyo [2,*]

[1] Department of Civil Engineering, Kyungdong University, 27 Gyeongdongdaehak-ro, Yangju-Si 11458, Korea; csomy1113@kduniv.ac.kr

[2] Department of Urban and Environmental Engineering, Ulsan National Institute of Science and Technology (UNIST), 50 UNIST-gil, Ulju-gun, Ulsan 44919, Korea

* Correspondence: shpyo@unist.ac.kr; Tel.: +82-52-217-2827

Received: 4 September 2020; Accepted: 29 September 2020; Published: 30 September 2020

Abstract: This experimental research investigated the applicability of the liquid crystal display (LCD) by-product of the refining process as a sustainable and alternative alkali activator for ground granulated blast-furnace slag (GGBFS) blended cement concrete. Three levels of binder replacement using the industrial by-product, and four water/binder ratios were considered in order to evaluate the effects of the replacement in fresh and hardened properties of the blended concrete. XRD and TG analyses confirmed that the by-product that contains abundant alkali compounds promotes the reactivity of GGBFS. The test results indicated that the incorporation of the by-product results in delayed setting and degraded workability due to the highly porous nature of the by-product, yet shows rapid early-age strength development of the blended concrete as conventional alkaline activators for GGBFS. These characteristics shed light on a simple yet effective and practical means of reusing the industrial by-product as an alternative alkaline activator.

Keywords: alternative alkali-activated material; ground granulated blast-furnace slag; strength development; setting time; workability

---

1. Introduction

As one of the most sustainable approaches to reducing carbon dioxide emissions from the production of cement-based construction materials, various types of industrial by-products are currently used worldwide as Portland cement replacements, called supplementary cementitious materials (SCMs), e.g., ground granulated blast-furnace slag (GGBFS), fly ash, and silica fume. It is well known that the replacement level of cement with SCMs depends on the reactivity, local availability, and legislation [1]. For the strength evolution of SCMs with cement, the pozzolanic reaction of SCMs is an essential chemical process that uses high alkalinity compounds, such as $Ca(OH)_2$, from cement hydration products [2]. Although blending cement with SCMs has many advantages—such as reducing carbon dioxide emissions by saving cement, and improving long-term strength, durability, and chemical stability [3–7]—the setting time and early-age strength development can be drastically delayed if the amount SCMs in the blend is excessive, without adequate alkali activators [8,9].

Various alkali activators have been suggested for adequately promoting strength of blended concrete using SCMs to either partially or fully replace ordinary Portland cement (OPC). For example, it is well known that sodium hydroxide (NaOH), sodium silicate ($Na_2O{\cdot}rSiO_2$), sodium carbonate (Na_2CO_3), and sodium sulphate (Na_2SO_4) are effective and common ones used to activate GGBFS [10,11]. These alkali activators are effective in promoting the initial reactivity of GGBFS and lead to rapid strength development [3]. However, the radical chemical reaction can also dramatically accelerate the reaction of GGBFS, which can significantly affect the workability of concrete—such as its setting time and flowability—depending on the alkali content and the slag/activator ratio [10,12]. In addition, the manufacturing process of alkali activators is usually both energy intensive and costly [13], which could limit wider application of alkali activated slag concrete. Alternative studies have been attempted to evaluate sustainable alkali activators for SCMs using industrial by-products [14]. For example, Maraghechi et al. [1] tested recycled glass powder as an alkali activator for binary mixtures with OPC, slag, and fly ash. The findings suggested that elevated temperature curing (60 °C) is preferable to effectively consuming glass powder for alkali activated slag mortar. However, limited work has been conducted on the use of industrial by-products as alkali activators for slag cement and concrete.

The current mainstream liquid crystal display (LCD) design primarily consists of color filter (CF) substrate glass on which RGB pixels are deposited and of thin-film-transistor (TFT) substrate glass that is painted with thin film circuitry that delivers signals to liquid crystal. In order to maintain a high quality and a high resolution, TFT uses an alkali-free glass that is obtained through a high refining process that involves the generation of by-product. Because the by-product of this refining process contains a large amount of alkali ingredients—such as SO_3, Na_2O, and K_2O—it is expected to be used as an alternative alkali activator for GGBFS blended cement concrete. Although the LCD by-product that results from the refining process does not require an additional treatment because it is in the form of powder, it is necessary to investigate its fundamental properties in order to verify its applicability as an alternative alkali activator for GGBFS blended cement concrete. However, to the best of the authors' knowledge, alkaline activation of LCD by-product has not been investigated as a means of developing sustainable construction material that can be utilized for the production of high strength concrete. In order to adapt the alkali industrial by-product to slag concrete, it is necessary to find suitable combinations of base binders by investigating the reactivity of the by-product with existing binding materials. Therefore, the effects of the alkali by-product on the fundamental properties of concrete—such as workability, setting time, and strength enhancement under various mixing conditions—need to be identified. This research gap motivated the study presented in this paper. It should be pointed out, however, that multiple attempts have been conducted to use thin-film transistor liquid-crystal display (TFT-LCD) waste glass to partially replace OPC [15–18]. For example, Lin et al. [15] tested up to 40% of TFT-LCD waste glass to replace OPC, and concluded that as the amount of TFT-LCD waste glass increases the strength of the paste distinctly decreases. Jang et al. [16] used additional activator to promote reactivity of TFT-LCD waste glass with OPC, and concluded that the pozzolanic reaction between the waste glass and the activator leads to enhance compressive strength of high strength concrete products. Kim et al. [18] found that smaller particle size of ground TFT-LCD waste glass would lead to the decreased porosity of TFT-LCD waste glass concrete, which is expected to enhanced durability and permeability.

In this study, the applicability of the LCD by-product of the refining process as an alkali activator was evaluated by characterizing the reactivity of GGBFS with OPC under a normal curing condition with the aim of developing practical applications. The binary paste was prepared by mixing GGBFS and OPC with the LCD by-product-based activator (LCDBA) and the variation of the hydration products in the paste, according to the curing age, was characterized through X-ray diffraction (XRD) as well as through thermogravimetry and differential thermal analysis (TG-DTA). In addition, fresh and hardened properties of OPC-slag concrete, activated by the alkali by-product, were assessed by examining the slump, bleeding, setting time, and compressive strength of concrete.

2. Experimental Program

The applicability of the LCDBA was investigated with the aim of developing practical applications as a sustainable and alternative alkali activator for GGBFS blended cement concrete. The activation effects of the LCDBA were characterized using GGBFS blended cement paste. In addition, the fresh and hardened properties of GGBFS blended cement concrete incorporating LCDBA were investigated.

2.1. Raw Materials

OPC (ASTM C 150 Type I) and GGBFS were used as binders to prepare the paste and concrete and their chemical compositions are summarized in Table 1. The OPC and GGBFS that were used had a Blaine specific surface of 339 m^2/kg and 449 m^2/kg and a density of 3150 kg/m^3 and 2910 kg/m^3, respectively. The major chemical components of GGBFS were CaO, SiO_2, and Al_2O_3, as shown in Table 1. The basicity coefficient (Kb = (CaO + MgO)/(SiO_2 + Al_2O_3)) was 0.977, which is similar to the neutral value of 1.0 for ideal alkali activation [19]. The hydration modulus, according to a formula proposed in the literature (HM = (CaO + MgO + Al_2O_3)/SiO_2) of GGBFS, was 1.92. This was higher than the value of 1.4, which is required for good hydration properties of GGBFS [19]. The LCD by-product of the refining process (see Figure 1), LCDBA—the chemical composition of which is given in Table 1—had a Blaine specific surface of 770 m^2/kg and a density of 2540 kg/m^3. XRD analysis was performed to determine the nature of LCDBA, as shown in Figure 2, which indicated that this was crystallized mainly with K_2SO_4 and Na_2SO_4. The particle size distributions of GGBFS and LCDBA are shown in Figure 3, which were measured using a laser particle size analyzer (PSA, Beckman Coulter LS 13 320, Brea, CA, USA). The overall particle size distribution of LCDBA was similar to that of GGBFS, showing a narrow distribution with the peak around 10 µm. Although LCDBA had about 1.7 times higher specific surface in comparison to GGBFS, the particle size of LCDBA was slightly larger than that of GGBFS, which indicated that LCDBA has porous microstructures resulting from the refining process.

Table 1. Chemical composition of OPC, GGBFS, and LCDBA (wt.%).

Element	SiO_2	Al_2O_3	Fe_2O_3	CaO	MgO	SO_3	K_2O	Na_2O	LOI
OPC	19.5	5.6	3.5	61.5	3.8	2.5	1.1	0.1	2.5
GGBFS	32.6	15.5	0.5	42.2	4.6	3.3	0.5	0.2	-
LCDBA	4.3	-	0.2	10.0	0.5	47.4	4.3	30.4	3.0

Figure 1. Image of LCDBA used in this study.

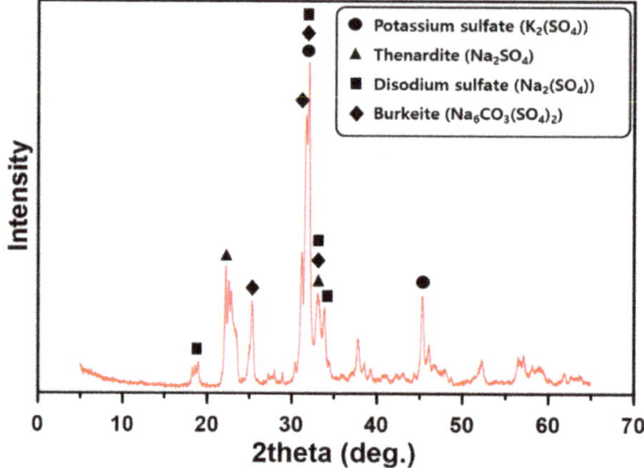

Figure 2. XRD patterns of LCDBA.

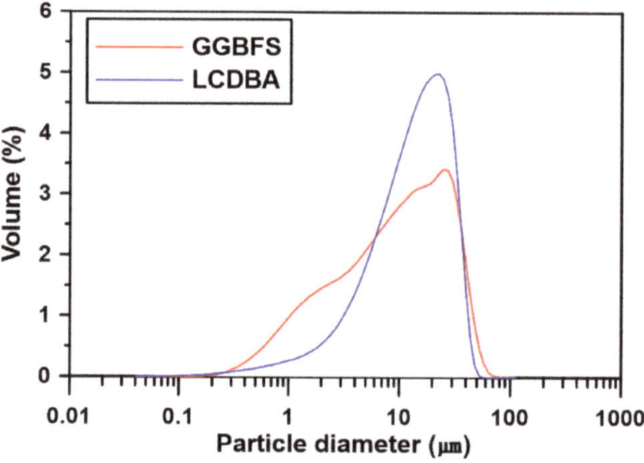

Figure 3. Particle size distribution of GGBFS and LCDBA.

Non-reactive river sand, with a specific gravity of 2.62, a fineness modulus of 2.77, and an absorption capacity of 1.1%, was used in preparation of all mortars and concrete. The crushed basalt aggregates were adopted as the coarse aggregate for the concrete mix in which the maximum size, specific gravity, absorption capacity, and fineness modulus were 25 mm, 2.63, 0.8%, and 6.4, respectively. In addition, a polycarboxylate-based superplasticizer with 17% solid content by weight was used to enhance particle dispersions within the mixture.

2.2. Mix Proportions

The characterization of the reactivity of the binders was carried out based on XRD and TG-DTA tests using GGBFS blended cement paste with the OPC/GGBFS ratio of 55:45 and the water/binder ratio of 30%. LCDBA replaced the binder by 0%, 3%, and 5%. Table 2 shows the concrete mix proportions according to the usage of LCDBA. As in the paste mix design, the ratio of OPC and GGBFS was set to 55:45. In addition, the binder substitution ratio of LCDBA was set to 0%, 3%, and 5% in order to evaluate the effects of the LCDBA substitution level on the fresh and hardened properties of GGBFS blended cement concrete. The GGBFS blended cement concrete was designed to have the total volume of mortar in the 615 ± 5 L range and the slump of concrete in the 180 ± 25 mm range by controlling the amount of superplasticizer and the sand to total aggregate volume ratio (s/a).

Table 2. Mix proportions of concrete.

Mixture	W/B (%)	s/a (%)	Unit Weight (kg/m^3)						Super-Plasticizer (%)
			Water	Binder			Fine Aggregate	Coarse Aggregate	
				OPC	GGBFS	LCDBA			
35/L0	35	47.5	160	251	206	-	814	903	0.70
35/L3	35	46.0	160	244	200	14	788	929	0.80
35/L5	35	45.0	160	239	195	23	770	945	1.00
40/L0	40	49.0	160	220	180	-	864	903	0.70
40/L3	40	48.0	160	213	175	12	846	920	0.80
40/L5	40	47.5	160	209	171	20	837	928	0.97
45/L0	45	50.0	160	196	160	-	901	904	0.70
45/L3	45	49.0	160	190	155	11	882	922	0.75
45/L5	45	48.0	160	186	152	18	864	940	0.80
50/L5	50	49.0	160	167	137	16	897	937	0.70

2.3. Experimental Methods

The reactivity of the paste with different LCDBA substitution levels was investigated through XRD and TG analysis. GGBFS blended cement pastes were cured at 20 °C and 90 ± 2% relative humidity for a predetermined curing duration (3, 7, and 28 days). Thereafter, the samples were ground and immersed in acetone to stop hydration and were suction filtered using an aspirator. The crushed samples that stopped hydration were ground further and powdered to particles smaller than 106 μm for the XRD and TG measurements. XRD was conducted using a D/MAX 2500V/PC (Rigaku, Tokyo, Japan) with a scan range of 5°–65° 2θ. TG was conducted using a NETZSCH STA 409 C/CD (NETZSCH, Selb, Bavaria, Germany) with a heating rate of 5 °C/min in the 20–1000 °C range.

The fresh properties of GGBFS blended cement concrete were measured using the slump, bleeding, and setting time. The slump of fresh concrete were measured in accordance with ASTM C 94 [20]. The bleeding test of fresh concrete was measured in accordance with ASTM C 232 [21] by drawing off the bleed water until cessation of bleeding. The initial and final setting times were measured using a penetration resistance apparatus in accordance with ASTM C 803 [22]. The measurements were conducted on sieved mortar samples from the mixed concrete. The fresh concrete was cast into cylindrical molds (100 × 200 mm) for the compression test. Following 24 h of air curing, these cylindrical samples were demolded and immersed into water at a temperature of 20 ± 1 °C for additional curing. Compressive strength tests were carried out in accordance with ASTM C 39 [23]. The strength was determined at 3, 7, 28, and 91 days of curing by averaging the tested values of the three replicates.

3. Test Results and Discussions

3.1. XRD Analysis

The diffraction patterns of the hardened pastes were analyzed by characterizing the crystal phases in order to investigate the applicability of LCDBA with high alkali content for stimulating GGBFS. The activation effect of LCDBA for GGBFS can be demonstrated by investigating the generation of hydration products for different curing ages. Figure 4 shows the results of an XRD analysis of the pastes, according to the LCDBA substitution level.

As can be seen in Figure 3, all specimens included akermanite ($Ca_2Mg(Si_2O_7)$) with strong peaks in the XRD analysis. It has been reported that akermanite exists in a crystalline form in raw materials and hardened GGBFS pastes [24,25] and that it can be identified more clearly for cases with LCDBA. In particular, a strong akermanite peak was observed for the 5% LCDBA series at 28 days of curing, indicating that LCDBA affects the hydration of GGBFS in the long term.

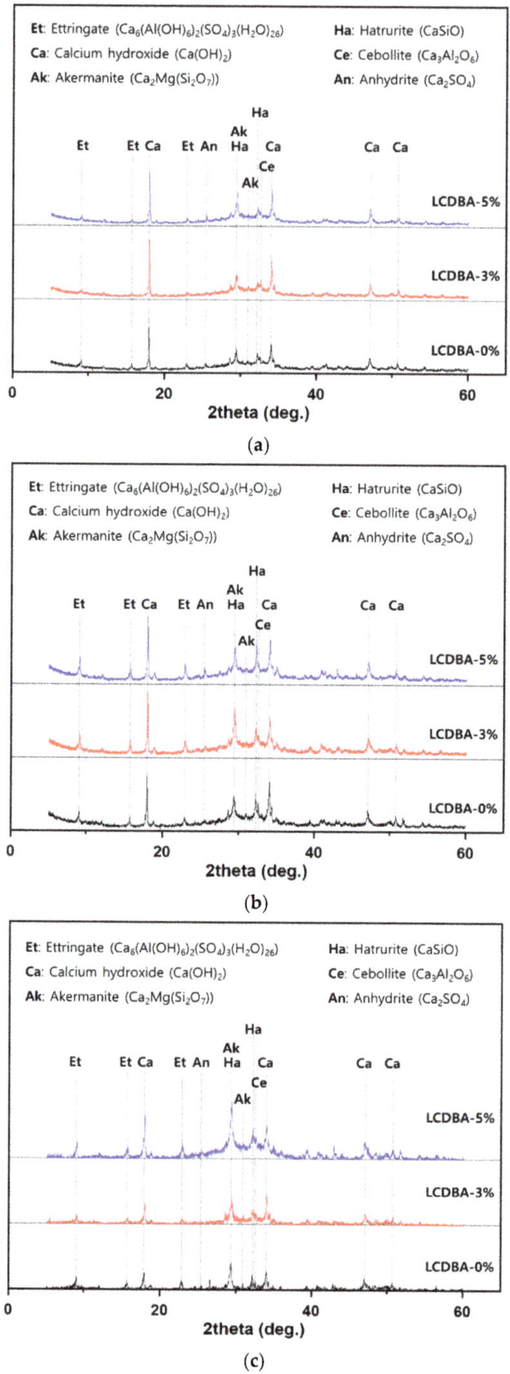

Figure 4. XRD results of the blended pastes at different curing ages: (**a**) 3 days, (**b**) 7 days, and (**c**) 28 days.

In addition, the strong alkali compound of the cement hydration product—calcium hydroxide ($Ca(OH)_2$)—is known to play a role in promoting the activation reaction of GGBFS [2]. This effect is evidenced by the XRD pattern at 3 and 28 days. For example, it can be clearly seen in the LCDBA-0% series that the higher peak of the $Ca(OH)_2$ at 3 days decreases as the curing age increases through the consuming necessary for the activation of the GGBFS to form calcium-silicate-hydrate (C-S-H) gel. The pastes of the two series containing LCDBA also had high $Ca(OH)_2$ at their early age and the amount decreased as the age increased. However, the decrease of $Ca(OH)_2$ was smaller than that of LCDBA-0% series, although the C-S-H gel peak increased significantly. This phenomenon was the likely cause of the high alkaline LCDBA activator. In the section where 2θ is around 9°, it is known that ettringite peaks result from the hydration of cement [26]. It should be noted that the ettringite peak was less affected by the incorporation of LCDBA at 3 days, increasing with the higher LCDBA amount at 28 days. This implies that the production of ettringite is increased by the supply of SO_3, as with the high SO_3 content in LCDBA, at 47.1%. In addition, Mohammed and Safiullah [27] revealed that the amount of ettringite formation highly correlates with the amount of SO_3.

3.2. TG Analysis

The degrees of hydration of GGBFS blended cement pastes were evaluated using the thermogravimetry method and the results are presented in Figure 5. The weight loss at around 100 °C (refer to Section I afterward) can mostly be attributed to the decomposition of ettringite and C-S-H gel [28]. The weight loss at around 450 °C (refer to Section II afterward) can primarily be attributed to the decomposition of $Ca(OH)_2$ to CaO [29].

From the results for Sections I and II, it can be seen that the weight losses of the LCDBA-3% and LCDBA-5% series are more pronounced than those of the LCDBA-0% series for all curing ages. Therefore, the large mass change in Section I implies that a large amount of C-S-H gel initially forms due to the incorporation of LCDBA that contains abundant alkali compounds that promote the reactivity of GGBFS. The large mass change in Section II can be explained that the cement hydration product, $Ca(OH)_2$, was relatively less consumed in the LCDBA-3% and LCDBA-5% mixtures due to the additional supply of alkaline compounds by LCDBA. In the meantime, it should be noted that the differences in mass change between the pastes with LCDBA and without LCDBA at 28 days were relatively large in Section I in comparison to those in Section II. This might be attributed to the fact that, as curing age increases, the alkali components supplied for the pozzolanic reaction of GGBFS in the LCDBA-3% and LCDBA-5% series were also consumed, resulting in a $Ca(OH)_2$ decrease and reaching similar remaining levels of $Ca(OH)_2$ as those of the LCDBA-0% series at 28 days of curing.

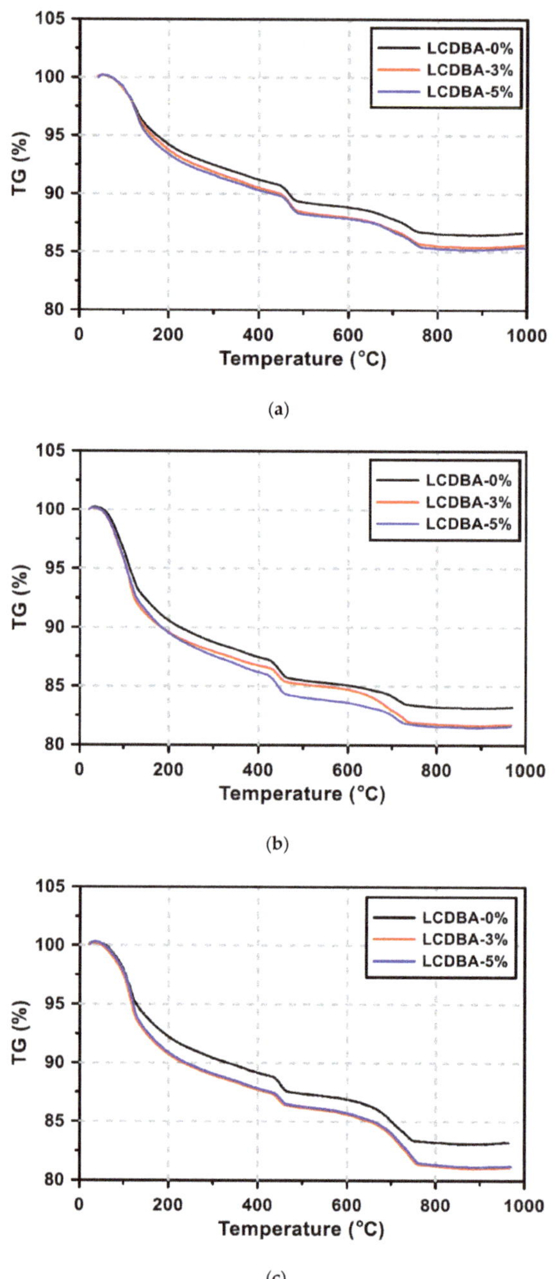

Figure 5. TG curves of the blended pastes at different curing ages: (**a**) 3 days, (**b**) 7 days, and (**c**) 28 days.

3.3. Workability

The slump test is widely used and the most well-known test to assess the workability of concrete. In this study, the effect of the LCDBA usage on workability was evaluated using the slump test. Figure 6 shows the slump values and the superplasticizer dosage of the blended concrete, indicating that all tested series met the target slump level of 180 ± 25 mm. For all concrete series without LCDBA, the superplasticizer amount was fixed at 0.7%. The figure shows that as the water/binder ratio increases the slump increases by 10 mm per 0.05 of water/binder ratio. The blended concrete that contained 3% and 5% LCDBA tended to decrease in slump even though the dosage of superplasticizer increased. For mixtures with 5% LCDBA, all slump results were the same, 160 mm, and the corresponding superplasticizer amounts were 1.0%, 0.9%, and 0.8%, for the water/binder ratios of 0.35, 0.40, and 0.45, respectively. Therefore, the increase in the required superplasticizer amount for maintaining a similar level of workability implies that the porous LCDBA absorbs the mix water during the mixing and thus degrades the workability of the blended concrete.

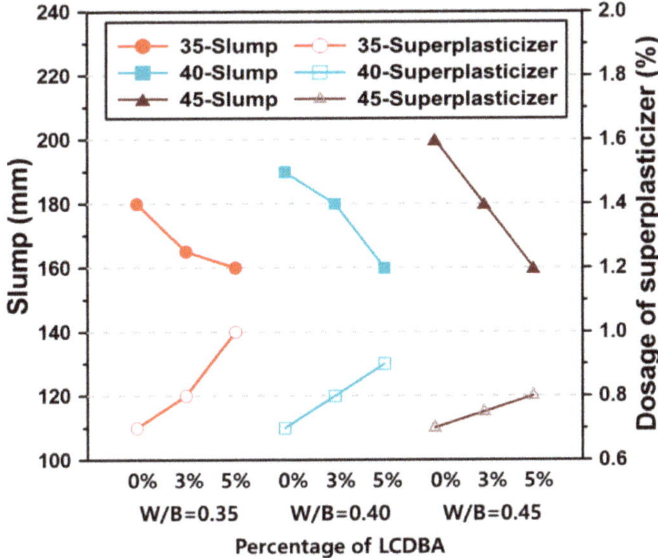

Figure 6. Slump of the blended concrete with different water/binder ratios and LCDBA amount (solid markers and hollow markers stand for slump values and dosage of superplasticizer, respectively).

3.4. Bleeding

The bleeding of concrete, a necessary part of the life of concrete, was considered to occur when the mix water would raise to the surface of freshly placed concrete. It has been reported that the bleed rate and capacity of GGBFS blended cement concrete highly depends on the GGBFS replacement level and the water/binder ratio [30]. Figure 7 shows the development of bleeding over elapsed time of concrete for the water/binder ratio of 0.40, as a function of the LCDBA replacement level. The ending point and the slope of the curves indicate the bleeding capacity of the mixtures and the bleeding rate, respectively. It should be pointed out from the figure that the bleeding capacity decreases as the LCDBA replacement level increases. In particular, the mixture without LCDBA (the black curve) showed the highest bleeding rate for the first four hours and then it quickly reached the highest value. On the other hand, the mixtures with LCDBA (the blue and red curves) showed that as the LCDBA replacement level increases the bleeding rate becomes slower for the first three hours. After the first three hours, all mixtures showed similar bleeding rates until reach the bleeding capacity. The delayed endings of the bleeding were observed for both mixtures with LCDBA in comparison to the one without LCDBA. This can be attributed to the fact that the porous LCDBA absorbed excess water during mixing because of its high specific surface area, which reduced the initial bleeding rate. In addition, the bleeding time increased as the water absorbed by the LCDBA was slowly released to the fresh mixture.

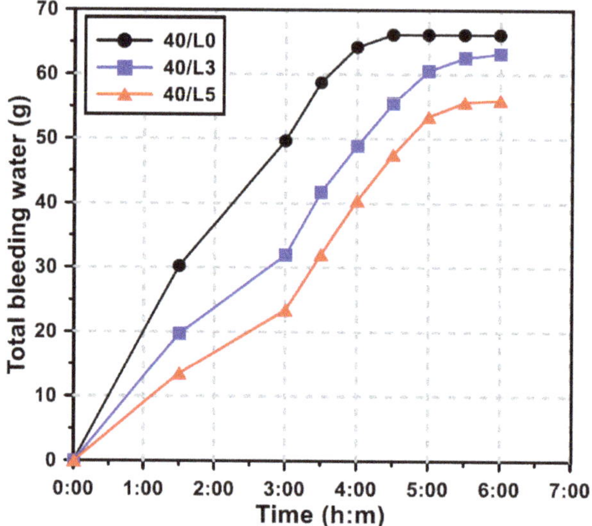

Figure 7. Effect of the LCDBA replacement level on bleeding (W/B = 0.40).

3.5. Setting Time

The effect of the LCDBA replacement level on the initial and final setting times of GGBFS blended cement concrete is shown in Figure 8. The setting times were defined when the penetration resistance reached the pre-defined criteria in accordance with ASTM C 803 [22]. The initial setting time of the mixture without LCDBA was 340 min but the mixtures incorporating LCDBA were delayed for more than 80 min. The delay in the initial setting time can be explained by the delay in the bleeding. Similar to the results of the initial setting time, the mixture without LCDBA showed the fastest final setting time, at 483 min, and 20- and 40-min delays were observed for the mixtures with 3% and 5% LCDBA replacement, respectively. It should be pointed out, however, that the differences of final setting time between the mixtures with and without LCDBA were smaller than those of the initial setting time. This can be attributed to the activation of GGBFS promoted by the alkali supply of LCDBA after initial setting.

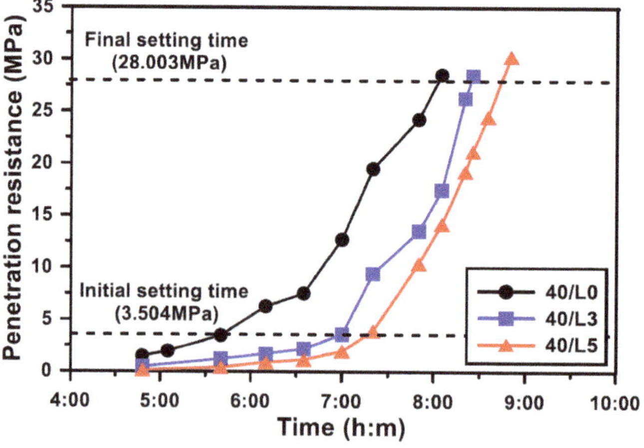

Figure 8. Effect of the LCDBA replacement level on penetration resistance (W/B = 0.40).

3.6. Compressive Strength

The compressive strength development of GGBFS blended cement concrete with different LCDBA replacement levels and with water/binder ratios at 3, 7, 28, and 91 days is shown in Figure 9. The concrete mixtures that incorporate LCDBA showed a generally higher strength than the mixture without LCDBA. In particular, the incorporation of LCDBA was more effective for early strength development and for mixtures with lower water/binder ratios.

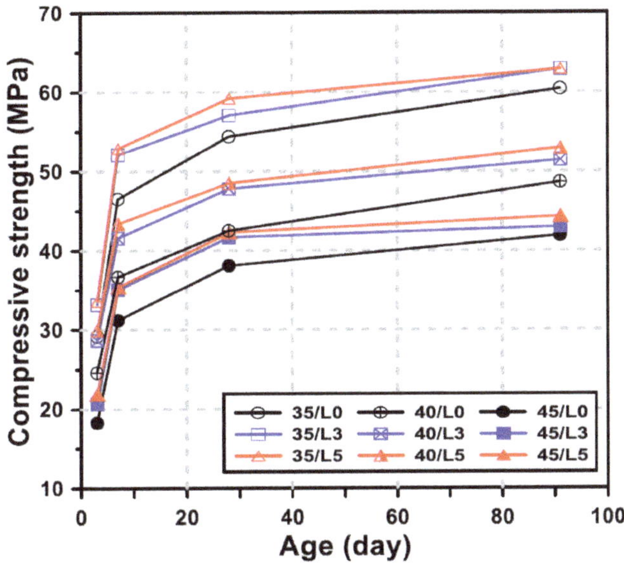

Figure 9. Effect of the LCDBA replacement level on compressive strength.

The rates of increase in the compressive strength of the concrete incorporating LCDBA relative to the strength of concrete without LCDBA were evaluated for each curing age, with the results summarized in Table 3. At 3 days of curing, the strength improvement rates of the concrete with 3% and 5% replacement were 13.1–16.7% and 15.1–22.0%, respectively. On the other hand, the strength improvement rates of the concrete with 3% and 5% replacement at 91 days of curing were 1.4–5.8% and 4.1–8.8%, respectively. Therefore, as the replacement level of LCDBA increased, the strength increase rate became higher, while the strength increase rate gradually decreased as the curing age increased. This tendency is similar to the typical pattern of rapid strength development of GGBFS promoted by alkaline activators (see also [31–33]). It should be pointed out, therefore, that LCDBA could be a sustainable and alternative activator for GGBFS blended cement concrete, with similar effects as those of conventional alkaline activators, which is effective for the early-age strength development of concrete.

Table 3. Rate of increase in the compressive strength of concrete incorporating LCDBA.

Mixture	Curing Age (%)			
	3 Days	7 Days	28 Days	91 Days
35/L3	13.7	12.0	5.0	4.1
35/L5	15.1	13.8	8.8	4.1
40/L3	16.7	13.4	12.5	5.8
40/L5	22.0	18.3	14.1	8.8
45/L3	13.1	9.0	1.5	1.4
45/L5	19.7	9.9	2.9	4.5

In order to quantitatively analyze the strength developing characteristics of GGBFS blended cement concrete incorporating LCDBA, the compressive strength prediction model suggested in ACI 209 was used. The ACI Committee 209 [34] recommends the following equation for predicting the compressive strength of concrete with time

$$(f_c')_t = \frac{t}{a + b \times t}(f_c')_{28}, \tag{1}$$

where a and b are material constants considering the type of binders and curing methods, calculated in this research based on regression analysis using measured values. In addition, $(f_c')_t$ and $(f_c')_{28}$ are compressive strength at the age of t and 28 days, respectively.

The concrete mixtures with 5% LCDBA replacement and without LCDBA were considered to quantitatively illustrate the effect of LCDBA on the development of the compressive strength of concrete. Figure 10 shows the compressive strength results, according to various curing ages and water/binder ratios for the concrete mixtures with 0% or 5% of LCDBA and with the regression curves based on the ACI 209 model.

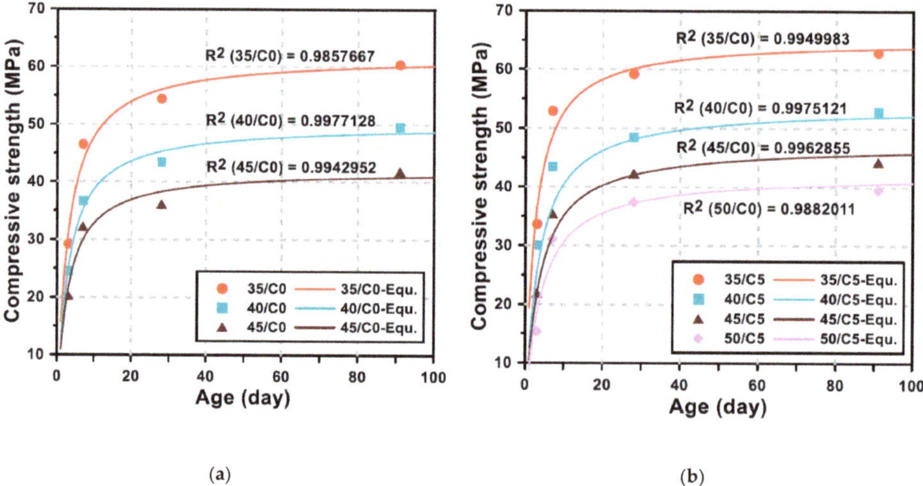

Figure 10. Comparison between the measured compressive strength and the regression curves for 0% or 5% LCDBA replacement series; (**a**) LCDBA-0%, (**b**) LCDBA-5%.

As a result of the regression analysis of the compressive strength, according to the ACI 209 model, the R-squared values for the mixtures with and without LCDBA were higher than 0.985, indicating a high goodness of fit. With the high goodness of fit, the material constants a and b were used to further analyze the strength development patterns of the tested concrete. The calculated material constants a and b are illustrated in Figure 11, where the constants a and b considerably correlate with the strength development of concrete at early age and at long-term age, respectively [35]. Mathematically, a is inversely proportional to the initial strength development and b is inversely proportional to the increase in compressive strength according to the curing age. As shown in the figure, a and b tend to decrease and increase, respectively, as the water/binder ratio increases. As the water/binder ratio decreases, the initial strength development rate becomes high, which leads to a decrease in a. On the other hand, the material constant b is related to the long-term strength development, where the lower b value stands for the higher long-term strength increment. The constant b values for the mixtures with LCDBA were similar, regardless of the water/binder ratio, and higher than those for the mixtures without LCDBA. The higher b values for the mixture with LCDBA were attributed to the higher initial strength development activated by LCDBA and to relatively less long-term strength development.

This is similar to the strength development characteristics of GGBFS and/or fly ash blended cement concrete activated using conventional alkali activators. Therefore, it can be concluded that LCDBA is an effective and sustainable alternative to alkali activators for GGBFS blended cement concrete.

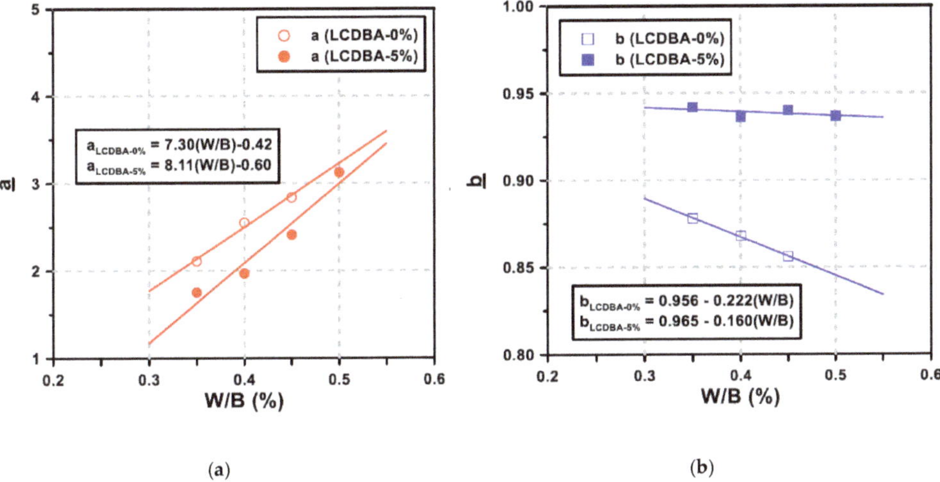

(**a**) (**b**)

Figure 11. Relationship between the material constants and the W/B for 0% or 5% LCDBA replacement series, corresponding to Figure 9; (**a**) constant *a* (**b**) constant *b*.

4. Conclusions

This experimental study evaluates the LCD by-product of the refining process, LCDBA, as an alternative and sustainable alkaline activator for GGBFS blended cement concrete. To investigate the applicability of this the alternative activator, the tested experimental parameters were set at three LCDBA replacement levels and four water/binder ratios. The activation effects were characterized based on XRD and TG analyses using GGBFS blended cement paste. The fresh and hardened properties of GGBFS blended cement concrete incorporating LCDBA were investigated using slump, bleeding, setting time, and compressive strength tests. The key observations and findings of this research can be summarized as follows:

(1) The effectiveness of LCDBA as an alternative and sustainable alkaline activator for GGBFS blended cement concrete was demonstrated for the first time in this study. The formations of akermanite and ettringite, as well as the consumption of $Ca(OH)_2$ due to the stimulation of GGBFS hydrations with LCDBA as days of curing increase, were characterized using XRD analysis. This finding highlights the role of LCDBA as an activator for GGBFS blended cement concrete by providing the essential chemical evidence.

(2) The thermogravimetric analysis was employed to evaluate the degrees of hydration. The results highlight that a large amount of C-S-H gel was initially formed and that the cement hydration product, $Ca(OH)_2$, was relatively less consumed due to the incorporation of LCDBA that contains abundant alkali compounds that promote the reactivity of GGBFS. As curing age increased, the remaining amount of $Ca(OH)_2$ in the pastes with LCDBA became similar as to the one in the pastes without LCDBA, which is another implication of the relatively less long-term strength development.

(3) It is identified from this research that by incorporating LCDBA, the fresh GGBFS blended cement concrete showed degraded workability, delayed bleeding ends, reduced bleeding capacity, and delayed setting times, attributed to the porous nature of LCDBA that led to mix water absorption.

(4) The series of compressive strength tests conducted in this research concluded that LCDBA was an effective alkaline activator for GGBFS blended cement concrete, showing early-age strength developing characteristics, especially for mixtures with lower water/binder ratios. The compressive strength model, suggested by the ACI Committee 209, also highlighted the early-age strength developing characteristics of the blended concrete activated with LCDBA.

The results obtained in this study provide a simple, yet effective and practical, means of reusing an industrial by-product as an alternative alkaline activator for GGBFS blended concrete. Further studies are, however, necessary to determine its long-term durability and dimensional stabilities, such as shrinkage and creep.

Author Contributions: Conceptualization, S.C.; methodology, S.C.; software, S.C.; validation, S.C.; formal analysis, S.C. and S.P.; investigation, S.C.; resources, S.P.; data curation, S.C.; writing—original draft preparation, S.C.; writing—review and editing, S.P.; visualization, S.C.; supervision, S.P.; project administration, S.P.; funding acquisition, S.P. Both authors have read and agreed to the published version of the manuscript.

Funding: This work was supported by the National Research Foundation of Korea (NRF) grant funded by the Korea government (MSIT) (no. NRF-2019R1F1A1060906). The research described herein was also supported by the 2019 Research Fund (1.190015.01) of UNIST (Ulsan National Institute of Science and Technology).

Conflicts of Interest: The authors declare no conflict of interest.

References

1. Maraghechi, H.; Salwocki, S.; Rajabipour, F. Utilisation of alkali activated glass powder in binary mixtures with Portland cement, slag, fly ash and hydrated lime. *Mater. Struct.* **2017**, *50*, 16. [CrossRef]
2. Sajedi, F.; Razak, H.A. The effect of chemical activators on early strength of ordinary Portland cement-slag mortars. *Constr. Build. Mater.* **2010**, *24*, 1944–1951. [CrossRef]
3. Ding, Y.; Dai, J.G.; Shi, C.J. Mechanical properties of alkali-activated concrete: A state-of-the-art review. *Constr. Build. Mater.* **2016**, *127*, 68–79. [CrossRef]
4. Kim, H.; Koh, T.; Pyo, S. Enhancing flowability and sustainability of ultra high performance concrete incorporating high replacement levels of industrial slags. *Constr. Build. Mater.* **2016**, *123*, 153–160. [CrossRef]
5. Kang, H.; Kang, S.H.; Jeong, Y.; Moon, J. Quantitative analysis of hydration reaction of GGBFS using X-ray diffraction methods. *J. Korea Concr. Inst.* **2020**, *32*, 241–250. [CrossRef]
6. Lee, H.K.; Jeon, S.M.; Lee, B.Y.; Kim, H.K. Use of circulating fluidized bed combustion bottom ash as a secondary activator in high-volume slag cement. *Constr. Build. Mater.* **2020**, *234*, 117240. [CrossRef]
7. Lee, B.Y.; Jeon, S.M.; Cho, C.G.; Kim, H.K. Evaluation of time to shrinkage-induced crack initiation in OPC and slag cement matrices incorporating circulating fluidized bed combustion bottom ash. *Constr. Build. Mater.* **2020**, *257*, 119507. [CrossRef]
8. Kang, C.; Kim, T. The strength properties of alkali-activated cement with blended ground granulated blast furnace slag. *J. Korea Concr. Inst.* **2018**, *30*, 315–324.
9. Suh, J.I.; Yum, W.S.; Song, H.; Park, H.G.; Oh, J.E. Influence of calcium nitrate and sodium nitrate on strength development and properties in quicklime (CaO)-activated Class F fly ash system. *Mater. Struct.* **2019**, *52*, 115. [CrossRef]
10. Humad, A.M.; Provis, J.L.; Cwirzen, A. Alkali activation of a high MgO GGBS–fresh and hardened properties. *Mag. Concr. Res.* **2018**, *70*, 1256–1264. [CrossRef]
11. Mohamed, O.A. A review of durability and strength characteristics of alkali-activated slag concrete. *Materials* **2019**, *12*, 1198. [CrossRef] [PubMed]
12. Qureshi, M.N.; Ghosh, S. Workability and setting time of alkali activated blast furnace slag paste. *Adv. Civ. Eng. Mater.* **2013**, *2*, 62–77. [CrossRef]
13. Turner, L.K.; Collins, F.G. Carbon dioxide equivalent (CO_2-e) emissions: A comparison between geopolymer and OPC cement concrete. *Constr. Build. Mater.* **2013**, *43*, 125–130. [CrossRef]
14. Liu, Y.; Shi, C.; Zhang, Z.; Li, N. An overview on the reuse of waste glasses in alkali-activated materials. *Resour. Conserv. Recycl.* **2019**, *144*, 297–309. [CrossRef]

15. Lin, K.L.; Huang, W.J.; Shie, J.L.; Lee, T.C.; Wang, K.S.; Lee, C.H. The utilization of thin film transistor liquid crystal display waste glass as a pozzolanic material. *J. Hazard. Mater.* **2009**, *163*, 916–921. [CrossRef] [PubMed]
16. Jang, H.S.; Jeon, S.H.; So, H.S.; So, S.Y. A study of the possibility of using TFT-LCD waste glass as an admixture for steam-cured PHC piles. *Mag. Concr. Res.* **2014**, *66*, 196–208. [CrossRef]
17. Wang, H.Y.; Zeng, H.H.; Wu, J.Y. A study on the macro and micro properties of concrete with LCD glass. *Constr. Build. Mater.* **2014**, *50*, 664–670. [CrossRef]
18. Kim, S.K.; Kang, S.T.; Kim, J.K.; Jang, I.Y. Effects of particle size and cement replacement of LCD glass powder in concrete. *Adv. Mater. Sci. Eng.* **2017**, *2017*, 3928047. [CrossRef]
19. Chang, J.J. A study on the setting characteristics of sodium silicate-activated slag pastes. *Cem. Concr. Res.* **2003**, *33*, 1005–1011. [CrossRef]
20. ASTM International. *ASTM C94/C94M-19a. Standard Specification for Ready-Mixed Concrete*; ASTM International: West Conshohocken, PA, USA, 2019.
21. ASTM International. *ASTM C232/C232M-14. Standard Test Method for Bleeding of Concrete*; ASTM International: West Conshohocken, PA, USA, 2019.
22. ASTM International. *ASTM C803/C803M-18. Standard Test Method for Penetration Resistance of Hardened Concrete*; ASTM International: West Conshohocken, PA, USA, 2018.
23. ASTM International. *ASTM C39/C39M-18. Standard Test Method for Compressive Strength of Cylindrical Concrete Specimens*; ASTM International: West Conshohocken, PA, USA, 2018.
24. Kim, M.S.; Jun, Y.; Lee, C.; Oh, J.E. Use of CaO as an activator for producing a price-competitive non-cement structural binder using ground granulated blast furnace slag. *Cem. Concr. Res.* **2013**, *54*, 208–214. [CrossRef]
25. Yum, W.S.; Suh, J.I.; Jeon, D.; Oh, J.E. Strength enhancement of CaO-activated slag system through addition of calcium formate as a new auxiliary activator. *Cem. Concr. Compos.* **2020**, *109*, 103572. [CrossRef]
26. Colombo, A.; Geiker, M.; Justnes, H.; Lauten, R.A.; De Weerdt, K. The effect of calcium lignosulfonate on ettringite formation in cement paste. *Cem. Concr. Res.* **2018**, *107*, 188–205. [CrossRef]
27. Mohammed, S.; Safiullah, O. Optimization of the SO3 content of an Algerian Portland cement: Study on the effect of various amounts of gypsum on cement properties. *Constr. Build. Mater.* **2018**, *164*, 362–370. [CrossRef]
28. Song, H.; Jeong, Y.; Bae, S.; Jun, Y.; Yoon, S.; Oh, J.E. A study of thermal decomposition of phases in cementitious systems using HT-XRD and TG. *Constr. Build. Mater.* **2018**, *169*, 648–661. [CrossRef]
29. Mirghiasi, Z.; Bakhtiari, F.; Darezereshki, E.; Esmaeilzadeh, E. Preparation and characterization of CaO nanoparticles from $Ca(OH)_2$ by direct thermal decomposition method. *J. Ind. Eng. Chem.* **2014**, *20*, 113–117. [CrossRef]
30. Siddique, R.; Bennacer, R. Use of iron and steel industry by-product (GGBS) in cement paste and mortar. *Res. Conserv. Recycl.* **2012**, *69*, 29–34. [CrossRef]
31. Barnett, S.J.; Soutsos, M.N.; Millard, S.G.; Bungey, J.H. Strength development of mortars containing ground granulated blast-furnace slag: Effect of curing temperature and determination of apparent activation energies. *Cem. Concr. Res.* **2006**, *36*, 434–440. [CrossRef]
32. Deb, P.S.; Nath, P.; Sarker, P.K. The effects of ground granulated blast-furnace slag blending with fly ash and activator content on the workability and strength properties of geopolymer concrete cured at ambient temperature. *Mater. Des.* **2014**, *62*, 32–39. [CrossRef]
33. Fang, G.; Ho, W.K.; Tu, W.; Zhang, M. Workability and mechanical properties of alkali-activated fly ash-slag concrete cured at ambient temperature. *Constr. Build. Mater.* **2018**, *172*, 476–487. [CrossRef]
34. American Concrete Institute (ACI). *ACI Committee 209.2R-08. Guide for Modeling and Calculating Shrinkage and Creep in Hardened Concrete*; American Concrete Institute (ACI): Farmington Hills, MI, USA, 2008.
35. Yang, K.H.; Song, J.K.; Ashour, A.F.; Lee, E.T. Properties of cementless mortars activated by sodium silicate. *Constr. Build. Mater.* **2008**, *22*, 1981–1989. [CrossRef]

Article

A Study on Initial Setting and Modulus of Elasticity of AAM Mortar Mixed with CSA Expansive Additive Using Ultrasonic Pulse Velocity

Gum-Sung Ryu [1,*], Sung Choi [2], Kyung-Taek Koh [1], Gi-Hong Ahn [1], Hyeong-Yeol Kim [1] and Young-Jun You [1]

1 Department of Infrastructure Safety Research, Korea Institute of Civil Engineering and Building Technology, 283 Goyangdae-Ro, Ilsanseo-Gu, Goyang-Si 10223, Korea; ktgo@kict.re.kr (K.-T.K.); agh0530@kict.re.kr (G.-H.A.); hykim1@kict.re.kr (H.-Y.K.); yjyou@kict.re.kr (Y.-J.Y.)

2 Department of Civil Engineering, KyungDong University 27, Gyeongdongdaehak-Ro, Yangju-Si 11458, Korea; csomy1113@kduniv.ac.kr

* Correspondence: ryu0505@kict.re.kr; Tel.: +82-31-910-0050

Received: 31 August 2020; Accepted: 28 September 2020; Published: 5 October 2020

Abstract: This study investigated the hardening process of alkali-activated material (AAM) mortar using calcium sulfoalumiante (CSA) expansive additive (CSA EA), which accelerates the initial reactivity of AAMs, and subsequent changes in ultrasonic pulse velocity (UPV). After the AAM mortar was mixed with three different contents of CSA EA, the setting and modulus of elasticity of the mortar at one day of age, which represent curing steps, were measured. In addition, UPV was used to analyze each curing step. The initial and final setting times of the AAM mortar could be predicted by analyzing the UPV results measured for 14 h. In addition, the dynamic modulus of elasticity calculated using the UPV results for 24 h showed a tendency similar to that of the static modulus of elasticity. The test results showed that the use of CSA EA accelerated the setting of the AAM mortar and increased the modulus of elasticity, and these results could be inferred using UPV. The proposed measurement method can be effective in evaluating the properties of a material that accelerates the initial reactivity.

Keywords: alkali-activated material; CSA expansive additive; ultrasonic pulse velocity; setting time; modulus of elasticity

---

1. Introduction

Alkali-activated materials (AAMs) are binders that accelerate the reaction of mineral admixtures, such as ground granulated blast furnace slag (GGBFS) and fly ash (FA), using strong alkali activators with high pH (>12). As AAMs are highly reactive, the hydrates of mineral admixtures are generated earlier. This lowers the fluidity of the mortar and causes setting and hardening at the same time. This process occurs quite suddenly [1–4]. The acceleration of the mortar setting has a significant impact on the physical properties of mortar at early ages [5–9], especially the initial shrinkage stress of the AAM mortar. According to the studies of many researchers, the shrinkage of mortar and concrete is evaluated after the occurrence of the final setting [10–14]. This is because the final setting means that the AAM matrix has been hardened and the deformation that occurs after the hardening of the AAM matrix acts as stress. Therefore, if the final setting of mortar that uses AAMs as binders is accelerated, shrinkage stress occurs earlier. This may cause significant initial shrinkage of the AAM mortar and high shrinkage stress. In addition, as the hydration reaction of AAMs is fast even after the final set, the strength and modulus of elasticity of the AAM mortar also significantly increase at early ages. The increase in the modulus of elasticity of the mortar causes a larger initial shrinkage stress. In general, additives, such as expansive additives (EAs) and shrinkage reducing agents, are used to control the shrinkage of mortar. These additives are effective in reducing shrinkage, but shrinkage stress can be evaluated differently if the setting and modulus of elasticity of mortar are considered [15]. In particular, when a calcium sulfoalumiante (CSA) expansive additive (CSA EA) is used, the formation of ettringite and calcium monosulfoaluminate at the initial stage can compensate for shrinkage, but it can also increase the modulus of elasticity and compressive strength [16–19]. If the setting is accelerated or the modulus of elasticity increases despite the reduction in shrinkage, the initial shrinkage stress of mortar can be evaluated to be larger [20]. Therefore, mortar and concrete that used AAMs require a comprehensive evaluation of the properties that can calculate the stress caused by deformation, such as setting, modulus of elasticity, and shrinkage at early ages.

In general, it is possible to measure the shrinkage of mortar before setting using an embedded strain gauge and a data logger. However, to measure the shrinkage stress of mortar, the modulus of elasticity at the time of the occurrence of shrinkage must also be considered.

The elastic modulus is determined from the load–displacement curve, as it is the slope of the curve. Nevertheless, the strength of concrete and mortar specimens is too low to be measured with mechanical tests, making it difficult to determine elastic modulus. In addition, as the strength and modulus of elasticity vary dramatically over time at early ages, the limited number of specimens to perform continuous monitoring may cause larger experimental errors [21–23]. The use of the ultrasonic pulse velocity (UPV), a type of non-destructive testing, can detect the physical properties of a specimen continuously and successively without causing damage. With respect to the analysis of the properties of concrete, UPV has been verified by many researchers [24–27]. Öztürk et al. detected the early hydration status of cement materials through ultrasonic reflectance measurements. They showed that changes in reflected waves responded well to various stages of hydration [28]. Voigt et al. conducted studies on changes in microstructure during cement hydration using ultrasonic reflection and transmission techniques [29,30].

In addition, the use of UPV makes it possible to quantitatively analyze the hydration process of concrete. In particular, many researchers have reported that the setting time can be predicted using changes in the UPV curve [31–33]. Reinhardt and Grosse developed an ultrasonic device for concrete quality control testing that can continuously observe the characteristics of concrete. The results obtained using this device reported that the initial setting time can be determined by UPV [34]. Krauß and Hariri reported that the degree of hydration of concrete at the initial stage can be analyzed using UPV and proposed a method to predict the setting time [35]. Belie et al. conducted research on the setting and hardening of shotcrete mortar, for which the hydration reaction of the binder is significantly fast. They found that shotcrete mortar was immediately hardened in the fluid state, and they could analyze the sudden setting and hardening process of the mortar using UPV [36].

The use of UPV makes it possible to analyze the overall hardening process of concrete and mortar, and UPV has been used to analyze the static and dynamic elasticity characteristics as well as the tensile and compressive strengths of concrete and mortar [35,37–41]. Rajagopalan et al. reported that they identified highly reliable concrete characteristics at early ages using the relationship between the UPV and compressive strength of concrete at early ages [42]. Anderson and Seals conducted research to predict the long-term compressive strength of concrete through non-destructive testing using UPV and proposed a non-destructive test method for predicting the long-term strength [43]. Abdel-Jawed and Afaneh investigated various factors in concrete that affect the ultrasonic pulse. They mentioned that the compressive strength characteristics over time according to the water/binder (W/B) ratio can be analyzed using UPV [44]. Trtnik et al. investigated material and mix properties that affect the UPV of concrete and verified the relationship between UPV and the static/dynamic modulus of elasticity based on the results [39].

There have been attempts of measuring UPV as a means of exploring the fresh state and setting properties of ordinary Portland cement (OPC)-based concrete and mortar, while it has been rarely applied to the binders that set very rapidly. Determination of elastic modulus after the setting has occurred is particularly important for controlling the shrinkage-induced stress and crack formation. This study attempted to evaluate the initial physical properties of the AAM mortar using CSA EA, which affects the setting and modulus of elasticity of mortar despite its shrinkage compensation effect. It was possible to find the initial and final setting times of the AAM mortar in the UPV inflection section by measuring the UPV for 48 h and to identify changes in its modulus of elasticity from the final setting to 48 h through the relationship between the UPV and dynamic modulus of elasticity. Based on this, the relationship between the setting and modulus of elasticity of the AAM mortar was analyzed according to the content of CSA EA, which is used for shrinkage reduction.

2. Materials and Methods

2.1. Materials and Mixture Proportions of AAM Mortar

Table 1 shows the physical properties and chemical compositions of the ground granulated blast furnace slag (GGBFS), fly ash (FA), and calcium sulfoalumiante expansive additive (CSA EA) used in AAM mortar. GGBFS satisfied the third type of KS F2563 (density: >2.8 g/cm^3, fineness: 4000–6000 cm^2/g), and FA met the second type of KS L5405 (density: >1950 g/cm^3, fineness: >3000 cm^2/g). Both GGBFS and FA were produced by the company Sampyo in South Korea [45,46]. The CSA EA used in this study was POWER CSA TYPE from Denka in Japan. It was used to compensate for the shrinkage of the AAM mortar and was added on the basis of the binder mass. The main chemical components of GGBFS were CaO (41.9%), SiO_2 (13.8%), and Al_2O_3 (4.9%). Its basicity coefficient (K_b = (CaO + MgO)/(SiO_2 + Al_2O_3)) and hydration modulus (HM = (CaO + MgO + Al_2O_3)/SiO_2) were 0.99 and 1.82, respectively. K_b was close to 1.0, which is a neutral value for ideal alkali activation, and HM was higher than 1.4, which is a value for excellent hydration reaction. FA was composed of SiO_2 (56.8%), Al_2O_3 (22.8%), Fe_2O_3 (6.9%), and M_2O(K_2O+Na_2O) (1.9%). The main components of CSA EA were lime, gypsum, and bauxite, and they were composed of CaO (34.6%), SiO (30.2%), and Al_2O_3 (24.2%). The alkali activator was used to accelerate the reaction of the binder. The alkali activator was in the form of a white powder with a molar ratio of 0.95. In addition, alkali activators are manufactured separately by adjusting the chemical components. The SiO_2/Na_2O ratio of the alkali activator used in this study was 0.92. The fine aggregate used was river sand with a density of 2.53, water absorptivity of 1.08, and fineness modulus of 2.77.

Table 1. Physical properties and chemical composition of alkali-activated material (AAM) binder and calcium sulfoalumiante expansive additive (CSA EA).

Type	Density (g/cm^3)	Blaine Specific Surface (cm^2/g)	CaO	SiO_2	Al_2O_3	Fe_2O_3	SO_3	MgO	K_2O	Na_2O
GGBFS	2.9	4680	43.4	34.6	14.3	0.6	5.0	5.1	0.5	0.2
FA	2.2	3220	3.5	56.8	22.8	6.9	0.5	1.8	1.1	0.8
CSA expansive additive	2.9	3750	36.4	30.2	24.2	1.7	5.3	1.4	0.5	0.3
Alkali activator	1.0	-	-	46.2	-	-	-	-	-	50.2

Table 2 summarizes the mix proportions of the AAM mortar. A binary blended binder in which GGBFS and FA were mixed at a ratio of 7:3 was used, and 24% alkali activator was used compared to the unit water content. The contents of CSA EA used were 0%, 2.5%, 5.0%, and 7.5% compared to the amount of the binder. The water/binder (W/B) ratio was 45.1%, and the sand/binder (S/B) ratio was 1.2. For the mixing method, the basic binder, alkali activator, and CSA EA were placed in a test container and dry mixing was performed for 30 s. Then, water was added, and mixing was performed for ten min at 150 rpm to produce paste. In addition, fine aggregate was added, and mixing was performed for 90 s at 300 rpm.

Table 2. Mix proportions of alkali-activated material (AAM) mortar.

Type	Water	Binder	Activator	Sand	CSA
	(g)	(g)	(g)	(g)	(g)
0 EA	0.451	1.000	0.108	1.200	0.000
2.5 EA	0.451	1.000	0.108	1.200	0.250
5.0 EA	0.451	1.000	0.108	1.200	0.500
7.5 EA	0.451	1.000	0.108	1.200	0.750

2.2. Test Methods

2.2.1. Setting Time Test

To evaluate the setting characteristics of the AAM mortar, the mortar Vicat test was conducted in accordance with ASTM C191-18a [47]. The initial and final settings of the AAM mortar were determined using the penetration depth of the Vicat needle (diameter: 1.00 ± 0.05 mm, length: 50 mm or more) in a mortar for 30 s. The Vicat needle penetration tests were conducted every five min, and the time at which no trace of the Vicat needle was observed on the AAM mortar surface was determined as the final setting time.

2.2.2. Compressive Strength Test

The compressive strength test was conducted in accordance with ASTM C109-16a [48]. The diameter and height of the cylindrical specimens were 100 and 200 mm, respectively, and the tests were conducted at curing ages of 1, 7, and 28 days. The average compressive strengths of the three specimens were used. The specimens were cured in a constant temperature and humidity chamber at a temperature of 23 ± 2 °C and a relative humidity of 90 ± 2%.

2.2.3. Modulus of Elasticity Test

The concrete modulus of elasticity test was conducted in accordance with ASTM C469M-14 [49]. The diameter and height of the cylindrical specimens were 100 and 200 mm, respectively. Two strain gauges were attached to the side of each specimen to measure the modulus of elasticity. A load of up to 40% of the ultimate load was applied at a rate of 0.25 MPa/s, and the modulus of elasticity was calculated through regression based on the interpolation of strain measurements for the load. The modulus of elasticity test results was obtained by averaging the modulus of elasticity measurement results of three specimens. The specimens were cured in a constant temperature and humidity chamber at a temperature of 23 ± 2 °C and a relative humidity of 90 ± 2%.

2.2.4. Ultrasonic Pulse Value (UPV) Test

The system used for monitoring the UPV in the AAM mortar specimens is shown in Figure 1. A hole was made in the Styrofoam container, and an oscillator and a receiver were placed so that they could face each other at a distance of 30 mm. After filling the container with mortar, the transit time of ultrasonic waves from the oscillator to the receiver was measured every 30 s. The Pundit-2 model from PROCEQ was used to measure the UPV. UPV (V_c) was obtained by dividing the length of the specimen (L) by the transit time (T), and it has the following relationship with the dynamic modulus of elasticity (E_d) and density (ρ) [50–52].

$$V_c = \frac{L}{T} = \sqrt{\frac{E_d}{\rho}} \tag{1}$$

Figure 2 shows the typical evolution curve of the UPV of the AAM mortar, which was altered with three sections. According to the curve, points A and B can be determined in terms of altered time points. The UPV begins to increase at point A, and it begins to slow down and converges to a certain value at point B [53–55].

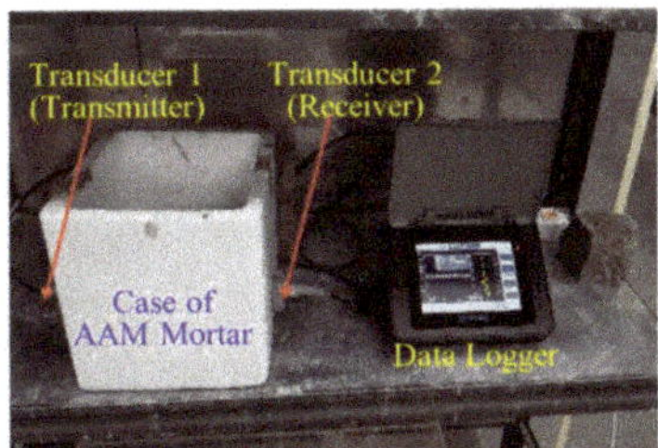

Figure 1. Schematic of ultrasonic pulse velocity (UPV) monitoring.

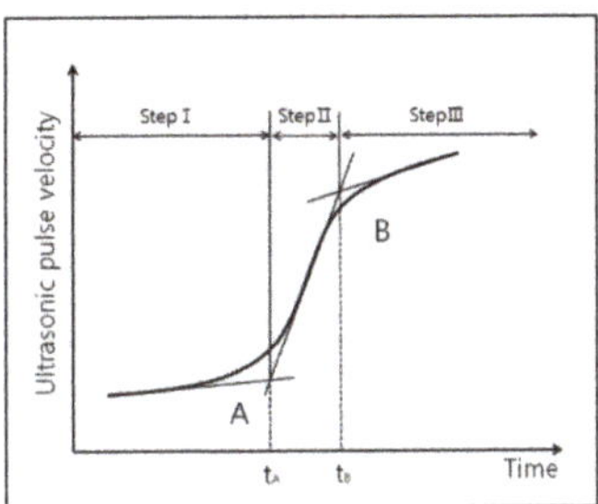

Figure 2. Typical evolution curve of ultrasonic pulse velocity.

3. Results and Discussion

3.1. Setting Time

The setting of AAM mortar is affected by the conditions of the alkali activator, and it is generally faster than that of ordinary Portland cement (OPC) [56,57]. Table 3 shows the setting time of the AAM mortar according to the content of EA. For 0 EA without EA, the initial setting time (IST) was 101 min and the final setting time (FST) was 292 min. As for paste (w/c = 0.5) using ordinary Portland cement (OPC), the initial setting is within 6 to 7 h according to the literature [58]. Under the same mixing conditions, the AAM mortar exhibited a faster setting compared to mortar that used cement as a binder. This is because the alkali activator stimulated the binder quite early [59].

Table 3. Setting time and strength characteristics of the AAM mortar mixed with EA.

Type	Setting Time (min)		Compressive Strength (MPa)		Modulus of Elasticity (E_s, GPa)	
	IST	FST	1D	28D	1D	28D
0 EA	101	292	3.2	43.6	1.7	19.3
2.5 EA	93	238	5.2	49.1	2.1	20.1
5.0 EA	70	223	5.9	49.6	2.3	20.2
7.5 EA	55	171	5.7	47.3	2.5	20.2

The setting time was accelerated as the CSA EA content increased. For 7.5 EA with the highest CSA EA content, the initial setting time and final setting time were 55 and 171 min, respectively, which were 45.4% and 41.4% shorter compared to those of 0 EA. When a CaO-based CSA EA is used in mortar, Ca^{2+} ions diffuse and generate hydrates ($Ca(OH)_2$). The setting characteristic of OPC is highly associated with the hydration of C_3S (Ca_3SiO_5, alite). An OPC paste reaches its initial setting state once C_3S has been sufficiently hydrated, and the final setting occurs when a sufficient amount of C-S-H has precipitated. On the other hand, the main constituent of CSA expansive additives is C_4A_3s (Ye'elimite), which dissolves in contact with water and forms ettringite ($Ca_6Al_2(SO_4)_3(OH)_{12}{\cdot}26H_2O$) at a much faster rate than C_3S does.

This action, which is similar to the alkali activation reaction of AAM, accelerates the reaction of AAM by increasing the dissolution of Ca^{2+} ions in the mixing water, thereby further accelerating setting [59,60].

3.2. Compressive Strength and Modulus of Elasticity

When a CSA EA is used in mortar, the mortar temporarily expands at the initial stage owing to the hydrates of EA, and its internal structure becomes dense. This increases the strength and modulus of elasticity of the mortar [1]. Table 3 shows the compressive strength and modulus of elasticity of AAM mortar mixed with CSA EA.

The average compressive strength of the three AAM mortars exceeded 40 MPa, which was the target strength set during the mix design. The AAM mortar without EA (0 EA) exhibited high initial strength, with a compressive strength of 3.2 MPa and modulus of elasticity of 1.2 GPa at 1 day of age. However, for the AAM mortar mixed with CSA EA, the compressive strength was higher than 5 MPa, and the modulus of elasticity was higher than 2 GPa at 1 day of age, which were 63–87% and 76.3–103.5% higher, respectively, compared to those of 0 EA. AAM mortar containing CSA EA exhibited higher compressive strength values at 28 days of age compared to 0 EA. At one day of age, the strength and modulus of elasticity increased as the EA content increased. However, at 28 days of age, the compressive strength of 5.0 EA was the highest (49.6 MPa), and the compressive strength of 7.5 EA was 47.3 MPa, which was lower than that of 2.5 EA.

When the modulus of elasticity of the AAM mortar at 28 days of age was compared, the modulus of elasticity of the AAM mortar containing CSA EA was higher than that of 0 EA, but there was no clear improvement in the modulus of elasticity due to EA. This is because the modulus of elasticity was not significantly affected by the binder, as it was also affected by factors other than the cement matrix, such as fine aggregate.

3.3. Ultrasonic Pulse Velocity

UPV can measure changes in the internal structure of mortar caused by a hydration reaction without physical damage using the speed of sound waves, and it is possible to identify sudden changes in the internal structure of mortar through the monitoring of UPV. According to Lee et al., the UPV of mortar and concrete has three different sections [61]. The initial fluid state section (step 1) represents the unhardened state of the mortar. In this instance, the UPV exhibits a level similar to that of the fluid state and ranges from 300 to 500 m/s. In step 2, pores are filled with the hydration products of cement, and moisture unsaturated porous solid structures are connected, thereby increasing the UPV. In the step 3 section, high UPV is maintained because the hydrates are connected to each other, and mortar begins to develop strength owing to the curing process. Micropores are filled with hydration products, and UPV continuously increases, even though it does not increase as rapidly as in step 2.

Figure 3 shows the UPV curves of the AAM mortar according to the EA content. The UPV of the AAM mortar incorporating CSA EA was 348–412 m/s, corresponding to the values typically shown by a mixture in the fluid state. An increase in the UPV value was observed after 0.5 h by 7.5 EA, which incorporated the highest dosage of CSA EA and was the first to enter step 2, followed by 2.5 EA and 0 EA after 1.5 h. It was observed that the UPV rapidly increased in step 2, while its increasing rate decreased after 3–8 h, reaching step 3. Similarly, the time taken to reach step 3 was the shortest in 7.5 EA. The rate at which the UPV increased tended to gradually decrease 48 h after the specimens reached step 3.

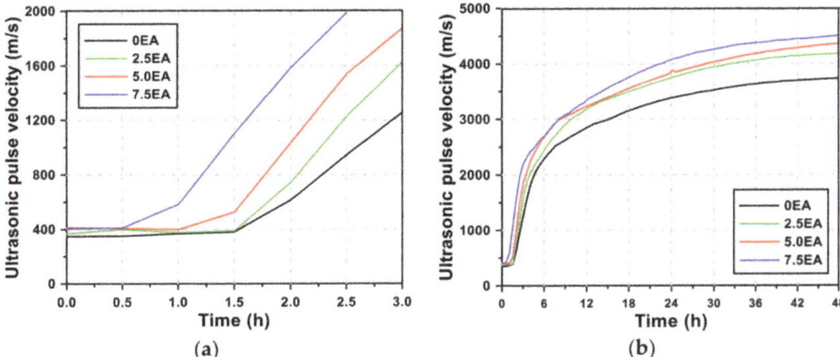

Figure 3. AAM mortar of ultrasonic pulse velocity curve: (**a**) Time: 3 h, (**b**) Time: 48 h.

The curves show shapes similar to those of the UPV curves of other researchers, but there are some differences. In a study by Lee et al. [61], the duration of step 1 for OPC mortar was 6–10 h, whereas that for the AAM mortar was 0.5–1.5 h, suggesting that the duration of step 1 was much shorter for AAM. In addition, while step 2 was reached after a sudden increase from step 1 for the UPV curves of the AAM mortar, the UPV curves of the OPC mortar continuously increased without clear boundary points to reach step 2. This is because the rapid hydration reaction occurred as AAM mortar-stimulated GGBFS using the alkali activator. In addition, CSA EA further reduced step 1 as it accelerated the hydration reaction. Lee et al. [61] suggested that OPC mortar reaches step 3 after 16–20 h, while it was shown that AAM took 3–8 h to reach step 3, showing a much faster rate than OPC mortar. This is because the rapid reaction of AAM mortar occurred as described above. The AAM mortar containing EA exhibited a higher UPV at the same age. This is due to the generation of ettringite, which is the main hydration product of EA.

3.4. Setting Time in UPV Test

The UPV curve can reflect changes in the internal structure of mortar caused by the hydration reaction. Pessiki and Carino et al. analyzed the relationship between UPV and setting or strength using the UPV curve [62]. According to a report by Lee et al., the analysis of the relationship between the UPV and the setting of concrete revealed that UPV increased when setting began. It was confirmed that initial and final settings occurred when UPV reached certain values [61,63,64].

The UPV measurement for AAM according to each step defined in Figure 2 is analyzed in Figure 4. As can be seen from the figure, the UPV curves are divided into three sections. According to a report by Lee, the intersections of the straight lines created by connecting UPV values in each section are related to setting. As the UPV curves of the AAM mortar have three different sections, straight lines were created in each section by connecting the data, and the intersections of the lines were compared with the setting time by the ASTM C191-18a test method.

Step 1 lasted for approximately 60 min (30 min for EA7.5), and the UPV results were used to create a straight line (black curve in Figure 4). The UPV sharply increased afterwards, and step 2 began approximately 90–120 min after mixing. In this instance, the use of EA advanced the time at which step 2 began. During step 2, the UPV results for an hour were used to create a straight line (blue curve in Figure 4). Step 3 began approximately eight hours after mixing. Mortar mixtures tend to reach step 3 within 3–8 h after setting. The point at which the UPV converges has been determined by regression analysis (blue curve in Figure 4).

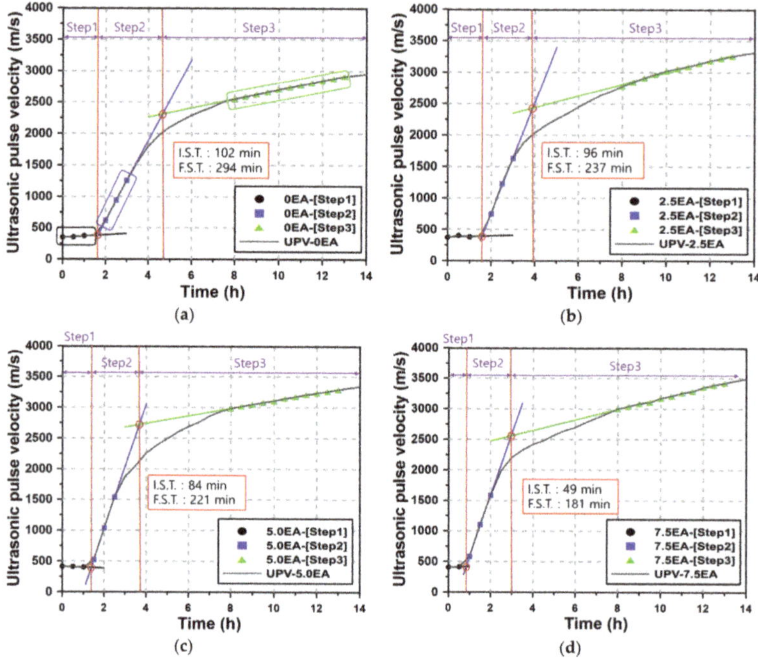

Figure 4. UPV curves of AAM mortar: (**a**) 0 EA, (**b**) 2.5 EA, (**c**) 5.0 EA, (**d**) 7.5 EA.

To determine the setting time of the AAM mortar using the UPV curves, the intersections of the straight lines created in each section of the curves were examined. The intersection of the straight lines of the step 1 and step 2 sections was defined as the initial setting, and that of the straight lines of the step 2 and step 3 sections was defined as the final setting. The analysis results showed that the initial setting time of 0 EA was 102 min, and its final setting time was 294 min. In the same manner, the initial setting and final setting of 2.5, 5.0, and 7.5 EA are shown in each graph. The initial and final setting times determined using UPV became shorter as the EA content increased. Table 4 shows the differences between the initial and final setting times determined by the UPV curves and those determined by ASTM C 191-18a. The difference in initial setting time ranged from 1 to 14 min, and that in the final setting ranged from 1 to 10 min. In particular, when the EA content was 2.5% or less, the difference in setting time was 3 min or less, resulting in no significant difference in the results of the two test methods.

Table 4. Differences in setting time between the ASTM C 191-18a test method and UPV.

Type	ASTM C 191-18a—UPV (min)			
	0 EA	2.5 EA	5.0 EA	7.5 EA
IST *	1	3	14	6
FST **	2	1	2	10

IST *: Initial setting time; FST **: Final setting time.

3.5. Elasticity in UPV Test

The dynamic modulus of elasticity of mortar can be obtained through an empirical equation if dynamic characteristics, such as UPV, are used.

$$E_d = \rho V_c^2 \quad (2)$$

where E_d is the dynamic modulus of elasticity (GPa), ρ is the density of the specimen (ton/m^3), and V_c is UPV (m/s). Table 5 shows the UPV results of the AAM mortar at one day of age, the E_d values obtained using the results, and the static modulus of elasticity (E_s) values at one day of age obtained through an experiment. The density of the AAM mortar used was 2.4 ton/m^3.

Table 5. UPV and dynamic and static moduli of elasticity of the AAM mortar at one day of age.

Type	By UPV Test (1D)			By ASTM C469M-14 (1D)		E_d/E_s
	UPV (m/s)	E_d (GPa)	Increase Rate (0 EA)	E_s (GPa)	Increase Rate (0 EA)	
0 E	3381.9	2.68	100.0	1.65	100.0	1.624
2.5 E	3753.5	3.31	123.5	2.13	129.1	1.554
5.0 E	3883.9	3.54	132.1	2.27	137.6	1.560
7.5 E	4081.1	3.91	145.9	2.46	149.1	1.589

As the EA content increased at 1 day of age, both E_d and E_s increased. As the content of CSA EA increased to 2.5, 5.0, and 7.5%, E_s increased by 23.5, 32.1, and 45.9% compared to that of 0 EA and E_d increased by 29.1, 37.6, and 49.1%. Here, the difference between the increase rates of E_d and E_s of the AAM mortar that used CSA EA based on 0 EA ranged from 3.2% to 5.6%, indicating that E_d and E_s showed relatively similar tendencies. This means that E_s can be indirectly calculated through UPV because E_d is obtained using the equation of UPV. This approach has been mainly used to analyze the characteristics of rocks, and the correlation between the static and dynamic characteristics of rocks has been verified by several researchers [65–69].

In general, the dynamic modulus of elasticity tends to be higher than the static modulus of elasticity [65,70–72]. This can be because an increase in the crack and pore volume in concrete and mortar decreases their static modulus of elasticity to a larger extent [65,73]. In the test results, E_d by UPV exhibited higher results than E_s. It was found that E_d was 62.4–55.4% higher than E_s.

Table 6 shows the static modulus of elasticity of the AAM mortar at the final setting time obtained using UPV. The UPV at the accurate final setting time was unknown because it was measured every 30 min. Therefore, the dynamic modulus of elasticity was calculated using UPV at a time point similar to the final setting time. At the final setting time of the AAM mortar, UPV exhibited similar values, ranging from 2039.9 to 2205.0 m/s. Based on this, the dynamic modulus of elasticity was measured to be between 0.98 and 1.14 GPa. The AAM mortar developed stiffness as it was cured, and UPV and the dynamic modulus of elasticity were found to be constant at the final setting time. They were similar regardless of EA content.

Table 6. UPV and dynamic modulus of elasticity at the final setting time of the AAM mortar.

Type	By ASTM C 191-18a Final Setting (h:m)	Time	By UPV Test UPV(m/s)	E_s (GPa)
0 EA	4:52	5:00	2105.1	1.04
2.5 EA	3:58	4:00	2039.9	0.98
5.0 EA	3:43	3:30	2056.6	0.99
7.5 EA	2:51	3:00	2205.0	1.14

4. Conclusions

In this study, the initial hardening characteristics of alkali-activated material (AAM) mortar were analyzed using the four levels of CSA expansive additive (CSA EA) content. For the analysis of the initial hardening characteristics using ultrasonic pulse velocity (UPV), the results of the Vicat setting test, a conventional test method, were compared with those of the modulus of elasticity test. The main observations and findings of this study can be summarized as follows:

1. CSA EA is used to compensate for initial shrinkage, but it accelerates the initial reactivity and shortens the setting time, thereby causing the final setting, which is the basis of shrinkage stress. The dynamic modulus of elasticity of the AAM mortar obtained using UPV at the final setting time was approximately 1 GPa, which was constant regardless of the CSA EA content. As the modulus of elasticity of the AAM mortar increases over time, CSA EA that accelerates the final setting may cause high shrinkage stress at early ages.
2. The compressive strength of the AAM mortar tended to increase with the CSA EA up to the dosage level of 5.0%, whereas it decreased at a dosage of 7.5%. Therefore, it is concluded that the maximum dosage of CSA EA should be limited to 5.0%, considering its effect on the compressive strength development.
3. The initial and final setting times obtained by the UPV test were similar to those obtained by the Vicat test. The dynamic modulus of elasticity derived through UPV increased as the CSA EA content increased, and it exhibited increase rates similar to those of the static modulus of elasticity. Therefore, UPV can reflect the curing, initial setting, and final setting processes of AAM mortar caused by the initial alkali activation reaction, and it is possible to evaluate the degree of curing of AAM mortar through the dynamic modulus of elasticity.
4. The use of a UPV monitoring system makes it possible to continuously observe the curing process of mortar and to identify changes in the stiffness of AAM mortar using the dynamic modulus of elasticity at early ages when it is difficult to conduct the static modulus of elasticity test. These data are expected to be useful in evaluating the initial shrinkage stress of AAM mortars.
5. In addition, the relationship between UPV and the static modulus of elasticity was established through various tests according to the material properties. Based on this relationship, it will be possible to evaluate the stress acting on mortar due to the displacement caused by various causes at early ages.

Author Contributions: Conceptualization, G.-S.R., S.C.; setting time test, S.C., G.-H.A.; data analysis, S.C.; compressive strength & modulus of elasticity test, G.-H.A., K.-T.K.; data analysis, G.-S.R., H.-Y.K. Y.-J.Y.; UPV test & data analysis, S.C., G.-S.R.; writing, G.-S.R., S.C. All authors have read and agreed to the published version of the manuscript.

Funding: This research received no external funding.

Acknowledgments: This work was supported by the Korea Institute of Civil Engineering and Building Technology (Project No. 2020-60).

Conflicts of Interest: The authors declare no conflict of interest

References

1. Lee, N.K.; Lee, H.K. Setting and mechanical properties of alkali-activated fly ash slag concrete manufactured at room temperature. *Constr. Build. Mater.* **2013**, *47*, 1201–1209. [CrossRef]
2. Brough, A.R.; Holloway, M.; Sykes, J.; Atkinson, A. Sodium silicate-based alkali-activated slag mortars: Part II. The retarding effect of additions of sodium chloride or malic acid. *Cem. Concr. Res.* **2000**, *30*, 1375–1379. [CrossRef]
3. Ravikumar, D.; Neithalath, N. Reaction kinetics in sodium silicate powder and liquid activated slag binders evaluated using isothermal calorimetry. *Thermochim. Acta* **2012**, *546*, 32–43. [CrossRef]
4. Ravikumar, D.; Neithalath, N. Effects of activator characteristics on the reaction product formation in slag binders activated using alkali silicate powder and NaOH. *Cem. Concr. Comp.* **2012**, *34*, 809–818. [CrossRef]
5. Garnier, V.; Corneloup, G.; Sprauel, J.M.; Perfumo, J.C. Setting time study of roller compacted concrete by spectral analysis of transmitted ultrasonic signals. *NDT E Int.* **1995**, *28*, 15–22. [CrossRef]
6. Li, Z.; Xiao, L.; Wei, X. Determination of concrete setting time using electrical resistivity measurement. *J. Mater. Civ. Eng.* **2007**, *19*, 423–427. [CrossRef]
7. Coppola, L.; Coffetti, D.; Crotti, E.; Candamano, S.; Crea, F.; Gazzaniga, G.; Pastore, T. The combined use of admixtures for shrinkage reduction in one-part alkali activated slag-based mortars and pastes. *Constr. Build. Mater.* **2020**, *248*, 118682. [CrossRef]

8. Coppola, L.; Coffetti, D.; Crotti, E.; Gazzaniga, G.; Pastore, T. The Durability of One-Part Alkali-Activated Slag-Based Mortars in Different Environments. *Sustainability* **2020**, *12*, 3561. [CrossRef]
9. Tataranni, P.; Besemer, G.M.; Bortolotti, V.; Sangiorgi, C. Preliminary research on the physical and mechanical properties of alternative lightweight aggregates produced by alkali-activation of waste powders. *Materials* **2018**, *11*, 1255. [CrossRef]
10. Cartwright, C.; Rajabipour, F.; Radlińska, A. Shrinkage characteristics of alkaliactivated slag cements. *J. Mater. Civ. Eng.* **2015**, *27*, B4014007. [CrossRef]
11. Collins, F.; Sanjayan, J.G. Effect of pore size distribution on drying shrinking of alkali-activated slag concrete. *Cem. Concr. Res.* **2000**, *30*, 1401–1406. [CrossRef]
12. Ye, H.; Radlińska, A. Shrinkage mechanisms of alkali-activated slag. *Cem. Concr. Res.* **2016**, *88*, 126–135. [CrossRef]
13. Ma, Y.; Ye, G. The shrinkage of alkali activated fly ash. *Cem. Concr. Res.* **2015**, *68*, 75–82. [CrossRef]
14. Hojati, M.; Rajabipour, F.; Radlińska, A. Drying shrinkage of alkali activated fly ash: Effect of activator composition and ambient relative humidity. In Proceedings of the 4th International Conference on Sustainable Construction Materials and Technologies (SCMT4), Las Vegas, NV, USA, 7–11 August 2016.
15. Nguyen, T.B.T.; Chatchawan, R.; Saengsoy, W.; Tangtermsirikul, S.; Sugiyama, T. Influences of different types of fly ash and confinement on performances of expansive mortars and concretes. *Constr. Build. Mater.* **2019**, *209*, 176–186. [CrossRef]
16. Sharp, J.H.; Lawrence, C.D.; Yang, R. Calcium sulfoaluminate cements—Low-Energy cements, special cements or what? *Adv. Cem. Res.* **1999**, *11*, 3–13. [CrossRef]
17. Lan, W.; Glasser, F.P. Hydration of calcium sulphoaluminate cements. *Adv. Cem. Res.* **1996**, *8*, 127–134. [CrossRef]
18. Ioannou, S.; Reig, L.; Paine, K.; Quillin, K. Properties of a ternary calcium sulfoaluminate-calcium sulfate-fly ash cement. *Cem. Concr. Res.* **2014**, *56*, 75–83. [CrossRef]
19. Winnefeld, F.; Martin, L.H.; Müller, C.J. Using gypsum to control hydration kinetics of CSA cements. *Constr. Build. Mater.* **2017**, *155*, 154–163. [CrossRef]
20. Choi, S.; Ryu, G.S.; Koh, K.T.; An, G.H.; Kim, H.Y. Experimental Study on the Shrinkage ehavior and Mechanical Properties of AAM Mortar Mixed with CSA Expansive Additive. *Materials* **2019**, *12*, 3312. [CrossRef]
21. Sant, G. Examining volume changes, stress development and cracking in cement based systems. Master's Thesis, Purdue University, West Lafayette, IN, USA, 2007.
22. Sant, G.; DehaDrai, M.; Bentz, D.; Lura, P.; Ferraris, C.F.; Bullard, J.W.; Weiss, J. Detecting the fluid-to-solid transition in cement pastes. *Concr. Int.* **2009**, *31*, 53–58.
23. Stark, J. Recent advances in the field of cement hydration and microstructure analysis. *Cem. Concr. Res.* **2011**, *41*, 666–678. [CrossRef]
24. Carrión, A.; Genovés, V.; Gosálbez, J.; Miralles, R.; Payá, J. Ultrasonic signal modality: A novel approach for concrete damage evaluation. *Cem. Concr. Res.* **2017**, *101*, 25–32. [CrossRef]
25. Sepehrinezhad, A.; Toufigh, V. The evaluation of distributed damage in concrete based on sinusoidal modeling of the ultrasonic response. *Ultrasonics* **2018**, *89*, 195–205. [CrossRef]
26. Wu, J.; Feng, M.; Ni, X.; Mao, X.; Chen, Z.; Han, G. Aggregate gradation effects on dilatancy behavior and acoustic characteristic of cemented rockfill. *Ultrasonics* **2019**, *92*, 79–92. [CrossRef]
27. Cao, S.; Yilmaz, E.; Song, W.; Xue, G. Assessment of acoustic emission and triaxial mechanical properties of rock-cemented tailings matrix composites. *Adv. Mater. Sci. Eng.* **2019**, 1–12. [CrossRef]
28. Öztürk, T.; Kroggel, O.; Grübl, P.; Popovics, J.S. Popovics, Improved ultrasonic wave reflection technique to monitor the setting of cement-based materials. *NDT E Int.* **2006**, *39*, 258–263. [CrossRef]
29. Voigt, T.; Grosse, C.U.; Sun, Z.; Shah, S.P.; Reinhardt, H.W. Comparison of ultrasonic wave transmission and reflection measurements with P- and S waves on early age mortar and concrete. *Mater. Struct.* **2005**, *38*, 729–738. [CrossRef]
30. Voigt, T.; Ye, G.; Sun, Z.; Shah, S.P.; Van Breugel, K. Early age microstructure of Portland cement mortar investigated by ultrasonic shear waves and numerical simulation. *Cem. Concr. Res.* **2005**, *35*, 858–866. [CrossRef]
31. Zhu, J.; Cao, J.N.; Bate, B.; Khayat, K.H. Determination of mortar setting times using shear wave velocity evolution curves measured by the bender element technique. *Cem. Concr. Res.* **2018**, *106*, 1–11. [CrossRef]

32. Renteria-Marquez, I.A.; Renteria-Marquez, A.; Tseng, B.T.L. A novel contact model of piezoelectric traveling wave rotary ultrasonic motors with the finite volume method. *Ultrasonics* **2018**, *90*, 5–17. [CrossRef]
33. Carette, J.; Staquet, S. Monitoring the setting process of mortars by ultrasonic P and S-wave transmission velocity measurement. *Constr. Build. Mater.* **2015**, *94*, 196–208. [CrossRef]
34. Reinhardt, H.W.; Grosse, C.U. Continuous monitoring of setting and hardening of mortar and concrete. *Constr. Build. Mater.* **2004**, *18*, 145–154. [CrossRef]
35. Krauß, M.; Hariri, K. Determination of initial degree of hydration for improvement of early-age properties of concrete using ultrasonic wave propagation. *Cem. Concr. Compos.* **2006**, *28*, 299–306. [CrossRef]
36. De Belie, N.; Grosse, C.U.; Kurz, J.; Reinhardt, H.W. Ultrasound monitoring of the influence of different accelerating admixtures and cement types for shotcrete on setting and hardening behavior. *Cem. Concr. Res.* **2005**, *35*, 2087–2094. [CrossRef]
37. Yoo, D.Y.; Park, J.J.; Kim, S.W.; Yoon, Y.S. Early age setting, shrinkage and tensile characteristics of ultra high performance fiber reinforced concrete. *Constr. Build. Mater.* **2013**, *41*, 427–438. [CrossRef]
38. Shariq, M.; Prasad, J.; Masood, A. Studies in ultrasonic pulse velocity of concrete containing GGBFS. *Constr. Build. Mater.* **2013**, *40*, 944–950. [CrossRef]
39. Trtnik, G.; Kavčič, F.; Turk, G. Prediction of concrete strength using ultrasonic pulse velocity and artificial neural networks. *Ultrasonics* **2009**, *49*, 53–60. [CrossRef]
40. Keating, J.; Hannant, D.J.; Hibbert, A.P. Correlation between cube strength, ultrasonic pulse velocity and volume change for oil well cement slurries. *Cem. Concr. Res.* **1989**, *19*, 715–726. [CrossRef]
41. Bogas, J.A.; Gomes, M.G.; Gomes, A. Compressive strength evaluation of structural lightweight concrete by non-destructive ultrasonic pulse velocity method. *Ultrasonics* **2013**, *53*, 962–972. [CrossRef]
42. Rajagopalan, P.R.; Prakash, J.; Narasimhan, V. Correlation between ultrasonic pulse velocity and strength of concrete. *Indian Concr. J.* **1973**, *47*, 416–418.
43. Anderson, D.A.; Seals, R.K. Pulse velocity as a predictor of 28- and 90-day strength. *J. Proc.* **1981**, *78*, 116–122.
44. Abel-Jawad, Y.A.; Afaneh, M. Factors affecting the relationship between ultrasonic pulse velocity and concrete compressive strength. *Indian Concr. J.* **1997**, *71*, 373–378.
45. KS F 2563. *Ground Granulated Blast-Furnace Slag for Use in Concrete*; Korean Standards: Seoul, Korea, 2019.
46. KS L5405. *Fly Ash*; Korean Standards: Seoul, Korea, 2018.
47. ASTM, C191-18a. *Standard Test Methods for Time of Setting Hydraulic Cement by Vicat Needle*; ASTM International: West Conshohocken, PA, USA, 2018.
48. ASTM C109-16a. *Standard Test Method for Compressive Strength of Hydraulic Cement Mortars (Using 2-in. or [50-mm] Cube Specimens)*; ASTM International: West Conshohocken, PA, USA, 2016.
49. ASTM C469M-14. *Standard Test Method for Static Modulus of Elasticity and Poisson's Ratio of Concrete in Compression*; ASTM International: West Conshohocken, PA, USA, 2014.
50. Komlos, K.; Popovics, S.; Nürnbergerová, T.; Babal, B.; Popovics, J.S. Ultrasonic pulse velocity test of concrete properties as specified in various standards. *Cem. Concr. Compos.* **1996**, *18*, 357–364. [CrossRef]
51. Qixian, L.; Bungey, J.H. Using compression wave ultrasonic transducers to measure the velocity of surface waves and hence determine dynamic modulus of elasticity for concrete. *Constr. Build. Mater.* **1996**, *10*, 237–242. [CrossRef]
52. Lu, X.; Sun, Q.; Feng, W.; Tian, J. Evaluation of dynamic modulus of elasticity of concrete using impact-echo method. *Constr. Build. Mater.* **2013**, *47*, 231–239. [CrossRef]
53. Chotard, T.; Gimet-Breart, N.; Smith, A.; Fargeot, D.; Bonnet, J.P.; Gault, C. Application of ultrasonic testing to describe the hydration of calcium aluminate cement at the early age. *Cem. Concr. Res.* **2001**, *31*, 405–412. [CrossRef]
54. Smith, A.; Chotard, T.; Gimet-Breart, N.; Fargeot, D. Correlation between hydration mechanism and ultrasonic measurements in an aluminous cement: Effect of setting time and temperature on the early hydration. *J. Eur. Ceram. Soc.* **2002**, *22*, 1947–1958. [CrossRef]
55. Boumiz, A.; Vernet, C.; Tenoudji, F.C. Mechanical properties of cement pastes and mortars at early ages: Evolution with time and degree of hydration. *Adv. Cem. Res.* **1996**, *3*, 94–106. [CrossRef]
56. Chang, J.J. A study on the setting characteristics of sodium silicate-activated slag pastes. *Cem. Concr. Res.* **2003**, *33*, 1005–1011. [CrossRef]
57. Choi, S.; Lee, K.M. Influence of Na_2O Content and Ms (SiO_2/Na_2O) of Alkaline Activator on Workability and Setting of Alkali-Activated Slag Paste. *Materials* **2019**, *12*, 2072. [CrossRef]

58. Nedeljković, M.; Li, Z.; Ye, G. Setting, strength, and autogenous shrinkage of alkali-activated fly ash and slag pastes: Effect of slag content. *Materials* **2018**, *11*, 2121. [CrossRef] [PubMed]
59. Chatterji, S. Mechanism of expansion of concrete due to the presence of dead-burnt CaO and MgO. *Cem. Concr. Res.* **1995**, *25*, 51–56. [CrossRef]
60. Shi, C.; Day, R.L. A calorimetric study of early hydration of alkali-slag cements. *Cem. Concr. Res.* **1995**, *25*, 1333–1346. [CrossRef]
61. Lee, H.K.; Lee, K.M.; Kim, Y.H.; Yim, H.; Bae, D.B. Ultrasonic in-situ monitoring of setting process of high-performance concrete. *Cem. Concr. Res.* **2004**, *34*, 631–640. [CrossRef]
62. Pessiki, S.P.; Carino, N.J. Setting Time and Strength of Concrete Using the Impact-Echo Method. *Mater. J.* **1988**, *85*, 389–399.
63. Wei, S.; Yunsheng, Z.; Jones, M.R. Using the ultrasonic wave transmission method to study the setting behavior of foamed concrete. *Constr. Build. Mater.* **2014**, *51*, 62–74. [CrossRef]
64. Zhang, S.; Zhang, Y.; Li, Z. Ultrasonic monitoring of setting and hardening of slag blended cement under different curing temperatures by using embedded piezoelectric transducers. *Constr. Build. Mater.* **2018**, *159*, 553–560. [CrossRef]
65. King, M.S. Static and dynamic elastic properties of igneous and metamorphic rocks from the canadian shield. *Int. J. Rock Mech. Min. Sci.* **1983**, *20*, 237–241. [CrossRef]
66. Christaras, B.; Auger, F.; Mosse, E. Determination of the moduli of elasticity of rocks. Comparison of the ultrasonic velocity and mechanical resonance frequency methods with direct static methods. *Mater. Struct.* **1994**, *27*, 222–228. [CrossRef]
67. Ameen, M.S.; Smart, B.G.; Somerville, J.M.; Hammilton, S.; Naji, N.A. Predicting rock mechanical properties of carbonates from wireline logs (A case study: Arab-D reservoir, Ghawar field, Saudi Arabia). *Mar. Pet. Geol.* **2009**, *26*, 430–444. [CrossRef]
68. Martínez-Martínez, J.; Benavente, D.; García-del-Cura, M.A. Comparison of the static and dynamic elastic modulus in carbonate rocks. *Bull. Eng. Geol. Environ.* **2012**, *71*, 263–268. [CrossRef]
69. Sone, H.; Zoback, M.D. Mechanical properties of shale-gas reservoir rocks-Part 1: Static and dynamic elastic properties and anisotropy. *Geophysics* **2013**, *78*, 381–392. [CrossRef]
70. Simmons, G.; Brace, W.F. Comparison of static and dynamic measurements of compressibility of rocks. *J. Geophys. Res.* **1965**, *70*, 5649–5656. [CrossRef]
71. Eissa, E.A.; Kazi, A. Relation between static and dynamic young's moduli of rocks. *Int. J. Rock Mech. Min. Sci.* **1988**, *25*, 479–482. [CrossRef]
72. Ciccotti, M.; Mulargia, F. Differences between static and dynamic elastic moduli of a typical seismogenic rock. *Geophys. J. Int.* **2004**, *157*, 474–477. [CrossRef]
73. Cheng, C.H.; Johnston, D.H. Dynamic and static moduli. *Geophys. Res. Lett.* **1981**, *8*, 39–42. [CrossRef]

materials

Article

The Influence of Ambient Temperature on High Performance Concrete Properties

Alina Kaleta-Jurowska * **and Krystian Jurowski**

Faculty of Civil Engineering and Architecture, Opole University of Technology, Katowicka 48, 45-061 Opole, Poland; k.jurowski@po.edu.pl

* Correspondence: a.kaleta-jurowska@po.edu.pl

Received: 2 September 2020; Accepted: 12 October 2020; Published: 18 October 2020

Abstract: This paper presents the results of tests on high performance concrete (HPC) prepared and cured at various ambient temperatures, ranging from 12 °C to 30 °C (the compressive strength and concrete mix density were also tested at 40 °C). Special attention was paid to maintaining the assumed temperature of the mixture components during its preparation and maintaining the assumed curing temperature. The properties of a fresh concrete mixture (consistency, air content, density) and properties of hardened concrete (density, water absorption, depth of water penetration under pressure, compressive strength, and freeze–thaw durability of hardened concrete) were studied. It has been shown that increased temperature (30 °C) has a significant effect on loss of workability. The studies used the concrete slump test, the flow table test, and the Vebe test. A decrease in the slump and flow diameter and an increase in the Vebe time were observed. It has been shown that an increase in concrete curing temperature causes an increase in early compressive strength. After 3 days of curing, compared with concrete curing at 20 °C, an 18% increase in compressive strength was observed at 40 °C, while concrete curing at 12 °C had a compressive strength which was 11% lower. An increase in temperature lowers the compressive strength after a period longer than 28 days. After two years of curing, concrete curing at 12 °C achieved a compressive strength 13% higher than that of concrete curing at 40 °C. Freeze–thaw performance tests of HPC in the presence of NaCl demonstrated that this concrete showed high freeze–thaw resistance and de-icing materials (surface scaling of this concrete is minimal) regardless of the temperature of the curing process, from 12 °C to 30 °C.

Keywords: concrete; temperature; high performance concrete (HPC)

1. Introduction

1.1. Temperature Influence on the Hydration Process of Portland Cement

Temperature is an important factor which influences the hydration process of cement and the properties of concrete mixture and hardened concrete. It is known that the rate of reaction of cement hydration grows with increasing temperature. The consequence of this is a faster increase in the strength of concrete in the early stage of maturation [1–3].

The influence of temperature on the cement hydration process has been the subject of many studies. It has been found that, in the early stages of maturation, the rate of hydration of the alite significantly increases along with an increase in temperature, but later (from 28 to 90 days) it decreases depending on the type of cement [4]. After a year of maturation, the highest degree of hydration was observed in cement pastes cured at 10 °C and the lowest in those at 60 °C. Furthermore, it was found that, in cement paste curing at 10 °C, almost all the cement grains were hydrated, while at 60 °C cement grains which were only partially hydrated could be observed.

Studies on the microstructure of hydrated cement phases at different temperatures have shown that temperature also influences the morphology, type, and number of hydrate phases formed. At higher temperatures, the more heterogeneous distribution of hydrate phases and formations of shorter needle-shaped ettringite crystals are observed [5]. Moreover, the results indicate that at elevated temperatures the hydration rate of alite and belite is higher.

The results of the authors of the study [6] indicate that the apparent density of cement paste increases with temperature (in the range from 5 °C to 60 °C). According to the authors, this is due to the reduction of bonded water. This results in a more porous microstructure of the cement paste and a reduction in the volume occupied by the C-S-H phase. Higher porosity of cement pastes cured at elevated temperatures was also found by the authors [7]. The result is lower strength of the paste and lower durability of the resulting material.

The studies presented in the paper [8] indicate that at elevated temperatures (40 °C and 50 °C) the formation of the C-S-H phase with higher density, more heterogeneous distribution of hydration products and higher porosity were observed. At 50 °C, calcium monosulphate was observed in the initial period, while the amount of ettringite significantly decreased. This was also confirmed by the authors of other works [9,10]. Due to the increase in porosity, the strength decreases later on. An increase in the porosity of cement pastes cured at elevated temperatures was also observed in binders containing granulated blast-furnace slag [11–14].

Cement pastes with the addition of fly ash, volcanic ash or granulated blast-furnace slag, cured in the temperature range from 10 °C to 60 °C were tested in the study [11]. It was found that blast furnace slag was the only additive that positively influenced the strength (in relation to the strength of the cement paste without additives), especially at 60 °C. According to these authors, the microstructure of cement pastes cured at 60 °C showed higher porosity than the microstructure of grouts cured at 10 °C.

In [15], it was found that the microstructure of cement paste with silica fume, cured at 23 °C, is homogeneous. This cement paste has a much less porous structure compared to a cement paste without an additive, with the same degree of cement hydration. On the other hand, cement pastes cured at 30 °C and 70 °C differ from cement pastes cured at 23 °C with their $Ca(OH)_2$ concentration. While the distribution of hydration products is still relatively homogenous, there are larger continuous pores between the cement grains. The authors found that the temperature of curing has a greater impact on the microstructure of a cement paste with silica fume than cement paste without this additive.

Tests of cement pastes cured at temperatures ranging from 5 °C to 50 °C carried out after a longer period of time (up to 91 days) showed that the cement pastes cured at the lowest temperature was hydrated to the greatest extent [16]. These authors have also shown that at a higher curing temperature in a cement paste, the distribution of hydration products is uneven, resulting in a lower compressive strength of these cement pastes after a longer curing time [17,18].

Summing up the results of the research conducted by various authors, it should be stated that the increase in temperature leads to the acceleration of the hydration process of Portland cement, with the distribution of hydration products being more irregular. This results in increased compressive strength in the early stages of curing. Increased temperature also makes the distribution of cement hydration products uneven and increases the porosity of the resulting structure. The consequence of this is a reduction in compressive strength after a longer curing period. This also applies to cement pastes containing mineral additives, although, in the case of additives, such as fly ash or granulated blast-furnace slag, the scale of the phenomenon is smaller, which can be explained by the reduction in hydration heat of binders with these mineral additives.

1.2. Influence of Temperature on the Properties of a Fresh Concrete Mixture and Hardened Concrete

The influence of temperature on the cement hydration is reflected in the properties of the concrete mixture and hardened concrete. The production of concrete mixtures at elevated temperatures causes many problems due to the accelerated hydration process of the cement. In addition, the concrete mix has a higher water demand due to evaporation. The influence of temperature on the workability of normal strength concrete is well recognised—increasing temperature leads to workability deterioration [19,20]. The authors of the paper [21] also stated that there is an optimal temperature (about 20 °C) allowing them to obtain a concrete mixture with the most advantageous workability. Klieger [22] found that with the temperature increase of 11 °C, the slump decreases by 25 mm, the result of which it is necessary to increase the water content to maintain its consistency.

The consistency of the concrete mixture also depends on the effectiveness of chemical admixtures at elevated temperatures. Schmidt et al. [23] demonstrated that the behaviour of Self-Compacting Concrete (SCC), containing a superplasticizer, at different temperatures, is different from that of normal concrete. Superplasticizers in a concrete mixture, depending on their chemical structure, have different effects on rheological properties of the concrete mixture. A linear relationship between the temperature and the yield stress of the concrete mixture was shown. The higher the temperature, the faster the yield stress increases [24].

The paper in [25] shows that the temperature of concrete mixture also has an influen on the initian and final setting time of cement. The difference between the initial and the final setting time of the cement decreases as the ambient temperature increases. Moreover, the study [26] shows that an increase in the cement content results in an increase in the temperature of the concrete mixture, as well as a shortening of the setting time.

An increase in ambient temperature generally results in a loss of workability of the concrete mixture. The reason for this phenomenon is both the acceleration of the cement setting process and the faster evaporation of the mixing water at higher temperatures.

The influence of temperature on the properties of hardened concrete is similar to that of cement pastes [27]. Increasing the temperature of concrete curing results in higher early concrete strength; however, the strength decreases after as time goes on. An increase in temperature also reduces the corrosion resistance of concrete [5,28]. This effect is most evident when the concrete mixture is exposed to high temperatures immediately after casting.

The most susceptible to excessive heating are massive elements, where the cooling surface is small in relation to the mass of the concrete mixture being caste. The negative phenomena caused by excessive heating can be minimized by proper selection of binder composition [29].

There are methods to minimise the adverse effects of increased temperature on concrete properties. These include: reducing the cement content in concrete; partial replacement of the cement by mineral pozzolanic and hydraulic additives; the use of cement with low hydration heat; thermal control of aggregates; the use of cool water or the addition of crushed ice to the concrete mixture. In practice, good results are achieved by introducing granulated blast-furnace slag and fly ash into the cement [30–32].

The influence of temperature on the properties of normal strength concrete is widely recognised. An increase in the curing temperature also increases the early strength; however, it reduces the strength of the concrete later on and has a negative impact on its durability, which is related to the cement hydration process. However, it should be noted that the influence of temperature on the properties of High Performance Concrete (HPC), which is particularly sensitive to temperature changes due to the relatively low w/c ratio and the use of high range water reducing admixtures (HRWR), is much less well-known.

Apart from the heat generated by the hydration reaction, the temperature of the concrete mixture is also influenced by the temperature of the mixture components, the ambient temperature, and the heat generated by friction as a result of mixing. The temperature of the aggregate is of particular importance, because its content in concrete is relatively high. The temperature of aggregate and water generally corresponds to the ambient temperature, while the temperature of cement stored in silos can be much higher, which further increases the temperature of the concrete mixture.

This paper presents the results of a study on the influence of temperature on the properties of a fresh concrete mixture and hardened HPC containing a polycarboxylate superplasticizer and an addition of silica fume. The tests were carried out at both increased (30 °C) and lowered (12 °C) concrete curing temperatures, but within the range of the practical applicability of concrete. The compressive strength and concrete mix density were tested at temperatures of 12 °C, 20 °C, 30 °C and 40 °C. Special attention was paid to achieving the desired temperature of the mixture components and maintaining this temperature while making the mixture and curing the concrete.

2. Materials

The concrete (HPC) was made with Portland cement CEM I 42.5 R (CEM I), with a specific surface area (Blaine) of 440 m^2/kg. The chemical composition of the cement is shown in Table 1. The results of particle size distribution tests performed with a laser grain analyser are presented in Figure 1.

Table 1. Chemical composition of Portland cement CEM I 42.5 R (% mas.).

Cement	SiO_2	Al_2O_3	Fe_2O_3	CaO	MgO	Cl^-	Na_2O_{eq}	SO_3	K_2O
CEM I	21.9	5.8	2.9	63.1	1.2	0.01	0.7	2.1	0.5

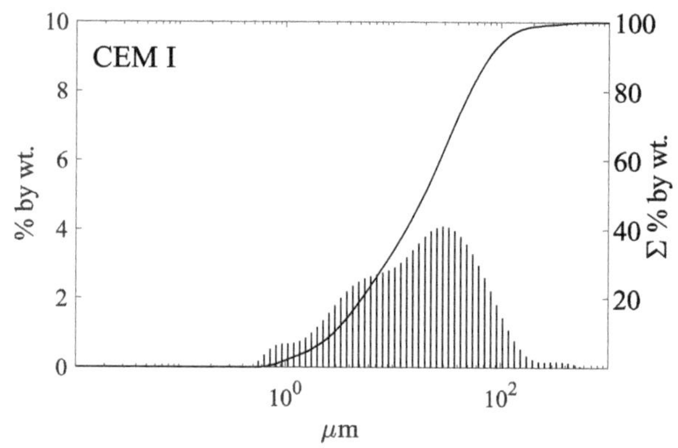

Figure 1. Particle size distribution of Portland cement CEM I 42.5 R (CEM I).

A superplasticizer based on polycarboxylates (SP) was used as a HRWR. The SP was added in the amount of 1.5% in relation to the mass of cement.

Silica fume (SF) was used as the mineral additive, an amount of 10% in relation to the mass of cement. According to the manufacturer's specification, the chemical composition of SF is as follows: SiO_2 (min. 85%), Fe_2O_3 (max. 2.5%), CaO (max. 1.0%) and Al_2O_3 (max. 1.5%).

The concrete mixture was made of natural fine aggregate (0/2 mm fraction) and crushed basalt aggregate (fractions 2/8 and 8/16 mm). The particle size distribution of individual aggregate fractions is shown in Figure 2.

The results of the physical properties of aggregates, such as bulk density, specific density and water absorption are presented in Table 2.

The particle size distribution was selected using the iterative method described by Kuczyński [33,34]. When composing the particle size distribution from several different aggregate fractions, they were put together in such a way as to ensure the greatest possible tightness, with the lowest possible water demand.

The composition of the HPC mixture was designed using the experimental method, assuming the particle size distribution, as defined above, and the amount and type of cement with silica fume added (Table 3). The w/c ratio was selected to obtain concrete with a compressive strength of more than 100 MPa. The designed concrete mixture composition is presented in Table 3. The consistency of the concrete mixture was regulated by adding an appropriate amount of the SP.

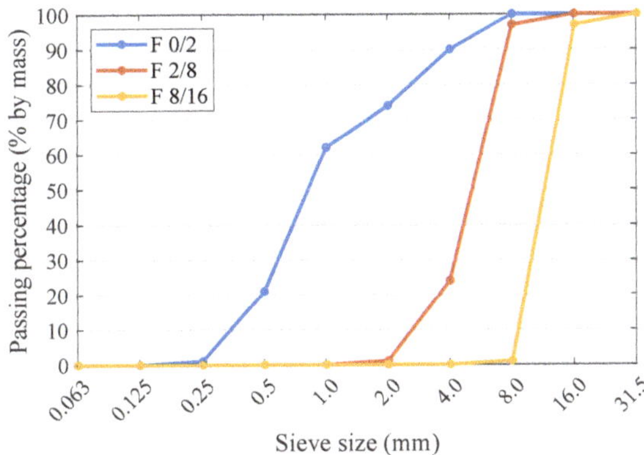

Figure 2. Particle size distribution of aggregates used in the production of HPC.

Table 2. Physical properties of aggregates.

Fraction	Average Value		
	Bulk Density (kg/dm^3)	**Specific Density (kg/dm^3)**	**Water Absorption (%)**
0/2	1.72	2.64	1.4
2/8	1.64	3.12	1.6
8/16	1.60	3.18	0.5

Table 3. Concrete mixture composition.

Ingredient	Content (kg/m^3)
cement	500
0/2 fraction fine aggregate	656
2/8 fraction coarse aggregate	592
8/16 fraction coarse aggregate	740
water	150
silica fume	50
superplasticizer	7.5

The ingredients were mixed in a forced circulation mixer by Zyklos Mixer ZK 150 HE (Pemat, Freisbach, Germany). The same procedure for adding ingredients to the mixer and a constant mixing time of the concrete mixture at all temperatures were applied. The mixing procedure used is shown in Table 4.

Table 4. Mixing procedure for concrete mixture ingredients.

Time (min)	Activity Performed
0–2	mixing the coarse aggregate fractions (2/8 and 8/16 mm)
2–4	adding the fine aggregate fraction (0/2 mm)
4–6	adding the cement and silica fume
6–8	adding 1/2 the water amount
8–14	adding 1/2 the water amount with the superplasticizer
14	mixing completion

Concrete mixtures were made at temperatures of 12 °C, 20 °C, 30 °C and 40 °C. In order to stabilise the temperature of the components, the cement, silica fume, aggregate and water were kept at a controlled, assumed temperature for at least 72 h before the concrete mix was made, using climate chambers.

Every effort was made to keep the temperature in the room where the concrete mix was prepared at the assumed level. The temperature was increased using an appropriate heating system and air heater units. Tests at lowered temperatures were carried out in the winter period, which made it possible to maintain the assumed temperature.

3. Methods

3.1. Tests of the Properties of the Materials Used

The particle size distribution of cement was tested with the use of the Mastersizer 3000 laser particle analyser (Malvern, UK), and the wet method. Isopropanol was used as the dispersant. The test was carried out with an obscuration range of 10–15%. The presented test results are an average of at least 3 measurements.

Physical properties of aggregates such as bulk density, specific density and water absorption were tested in accordance with standards EN 1097-3:2000, EN 1097-6:2013 and EN 1097-6:2013, respectively.

3.2. Concrete Mixture Testing

Consistency testing by the concrete slump test was performed according to EN 12350-2:2011. This test consists of placing and compacting the concrete mixture in a truncated cone shaped mould. The result of the test is the decrease in the height of the concrete mixture immediately after removing the mould.

Consistency determination with the flow table test was carried out according to EN 12350-5:2011. This test consists of placing the mixture in a truncated cone on a plate and then, after lifting the mould, 15 cycles of lifting and free falling of the upper table plate are performed. The result of the test is the flow diameter of concrete mixture.

Consistency determination with the Vebe test was carried out according to EN 12350-3:2011. The result of the test is the time of vibration of the mixture placed in the cylinder until the level of the mixture in the cylinder is completely consolidated.

Testing of the consistency of the concrete mixture using the concrete slump test, the flow table test and the Vebe test was carried out immediately after mixing the ingredients, as well as after 30 and 60 min. The concrete mixture was mixed at slow speed until testing.

The air content of the concrete mixture was determined in accordance with EN 12350-7:2011 by means of a pressure gauge. Testing using a pressure gauge is based on Boyle–Mariotte's law and consists in the fact that a known volume of air at a certain pressure combines in a tightly closed container with an unknown volume of air contained in a concrete mixture sample.

The density of the concrete mixture was determined according to EN 12350-6:2011. This method consists in determining the weight of a concrete mixture, completely filling a container with a known volume.

3.3. Testing on Hardened Concrete

The density of the concrete was determined according to EN 12390-7:2011, hydrostatic method. The method consists in determining the mass and volume of a concrete sample by determining the mass of displaced water. The test was carried out on cubic samples of 100 mm side.

Tests of concrete water absorption were carried out using the method described in the PN-B-06250:1988 standard, after 28 days of concrete curing. Each time the test was performed on 3 cubic samples with a side dimension of 100 mm. The samples were first saturated in water to constant mass and then dried at 105 °C to constant mass.

The compressive strength was tested according to EN 12390-3:2011. The test was carried out using the Controls MCC8 strength press (Controls Group, Liscate, Italy). The test was carried out on cubic samples of 100 mm side. The concrete samples cured for more than 28 days were taken out of water and cured at laboratory temperature (20 ± 2) °C until the test.

Testing of the depth of penetration of water under pressure of the concrete was carried out in accordance with EN 12390-8:2011. Each time, the test was carried out on 6 cubic samples of concrete with a side of 150 mm, cured before the test for 28 days in water.

The tests of concrete resistance to cyclic freezing and thawing in the environment of de-icing salt were performed according to EN 12390-9:2007, using the "slab test". This test consists in determining the weight of scaled material from the concrete sample surface after 7, 14, 28, 42 and 56 freezing and thawing cycles in the presence of 3% NaCl solution. Then, 4 cubic samples of concrete with a side of 150 mm were used for the test, which were cured for 7 days at 12 °C, 20 °C or 30 °C. The remaining curing and preparation period of the test was carried out in accordance with EN 12390-9:2007.

The freeze–thaw durability was also tested according to Polish standard PN-B-06250:1988. The test is carried out on 12 cubic concrete samples with a side dimension of 100 mm, 6 of which undergo 300 freeze/thaw cycles. The result of the test is a loss of compressive strength of the tested samples, compared to the other 6 "witness samples". Before testing, the samples were cured for 28 days at 12 °C, 20 °C or 30 °C water temperatures.

All test results presented in this paper are average values for a minimum of three measurements. The uncertainty values given are the expanded uncertainty of measurement with an expansion probability of approximately 95% and a corresponding expansion factor of k = 4.30 (for 3 samples) k = 3.18 (for 4 samples) and k = 2.57 (for 6 samples).

4. Results and Discussion

4.1. Temperature Influence on the Properties of the Concrete Mixture

Consistencies in the concrete mixture were tested by three methods. This was necessary because consistency differed significantly from one temperature to another and only one of the methods give results outside the range of applicability of this method.

Figure 3 shows the slump depending on the time of the test and the ambient temperature at which the concrete mixture was prepared from ingredients that had previously obtained the same temperature as the ambient temperature. It is clear that the concrete mixture preparation temperature in the range from 12 °C to 30 °C has a significant impact on the consistency. At 30 °C, immediately after mixing the ingredients, a minimum slump (20 mm) is observed, while at 12 °C this value is as high as 280 mm, which is beyond the applicability range of this test method.

Similar results were obtained using the flow table test and the Vebe test. The flow diameters and Vebe time depending on the temperature as well as the time of the test are shown in Figures 4 and 5, respectively.

The results of the consistency test using the flow table test (Figure 4) showed a decrease in the flow diameter of the concrete mixture as the temperature and time increase. The Vebe time needed for compaction of the concrete mixture also increased with increasing temperature (Figure 5).

The workability of the concrete mixture has also decreased significantly in up to 60 min. Moreover, at 30 °C, the decrease in flowability was so great that after only 30 min correct measurement was impossible (both the Vebe test and the flow table test).

The presented results of the concrete mixture consistency tests showed that an increase in temperature causes a loss of workability. This effect is particularly visible at 30 °C. At this temperature, the workability of the concrete mixture is lost after only 30 min. The observed phenomenon is mainly the effect of increasing the rate of water evaporation from the concrete mixture at elevated temperature and accelerated cement hydration process [19,21], as well as from low ratio $w/b = 0.27$ in HPC. The results of consistency tests over time are difficult to assess as no similar tests of an HPC concrete mixture were found in the literature. The loss of workability can also be caused by a decrease in the effectiveness of the HRWR admixture at elevated temperatures [23,35]. In practice, it would be necessary to use more of the HRWR admixture. In this paper, the proportions of ingredients have been retained to ensure the comparability of test results.

The results of air content tests in the concrete mixture depending on the curing temperature and testing time are presented in Figure 6.

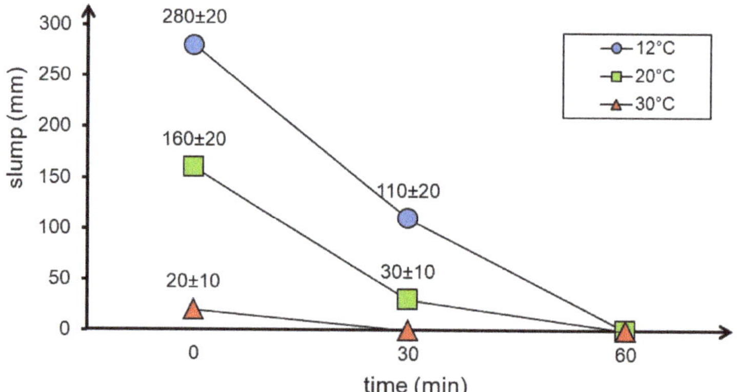

Figure 3. Results of concrete mixture consistency tests with the use of the slump test depending on the temperature and time of testing.

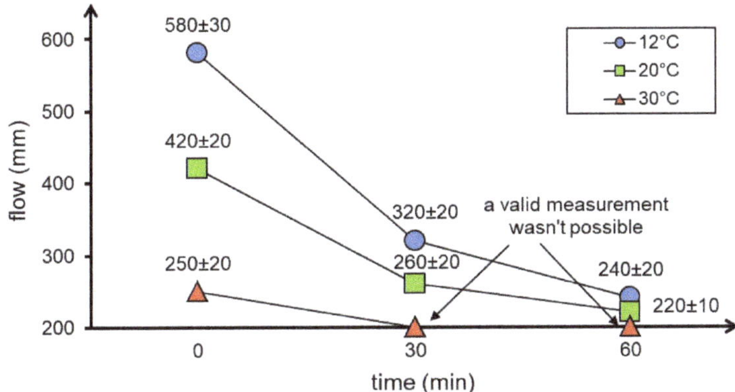

Figure 4. Results of concrete mixture consistency tests with the use of the flow table test depending on the temperature and time of testing.

As shown in Figure 6, the air content in the concrete mixture generally increases with the mixture temperature and does not change significantly over time. It should be noted that the results obtained at 30 °C may be unreliable as the consolidation of concrete mix was difficult at this temperature.

The concrete mixture density tests showed that the temperature in the tested range (from 12 °C to 40 °C) does not have a significant influence on this parameter. Differences in the results of density determination did not exceed the measurement error (Table 5). Similar results were obtained by the authors in the paper [36], who demonstrated that the differences in the density of the concrete mixture tested, at different temperatures, are negligible.

Table 5. Temperature influence on the density of the concrete mixture.

Ambient Temperature (°C)	Concrete Mixture Temperature (°C)	Compaction Method	Density (kg/m^3)
12	17.3	rodding/manual	2620 ±10
20	23.5	vibrating table/mechanical	2630 ±10
30	32.7	vibrating table/mechanical	2630 ±10
40	41.6	vibrating table/mechanical	2630 ±10

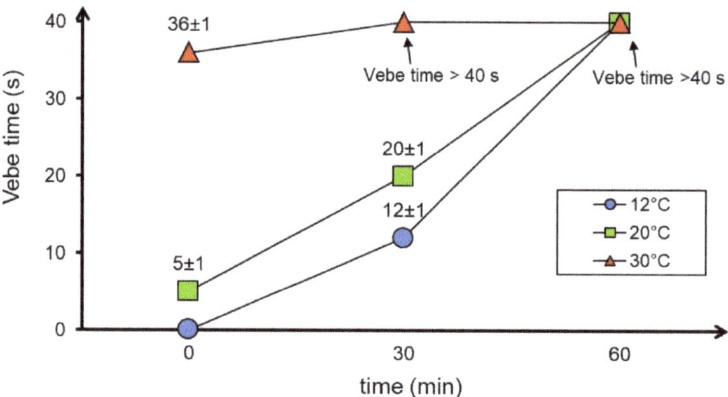

Figure 5. Results of concrete mixture consistency tests with the use of the Vebe test depending on the temperature and time of testing.

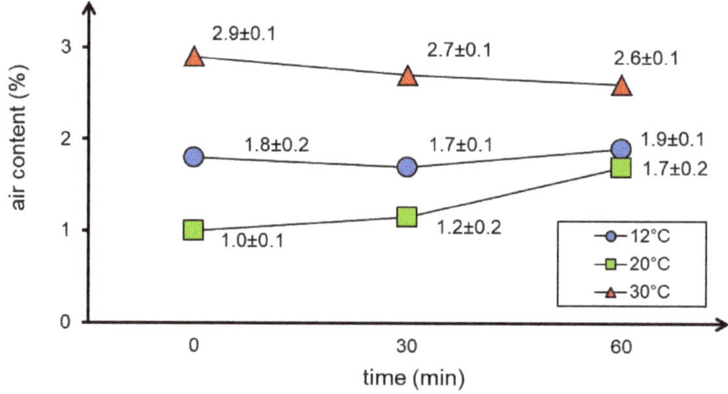

Figure 6. The results of air content tests in the concrete mixture depending on the curing temperature and time.

4.2. Temperature Influence on the Properties of Hardened Concrete

The concrete density tests showed that in the range from 12 °C to 30 °C, an increase in temperature causes a slight increase in the concrete density. However, it should be noted that these differences are minimal, slightly exceeding the error of measurement of the test method (Table 6).

The results of the water absorption tests of the samples of concrete cured at different temperatures showed that in all cases the obtained values were at a very low level. The least absorbent is the concrete cured at 20 °C, and the largest is the concrete cured at 30 °C (Figure 7), however the presented differences should be considered as small. In addition, at elevated temperatures the water absorption rate increases more than at the lower curing temperature (12 °C). Similar results were obtained by the authors of the paper [37] in testing mortars.

Table 6. Influence of ambient temperature on the density of concrete samples.

Curing Temperature (°C)	Concrete Density (kg/m^3)
12	2650 ± 20
20	2660 ± 10
30	2680 ± 20

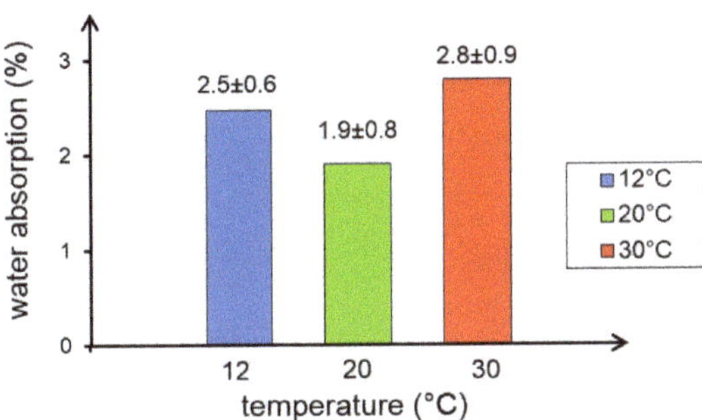

Figure 7. Water absorption of concrete samples depending on concrete curing temperature.

The results of compressive strength testing of concrete samples depending on curing temperature (12 °C, 20 °C, 30 °C and 40 °C) are shown in Table 7. The increase in compressive strength over time of these concrete is shown in Figure 8.

Table 7. Results of compressive strength tests (MPa) of concrete samples cured at 12 °C, 20 °C, 30 °C and 40 °C.

Time (days)	Curing Temperature			
	12 °C	20 °C	30 °C	40 °C
3	74.5 ± 5.3	84.0 ±1.3	87.9 ± 2.5	99.0 ± 9.1
7	87.8 ± 6.1	93.1 ± 4.0	97.0 ± 11.0	103.3 ± 4.6
28	106.6 ± 2.9	116.9 ± 7.6	107.6 ± 6.3	108.8 ± 6.2
90	128.3 ± 10.4	127.3 ± 7.1	122.7 ± 6.9	119.9 ± 10.7
365	127.7 ± 4.8	131.9 ± 1.8	126.8 ± 0.6	119.0 ± 7.0
730	135.6 ± 6.9	134.8 ± 7.9	124.5 ± 3.7	119.7 ± 6.8

As can be seen from the data in Figure 8, the early strength (after 3 and 7 days) of concrete samples cured at elevated temperatures (30 °C and 40 °C) is higher than that of concrete samples cured at lower temperatures (12 °C and 20 °C). However, after a longer curing period (from 90 days to 2 years), the highest strength was demonstrated for concrete cured at lower temperatures. The data presented in Figure 8 show that for concrete curing at almost all tested temperatures, the strength is similar at age of about 40 days.

The increase in the compressive strength of concrete in the initial period of time (up to 7 days), with an increase in the curing temperature, can be explained by the acceleration of the cement hydration process [38]. However, as the results of the research [25] show, an increase in the rate of cement hydration contributes to the formation of a weaker, more porous microstructure, which in effect results in lower compressive strength after a longer curing period (over 28 days).

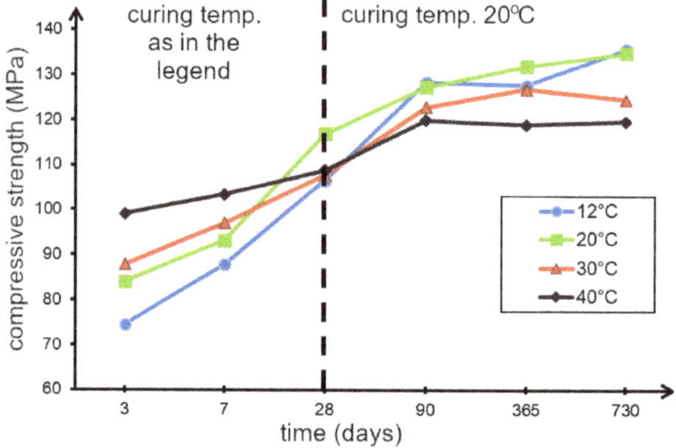

Figure 8. Results of compressive strength tests of concrete cured at 12 °C, 20 °C, 30 °C and 40 °C.

Concrete cured at 12 °C and 20 °C achieved similar strength values at the age of 2 years (about 130 MPa), while concrete made at 20 °C achieved this strength much faster. It can therefore be concluded that a temperature of 20 °C is the most favourable, due to the increase in strength over time. Decreased early concrete strength at low temperatures is also confirmed by the results of the research presented, for example, in [22].

The results of penetration depth tests of the concrete samples cured at 12 °C, 20 °C and 30 °C showed that no significant differences were observed in the values of this parameter (Figure 9) which is in the 9–11 mm range. The above shows a very high tightness of the tested concrete and indicates a properly selected composition of the concrete mixture, including a tight particle size distribution.

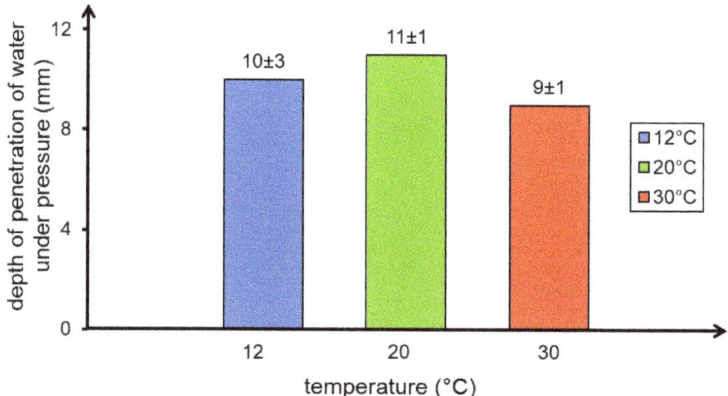

Figure 9. Depth of penetration of water under pressure in concrete samples depending on the curing temperature.

The high concrete tightness shown in the test for the depth of penetration of water under pressure contributes to the durability of the concrete. This is confirmed by the results of the freeze–thaw durability tests carried out with two methods: the method according to PN-B-06250 and the method according to EN-12390-9.

The results of concrete freeze–thaw durability tests using the method according to PN-B-06250 are presented in Table 8. The tests have shown that, regardless of the concrete curing temperature (12 °C, 20 °C and 30 °C), this concrete can be classified as having the highest degree of freeze–thaw durability (*F*300 according to PN-B-06250).

The HPC subjected to 300 freezing and thawing cycles shows no cracking or significant mass loss. The decrease in compressive strength is highest for the concrete prepared at 12 °C (8.6%) and lowest for the concrete prepared at 30 °C (1.7%). The reason for the observed phenomenon may be the fact that, on the day of the beginning of freezing (28th day of maturation), the concrete samples curing at lower temperatures had not yet obtained sufficiently high strength.

Table 8. Results of the tests of resisting of the concrete samples to cyclic freezing and thawing determined by the method according to PN-B-06250.

Curing Temperature	12 °C	20 °C	30 °C
loss of mass of samples after the test (%)	0.01	0.00	0.03
decrease in strength of samples after the test (%)	8.6	4.3	1.7

The results of tests of resistance of the concrete samples matured at 12, 20 and 30 °C to cyclic freezing and thawing in the presence of de-icing salt are shown in the table (Table 9).

Table 9. Results of the tests of resisting of the concrete samples to cyclic freezing and thawing in the presence of de-icing salt (3% NaCl), cured at 12, 20, and 30 °C.

Curing Temperature	Average Value of the Sample Scaling Mass (kg/m^2)				
	After 7 Cycles	After 14 Cycles	After 28 Cycles	After 42 Cycles	After 56 Cycles
12 °C	0.00 ± 0.02	0.00 ± 0.02	0.02 ± 0.02	0.02 ± 0.02	0.02 ± 0.02
20 °C	0.00 ± 0.02	0.02 ± 0.02	0.02 ± 0.02	0.04 ± 0.02	0.04 ± 0.02
30 °C	0.00 ± 0.02	0.02 ± 0.02	0.02 ± 0.02	0.04 ± 0.02	0.04 ± 0.02

The scaling masses under the influence of cyclical freezing and thawing of concrete with simultaneous action of salt solution were minimal. Thus, the influence of the concrete curing temperature on the freeze–thaw durability result of this method has not been demonstrated. At the same time, the assessment according to the Borås criterion proves that the designed concrete is of very good quality, regardless of the curing temperature applied.

5. Conclusions

HPC is increasingly used in civil engineering. The assessment of the influence of temperature on the properties of concrete is particularly important because of the application of HPC concrete at both elevated and lowered temperatures. This paper presents the results of research on the influence of ambient and curing temperature on the properties of the concrete mixture and hardened HPC.

Consistency tests carried out using various methods have shown that HPC mixtures made at different ambient temperatures (within the range of 12 °C to 30 °C) have extremely different consistencies. The slump of the concrete mixture made at 20 °C (160 mm) was almost twice as high as at 12 °C (280 mm) and almost five times lower than at 30 °C (20 mm). This dependence was demonstrated by keeping the mixture ingredients for 72 h before executing the mixture at the assumed test temperature.

The results of the fresh concrete mixture consistency tests confirm that the HPC mixture is very sensitive to temperature increases. The reason for this could be the faster evaporation of water which, combined with the low water–binder ratio, results in a clear decrease in workability. This effect is also influenced by the acceleration of the cement hydration reaction. Losing workability of a concrete mixture at elevated temperatures can significantly hinder its application. This is indicated by the results of consistency tests as well as air content in the concrete mixture at 30 °C.

Temperature has been shown not to have a significant effect on properties, such as density, water absorption, and depth of water penetration under pressure. The indicated parameters for the concrete samples prepared at different temperatures differed slightly.

Compressive strength tests carried out over a period of 3 days to 2 years on concrete samples prepared at temperatures of 12, 20, 30, and 40 °C showed that the rate of compressive strength growth increases with increases in temperature. The concrete prepared at 40 °C reached 99 MPa after only three days (i.e., 91% of the 28-day strength), while at 12 °C the concrete reached 74.5 MPa (i.e., 70% of the 28-day strength). This confirms that at elevated temperatures the cement hydration rate increases, resulting in a faster increase in concrete compressive strength in the first 28 days of curing.

After 28 days, the highest compressive strength was achieved by the concrete maturing at lower temperatures. At the age of two years, concrete produced at temperatures of 12, 20, 30, and 40 °C reached compressive strengths of 135.6, 134.8, 124.5, and 119.7 MPa, respectively.

As shown, an increase in the temperature of concrete curing results in a decrease in its compressive strength after a long period of time, compared with concrete cured at lower temperatures. This may be due to the fact that a faster rate of hydration results in hydration products with a more irregular structure and higher porosity, which negatively affect compressive strength after a longer curing time.

The results of the freeze–thaw durability tests performed using the Polish standard method and the "slab test" show that HPC is resistant to freeze–thaw. The decrease in compressive strength after 300 freezing and thawing cycles was relatively small and amounted to a maximum of 8.6%, compared with samples not subjected to cyclic freezing and thawing, in the case of the concrete prepared at the lowest temperature of 12 °C. For the concrete cured at 20 °C and 30 °C, the strength drop was 4.3% and 1.7%, respectively. This may be due to the fact that, at the start of the test (28th day of curing), the concrete stored at lower temperatures reached a lower compressive strength than the concrete stored at higher temperatures.

The freeze–thaw durability tests in the presence of a NaCl solution showed minimal scaling of the concrete surface, which is proof of very good freeze–thaw durability of the tested HPC, regardless of its preparation temperature.

This study shows a significant influence of lowered and increased temperature on the properties of the concrete mix and hardened HPC concrete, especially in terms of consistency and compressive strength. This points to directions for further research, which should include studies of the rheological parameters of the HPC concrete mix and its changes over time. Research aimed at demonstrating the influence of curing temperature on the cement hydration process for binders used in HPC concrete (with a low water–binder ratio) should also be carried out. Future research should include changes of hydrate phases over time.

Author Contributions: Conceptualization, A.K.-J. and K.J.; methodology, A.K.-J; formal analysis, A.K.-J. and K.J.; investigation, A.K.-J.; writing—original draft preparation, A.K.-J.; writing—review and editing, A.K.-J. and K.J.; visualization, A.K.-J.; supervision, A.K.-J. and K.J. All authors have read and agreed to the published version of the manuscript.

Funding: The work accomplished within statutory research no NBS 16/2019 at the Department of Building Materials Engineering, Faculty of Civil Engineering and Architecture, Opole University of Technology, Opole, Poland.

Acknowledgments: We would like to express great appreciation to Professor Stefania Grzeszczyk for her valuable and constructive suggestions during the planning and development of this research work

Conflicts of Interest: The authors declare no conflict of interest

References

1. Elkhadiri, I.; Elkhadiri, M.; Puertas, F. Effect of curing temperature on cement hydration. *Ceram. Silik.* **2009**, *53*, 65–75.
2. Soutsos, M.; Kanavaris, F. The modified nurse-saul (MNS) maturity function for improved strength estimates at elevated curing temperatures. *Case Stud. Constr. Mater.* **2018**, *9*, 1–14.
3. Nasir, M.; Al-Amoudi, O.S.; Al-Gahtani, H.J.; Maslehuddin, M. Effect of casting temperature on strength and density of plain and blended cement concretes prepared and cured under hot weather conditions. *Constr. Build. Mater.* **2016**, *112*, 529–537. [CrossRef]
4. Escalante-Garcia, J.; Sharp, J. Effect of temperature on the hydration of the main clinker phases in portland cements: part i, blended cements. *Cem. Concr. Res.* **1998**, *28*, 1259–1274. [CrossRef]
5. Lothenbach, B.; Winnefeld, F.; Alder, C.; Wieland, E.; Lunk, P. Temperatureinfluss auf die Hydratation von Portland Zementen. In *16. Internationale Baustofftagung*; F.A. Finger-Institut fur Baustoffkunde: Weimar, Germany, 2006, pp. 401–408.
6. Gallucci, E.; Zhang, X.; Scrivener, K. Effect of temperature on the microstructure of calcium silicate hydrate (C-S-H). *Cem. Concr. Res.* **2013**, *53*, 185–195. [CrossRef]
7. Wang, Q.; Shi, M.; Wang, D. Influence of elevated curing temperature on the properties of cement paste and concrete at the same hydration degree. *J. Wuhan Univ. Technol.-Mater. Sci. Ed.* **2017**, *32*, 1344–1351. [CrossRef]
8. Lothenbach, B.; Winnefeld, F.; Alder, C.; Wieland, E.; Lunk, P. Effect of temperature on the pore solution, microstructure and hydration products of Portland cement pastes. *Cem. Concr. Res.* **2007**, *37*, 483–491. [CrossRef]
9. Thomas, J.J.; Rothstein, D.; Jennings, H.M.; Christensen, B.J. Effect of hydration temperature on the solubility behavior of Ca-, S-, Al-, and Si-bearing solid phases in Portland cement pastes. *Cem. Concr. Res.* **2003**, *33*, 2037–2047. [CrossRef]
10. Buck, A.D.; Burkes, J.P.; Poole, T.S. *Thermal Stability of Certain Hydrated Phases in Systems Made Using Portland Cement*; Technical Report; Department of the Army, Waterways Experiment Station, Corps of Engineers: Vicksburg, MS, USA, 1985.
11. Escalante-Garcia, J.I.; Sharp, J.H. The microstructure and mechanical properties of blended cements hydrated at various temperatures. *Cem. Concr. Res.* **2001**, *31*, 695–702. [CrossRef]
12. Escalante, J.; Gomez, L.; Johal, K.; Mendoza, G.; Mancha, H.; Mendez, J. Reactivity of blast-furnace slag in Portland cement blends hydrated under different conditions. *Cem. Concr. Res.* **2001**, *31*, 1403–1409.
13. Paul, M.; Glasser, F. Impact of prolonged warm (85C) moist cure on Portland cement paste. *Cem. Concr. Res.* **2000**, *30*, 1869–1877. [CrossRef]
14. Yang, H.M.; Kwon, S.J.; Myung, N.V.; Singh, J.K.; Lee, H.S.; Mandal, S. Evaluation of Strength Development in Concrete with Ground Granulated Blast Furnace Slag Using Apparent Activation Energy. *Materials* **2020**, *13*, 442. [CrossRef] [PubMed]
15. Cao, Y.; Detwiler, R.J. Backscattered electron imaging of cement pastes cured at elevated temperatures. *Cem. Concr. Res.* **1995**, *25*, 627–638. [CrossRef]
16. Kjellsen, K.O.; Detwiler, R.J. Reaction kinetics of portland cement mortars hydrated at different temperatures. *Cem. Concr. Res.* **1992**, *22*, 112–120. [CrossRef]
17. Kjellsen, K.O.; Detwiler, R.J.; Gjørv, O.E. Development of microstructures in plain cement pastes hydrated at different temperatures. *Cem. Concr. Res.* **1991**, *21*, 179–189. [CrossRef]
18. Verbeck, G.; Helmuth, R. Structures and Physical Properties of Cement Paste. In *5th International Congress on the Chemistry of Cement*; The Cement Association of Japan: Tokyo, Japan, 1969; pp. 1–44.
19. Neville, A.M. *Właściwości Betonu*; Polski Cement: Kraków, Poland, 2010; p. 874.

20. Ortiz, J.; Aguado, A.; Roncero, J.; Zermeno, M. Influencia de la temperatura ambiental sobre las propiedades de trabajabilidad y microestructurales de morteros y pastas de cemento. *Índice* **2009**, *1*, 2–24.
21. Ortiz, J.; Aguado, A.; Agulló, L.; García, T.; Zermeño, M. Influence of environmental temperature and moisture content of aggregates on the workability of cement mortar. *Constr. Build. Mater.* **2009**, *23*, 1808–1814. [CrossRef]
22. Klieger, P. Effect of Mixing and Curing Temperature on Concrete Strength. *Am. Concr. Inst.* **1958**, *54*, 54–62.
23. Schmidt, W.; Brouwers, H.; Kühne, H.C.; Meng, B. Influences of superplasticizer modification and mixture composition on the performance of self-compacting concrete at varied ambient temperatures. *Cem. Concr. Compos.* **2014**, *49*, 111–126. [CrossRef]
24. Petit, J.Y.; Khayat, K.H.; Wirquin, E. Coupled effect of time and temperature on variations of yield value of highly flowable mortar. *Cem. Concr. Res.* **2006**, *36*, 832–841. [CrossRef]
25. Burg, R.G. *The Influence of Casting and Curing Temperature on the Properties of Fresh and Hardened Concrete*; Portland Cement Association: Skokie, IL, USA, 1996; p. 18.
26. Marar, K.; Eren, Ö. Effect of cement content and water/cement ratio on fresh concrete properties without admixtures. *Int. J. Phys. Sci.* **2011**, *6*, 5752–5765.
27. Kjellsen, K.O.; Detwiler, R.J.; Gjørv, O.E. Backscattered electron imaging of cement pastes hydrated at different temperatures. *Cem. Concr. Res.* **1990**, *20*, 308–311. [CrossRef]
28. Rachel, J.D.; Jennifer, N.C.A.F. Use of Supplementary Cementing Materials to Increase the Resistance to Chloride Ion Penetration of Concretes Cured at Elevated Temperatures. *Mater. J.* **1994**, *91*, 63–66.
29. Kiernożycki, W.; Borucka-Lipska, J. Zmiany objętościowe twardniejącego betonu i ich następstwa. *Cem. Wapno Beton* **2004**, *9/71*, *1*, 19–25.
30. Barrow, R.S.; Carrasquillo, R.L. *The Effect of Fly Ash on the Temperature Rise in Concrete*; Technical Report; The University of Texas at Austin: Austin, TX, USA, 1988.
31. Soutsos, M.; Hatzitheodorou, A.; Kwasny, J.; Kanavaris, F. Effect of in situ temperature on the early age strength development of concretes with supplementary cementitious materials. *Constr. Build. Mater.* **2016**, *103*, 105–116. [CrossRef]
32. Wang, L.; Quan, H.; Li, Q. Evaluation of Slag Reaction Efficiency in Slag-Cement Mortars under Different Curing Temperature. *Materials* **2019**, *12*, 2875. [CrossRef]
33. Kuczyński, W. *Betony Konstrukcyjne. Projektowanie Metodą Kolejnych Przybliżeń (Iteracji)*; Budownictwo i Architektura: Warszawa, Poland, 1956; p. 219.
34. Bukowski, B.; Bastian, S.; Braun, K.; Gruener, M.; Kuczyński, W. *Technologia Betonu. cz. 2, Projektowanie Betonów*; Komitet Inżynierii: Warsaw, Poland, 1972 .
35. Grzeszczyk, S.; Sudoł, M. Wpływ temperatury na skuteczność działania superplastyfikatorów nowej generacji. *Cem. Wapno Beton* **2003**, *6*, 325–331.
36. Mannheimer, R. High temperature and high pressure rheology of oilwell cement slurries. In *Rheology of Fresh Cement and Concrete–Proceedings of an International Conference*; Banfill, P., Ed.; CRC Press: Liverpool, UK 1990; p. 384.
37. Kaczmarek, A. Wpływ zmiennej temperatury powietrza podczas kondycjonowania zapraw na ich parametry techniczne. *Mater. Bud.* **2018**, *5*, 12–13. [CrossRef]
38. Escalante-Garcia, J. Nonevaporable water from neat OPC and replacement materials in composite cements hydrated at different temperatures. *Cem. Concr. Res.* **2003**, *33*, 1883–1888. [CrossRef]

Article

Fibre Distribution Characterization of Ultra-High Performance Fibre-Reinforced Concrete (UHPFRC) Plates Using Magnetic Probes

Lufan Li [1], Jun Xia [1,2,*], Chee Chin [1,2] and Steve Jones [3]

1 Department of Civil Engineering, Xi'an Jiaotong-Liverpool University, Suzhou 215123, China; lufan.li@xjtlu.edu.cn (L.L.); chee.chin@xjtlu.edu.cn (C.C.)
2 Institute for Sustainable Material and Environment, Xi'an Jiaotong-Liverpool University, Suzhou 215123, China
3 Department of Civil Engineering, University of Liverpool, Liverpool L69 3GH, UK; Stephen.Jones@liverpool.ac.uk
* Correspondence: jun.xia@xjtlu.edu.cn

Received: 12 October 2020; Accepted: 5 November 2020; Published: 10 November 2020

Abstract: Ultra-high performance fibre reinforced concrete (UHPFRC) is an innovative cement-based engineering material. The mechanical properties of UHPFRC not only depend on the properties of the concrete matrix and fibres, but also depend on the interaction between these two components. The fibre distribution is affected by many factors and previous researchers had developed different approaches to test the fibre distribution. This research adopted the non-destructive C-shape ferromagnetic probe inductive test and investigated the straight steel fibre distribution of the UHPFRC plate. A simplified characterization equation is introduced with an attenuation factor to consider the different plate thicknesses. The effective testing depth of this probe was tested to be 24 mm. By applying this method, fibre volume content and the fibre orientation angle can be calibrated for the entire plate. The fibre volume content generally fulfilled the design requirement. The fibre orientation angle followed a normal distribution, with a mean value of 45.60°. By testing small flexural specimens cut from the plates, it was found out that the mechanical performance (peak flexural strength) correlates with the product of fibre volume content and cosine fibre orientation angle.

Keywords: C-shape magnetic probe test; fibre orientation angle; flexural test; attenuation factor

1. Introduction

Concrete has been the most significant construction material throughout history. Ultra-high performance concrete (UHPC) was developed as a cementitious composite material with far higher strength, durability, and resistance to external environments than traditional construction materials [1]. With the addition of fibres, the mechanical properties of ultra-high performance fibre reinforced concrete (UHPFRC) can be further improved especially the performance under tensile loading [2]. Moreover, the resistance to impact, chemical degradation, abrasion, and fire can also be improved [3]. UHPFRC has been widely used in airports [4], bridges, roofs, and cladding [5].

Comparing to the ascending branch of the stress-strain curve, the fibre addition always has a more significant impact on the cracking stage of UHPFRC. The fibre acts as a 'bridge' to bond the separated cracked concrete matrix and then being pulled out gradually, which is also known as the 'bridging effect' [6]. It compensates for the disadvantage of UHPC of being too brittle. However, if the fibres are distributed poorly inside concrete, for example, perpendicular to the loading direction, the bridging effect will be adversely affected.

Many factors can influence the fibre distribution, for example, fibre volume fraction [7,8], external vibration [9], and mixing sequence [10]. Previous researchers have developed various testing methods to test the fibre distribution inside the concrete matrix, for example, image analysis [10–12], AC-IS (AC-Impedance Spectroscopy) [13–15], X-ray scanning [16,17], and the magnetic core inductive method [18,19]. Compared to these testing methods, the C-shape ferromagnetic probe method is more economical and convenient, so this method is chosen to be the detecting method in this research.

The C-shape ferromagnetic probe test was proposed by Faifer et al. [20]. However, the main focus of their research was on the theoretical background modelling and derivation of the probe, for example, the simulation of flux density of the probe [21]. The application of this probe was also limited to the basic monitoring of fibre volume content and fibre orientation angle of steel fibre reinforced concrete [21]. In 2016, Nunes et al. made a simple and cheap C-shaped ferrite core with coils wound on it [22]. They further applied this testing method on thin UHPFRC plates. By placing the probe on 300 mm × 300 mm × 30 mm specimens in different directions and testing the inductance, the corresponding fibre distribution conditions can be identified. According to Nunes et al., a relative magnetic permeability $\mu_{r,mean}$, which reflects fibre content, can be calculated as:

$$\mu_{r,\varphi} = \frac{L_{\varphi}}{L_{air}} \quad (1)$$

$$\mu_{r,(90^\circ-\varphi)} = \frac{L_{(90^\circ-\varphi)}}{L_{air}} \quad (2)$$

$$\mu_{r,mean} = \frac{\mu_{r,\varphi} + \mu_{r,(90^\circ-\varphi)}}{2} \quad (3)$$

where L_{φ}, $L_{(90^\circ-\varphi)}$, and L_{air} represent the magnetic inductance at φ, $(90^\circ - \varphi)$, and in the air.

The relationship between the orientation indicator ρ_Δ and the relative magnetic permeability can be expressed as:

$$\rho_\Delta = \rho_{(90^\circ-\varphi)} - \rho_{\varphi} = \frac{\mu_{r,(90^\circ-\varphi)} - \mu_{r,\varphi}}{2\left(\mu_{r,mean} - 1\right)} \quad (4)$$

Nune et al.'s later research was mainly emphasizing on the detection of uniformly distributed UHPFRC by aligning the fibres with an intense magnetic field [23,24], but the intrinsic heterogeneity of fibre distribution had been proved to largely affect the tensile response of fibre reinforced concrete [25]. Moreover, recent studies were mainly focusing on the relationship between fibre content and flexural performance [26,27]. Therefore, both the local fibre spatial and orientation distribution should be taken into consideration, but this area still needs further investigation.

In previous research, the magnetic probe method was only applied on thin plates. If this method is going to be applied for the purpose of checking the structural safety, it is necessary to determine the effective depth of the probe. In order to fill the gap, the effective depth of the magnetic probe was firstly examined and an attenuation factor was obtained to further correct the fibre volume content data. Apart from this, a simplified analytical solution was derived to determine the relationship between the fibre orientation angle and orientation indicator. In addition, compressive tests and flexural tests had been conducted and the correlated equation between mechanical performance and the tested fibre distributions was derived. The corresponding theoretical explanation of strain-hardening behaviour was clarified.

2. Material Preparation and Concrete Mixing

2.1. Materials and Specimen List

Premix powder material provided from third party company (Beirong Circular Materials Co., Ltd., Yingtan, Jiangxi, China) were used to ensure the concrete grade in each batch. The premix powders contained the basic components needed for UHPFRC (cement, fine sand, silica fume, and quartz). Straight steel fibres (as shown in Figure 1) covered with a brass coating were added in the mix. The diameter and length of fibres were 0.25 mm and 12.5 mm respectively, given an aspect ratio of 50.

The uniaxial tensile strength of the fibres can go up to 2850 MPa. Polycarboxylate superplasticizer was added to ensure the workability. Detailed mix proportions are presented in Table 1.

Figure 1. Straight steel fibres used in research.

Table 1. Mix proportion for pre-mix ultra-high performance fibre reinforced concrete (UHPFRC; provided by third party company).

Material	Content (kg/m^3)
Dry-Mix	2200
Water	192
Steel Fibre	0, 78, 156, 195 (for 0%, 1%, 2%, and 2.5% vol.)
Superplasticizer	20.5
Total	2413, 2491, 2569, 2608 (for 0%, 1%, 2%, and 2.5% vol.)

Table 2 lists the dimensions and casting purpose of all specimens. Cubes were used to determine the compressive strength. Plates with different thicknesses were designed to detect the fibre orientation and spatial distribution, and the following flexural performances.

Table 2. Specimen list and casting purpose for different types of specimens.

Shape	Dimension (mm)	Casting Purpose
Cube	100 × 100 × 100	Compressive test
Plate	500 × 500 × 15	Flexural test, Magnetic probe test
Plate	500 × 500 × 20	Flexural test, Magnetic probe test
Plate	500 × 500 × 35	Flexural test, Magnetic probe test
Plate	500 × 500 × 50	Flexural test, Magnetic probe test

2.2. Concrete Mixing Process and Curing Condition

Table 3 shows the suggested mixing procedure and mixing time provided from the supply company. Considering the differences of the mixing environment and fibre amount, the mixing time varied by ±30 s.

Table 3. Mixing procedure and time for premix UHPFRC.

Procedure	Duration (min)
Mix Dry-Mix Material	1
Add Superplasticizer and 2/3 Water	6
Add 1/3 Water	3
Add Fibres	1

All the cube specimens were cast in three layers with minor hand vibration to expel the air trapped in the fresh concrete. For plates, concrete was poured from the centre and let it flow freely towards the four edges (see Figure 2). To avoid the segregation of fibres, no vibration was applied.

Figure 2. Plate casting direction.

All the fresh concrete specimens were covered with a plastic film to prevent the early shrinkage due to water evaporation. After 24 h of hardening, all specimens were demolded and moved to a steam curing tank. The curing duration included 2 h for increasing the curing temperature to 90 °C and then another 48 h constant curing at 90 °C.

3. C-Shape Magnetic Probe Test

3.1. Probe Specification

A magnetic probe was manufactured based on Nunes' research [22]. The probe (Figure 3a) was made of a high frequency inductive Mn-Zn ferrite core wrapped by 350 turns of 0.9 mm diameter enameled copper wire, then tightened by black insulated rubber tape to protect the safety of users. The ferrite core used in this research was 76 mm tall, 93 mm long, and 30 mm wide. Detailed dimensions of the magnetic probe can be seen in Figure 3b.

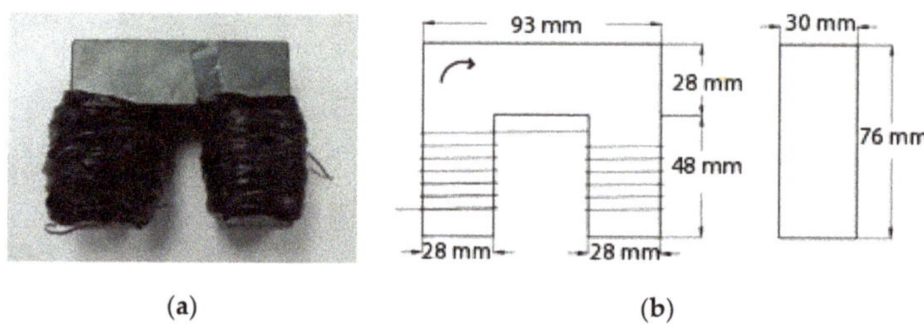

(a) (b)

Figure 3. (**a**) Appearance of the magnetic probe and (**b**) detailed dimensions of the ferrite core.

The inductive test was carried out after curing. The magnetic probe was placed on a smooth surface of the UHPFRC specimen and connected with a LCR meter with two clips. Tonghui TH2830 LCR meter was used to measure the magnetic inductance under 1 kHz with a test signal of 1 V. The variation of inductance of a single object was lower than ±0.01 mH under this testing condition.

3.2. Effective Depth Test Method

Non-ferromagnetic object like wood, plastic, and glass did not show any impact on the inductance data. By placing 100 mm × 150 mm non-ferromagnetic acrylic plates (Figure 4a) with different thicknesses (0, 2, 4, 8, 12, 15, 17, 20, 24, 28, 32, and 36 mm) between the specimens and the magnetic probe (Figure 4b), the magnetic inductance data can be measured and the relative magnetic permeability (RMP) μ can be calculated as:

$$\mu = \frac{L_0}{L_{air}} \tag{5}$$

where

L_0 magnetic inductance when placing the magnetic probe directly on the specimen;
L_{air} magnetic inductance when placing the magnetic probe in the air and far from any conductive object.

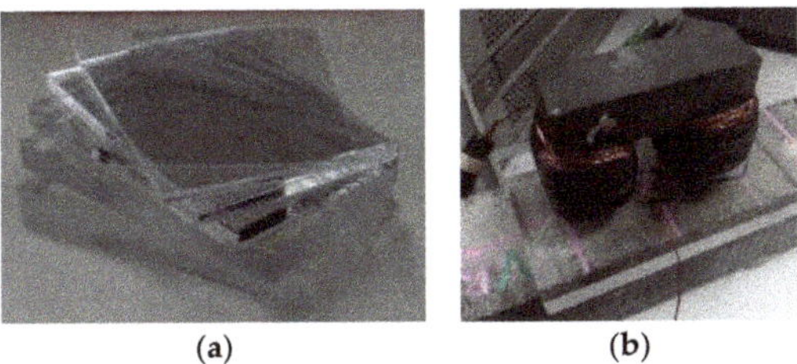

(a) (b)

Figure 4. (**a**) Acrylic plates used in the experiment and (**b**) experiment set up of an effective depth test.

However, since the initial relative magnetic permeability for each group was different, it was difficult to reflect on how the relative magnetic permeability decayed with the increase of thickness. Thus, an attenuation factor (AF) was introduced to describe the residual proportion of magnetic permeability. It can be calculated as:

$$AF_t = \frac{\mu_t - 1}{\mu_0 - 1} \times 100\% \tag{6}$$

where

μ_0 magnetic permeability when placing the probe directly on the specimen;
μ_t magnetic permeability at thickness t.

3.3. *Effective Depth Test Results*

The effective depth testing was conducted on 2% and 2.5% vol. UHPFRC. In total, 4 points (2 points for each group) were tested. Based on Equation (6), AF data of each testing point were calculated and the results are shown in Table 4. The relative magnetic permeability decreased with the increase of plate thickness. For specimens with lower fibre content, the relative magnetic permeability tended to drop quicker. All groups of AF data dropped below 10% for depths greater than 24 mm. This proved that the fibres 24 mm away from the testing surface had little effect on the relative magnetic permeability.

Table 4. Average attenuation factors calculated for each group.

t (mm)	2% vol.—A	2% vol.—B	2.5% vol.—A	2.5% vol.—B
0	100%	100%	100%	100%
2	75%	72%	77%	73%
4	57%	54%	57%	58%
8	40%	39%	39%	40%
12	23%	23%	24%	26%
15	19%	19%	16%	19%
17	14%	15%	13%	15%
20	10%	13%	12%	12%
24	6%	7%	7%	6%
32	3%	2%	4%	4%
36	3%	3%	3%	3%

The AF data presented in Figure 5 shows very similar exponential trends for all groups of testing points. It proved that the fibres closer to the testing surface had a more significant effect on the relative magnetic permeability, especially within the top 6 mm of the testing surface. After the top 6 mm, the AF dropped below 50%.

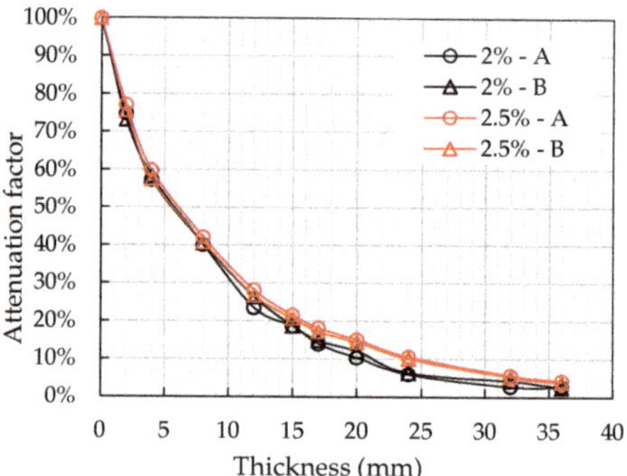

Figure 5. Attenuation factors of 2% and 2.5% vol. specimens at different thicknesses.

To get a theoretical expression of this relationship, a curve fitting analysis was conducted using MATLAB. The relationship between the testing depth and attenuation factor can be expressed by Equation (7) with an R-squared value equal to 0.995.

$$AF = e^{-0.115 \times t} \tag{7}$$

This test revealed the effective depth when using this particular probe. If any thin specimens are going to be tested in the future, AF can be used for correcting and calculating the real fibre volume content value.

3.4. Plate Test Method

The magnetic probe test of all the plates were tested right after finishing curing. The bottom surfaces during casting were used as the testing surface as they were smooth and flat. Considering the length of the magnetic probe was 93 mm, the 500 mm × 500 mm plates were labelled on a 9 × 9 grid from A1 to I9 at equal distances of 50 mm (Figure 6a). A paper testing map (Figure 6b) with 81 points highlighted in red was made to accurately locate the testing points.

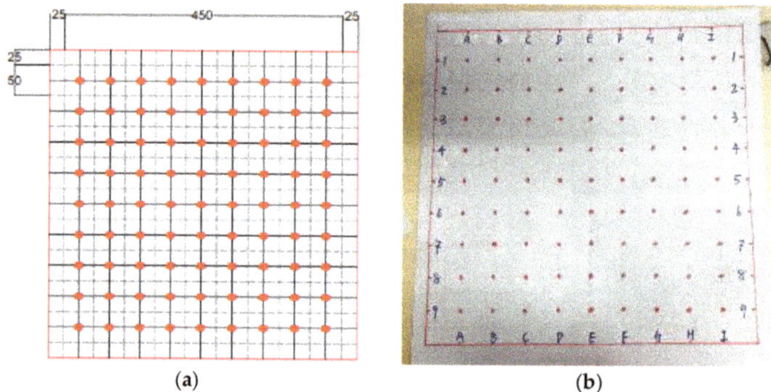

Figure 6. Testing area divisions of the UHPFRC plate. (**a**) Schematic graph (unit: mm) and (**b**) practical experimental design.

As shown in Figure 7, by placing the magnetic probe in two orthogonal directions (horizontally and vertically), the spatial distribution and orientation distribution at each red point were measured.

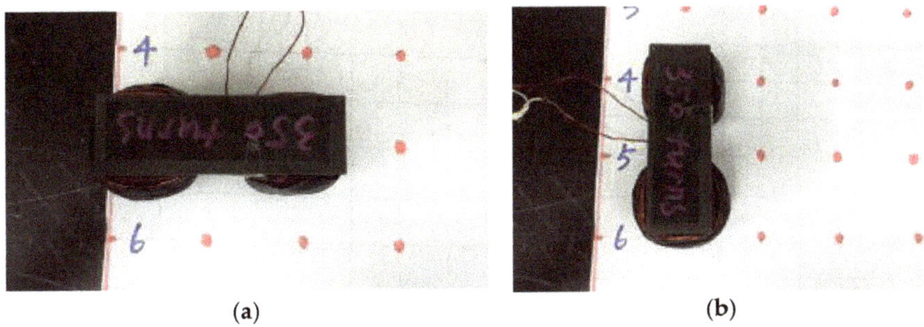

Figure 7. Testing directions. (a) Horizontal and (b) vertical.

The magnetic inductances of the red points in Figure 6 were recorded directly from the LCR meter. In total, 81 data points were collected for each plate. Air inductance was labelled as L_{air}. The magnetic inductance values measured in the horizontal and vertical directions were labelled as $L_{ij,x}$ and $L_{ij,y}$. All the magnetic inductance values were divided by the air inductance to get the relative magnetic permeability μ.

The average of relative magnetic permeability measured in two orthogonal directions is the indication of fibre volume content. Based on Equation (3), the average relative magnetic permeability on each red point can be calculated as Equation (8). Symbols i and j are used to represent the point location, e.g., $\mu_{11,x}$ represents the horizontally measured relative magnetic permeability at the point on the top left corner.

$$\mu_{ij,ave} = \frac{\mu_{ij,x} + \mu_{ij,y}}{2} \tag{8}$$

3.5. Plate Test Results

3.5.1. Fibre Spatial Distribution

Detailed mean and standard deviation (STD) of relative magnetic permeabilities for all plates are listed in Table 5. It can be seen that with the increase of fibre content, the relative magnetic permeability also increased. Due to the limited testing depth, the magnetic permeability only increased with the increase of plate thickness from 15 to 35 mm but no obvious difference between thicknesses from 35 to 50 mm.

Table 5. Mean and standard deviation (STD) of relative magnetic permeabilities of all plates.

Index	Fibre Volume Content	15 mm	20 mm	35 mm	50 mm
Mean	1%	1.031	1.035	1.038	1.043
	2%	1.059	1.060	1.074	1.075
	2.5%	1.069	1.078	1.095	1.088
Standard Deviation	1%	0.0004	0.0005	0.0002	0.0005
	2%	0.0005	0.0006	0.0006	0.0012
	2.5%	0.0016	0.0009	0.0008	0.0017

Combining the results from Table 5 and the results from Nunes et al. [22], a relationship between fibre volume content and theoretical relative magnetic permeability μ can be derived as shown in Figure 8. When fibre volume content is 0, the relative magnetic permeability value equals to 1, which represents the magnetic permeability of air. The R-squared value of 0.9987 shows a nearly perfect linear fitting.

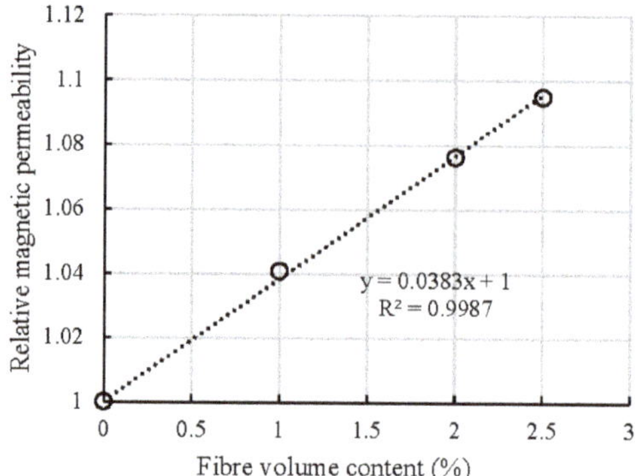

Figure 8. Relationship between fibre volume content and relative magnetic permeability.

Considering the effective depth of the magnetic probe, if relative magnetic permeability μ_{test} was obtained on a thin specimen, the attenuation factor should be applied to calculate the real fibre volume content. The relationship between real relative magnetic permeability μ_r and fibre volume content V_f is further developed into

$$\mu_r = \frac{\mu_{test} - 1}{1 - AF_t} + 1 \tag{9}$$

$$\mu_r = 0.0383 \times V_f + 1 \tag{10}$$

where AF_t represents the attenuation factor when the specimen's thickness is t. Combining Equations (9) and (10), the corrected relationship can be expressed as:

$$V_f = \frac{\mu_{test} - 1}{0.0383 \times (1 - AF_t)}, \tag{11}$$

By using Equation (11), real fibre volume contents can be derived and are indicated in Table 6. For all the 1% vol. plates, the fibre content fulfilled the designed requirement. For 2% and 2.5% vol., some plates possessed lower fibre volume content than the designed value.

Table 6. Mean and STD of fibre volume content of all plates.

Index	Fibre Volume Content	15 mm	20 mm	35 mm	50 mm
Mean	1%	1.0%	1.0%	1.0%	1.1%
	2%	1.9%	1.7%	1.9%	2.0%
	2.5%	2.2%	2.2%	2.5%	2.3%
Standard Deviation	1%	0.013%	0.016%	0.006%	0.014%
	2%	0.018%	0.018%	0.016%	0.030%
	2.5%	0.052%	0.026%	0.020%	0.045%

With the increase of fibre volume content, the standard deviation also increased, which revealed that fibre tended to distribute more non-uniformly. This effect was more severe with the 2.5% vol. UHPFRC plates. A possible reason was the fibre balling or gathering effect. Although there were variations within each plate, the differences were very small compared to the total fibre volume content. Therefore, the plate can be considered almost uniformly spatially distributed.

The first coloured contour plot in Figure 9a describes the fibre distribution of all plates at a unified scale. Colours ranging from blue to red represent the differences of fibre volume content. It can be seen directly that plates 2%–20 mm and 2.5%–15 mm have a lower fibre volume content than the designed fibre volume content. The detailed fibre distribution cannot be visualized, since the range of data was too wide in the coloured contour plots. Thus, a greyscale contour plot is given in Figure 9b as a comparison. The darker shading indicates a lower fibre volume content. It can be seen that there was no obvious fibre spatial distribution trend in the middle area of each plate, only the four boundaries appeared to have a lower fibre volume content. This mainly results from the limitation of the testing area (boundary effect).

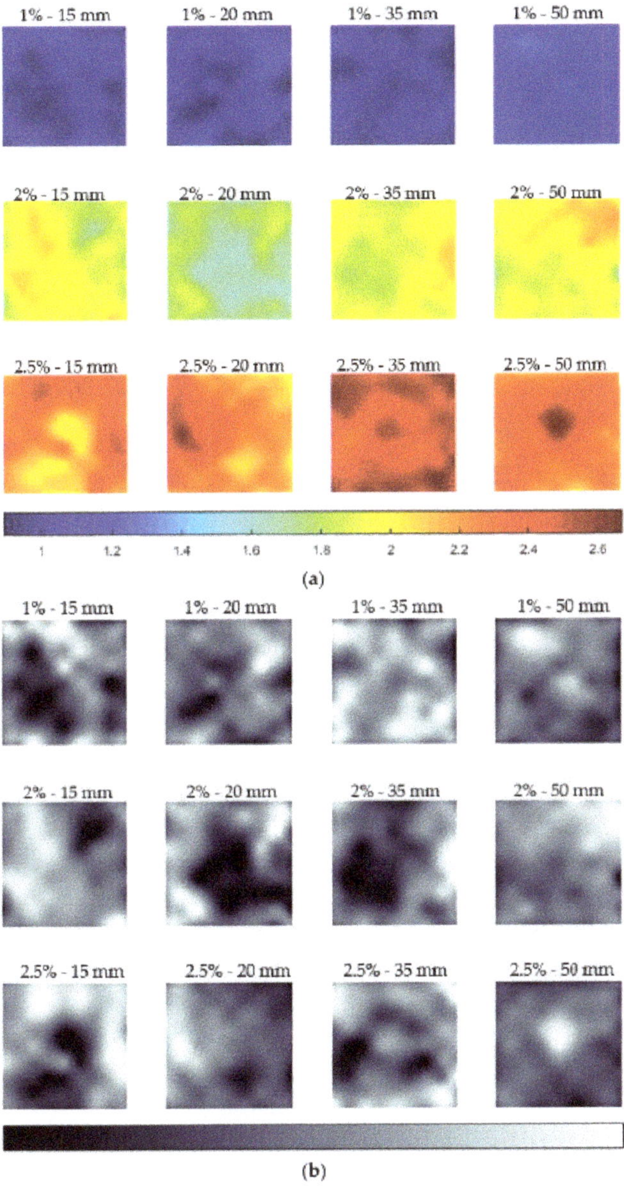

Figure 9. Fibre volume content distribution of all plates. (**a**) With a scale as the fibre volume content in percentage and (**b**) with a general non-unified scale.

3.5.2. Fibre Orientation Distribution

Based on previous literature by Nunes et al. [22], the fibre orientation was expressed by an orientation indicator ρ_Δ. The orientation indicator of the red points can be calculated as:

$$\rho_\Delta = \frac{\mu_{ij,y} - \mu_{ij,x}}{2\left(\mu_{ij,\ ave} - 1\right)} \quad (12)$$

Nunes et al. [22] found the orientation indicator had a sinusoidal relationship with the fibre orientation angle, but an analytical expression between the fibre orientation angle and orientation indicator ρ_Δ was not presented. Through further derivation, it was found that the orientation indicator is a function of polynomial terms of $\cos(\varphi)$. For example, the full expression of orientation indicator 1% vol. UHPFRC in terms of the orientation angle φ according to Nunes et al. [22] can be expressed as:

$$\rho_{\Delta,1\%} = \frac{-1076.68\cos^6(\varphi) + 1615.03\cos^4(\varphi) + 4.03\times10^7\cos^2(\varphi) - 2.02\times10^7}{\begin{array}{c}(-0.96\ \cos^6(\varphi) + 5.68\times10^5\ \cos^4(\varphi) - 5.68\times10^5\cos^2(\varphi)\\ -6.41\times10^7 + 0.48\cos^2(\varphi))\end{array}} \quad (13)$$

After numerical analysis, only the constant terms and the $\cos^2(\varphi)$ term on the numerator were found to be critical to the value of fibre orientation indicator ρ_Δ. Therefore, Equation (13) can be further simplified to:

$$\rho_{\Delta,1\%} \approx -0.63\ \times\cos^2(\varphi) + 0.315 = 0.315\times\ \cos(2\varphi) \quad (14)$$

Figure 10a shows the comparison between the original fibre orientation indicator calculated using Equation (13) and the simplified orientation indicator calculated using Equation (14). No obvious difference can be observed from 0 to 90°. Therefore, the original equation can be replaced with the simplified equation. There also works with other fibre volume percentages.

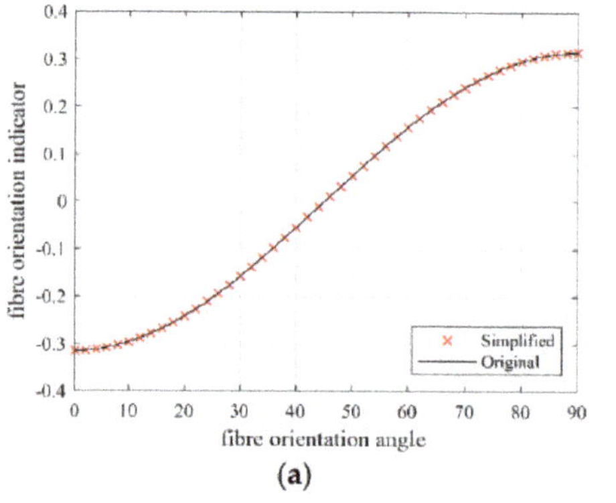

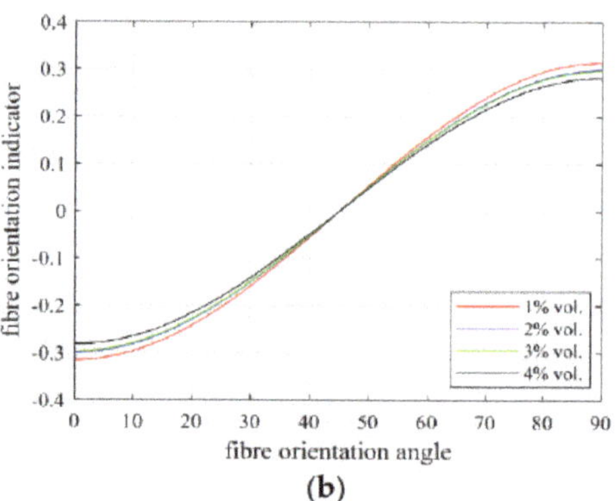

Figure 10. Relationship between fibre orientation angle (unit: degree) and fibre orientation indicator. (**a**) Comparison between simplified and original equation and (**b**) comparison between different fibre content.

The relationship for other fibre percentage is approximately equal ($\approx$) to:

$$\rho_{\Delta,2\%} \approx -0.299\times\ \cos(2\varphi) \quad (15)$$

$$\rho_{\Delta,3\%} \approx -0.297\ \times\cos(2\varphi) \quad (16)$$

$$\rho_{\Delta,4\%} \approx -0.282\ \times\cos(2\varphi) \quad (17)$$

This relationship was drawn in Figure 10b. It can be seen that the fibre volume content did not have a significant effect on the fibre orientation indicator, especially when the fibre orientation angle was around 45°. Generally, the fibre orientation indicator can be expressed as:

$$\rho_{\Delta} = a \times \cos(2\varphi) \tag{18}$$

where the fibre orientation indicator coefficient slightly ranges around −0.3 depending on the fibre volume content.

By using Equation (18), the fibre orientation distributions can be characterized. Figure 11 shows the fibre orientation distribution of all plates. Instead of contour plots, the fibre orientation angle is represented by dots in different colours. The fibre orientation ranged from 0 to 90° from the horizontal axis. It can be seen that fibres tended to orient at 0° at the top and bottom boundaries, while orienting at 90° along the left and right boundaries. There are two possible explanations: firstly, fibre tends to align their orientation to the mould due to the wall effect of the plate mould boundary; and secondly, fibre contents at the four boundaries were lower compared to other parts, which resulted from the testing method. The vertical testing value at the top and bottom points and the horizontal testing value at the furthest left and right points were lower.

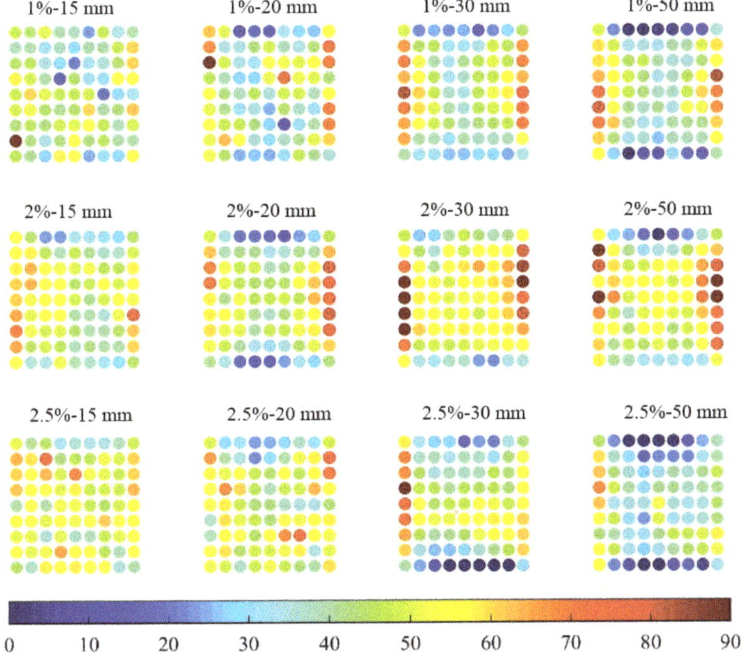

Figure 11. Fibre orientation angle distribution of all plates.

Figure 12 shows the distribution of fibre orientation angle based on 972 (9 × 9 × 12) sample points. It can be seen from the graph that under this specific casting method, the distribution of the fibre orientation angle generally follows a normal distribution and most fibres orient at an angle between 40° and 50°. The normal distribution function was calculated based on the mean and standard deviation values of fibre orientation angle, which can be represented as:

$$f(\varphi) = \frac{1}{\sqrt{2\pi}\sigma} \exp\left(\frac{(\varphi - \zeta)^2}{2\sigma^2}\right) \tag{19}$$

where the mean value $\sigma = 45.60°$ and standard deviation $\zeta = 15.32°$.

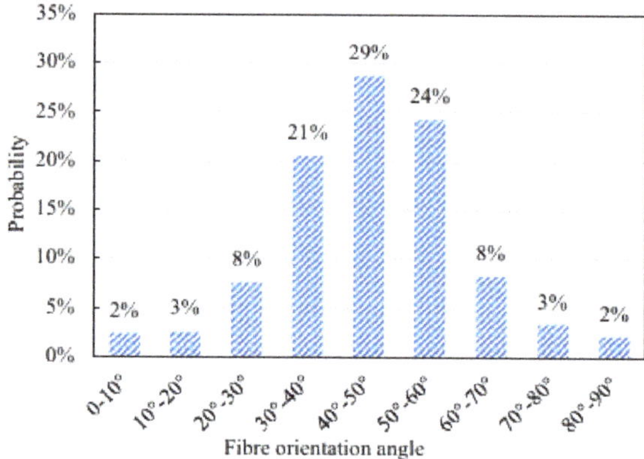

Figure 12. Distribution of fibre orientation angles.

Detailed mean and STD value of fibre orientation angles can be seen in Table 7. The increase of fibre content and fibre distribution do not have a direct relationship with the fibre orientation angle. Due to the wall effect, the coefficient of variance for the fibre orientation angle was higher comparing to the fibre volume content.

Table 7. Mean and STD of fibre orientation angles of all plates.

Index	Fibre Volume Content	15 mm	20 mm	35 mm	50 mm
Mean	1%	43	43	45	41
	2%	48	46	54	51
	2.5%	51	48	43	35
Standard Deviation	1%	2.8	3.1	3.0	2.6
	2%	2.2	2.4	4.0	3.0
	2.5%	2.0	1.2	3.3	3.7

4. Mechanical Test Results

4.1. Compressive Test

The compressive test was carried out on 0, 1%, 2%, and 2.5% specimens 5 days after curing. Six specimens were tested from each group. With the increase of fibre content, the compressive strength steadily increased in an almost linear trend. Detailed compressive test results can be seen in Table 8.

Table 8. Compressive strength of different groups of 100 mm cube specimens.

Fibre Volume Content	Average	STD	COV
0	87.8	4.7	5.4%
1%	134.1	5.3	4.0%
2%	141.4	4.4	3.1%
2.5%	149.5	5.4	3.6%

4.2. Flexural Test

All the 500 mm × 500 mm × 15 mm, 500 mm × 500 mm × 20 mm, 500 mm × 500 mm × 35 mm, and 500 mm × 500 mm × 50 mm plates were firstly cut into four sections by water blade to maintain the accuracy of dimension (Figure 13a,b). Then, the 250 mm × 250 mm × 15 mm, 250 mm × 250 mm × 20 mm,

250 mm × 50 mm × 35 mm, and 250 mm × 250 mm × 50 mm plates were further cut into 200 mm × 50 mm × 15 mm, 200 mm × 50 mm × 20 mm, 200 mm × 50 mm × 35 mm, and 200 mm × 50 mm × 50 mm. The detailed cutting scheme can be seen in Figure 13c. Beam No. 1–8 were tested in this research.

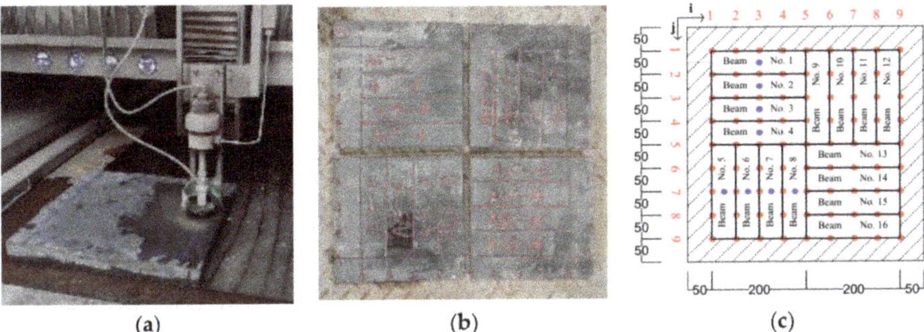

Figure 13. Concrete plate cutting. (**a**) Water blade cutting; (**b**) concrete plate after cutting; and (**c**) schematic graph.

The flexural test were conducted 5–7 days after curing. The experiment was carried out on a 3-ton universal testing machine at a constant deflection control speed of 0.3 mm/min. Taking a 15 mm thick beam as an example, the three-point bending test setup can be seen in Figure 14. The effective span was 150 mm. The experiment was terminated once the load dropped below 50% of the peak load. The displacement movement of the machine was used to plot the load–displacement diagram

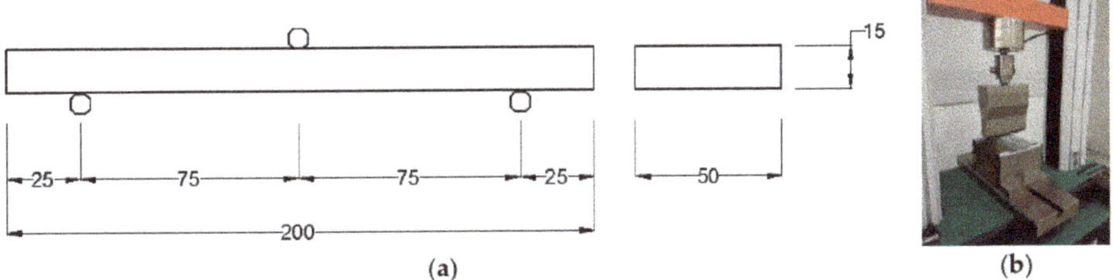

Figure 14. Three-point bending test for the UHPFRC beam. (**a**) Schematic graph (unit: mm) and (**b**) experimental set-up.

The flexural strength for a three-point bending test can be calculated from Equations (20) to (22).

$$M = \frac{F_f \times L}{4} \tag{20}$$

$$W = \frac{BH^2}{6} \tag{21}$$

$$f_f = \frac{M}{W} \tag{22}$$

where

F_f peak flexural load;
L distance between support and the nearby loading point;
M external moment;
W section modulus;

B width of the beam cross section;
H height of the beam cross section;
f_f flexural strength.

Average first crack flexural strength and peak flexural strength of eight beams of each plate can be seen in Table 9. The merged cell on the right side shows the average value of four plates. For UHPC, no micro or macrocracks were observed before reaching peak load. The first crack strength equalled the peak flexural strength.

Table 9. First crack flexural strength and peak flexural strength (unit: MPa).

Fibre Volume Content—Thickness	First Crack Flexural Strength		Peak Flexural Strength	
0–15	20.0	18.5	20.0	18.5
0–20	18.6		18.6	
0–35	15.0		15.0	
0–50	20.3		20.3	
1%–15	15.1	16.3	16.0	17.4
1%–20	13.9		16.0	
1%–35	20.2		21.0	
1%–50	15.8		16.6	
2%–15	18.5	18.6	22.4	22.6
2%–20	16.8		21.4	
2%–35	19.6		25.0	
2%–50	19.4		21.8	
2.5%–15	18.0	17.9	22.2	22.6
2.5%–20	16.2		20.9	
2.5%–35	18.3		24.1	
2.5%–50	18.9		23.1	

5. Correlation between Magnetic Probe Test and Mechanical Test

To correlate the results between the fibre distribution test and the mechanical test for each beam, the relative inductance values of the blue points in Figure 13c are also needed. The value is derived from the nearby two red points. For the blue points of Beams 1–4, the average relative magnetic permeability can be expressed as:

$$\mu_{ij,\,ave} = \frac{\mu_{ij,x} + \mu_{ij,y} + \mu_{i(j+1),x} + \mu_{i(j+1),y}}{4} \tag{23}$$

For Beams 5–8, it can be calculated by:

$$\mu_{ij,\,ave} = \frac{\mu_{ij,x} + \mu_{ij,y} + \mu_{(i+1)j,x} + \mu_{(i+1)j,y}}{4} \tag{24}$$

For the red points in Figure 13c, the fibre orientation indicator can be calculated straightly using Equation (12). For the blue points, the orientation indicator can be derived from the nearby two red points. For the blue points on Beams 1–4, the orientation indicator ρ_Δ can be derived as:

$$\rho_\Delta = \frac{\mu_{ij,y} + \mu_{i(j+1),y} - \mu_{ij,x} - \mu_{i(j+1),x}}{\mu_{ij,y} + \mu_{i(j+1),y} + \mu_{ij,x} + \mu_{i(j+1),x} - 4} \tag{25}$$

$$\rho_\Delta = \frac{\mu_{ij,y} + \mu_{(i+1)j,y} - \mu_{ij,x} - \mu_{(i+1)j,x}}{\mu_{ij,y} + \mu_{(i+1)j,y} + \mu_{ij,x} + \mu_{(i+1)j,x} - 4} \tag{26}$$

Figure 15 shows four typical load-deflection curves for 1% and 2% vol. UHPFRC. Both load-softening and load-hardening behaviours can be observed. Statistically, only 9 out of

32 1% vol. UHPFRC beams had obvious strain hardening behaviour. For 2% vol. UHPFRC, 27 out of 32 beams had load-hardening behaviour after the first crack. For 2.5% vol. UHPFRC beams, 29 out of 31 beams had load-hardening performance.

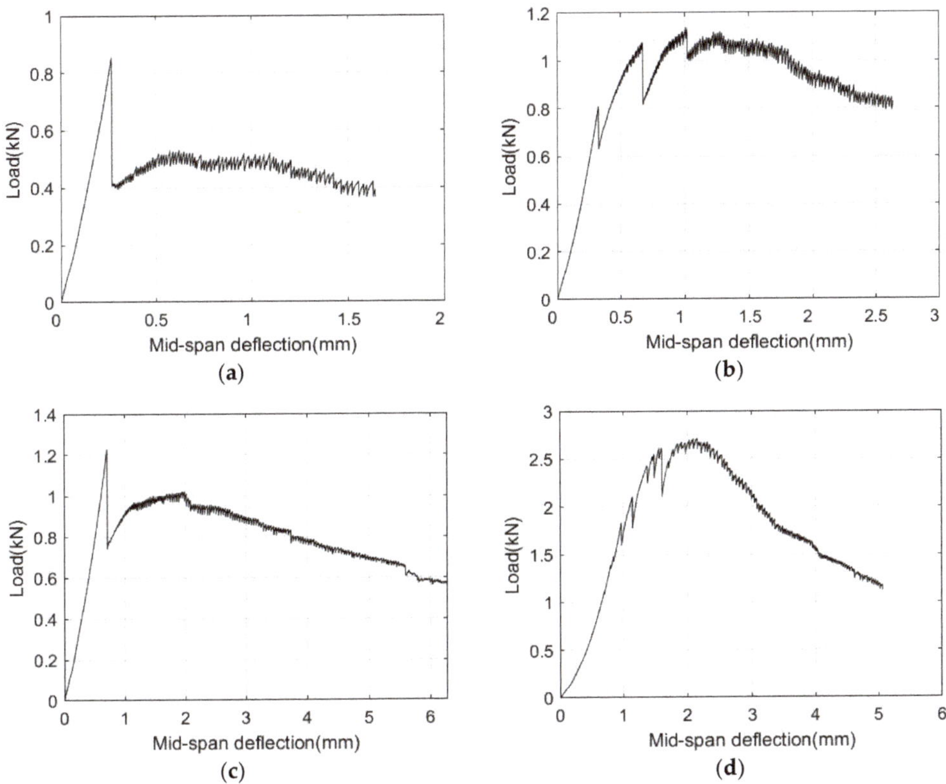

Figure 15. Load–deflection curves for 200 mm × 50 mm × 20 mm UHPFRC beams. (**a**) 1%-15-s6; (**b**) 1%-15-s4; (**c**) 2%-20-s8; and (**d**) 2%-20-s5.

Apart from the tensile property of the UHPC matrix, both the fibre orientation and fibre volume content can determine whether the beam performs a load-hardening behaviour or not. Initially in the uniaxial tensile test, fibres were fully bonded, and the tensile load was mainly carried by the concrete matrix. Thus, the first crack tensile strength mainly depended on the tensile properties of the concrete matrix, which agrees with previous researchers [28,29]. After the concrete cracks, concrete hardly sustained any loads, but fibres were not pulled-out yet due to the static frictional force τ between fibres and the concrete matrix. With the increase of uniaxial tensile load, the static frictional force also increased. Whether the beam had a load-hardening or load-softening performance depends on whether the whole fibre–concrete interfaces can provide enough frictional force, while the frictional force is related to the roughness of the interface, number of fibres (fibre volume content), and effective embedment length (fibre orientation). As can be seen in Figure 16, the effective static frictional force $f_{st,eff}$ carried by one fibre can be estimated as:

$$f_{st} = \pi \times d_f \times l_{em} \times \tau \tag{27}$$

$$f_{st,eff} = f_{st} \times \cos\varphi = \pi \times d_f \times l_{em} \times \tau \times \cos\varphi \tag{28}$$

where

f_{st} static frictional force;
$f_{st,eff}$ effective static frictional force;
d_f fibre diameter;
l_{em} effective embedment length;
τ static frictional strength;
φ fibre orientation angle.

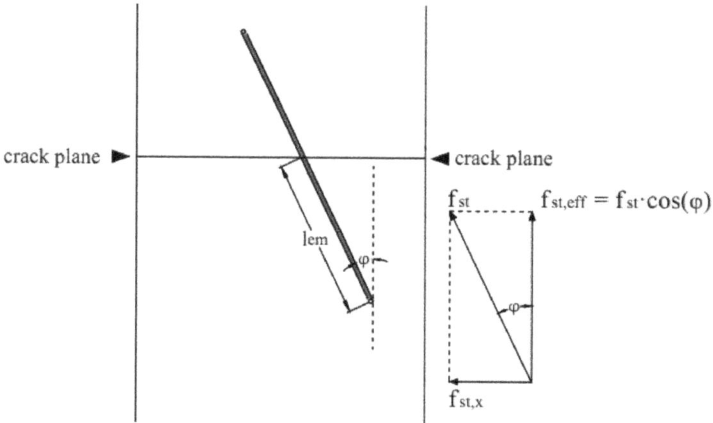

Figure 16. Effective embedment length of fibres inside the concrete matrix in the uniaxial tensile test in fibre activation stage.

The total effective static frictional force $f_{st,n}$ can be calculated as:

$$f_{st,n} = f_{st,eff} \times n_f = \pi \times d_f \times l_{em} \times \tau \times \cos\varphi \times n_f \tag{29}$$

Assuming the static frictional strength is constant in Equation (29), it can be seen that the total effective static frictional force has a linear relationship with $V_f \cos(\varphi)$ and number of fibres across the crack plane. Figure 17 shows the relationship between $V_f \cos(\varphi)$ and the peak flexural strength. For 1% UHPFRC, there is no obvious trend, mainly owing to the insufficient frictional force to support load-hardening performance, therefore, the peak flexural strength was determined by the properties of the concrete matrix. A linear relationship can be found for 2% and 2.5% UHPFRC. If the concrete matrix w consistent, the relationship between the uniaxial tensile strength $f_{t,\varphi 1}$, $f_{t,\varphi 2}$ at different fibre orientation angles φ_1, φ_2 can be calculated as:

$$\frac{f_{t,\varphi 1}}{\cos(\varphi_1) \times V_{f1}} = \frac{f_{t,\varphi 2}}{\cos(\varphi_2) \times V_{f2}} \tag{30}$$

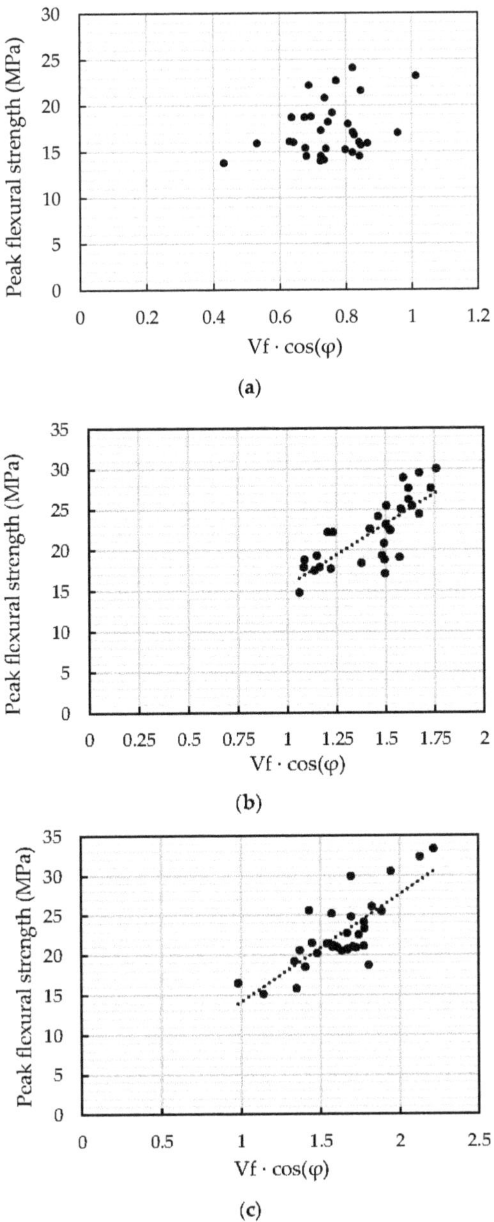

Figure 17. Relationship between $V_f \times \cos(\varphi)$ and peak flexural strength. (**a**) 1% vol. UHPFRC; (**b**) 2% vol. UHPFRC; and (**c**) 2.5% vol. UHPFRC.

6. Conclusions

This research focused on investigating fibre distribution of UHPFRC using C-shape magnetic probe. The following conclusions can be drawn.

- The effective testing depth of the C-shape magnetic probe was firstly determined by curve fitting analysis using MATLAB. An exponential equation was derived to alter the relative magnetic permeability value, and to correlate with the real fibre volume content for elements with different thicknesses.

- The magnetic permeability appeared to be linearly correlated with fibre volume content. With the increase of fibre volume content, the standard deviation of magnetic permeability also increased, which revealed a more non-uniform distribution of fibres.
- The analytical solution for deriving the fibre orientation indicator based on fibre orientation angle was derived. A fibre orientation indicator coefficient a was determined to be around −0.3.
- With a low fibre amount or large fibre orientation angle, the peak flexural strength and uniaxial tensile strength were more related to the tensile property of the concrete matrix. No strain hardening behaviour occurred after concrete cracks.
- With a high fibre amount or a small fibre orientation angle, the peak flexural strength and uniaxial tensile strength were closely affected by the static frictional force between the concrete matrix and fibres. Thus, strain hardening behaviour occurred. Results show that the uniaxial tensile strength had a linear relationship with the $V_f \cos(\varphi)$ value.

Author Contributions: Conceptualization, J.X. and L.L.; methodology, J.X. and L.L.; formal analysis, L.L. and J.X.; writing—original draft preparation, L.L.; writing—review and editing, J.X., C.C., and S.J.; supervision, J.X., C.C., and S.J.; All authors have read and agreed to the published version of the manuscript.

Funding: This research was funded by Suzhou Municipal Construction Research Centre, grant number 2019-13 and the APC was funded by Xi'an Jiaotong-Liverpool University.

Acknowledgments: The authors would like to thank Jiangxi Beirong Circular Materials Company for providing the dry-mix UHPFRC material in this investigation. The views and findings reported here are those of the writers alone, and not necessarily the views of sponsoring agencies.

Conflicts of Interest: The authors declare no conflict of interest.

References

1. Shi, C.; Mo, Y.L. (Eds.) *High-Performance Construction Materials: Science and Applications*; World Scientific: Singapore, 2008.
2. Fehling, E.; Schmidt, M.; Walraven, J.C.; Leutbecher, T.; Fröhlich, S. (Eds.) *Ultra-High Performance Concrete UHPC: Fundamentals —Design—Examples*; Ernst & Sohn: Berlin, Germany, 2014.
3. SINTEF Building and Infrastructure. *Ultra High Performance Fibre Reinforced Concrete (UHPFRC)—State of the Art*; COIN Project report; National Precast Concrete Association: Carmel, CA, USA, 2012.
4. Rebentrost, M.; Wight, G. *Experience and Applications of Ultra-High Performance Concrete*; Kassel University Press GmbH: Kassel, Germany, 2008; p. 10.
5. National Precast Concrete Association. *Ultra High Performance Concrete (UHPC) Guide to Manufacturing Architectural Precast UHPC Elements*; National Precast Concrete Association: Carmel, CA, USA, 2013.
6. Scott, D.A.; Long, W.R.; Moser, R.D.; Green, B.H.; O'Daniel, J.L.; Williams, B.A.; Graybeal, B.; Sritharan, S.; Wille, K. Influence of Steel Fiber Size and Shape on Quasi-Static Mechanical Properties and Dynamic Impact Properties of Ultra-High Performance Concrete. In Proceedings of the 1st International Interactive Symposium on UHPC, Des Moines, IA, USA, 18–20 July 2016. [CrossRef]
7. Li, V.C. Tailoring ECC for Special Attributes: A Review. *Int. J. Concr. Struct. Mater.* **2012**, *6*, 135–144. [CrossRef]
8. Yoo, D.-Y.; Kang, S.-T.; Yoon, Y.-S. Effect of fiber length and placement method on flexural behavior, tension-softening curve, and fiber distribution characteristics of UHPFRC. *Constr. Build. Mater.* **2014**, *64*, 67–81. [CrossRef]
9. Edgington, J.; Hannant, D.J. Steel fibre reinforced concrete. The effect on fibre orientation of compaction by vibration. *Mater. Struct.* **1972**, *5*, 41–44. [CrossRef]
10. Yang, L.; Yan, Y.; Ran, Z.; Liu, Y. A new method for generating random fibre distributions for fibre reinforced composites. *Compos. Sci. Technol.* **2013**, *76*, 14–20. [CrossRef]
11. Sun, X.; Lasecki, J.; Zeng, D.; Gan, Y.; Su, X.; Tao, J. Measurement and quantitative analysis of fiber orientation distribution in long fiber reinforced part by injection molding. *Polym. Test.* **2015**, *42*, 168–174. [CrossRef]

12. Švec, O.; Žirgulis, G.; Bolander, J.E.; Stang, H. Influence of formwork surface on the orientation of steel fibres within self-compacting concrete and on the mechanical properties of cast structural elements. *Cem. Concr. Compos.* **2014**, *50*, 60–72. [CrossRef]
13. Campo, M.; Woo, L.; Mason, T.; Garboczi, E. Frequency-Dependent Electrical Mixing Law Behavior in Spherical Particle Composites. *J. Electroceram.* **2002**, *9*, 49–56. [CrossRef]
14. Ozyurt, N.; Mason, T.O.; Shah, S.P.; Özyurt, N. Non-destructive monitoring of fiber orientation using AC-IS: An industrial-scale application. *Cem. Concr. Res.* **2006**, *36*, 1653–1660. [CrossRef]
15. Torrents, J.M.; Mason, T.O.; Peled, A.; Shah, S.P.; Garboczi, E.J. Analysis of the impedance spectra of short conductive fiber-reinforced composites. *J. Mater. Sci.* **2001**, *36*, 4003–4012. [CrossRef]
16. Bordelon, A.; Roesler, J. Spatial distribution of synthetic fibers in concrete with X-ray computed tomography. *Cem. Concr. Compos.* **2014**, *53*, 35–43. [CrossRef]
17. Ponikiewski, T.; Gołaszewski, J.; Rudzki, M.; Bugdol, M.N. Determination of steel fibres distribution in self-compacting concrete beams using X-ray computed tomography. *Arch. Civ. Mech. Eng.* **2015**, *15*, 558–568. [CrossRef]
18. Cavalaro, S.H.P.; López-Carreño, R.-D.; Torrents, J.M.; Aguado, A. Improved assessment of fibre content and orientation with inductive method in SFRC. *Mater. Struct.* **2015**, *48*, 1859–1873. [CrossRef]
19. Torrents, J.M.; Blanco, A.; Pujadas, P.; Aguado, A.; Juan-García, P.; Sánchez-Moragues, M. Ángel Inductive method for assessing the amount and orientation of steel fibers in concrete. *Mater. Struct.* **2012**, *45*, 1577–1592. [CrossRef]
20. Faifer, M.; Ottoboni, R.; Toscani, S. A compensated magnetic probe for steel fiber reinforced concrete monitoring. In Proceedings of the 2010 IEEE Sensors Conference, Kona, HI, USA, 1–4 November 2010; pp. 698–703.
21. Faifer, M.; Ottoboni, R.; Toscani, S.; Ferrara, L. Nondestructive Testing of Steel-Fiber-Reinforced Concrete Using a Magnetic Approach. *IEEE Trans. Instrum. Meas.* **2010**, *60*, 1709–1717. [CrossRef]
22. Nunes, S.; Pimentel, M.; Carvalho, A.S. Non-destructive assessment of fibre content and orientation in UHPFRC layers based on a magnetic method. *Cem. Concr. Compos.* **2016**, *72*, 66–79. [CrossRef]
23. Nunes, S.; Pimentel, M.; Ribeiro, F.; Milheiro-Oliveira, P.; Carvalho, A. Estimation of the tensile strength of UHPFRC layers based on non-destructive assessment of the fibre content and orientation. *Cem. Concr. Compos.* **2017**, *83*, 222–238. [CrossRef]
24. Abrishambaf, A.; Pimentel, M.; Nunes, S. A meso-mechanical model to simulate the tensile behaviour of ultra-high performance fibre-reinforced cementitious composites. *Compos. Struct.* **2019**, *222*, 110911. [CrossRef]
25. Bittencourt, T.; Frangopol, D.; Beck, A. Maintenance, Safety and Management, Maintenance, Monitoring, Safety, Risk and Resilience of Bridges and Bridge Networks. In Proceedings of the 8th International Conference on Bridge Maintenance, Safety and Management (IABMAS 2016), Foz do Iguaçu, Brazil, 26–30 June 2016; CRC Press: Boca Raton, FL, USA, 2016.
26. Meng, W.; Khayat, K.H. Effect of Hybrid Fibers on Fresh Properties, Mechanical Properties, and Autogenous Shrinkage of Cost-Effective UHPC. *J. Mater. Civ. Eng.* **2018**, *30*, 04018030. [CrossRef]
27. Zhang, L.; Liu, J.; Liu, J.; Zhang, Q.; Han, F. Effect of Steel Fiber on Flexural Toughness and Fracture Mechanics Behavior of Ultrahigh-Performance Concrete with Coarse Aggregate. *J. Mater. Civ. Eng.* **2018**, *30*, 04018323. [CrossRef]
28. Doo, D.-Y.; Kim, S.-W.; Park, J.-J. Comparative flexural behavior of ultra-high-performance concrete reinforced with hybrid straight steel fibers. *Constr. Build. Mater.* **2017**, *132*, 219–229. [CrossRef]
29. Park, J.-J.; Yoo, D.-Y.; Park, G.-J.; Kim, S.-W. Feasibility of Reducing the Fiber Content in Ultra-High-Performance Fiber-Reinforced Concrete under Flexure. *Materials* **2017**, *10*, 118. [CrossRef] [PubMed]

Article

Study of Combined Multi-Point Constraint Multi-Scale Modeling Strategy for Ultra-High-Performance Steel Fiber-Reinforced Concrete Structures

Zuohua Li [1,2], Zhihan Peng [1,2] and Jun Teng [1,2,*]

1 Shenzhen Key Lab of Urban & Civil Engineering Disaster Prevention & Reduction, Harbin Institute of Technology, Shenzhen 518055, China; lizuohua@hit.edu.cn (Z.L.); hayespeng@gmail.com (Z.P.)

2 School of Civil and Environment Engineering, Harbin Institute of Technology, Shenzhen 518055, China

* Correspondence: tengj@hit.edu.cn; Tel.: +86-755-26033806; Fax: +86-755-26033509

Received: 26 October 2020; Accepted: 22 November 2020; Published: 24 November 2020

Abstract: Compared with normal strength concrete (NSC), ultra-high-performance steel fiber-reinforced concrete (UHPFRC) shows superior performance. The concrete damage plasticity (CDP) model in ABAQUS can predict the mechanical properties of UHPFRC components well after calibration. However, the simulation of the whole structure is seriously restricted by the computational capability. In this study, a novel multi-scale modeling strategy for UHPFRC structure was proposed, which used a calibrated CDP model. A novel combined multi-point constraint (CMPC) was established by the simultaneous equations of displacement coordination and energy balance in different degrees of freedom of interface nodes. The advantage is to eliminate the problem of the tangential over-constraint of displacement coordination equation at the interface and to avoid stress iteration of the energy balance equation in the plastic stage. The expressions of CMPC equations of typical multi-scale interface connection were derived. The multi-scale models of UHPFRC components under several load cases were established. The results show that the proposed strategy can well predict the strain distribution and damage distribution of UHPFRC while significantly reducing the number of model elements and improving the computational efficiency. This study provides an accurate and efficient finite element modeling strategy for the design and analysis of UHPFRC structures.

Keywords: ultra-high-performance steel fiber-reinforced concrete; multiscale finite element modeling; multi-point constraint; multi-scale interface connection; concrete damage plasticity model; ABAQUS

---

1. Introduction

Concrete is currently the most widely used building material. Although many structures are built with concrete, the use of normal strength concrete (NSC) still has some limitations, such as low tensile strength and low ductility. Improving the mechanical properties of concrete to obtain higher strength and higher ductility has been widely of concern. Ultra-high-performance steel fiber concrete (UHPFRC) is a new type of fiber concrete, with high strength, fracture toughness, and ductility. Its compressive strength and tensile strength are generally over 150 MPa and 7 MPa, respectively [1–5], and even the tensile strength can reach 15 MPa [6]. Ultra-high-performance concrete has been applied and investigated in many kinds of engineering structures, such as concrete structures [7–9], seismic design [10,11], etc. For the material level, numbers of studies have been conducted on the influence of fiber types, fiber orientations, geometric shapes, dosages, and other factors on the mechanical properties of ultra-high-performance concrete [12–19]. Numerous experimental studies have been carried out on high-performance concrete components, which include full-size prestressed beams [20–22], reinforced beams [23–26], columns [27], slabs [28], etc.

Extensive tests are required at the material and structural levels in order to develop standard analytical procedures and design specifications for UHPFRC, which will take a lot of time and cost. Therefore, verifying the concrete material models in the existing finite element software by conducting a limited number of well-formulated tests on the material and structure levels is a way to save time and cost. The verified concrete material model and finite element modeling method can be used to establish extended analysis of various design parameters. In addition, the influences of changes in geometry, load cases, and reinforcement on mechanical properties of UHPFRC can be investigated. The finite element software ABAQUS is equipped with the concrete damage plasticity (CDP) model developed for NSC, which is a mature and reliable tool for predicting the mechanical behavior of NSC [29–31]. Compared with NSC, the material properties of UHPFRC have higher tensile strength and ductility, which makes the shape of a material constitutive curve substantially different from NSC. In order for CDP model to be used to simulate UHPFRC, the parameters of CDP model need to be calibrated. Tysmans et al. [32] used CDP model to simulate the behavior of high-performance fiber concrete composites under biaxial tension. Mahmud et al. [33] and Singh et al. [34] calibrated the CDP model through the UHPFRC material test and used the calibrated model to simulate the test results of the UHPFRC beam [24]. It was reported that the calibrated CDP model can accurately and effectively predict the load-displacement curves and plastic damage distributions of UHPFRC components.

Similar to the investigations of NSC, the investigations of UHPFRC need to be developed to the structural level as well as the material and component level. However, it is very expensive to establish a full-scale structural test, which is seriously restricted by the test conditions. When the calibrated CDP model is used to simulate a single UHPFRC component, the reduction of mesh size and the increase of number of elements will significantly increase the calculation time, while larger mesh size will lead to convergence problems [34]. Therefore, it is difficult to use solid elements to simulate all of the UHPFRC of the whole structure. Fortunately, the multi-scale finite element simulation strategy can solve this problem. The simulation strategy uses solid elements to simulate the key parts of the structure that need to be paid attention to, and adopts the macro-scale elements such as truss or beam elements for the other parts. Its advantage is to use limited computing resources to ensure the requirements for simulation accuracy and to improve computational efficiency. So the simulation strategy has been well applied in structural failure analysis, seismic design, optimization of structural system, etc. [35–39]. The key problem of the multi-scale finite element simulation strategy is to establish an accurate interface-coupling constraint relationship, so as to ensure the scientific and reasonable coordination between different scale elements. The multi-point constraint method is based on the relations of displacement coordination [40] or energy balance [41] between macro-scale and micro-scale elements at the interface, and the constraint equations containing the degrees of freedom of nodes of different scale elements are established at the interface [42,43]. However, a single multi-point constraint relation has the limitation that the stress state and deformation of the connection interface appear distorted after the material enters the plastic stage [44].

In order to promote the development of finite element simulation of UHPFRC structure, a novel multi-scale finite element modeling strategy was proposed in this study. A novel combined multi-point constraint (CMPC) based on displacement coordination and energy balance was established, aiming at the problems of the tangential over-constraint and the requirements for nonlinear stress iteration existing in the single multi-point constraint method. The nonlinear constitutive relationship of UHPFRC is considered. The multi-scale models of UHPFRC components under various load cases were established in the finite element software ABAQUS. The comparative analysis results show that the proposed multi-scale modeling strategy can well predict the strain distribution and damage distribution of UHPFRC components while significantly reducing the number of model elements and improving the computational efficiency. This study provides an accurate and efficient finite element modeling strategy for the design and analysis of UHPFRC structure, which can promote the application and development of UHPFRC in the construction industry.

2. Multi-Scale Modeling Strategy—Material Models

2.1. Calibrated Concrete Damage Plasticity (CDP) Model

The concrete damage plasticity (CDP) model is a concrete material model for NSC in the finite element software ABAQUS. It is a mature and reliable tool for predicting the mechanical behavior of NSC [29–31]. In order for CDP model to be used to simulate UHPFRC, the CDP model needs to be calibrated. Some studies have shown that the calibrated CDP model can accurately and effectively predict the mechanics characteristic of UHPFRC. In this study, the stress-strain curve for UHPFRC in compression proposed by Singh et al. [34], modified from Lu et al. [45], is used to calculate the data of compressive behavior in the CDP model. The stress-strain curve of UHPFRC specimen of the uniaxial tension test [34] is used to define the tensile behavior in the CDP model. The curves of compression damage and tension damage in the CDP model are defined according to the studies in [33,46], respectively. The parameters of the CDP model adopted in this study are shown in Table 1.

Table 1. Parameters of the concrete damage plasticity (CDP) model of ultra-high-performance fiber-reinforced concrete (UHPFRC).

Model Name	Young's Modulus	Compressive Strength	Tension Yield Stress	Tension Peak Stress
CDP	36.3 GPa	140 MPa	4.6 MPa	5.8 MPa
Dilation angle	Eccentricity	f_{b0}/f_{c0}	k_c	Viscosity Parameter
30	0.1	1.05	2/3	0.005

2.2. Validation of the Model

2.2.1. Test Specimens

In this study, the UHPFRC beams named as B25-1 and B25-2 [34] are chosen for the validation analysis. The cross section, spans, loading configuration details and reinforcement detail of the all the beams are given in Table 2, where the tensile reinforcement consisted of 20 mm diameter rebar with a yield strength and ultimate strength of 525 and 625 MPa, respectively. The four point bending test applies the same concentrated load symmetrically at a distance of 250 mm from the middle of the beams, resulting in pure bending stresses between the load points.

Table 2. Specimens details. [34].

Specimen Name	Cross Section (mm)	Effective Span (mm)	Loading Condition	Length of the Midspan without Stirrup (mm)
B25-1 & B25-2	250 × 250	3250	Four point bending	500
Top rebar diameter (mm)	Top rebar number	Bottom rebar diameter (mm)	Bottom rebar number	Stirrup diameter and spacing (mm)
10	2	20	3	D10@90

2.2.2. Finite Element Analysis (FEA) Model

According to the design diagram of UHPFRC beam specimen and the design of loading device, the corresponding finite element model of the test was established in ABAQUS with mesh size of 25 mm. The details of the reinforcement, mesh and load boundary condition of the finite element (FE) model are shown in Figure 1. The parameters of the CDP model adopted for the UHPFRC solid elements are shown in Table 1.

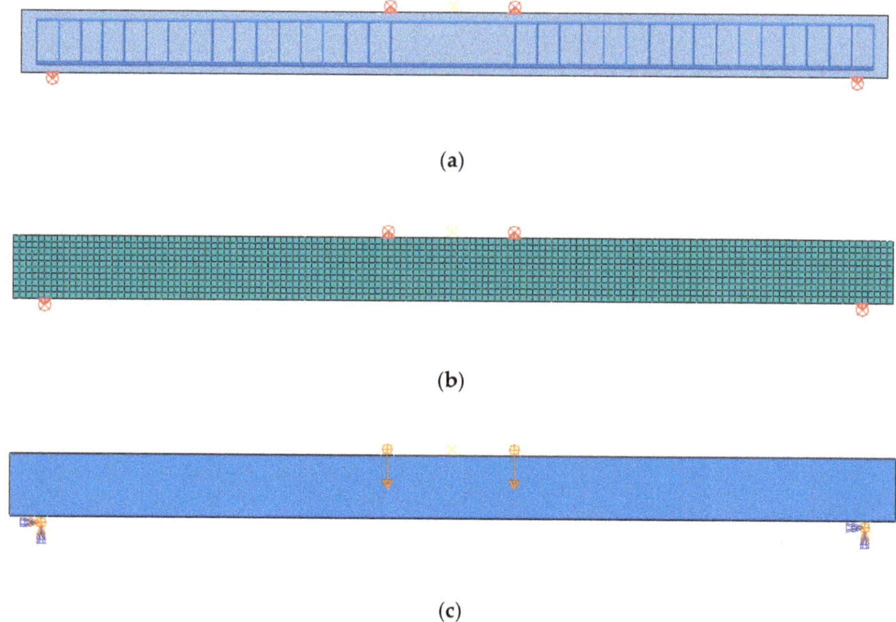

(a)

(b)

(c)

Figure 1. FE model of ultra-high-performance fiber-reinforced concrete (UHPFRC) beam test. (**a**) Reinforcement detail; (**b**) Mesh details; (c) Load boundary condition.

2.2.3. Results of FEA Simulation and Test

The test results and simulation results of the four point bending test of the two UHPFRC specimens are shown in Figure 2, where (a) is the relation curve between the mid-span displacement of the beam and the external load, (b) is the failure pattern of the specimen B25-1, and (c) is the tension damage distribution of UHPFRC in the finite element model. It can be seen that the finite element model can simulate the whole entire load-displacement curve, including the descending section after yielding. The finite element simulation results are in good agreement with the experimental results. The test results and the finite element results of peak load and corresponding displacement of UHPFRC specimens are shown in Table 3. The ultimate load capacity of specimens B25-1 and B25-2 predicted by the finite element model is 3% and 6% higher than the test results, respectively. It can be seen from Figure 2a,b, the damage distribution simulated by the finite element model is similar to the crack distribution of the specimen. Meanwhile, the validity of the parameters selected of the FEA model in this paper is proved so the FEA model with the same parameters can be taken as the standard for the extended study.

Table 3. Comparison of the FE model results with test results.

Specimen Name	Peak Load (kN)		Relative Error	Peak Displacement (mm)		Relative Error
	FE Model	Test		FE Model	Test	
B25-1	174	172	1.16%	40	59	−32%
B25-2	174	167	4.19%	40	39	2.56%

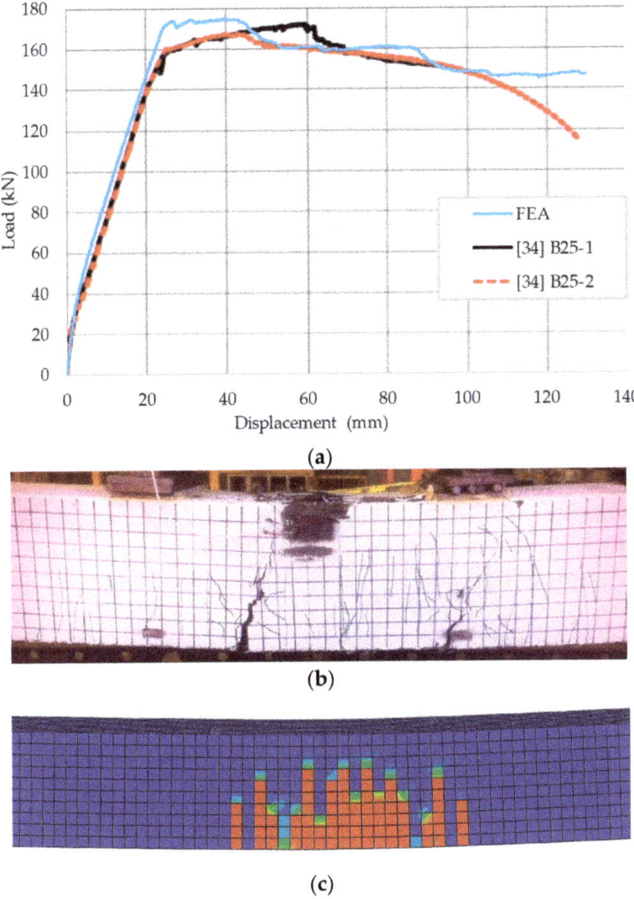

Figure 2. Test results and simulation results of beam test. (**a**) Load-displacement curve; (**b**) failure pattern of the specimen B25-1; (**c**) Tension damage distribution of UHPFRC in the FE model.

3. Multi-Scale Modeling Strategy—Interface Connection

3.1. Combined Multi-Point Constraint (CMPC) of Multi-Scale Model

3.1.1. Combine Multi-Point Constraint Relations

In the multi-scale model, the interface connection of different scale elements can be established by the constraint equations according to the degrees of freedom of interface nodes. The sketch of the interface connection of multi-scale model shown in Figure 3, where S_i (i = 1, 2, 3 . . . .) signifies micro element nodes with 3 degrees of freedom and B signifies macro element node with 6 degrees of freedom.

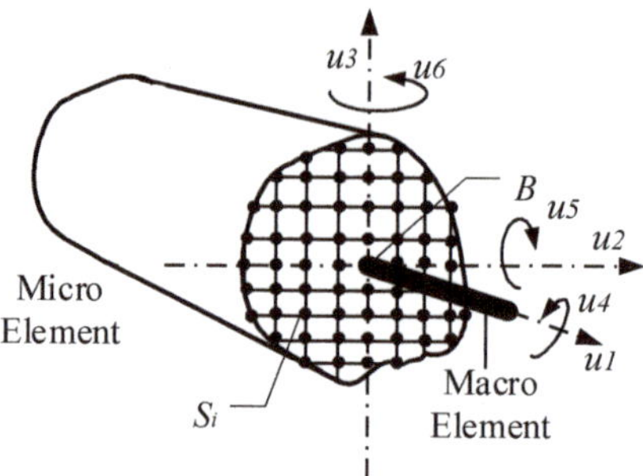

Figure 3. Sketch of the interface connection.

According to the coupling relation of degrees of freedom of nodes, the unified form of the constraint equations of the multi-scale interface connection is as follows:

$$c(u_{\mathrm{B}}, u_{\mathrm{S}i}) = u_{\mathrm{B}} - Cu_{\mathrm{S}i} = 0 \tag{1}$$

where u_{B} is the displacement vector of macro elements at the interface; $u_{\mathrm{S}i}$ is the displacement vector of micro elements at the interface; C is the coefficient matrix of interface constraint equations.

The accuracy of multi-scale simulation depends on the rationality of coefficient matrix C. If the constraint equations can effectively simulate the actual deformation coordination, a better effect of the coupling can be obtained.

The solution of multi-point constraint equations is usually based on the single constraint relation such as displacement coordination [40] or energy balance [41]. For the multi-scale simulation of UHPFRC structure, due to the nonlinear characteristics of the interface stress and deformation relation in the plastic stage, a single multi-point constraint method will lead to the over-constraint in tangential direction and the requirement of stress iteration in plastic stage. Therefore, the combined multi-point constraint (CMPC) is established in this study through the simultaneous equations of displacement coordination and energy balance. The equations form are as follows:

$$\begin{cases} u_{1\mathrm{S}i} - f_{1i}(u_{1\mathrm{B}}, u_{5\mathrm{B}}, u_{6\mathrm{B}}, b_i, h_i) = 0 \\ F_2 u_{2\mathrm{B}} = \int\limits_A \sigma_{2i,F2} u_{2\mathrm{S}i} \mathrm{d}A \\ F_3 u_{3\mathrm{B}} = \int\limits_A \sigma_{3i,F3} u_{3\mathrm{S}i} \mathrm{d}A \\ F_4 u_{4\mathrm{B}} = \int\limits_A (\sigma_{2i,F4} u_{2\mathrm{S}i} + \sigma_{3i,F4} u_{3\mathrm{S}i}) \mathrm{d}A \end{cases} \tag{2}$$

where $u_{1\mathrm{S}i}$ is the axial displacement of the node i of the micro element; $u_{2\mathrm{S}i}$, $u_{3\mathrm{S}i}$ are the tangential displacements of the node i of the micro element; $u_{1\mathrm{B}}$ is the axial displacement of the macro element node; $u_{2\mathrm{B}}$, $u_{3\mathrm{B}}$ are the tangential displacements of the macro element node; $u_{4\mathrm{B}}$, $u_{5\mathrm{B}}$, $u_{6\mathrm{B}}$ are the angular displacements of macro element node; F_j is the nodal force of macro element in the direction j; b_i, h_i are the distances from the node i of the micro element to the macro element node; $\sigma_{i2,Fj}$, $\sigma_{i3,Fj}$ are the nodal tangential stress of the node i of the micro element caused by F_j.

In the Equation set (2), the first equation is the axial and rotational constraint equation, which is established by displacement coordination. The last three equations are tangential constraint equations, which are established by energy balance. The aim is to eliminate the limitation of the single constraint relation in the plastic stage and improve the simulation accuracy of the multi-scale model.

3.1.2. Constraint of the Interface in Tangential Direction

Based on the multi-point constraint relation of displacement coordination, the tangential deformations of all micro element node at the interface are assumed to be consistent, and the displacement constraint relation of each node is established one by one. The deformation diagrams of displacement coordination are shown in Figure 4, and the constraint equations can be obtained as follows:

$$\begin{aligned} u_{2Si} - f_{2i}(u_{2B}, u_{4B}, b_i, h_i) = 0 \\ u_{3Si} - f_{3i}(u_{3B}, u_{4B}, b_i, h_i) = 0 \end{aligned} \tag{3}$$

In the equations, tangential displacements (u_{2Si}, u_{3Si}) of the micro element node are calculated from u_{2B}, u_{3B} and u_{4B}. There is no coupling relation between the degrees of freedom of different nodes of micro elements. Under this condition, when there is no nonzero tangential displacement or rotational displacement of macro element node, the tangential displacement of each node of micro elements along the interface is zero. It leads to the problem of over-constraint in tangential direction at the interface under the axial compression load.

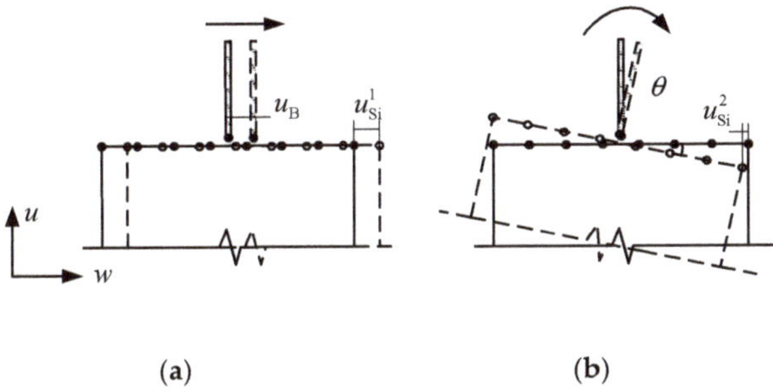

Figure 4. Deformation diagram of displacement coordination. (**a**) Tangential direction; (**b**) Rotational direction.

In the CMPC equation set, the displacement constraint equation in the tangential direction of the interface nodes can be obtained after the stress is eliminated by substituting the formula for the shear stress distribution:

$$u_B = f_2(u_{S1}, u_{S2} \cdots u_{Sn}) \tag{4}$$

where, the tangential displacements of each micro element node have a coupling relation with each other. When the tangential displacement of macro element node is zero, it can generate the relative displacements among micro element nodes and satisfy the constraint equation. The tangential deformation diagram of CMPC under the axial compression is shown in Figure 5.

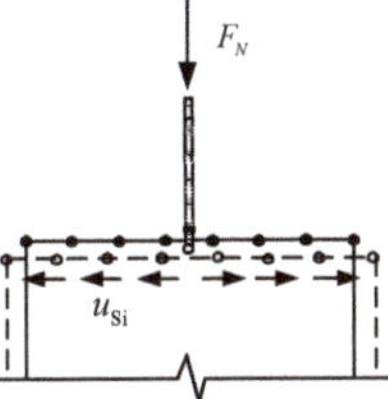

Figure 5. Tangential deformation diagram of CMPC under the axial compression.

The interface deformation under axial compression at the plasticity stage of the multi-scale model is shown in Figure 6. According to the Poisson ratio of UHPFRC, the uniform longitudinal stress causes the transverse strain of the section. And there is obvious transverse expansion deformation in the middle of the micro element model. The micro element nodes at the interface of displacement coordination model only produce vertical displacement with the macro element node with no tangential displacement, which is over-constraint compared with the micro model. The CMPC equation eliminates the over-constraint in tangential direction at the interface and conforms to the deformation relation of the interface nodes under the actual stress state.

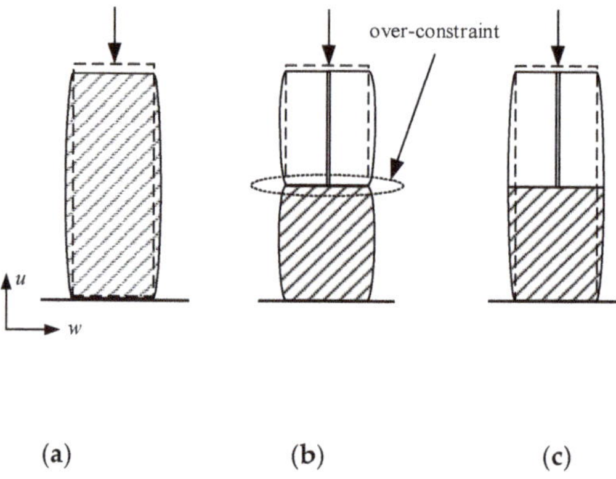

(a) (b) (c)

Figure 6. Interface deformation under axial compression. (**a**) Micro element model; (**b**) Multi-scale model of displacement coordination; (**c**) Multi-scale model of CPMC.

3.1.3. Constraint of the Interface in Rotational Direction

The multi-point constraint relation based on energy balance is established by the virtual work principle. It is assumed that the nodal force of the macro element and the nodal forces of the micro elements do equal work at the interface in the rotational direction. The equation is as follows:

$$\begin{aligned} F_5 u_{5\mathrm{B}} &= \int_A \sigma_{1i,F5} u_{1Si} \mathrm{d}A \\ F_6 u_{6\mathrm{B}} &= \int_A \sigma_{1i,F6} u_{1Si} \mathrm{d}A \end{aligned} \tag{5}$$

By substituting the formula for stress distribution under bending moment, the displacement constraint equation in rotational direction of the interface nodes can be obtained. Its precision depends on the rationality of the stress distribution of the stress formula.

The normal stress distribution of UHPFRC section under bending moment is shown in Figure 7. As in other studies [33,47], the stress distribution of UHPFRC section has underwent different stages. The first stage is the linear-elastic stage, in which the fiber and matrix show elasticity and the stress distribution is linear. With the increase of load, due to the strong bond between the high strength steel fiber and the matrix, the macro crack begins to expand slowly. The strain hardening phenomenon occurred is different from that of NSC, and the tensile stress is nonlinear distributed. This stage is called strain hardening stage and the formula for stress distribution in linear-elastic stage is no longer applicable. If the formula is not updated iteratively, the multi-point constraint equation at the interface based on energy balance will be distorted in the rotational direction in strain hardening stage.

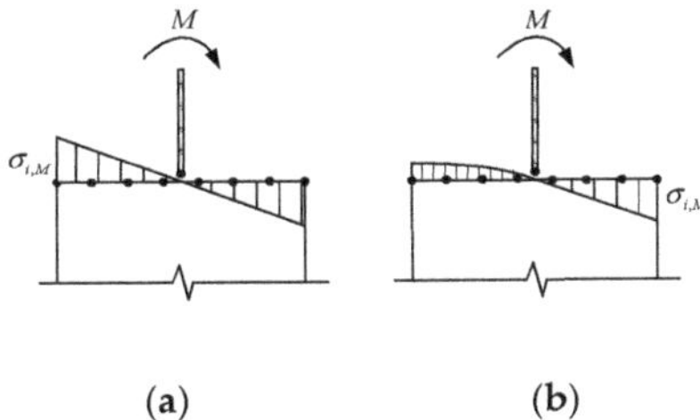

Figure 7. Stress distribution under bending moment. (**a**) Linear-elastic stage; (**b**) Strain hardening stage.

The CMPC method establishes the multi-point constraint equation of the interface in rotational direction based on the displacement coordination. The axial displacement of each node of the micro elements can be obtained through the s constraint equation (see the first equation in Equation set (2)). After entering the strain-hardening stage, this multi-point constraint equation avoids the problem that the formula for stress distribution needs to be updated iteratively. Therefore, the CMPC method proposed in this paper combines the advantages of displacement coordination method and energy balance method. The multi-point constraint equations conform to the transfer relations of displacement and stress between the interface nodes. It can achieve good constraint effect in axial, tangential, and rotational directions. It is applicable to the analysis of UHPFRC components under complex loads.

3.2. CMPC Equations of Multi-Scale Connection of Beam-Solid Element

According to Equation set (2), displacement vector $[u_B \ \ v_B \ \ w_B \ \ \theta_x \ \ \theta_y \ \ \theta_z \ ]$ of beam element and displacement vector $\left[\ u_{Si} \ \ v_{Si} \ \ w_{Si} \ \right]$ of solid element are substituted, and the multi-point constraint equations can be expressed as:

$$\begin{cases} w_{Si} - f_i\left(w_B, \theta_x, \theta_y, R_{xi}, R_{yi}\right) = 0 \\ F_x u_B = \int\limits_A \tau_{xi} u_{Si} \mathrm{d}A \\ F_y v_B = \int\limits_A \tau_{yi} v_{Si} \mathrm{d}A \\ T\theta_z = \int\limits_A (\tau_{xi} u_{Si} + \tau_{yi} v_{Si}) \mathrm{d}A \end{cases} \tag{6}$$

where R_{xi}, R_{yi} are the distances between the node i of the solid element and the beam element node at the interface in the x and y direction, respectively; F_x, F_y are the shear forces acting on the beam element node in the x and y direction, respectively; T is the torque acting on the beam element node; τ_{xi}, τ_{yi} are the shear stresses of the node i of the solid element in the x and y direction, respectively. The multi-scale connection of beam - solid element is shown in Figure 8.

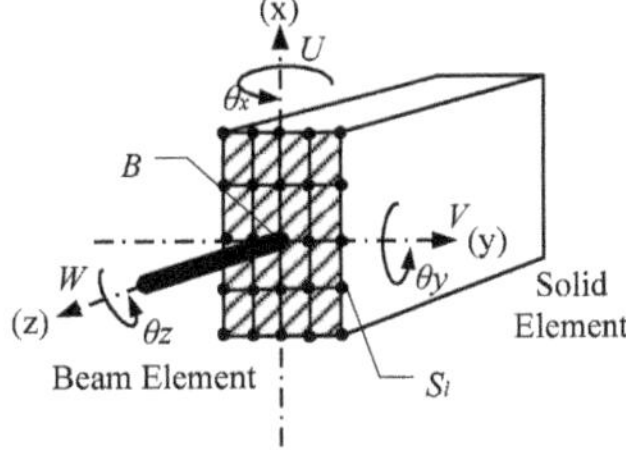

Figure 8. Multi-scale connection of beam - solid element.

In the Equation set (6), the first equation is the constraint equation of axial and rotational displacement, which can be solved according to the displacement coordination. The last three equations are the constraint equations of tangential displacements, which need to be solved by substituting the formula for stress distribution. For example, the formula for shear stress distribution of the rectangular section in the y direction is as follows:

$$\tau_{yi} = \frac{3F_y}{2bh}\left(1 - \frac{4R_{yi}^2}{h^2}\right) \tag{7}$$

where b and h are the width and height of the rectangular section, respectively.

After solving the Equation set (6), the following can be obtained:

$$\begin{cases} w_{\mathrm{S}i} = w_{\mathrm{B}} + R_{xi}\sin\theta_y + R_{yi}\sin\theta_x \\ u_{\mathrm{B}} = C_{u1}u_1 + C_{u2}u_2 + \cdots + C_{un}u_n \\ v_{\mathrm{B}} = C_{v1}v_1 + C_{v2}v_2 + \cdots + C_{vn}v_n \\ \theta_{\mathrm{Z}} = (u_1R_{y1} + \cdots + u_nR_{yn}) - (v_1R_{x1} + \cdots + v_2R_{x2}) \end{cases} \tag{8}$$

where C_{ui}, C_{vi} are the influence coefficients of the tangential displacements related to the section size and node position. An example of the Equation set (8) is given in Appendix A.

4. Multi-Scale Models of Ultra-High-Performance Steel Fiber-Reinforced Concrete

4.1. Multi-Scale Models Built-Up

The CMPC multi-scale modeling strategy with the same parameters selected above is adopted to establish the multi-scale models of reinforced UHPFRC components, as shown in Figures 9 and 10. Where (a) is the solid element model taken as the standard for comparison without experimental results, and (b), (c) and (d) are the multi-scale models of beam-solid element, whose interface connections are established by the displacement coordination method, the energy balance method and the CMPC method (Section 3.2. for the expressions) respectively. The height of the component is 3m, and the section size is 0.4 m × 0.4 m. The multi-scale interface is located at 1/3 height of the component with the height of 3 m. The parameters of the CDP model adopted for the UHPFRC solid elements are shown in Table 1.

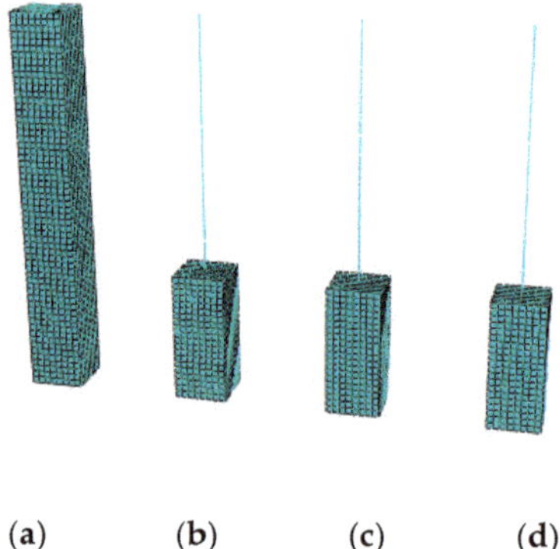

Figure 9. UHPFRC parts of the models. (**a**) Solid element model; (**b**) Displacement coordination model; (**c**) Energy balance model; (**d**) CMPC model.

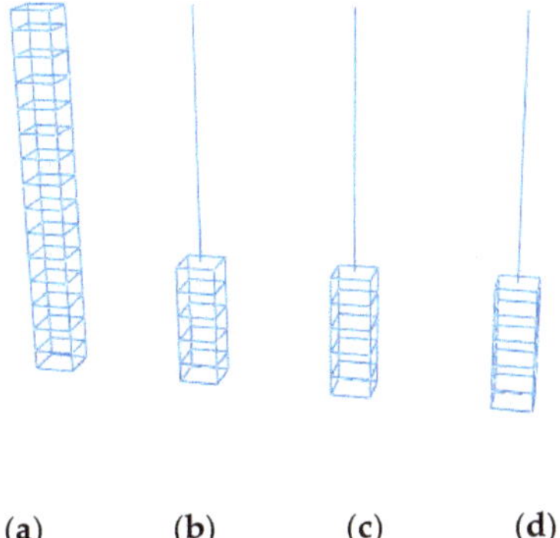

Figure 10. Reinforcement parts of the models. (**a**) Solid element model; (**b**) Displacement coordination model; (**c**) Energy balance model; (**d**) CMPC model.

In the micro element model, C3D8R elements are used to model UHPFRC with the calibrated CDP model which is the same as that in the second section. The reinforcement is simulated by T3D2 element. In the macro element model, B31 elements are used to simulate UHPFRC and the reinforcement net. The uniaxial stress-strain relationship of the material subroutine (UMAT) of UHPFRC is the same as that of the calibrated CDP model. The section size of the B31 element of the reinforcement net is calculated equivalent to total reinforcement area. The material model parameters of reinforcement are the same as those in Section 2. With a yield strength and ultimate strength of 525 and 625 MPa, respectively. The mesh size of the models is 0.05m. A fixed constraint is set at the bottom of the component and a loading point is set at the top. The number of model elements of the solid element model and the multi-scale model is shown in Table 4. It can be seen that the number of model elements in the multi-scale model is reduced by nearly 2/3 compared with that in the solid element model, which significantly improves the computational efficiency.

Table 4. Number of model elements.

Model	Element Type			Total
	C3D8R	T3D2	B31	
Solid element model	3840	616	0	4456
Multi-scale model	1280	220	80	1580

4.2. Unidirectional Load Cases

4.2.1. Axial Compression Load Case

Under the axial compression load, the stress distributions of UHPFRC of the multi-scale models and the connection interface are shown in Figure 11. By comparison, it can be seen that there is the phenomenon of stress concentration at the connection interface of the displacement coordination model whose stress distribution is different from that of the solid model. The overall stress distribution of the energy balance model is also different from that of the solid model, and the stress distribution at the interface connection is not uniform. The stress distribution obtained by the CMPC method is highly consistent with the solid element model. The constraint effect of the CMPC method is obviously better than that of the single multi-point constraint method.

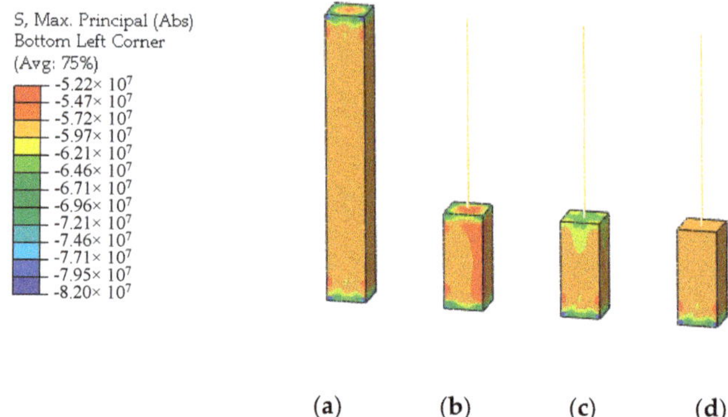

Figure 11. Stress distributions under the axial compression load (unit: Pa). (**a**) Solid element model; (**b**) Displacement coordination model; (**c**) Energy balance model; (**d**) CMPC model.

4.2.2. Bending Load Case

Unidirectional concentrated moment loads is applied to the loading point at the top of the models under the bending load case. The stress distribution and tensile damage distribution of UHPFRC of the models are shown in Figures 12 and 13. By comparison, it can be seen that the stress distribution at the connection interface in the plastic stage is nonlinear. The UHPFRC on the tensile side enters the strain hardening stage with tensile damage. Compared with the solid element model, the results of the displacement coordination model and the energy balance model show obvious stress distortion at the connection interface. As UHPFRC on the tension side enters the strain hardening stage, the formula for stress distribution of the energy balance method is no longer applicable. The original constraint equation needs to be balanced by over increasing the strain on the tension side, resulting in the distortion of the damage distribution at the connection interface. Due to the over-constraint in tangential direction mentioned above, the results of the displacement coordination model at the connection interface look distorted. The simulation results of CMPC model are highly consistent with that of the solid element model, and the stress and damage distribution of UHPFRC at the connection interface are simulated accurately.

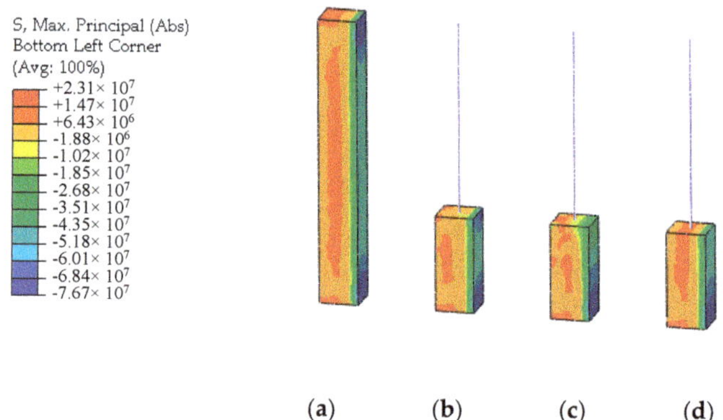

Figure 12. Stress distributions under the bending load (unit: Pa). (**a**) Solid element model; (**b**) Displacement coordination model; (**c**) Energy balance model; (**d**) CMPC model.

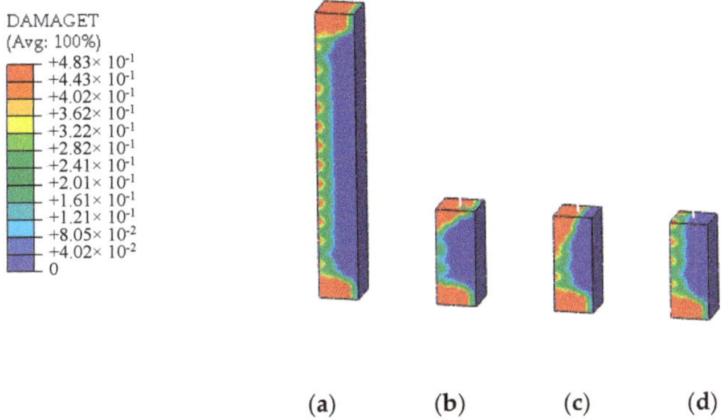

Figure 13. Tensile damage distributions under the bending load (unit: Pa). (**a**) Solid element model; (**b**) Displacement coordination model; (**c**) Energy balance model; (**d**) CMPC model.

4.2.3. Shear Load Case

In shear load case, a concentrated horizontal force is applied to the loading point at the top of the models. The stress distribution and tensile damage distribution of UHPFRC of the models are shown in Figures 14 and 15. It can be seen that after entering the strain hardening stage of UHPFRC, the stress distribution at the connection interface presents nonlinear. The multi-point constraint equation derived from the formula for elastic stress distribution is no longer applicable, and the simulation results of the stress distribution and damage distribution at the connection interface of the energy balance model are inaccurate. At the same time, due to the tangential over-constraint, the stress concentration occurs at the connection interface of the displacement coordination model. The damage distribution is distorted. However, the simulation results of the CMPC model still have good accuracy. The stress and damage distribution of UHPFRC at the connection interface are simulated accurately.

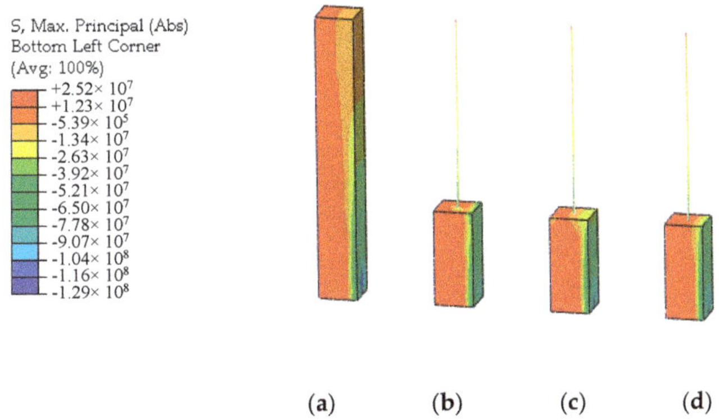

Figure 14. Stress distributions under the shear compression load (unit: Pa). (**a**) Solid element model; (**b**) Displacement coordination model; (**c**) Energy balance model; (**d**) CMPC model.

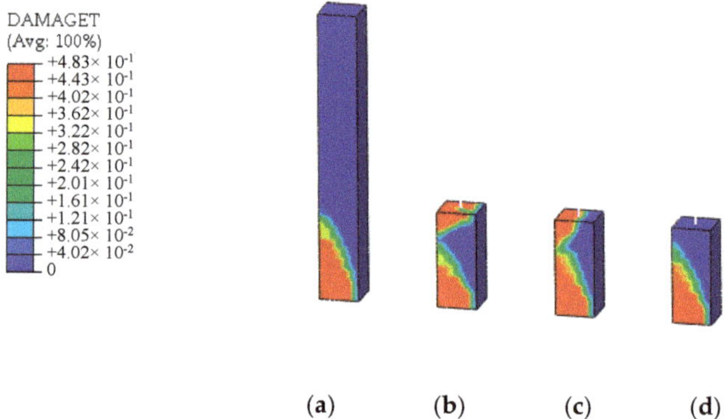

Figure 15. Stress distributions under the shear compression load (unit: Pa). (**a**) Solid element model; (**b**) Displacement coordination model; (**c**) Energy balance model; (**d**) CMPC model.

4.3. Multidirectional Composite Load Case

Under unidirectional load cases, the multi-scale model of UHPFRC component established according to the proposed multi-scale modeling strategy achieved good accuracy. The performance of this multi-scale model under multidirectional composite load case will be studied below.

In this load case, the axial compression force, the bidirectional moments and the bidirectional horizontal forces are applied composited at the loading point at the top of the models. The diagram of multidirectional composite load is shown in Figure 16. Where, the red arrow represents the bidirectional horizontal forces, the yellow arrow represents the axial compression force, and the purple double arrow represents the bidirectional moments.

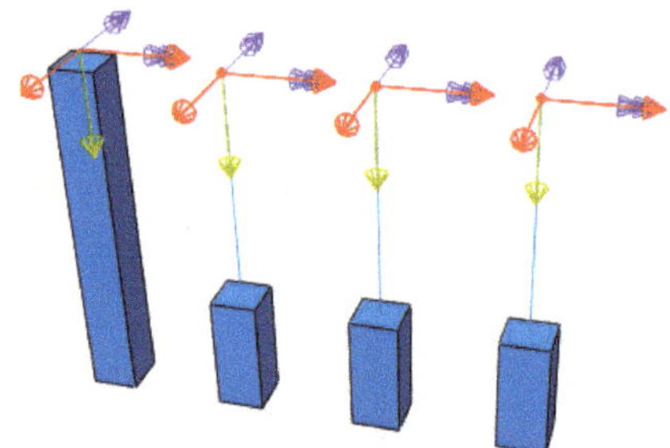

Figure 16. The diagram of multidirectional composite load.

Under the multidirectional composite load case, the stress distributions of UHPFRC of the multi-scale models and the connection interface are shown in Figure 17.

Through comparison, it can be seen that the CPMC equations established based on the proposed multi-scale modeling strategy achieve good connection effect under the multidirectional composite load case. The calculation accuracy of the CMPC model for UHPFRC is consistent with that of the solid element model, which is better than the displacement coordination model and the energy balance model. The multi-scale modeling strategy proposed in this study can be effectively applied to the multi-scale finite element analysis of UHPFRC structures with accuracy and efficiency.

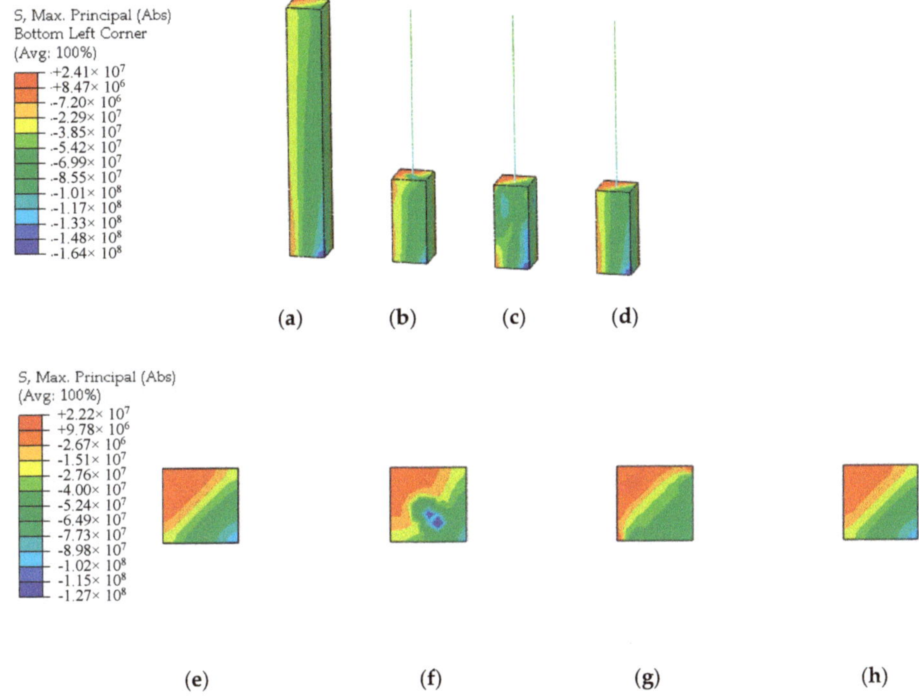

Figure 17. Stress distributions under the multidirectional composite loads case (unit: Pa). (**a**) Solid element model; (**b**) Displacement coordination model; (**c**) Energy balance model; (**d**) CMPC model; (**e**) Interface location of the solid element model; (**f**) Interface of the displacement coordination model; (**g**) Interface of the energy balance model; (**h**) Interface of CMPC model.

5. Conclusions

This study proposed a novel multi-scale modeling strategy for ultra-high-performance steel fiber-reinforced concrete (UHPFRC) structures. The main work and conclusions are summarized as follows:

1. The applicability of concrete damage plasticity (CDP) model in finite element software ABAQUS to UHPFRC was verified according to the four-point bending test results of reinforced UHPFRC beams. The simulation results show that the calibrated CDP model used in the multi-scale modeling strategy in this study can accurately and effectively predict the load-displacement curve and plastic damage distribution of UHPFRC components.
2. A novel combined multi-point constraint method was established by the simultaneous equations of the displacement coordination equation and energy balance equation in different directions of the interface. The CMPC method eliminates the problem of the tangential over-constraint of displacement coordination equation at the interface and avoids stress iteration of energy balance equation in the plastic stage. The multi-point constraint equations conform to the transfer relations of displacement and stress between the interface nodes.
3. The expression of the constraint equations of the multi-scale connection of beam-solid element by CMPC method was derived. The multi-scale model of the reinforced UHPFRC component was established in ABAQUS with this expression and the calibrated CDP model. The axial compression load case, bending load case, shear load case, and multidirectional composite load case are analyzed.

4. The simulation results of the multi-scale model under each load case show that the multi-scale model established by the CMPC method can significantly reduce the number of model elements and improve the calculation efficiency. The CMPC models have good simulation accuracy in the analysis of each load case compared with the displacement coordination model and the energy balance model. In the strain-hardening stage of UHPFRC, the CMPC method can still accurately simulate the stress distribution and damage distribution of the connection interface. It can be applied to multi-scale finite element analysis of UHPFRC structures with accuracy and efficiency.

Author Contributions: Conceptualization, Z.L. and J.T.; software, Z.P.; validation, Z.P.; writing—original draft preparation, Z.P.; writing—review and editing, Z.L.; supervision, Z.L.; project administration, J.T. All authors have read and agreed to the published version of the manuscript.

Funding: This study was funded by the National Key Research and Development Program of China under grant number 2016YFC0701102, the National Natural Science Foundations of China under grant number 51538003 and 51978224, China Major Development Project for Scientific Research Instrument under grant number 51827811, and Shenzhen Technology Innovation Program under grant number JCYJ20170811160003571 and JCYJ20180508152238111.

Conflicts of Interest: The authors declare that there are no conflicts of interests regarding the publication of this article.

Appendix A

An example of the multi-scale connection interface is established with section size 0.4 m × 0.4 m and mesh size 0.1 m shown in Figure A1, where the solid element node number on the connection interface is 1–25, and the beam element node number is 26.

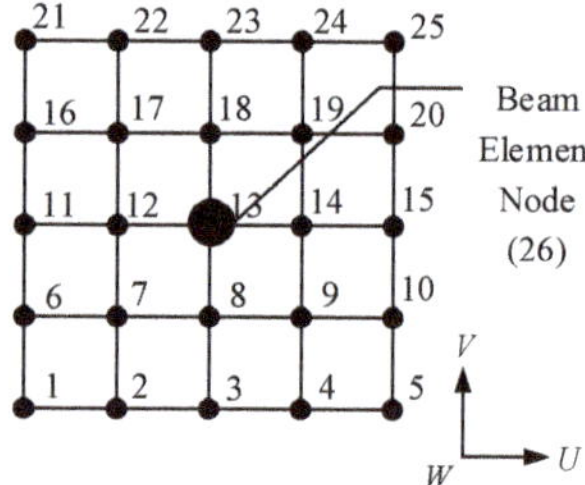

Figure A1. An example of the multi-scale connection interface.

The Equation set (8) of the multi-scale connection interface shown in Figure A1 can be obtained as follow. The multi-scale connection of CMPC model can be established by coding the '*EQUATION' keyword commands to input the constrained Equation set (8) into the '*.inp' file of ABAQUS. Similarly, this strategy can be implemented in the other FEA software by setting the multi-point constraint equations.

$$\left\{\begin{array}{l}
w_{Si} = w_B + R_{xi}\sin\theta_y + R_{yi}\sin\theta_x \\
1024u_{26} = [7(u_{25} + u_1 + u_5 + u_{21}) + 14(u_{20} + u_{15} + u_{10} + u_{16} + u_{11} + u_6) + 34(u_{24} + u_{22} + u_4 + u_2) + \\
\quad 68(u_{19} + u_{14} + u_9 + u_{17} + u_{12} + u_7) + 46(u_{23} + u_3) + 92(u_8 + u_{13} + u_{18})] \\
1024v_{26} = [7(v_{25} + v_1 + v_5 + v_{21}) + 14(v_{24} + v_{22} + v_4 + v_2 + v_{23} + v_3) + 34(v_{20} + v_{16} + v_{10} + v_6) + \\
\quad 68(v_{19} + v_{17} + v_9 + v_7 + v_{18} + v_8) + 46(v_{15} + v_{11}) + 92(v_{14} + v_{12} + v_{13})] \\
10\theta_{z26} = 2(u_1 + u_2 + u_3 + u_4 + u_5 + v_5 + v_{10} + v_{15} + v_{20} + v_{25}) + \\
\quad 2(-u_{21} - u_{22} - u_{23} - u_{24} - u_{25} - v_1 - v_6 - v_{11} - v_{16} - v_{21}) + \\
\quad (u_6 + u_7 + u_8 + u_9 + u_{10} + v_4 + v_9 + v_{14} + v_{19} + v_{24}) + \\
\quad (-u_{16} - u_{17} - u_{18} - u_{19} - u_{20} - v_2 - v_7 - v_{12} - v_{17} - v_{22}) \\
w_1 = w_{26} - 0.2\sin\theta_{x26} + 0.2\sin\theta_{y26} \\
w_2 = w_{26} - 0.2\sin\theta_{x26} + 0.1\sin\theta_{y26} \\
w_3 = w_{26} - 0.2\sin\theta_{x26} \\
w_4 = w_{26} - 0.2\sin\theta_{x26} - 0.1\sin\theta_{y26} \\
w_5 = w_{26} - 0.2\sin\theta_{x26} - 0.2\sin\theta_{y26} \\
w_6 = w_{26} - 0.1\sin\theta_{x26} + 0.2\sin\theta_{y26} \\
w_7 = w_{26} - 0.1\sin\theta_{x26} + 0.1\sin\theta_{y26} \\
w_8 = w_{26} - 0.1\sin\theta_{x26} \\
w_4 = w_{26} - 0.1\sin\theta_{x26} - 0.1\sin\theta_{y26} \\
w_{10} = w_{26} - 0.1\sin\theta_{x26} - 0.2\sin\theta_{y26} \\
w_{11} = w_{26} + 0.2\sin\theta_{y26} \\
w_{12} = w_{26} + 0.1\sin\theta_{y26} \\
w_{13} = w_{26} \\
w_{14} = w_{26} - 0.1\sin\theta_{y26} \\
w_{15} = w_{26} - 0.2\sin\theta_{y26} \\
w_{16} = w_{26} + 0.1\sin\theta_{x26} + 0.2\sin\theta_{y26} \\
w_{17} = w_{26} + 0.1\sin\theta_{x26} + 0.1\sin\theta_{y26} \\
w_{18} = w_{26} + 0.1\sin\theta_{x26} \\
w_{19} = w_{26} + 0.1\sin\theta_{x26} - 0.1\sin\theta_{y26} \\
w_{20} = w_{26} + 0.1\sin\theta_{x26} - 0.2\sin\theta_{y26} \\
w_{21} = w_{26} + 0.2\sin\theta_{x26} + 0.2\sin\theta_{y26} \\
w_{22} = w_{26} + 0.2\sin\theta_{x26} + 0.1\sin\theta_{y26} \\
w_{23} = w_{26} + 0.2\sin\theta_{x26} \\
w_{24} = w_{26} + 0.2\sin\theta_{x26} - 0.1\sin\theta_{y26} \\
w_{25} = w_{26} + 0.2\sin\theta_{x26} - 0.2\sin\theta_{y26}
\end{array}\right. \quad \text{(A1)}$$

References

1. De Larrard, F.; Sedran, T. Optimization of ultra-high-performance concrete by the use of a packing model. *Cem. Concr. Res.* **1994**, *24*, 997–1009. [CrossRef]
2. Rossi, P.; Arca, A.; Parant, E.; Fakhri, P. Bending and compressive behaviours of a new cement composite. *Cem. Concr. Res.* **2005**, *35*, 27–33. [CrossRef]
3. Graybeal, B. *Material Property Characterization of Ultra-High Performance Concrete*; No. Fhwa-Hrt-06-103; Federal Highway Administration: Washington, DC, USA, 2006; pp. 1–176.
4. Habel, K.; Viviani, M.; Denarié, E.; Brühwiler, E. Development of the mechanical properties of an Ultra-High Performance Fiber Reinforced Concrete (UHPFRC). *Cem. Concr. Res.* **2006**, *36*, 1362–1370. [CrossRef]
5. Magureanu, C.; Sosa, I.; Negrutiu, C.; Heghes, B. Mechanical properties and durability of ultra-high-performance concrete. *ACI Mater. J.* **2012**, *109*, 177–184.
6. Wille, K.; Naaman, A.E.; El-Tawil, S. Optimizing ultra-high-performance fiber-reinforced concrete. *Concr. Int.* **2011**, *33*, 35–41.
7. Triantafillou, T.C.; Papanicolaou, C.G. Shear strengthening of reinforced concrete members with textile reinforced mortar (TRM) jackets. *Mater. Struct. Mater. Constr.* **2006**, *39*, 93–103. [CrossRef]
8. Ortlepp, R.; Hampel, U.; Curbach, M. A new approach for evaluating bond capacity of TRC strengthening. *Cem. Concr. Compos.* **2006**, *28*, 589–597. [CrossRef]

9. Si Larbi, A.; Contamine, R.; Ferrier, E.; Hamelin, P. Shear strengthening of RC beams with textile reinforced concrete (TRC) plate. *Constr. Build. Mater.* **2010**, *24*, 1928–1936. [CrossRef]
10. Parra-Montesinos, G.J. High-performance fiber-reinforced cement composites: An alternative for seismic design of structures. *ACI Struct. J.* **2005**, *102*, 668–675.
11. Canbolat, B.A.; Parra-Montesinos, G.J.; Wight, J.K. Experimental study on seismic behavior of high-performance fiber-reinforced cement composite coupling beams. *ACI Struct. J.* **2005**, *102*, 159–166.
12. Barnett, S.J.; Lataste, J.F.; Parry, T.; Millard, S.G.; Soutsos, M.N. Assessment of fibre orientation in ultra high performance fibre reinforced concrete and its effect on flexural strength. *Mater. Struct. Mater. Constr.* **2010**, *43*, 1009–1023. [CrossRef]
13. Wille, K.; Kim, D.J.; Naaman, A.E. Strain-hardening UHP-FRC with low fiber contents. *Mater. Struct. Mater. Constr.* **2011**, *44*, 583–598. [CrossRef]
14. Kim, D.J.; Park, S.H.; Ryu, G.S.; Koh, K.T. Comparative flexural behavior of Hybrid Ultra High Performance Fiber Reinforced Concrete with different macro fibers. *Constr. Build. Mater.* **2011**, *25*, 4144–4155. [CrossRef]
15. Ryu, G.S.; Kim, S.H.; Ahn, G.H.; Koh, K.T. Evaluation of the direct tensile behavioral characteristics of UHPC using twisted steel fibers. In *Advanced Materials Research*; Inner Mongolia University of Technology, Korea Maritime University and Queensland University of Technology: Guangzhou, China, 2013; Volume 602–604, pp. 96–101.
16. Yoo, D.Y.; Lee, J.H.; Yoon, Y.S. Effect of fiber content on mechanical and fracture properties of ultra high performance fiber reinforced cementitious composites. *Compos. Struct.* **2013**, *106*, 742–753. [CrossRef]
17. Yoo, D.Y.; Kang, S.T.; Yoon, Y.S. Effect of fiber length and placement method on flexural behavior, tension-softening curve, and fiber distribution characteristics of UHPFRC. *Constr. Build. Mater.* **2014**, *64*, 67–81. [CrossRef]
18. Yoo, D.Y.; Shin, H.O.; Yang, J.M.; Yoon, Y.S. Material and bond properties of ultra high performance fiber reinforced concrete with micro steel fibers. *Compos. Part B Eng.* **2014**, *58*, 122–133. [CrossRef]
19. Choi, W.C.; Jung, K.Y.; Jang, S.J.; Yun, H.-D. The Influence of Steel Fiber Tensile Strengths and Aspect Ratios on the Fracture Properties of High-Strength Concrete. *Materials* **2019**, *12*, 2105. [CrossRef]
20. Graybeal, B.A. Flexural behavior of an ultrahigh-performance concrete I-girder. *J. Bridge Eng.* **2008**, *13*, 602–610. [CrossRef]
21. Chen, L.; Graybeal, B. Modeling Structural Performance of 2nd Generation Ultra-High Performance Concrete Pi-Girders. *J. Bridge Eng.* **2012**, *17*, 634–643. [CrossRef]
22. Yang, I.H.; Joh, C.; Kim, B.S. Flexural strength of large-scale ultra high performance concrete prestressed T-beams. *Can. J. Civ. Eng.* **2011**, *38*, 1185–1195. [CrossRef]
23. Yoo, D.Y.; Yoon, Y.S. Structural performance of ultra-high-performance concrete beams with different steel fibers. *Eng. Struct.* **2015**, *102*, 409–423. [CrossRef]
24. Yang, I.H.; Joh, C.; Kim, B.S. Structural behavior of ultra high performance concrete beams subjected to bending. *Eng. Struct.* **2010**, *32*, 3478–3487. [CrossRef]
25. Xia, J.; Mackie, K.R.; Saleem, M.A.; Mirmiran, A. Shear failure analysis on ultra-high performance concrete beams reinforced with high strength steel. *Eng. Struct.* **2011**, *33*, 3597–3609. [CrossRef]
26. Yoo, D.Y.; Banthia, N.; Kim, S.W.; Yoon, Y.S. Response of ultra-high-performance fiber-reinforced concrete beams with continuous steel reinforcement subjected to low-velocity impact loading. *Compos. Struct.* **2015**, *126*, 233–245. [CrossRef]
27. Astarlioglu, S.; Krauthammer, T. Response of normal-strength and ultra-high-performance fiber-reinforced concrete columns to idealized blast loads. *Eng. Struct.* **2014**, *61*, 1–12. [CrossRef]
28. Wu, C.; Oehlers, D.J.; Rebentrost, M.; Leach, J.; Whittaker, A.S. Blast testing of ultra-high performance fibre and FRP-retrofitted concrete slabs. *Eng. Struct.* **2009**, *31*, 2060–2069. [CrossRef]
29. Chen, J.F.; Chen, G.M.; Teng, J.G. Role of bond modelling in predicting the behaviour of RC beams shear-strengthened with FRP U-jackets, keynote. In Proceedings of the 9th International Symposium on Fibre Reinforced Polymer for Concrete Structures (FRPRCS9), Sydney, Australia, 15 July 2009.
30. Li, S.Q.; Chen, J.F.; Bisby, L.A.; Hu, Y.M.; Teng, J.G. Strain efficiency of FRP jackets in FRP-confined concrete-filled circular steel tubes. *Int. J. Struct. Stab. Dyn.* **2012**, *12*, 75–94. [CrossRef]
31. Chen, G.M.; Chen, J.F.; Teng, J.G. On the finite element modelling of RC beams shear-strengthened with FRP. *Constr. Build. Mater.* **2012**, *32*, 13–26. [CrossRef]

32. Tysmans, T.; Wozniak, M.; Remy, O.; Vantomme, J. Finite element modelling of the biaxial behaviour of high-performance fibre-reinforced cement composites (HPFRCC) using Concrete Damaged Plasticity. *Finite Elem. Anal. Des.* **2015**, *100*, 47–53. [CrossRef]
33. Mahmud, G.H.; Yang, Z.; Hassan, A.M.T. Experimental and numerical studies of size effects of Ultra High Performance Steel Fibre Reinforced Concrete (UHPFRC) beams. *Constr. Build. Mater.* **2013**, *48*, 1027–1034. [CrossRef]
34. Singh, M.; Sheikh, A.H.; Mohamed Ali, M.S.; Visintin, P.; Griffith, M.C. Experimental and numerical study of the flexural behaviour of ultra-high performance fibre reinforced concrete beams. *Constr. Build. Mater.* **2017**, *138*, 12–25. [CrossRef]
35. Fish, J.; Wagiman, A. Multiscale finite element method for a locally nonperiodic heterogeneous medium. *Comput. Mech.* **1993**, *12*, 164–180. [CrossRef]
36. Fish, J.; Belsky, V. Multi-grid method for periodic heterogeneous media Part 2: Multiscale modeling and quality control in multidimensional case. *Comput. Methods Appl. Mech. Eng.* **1995**, *126*, 17–38. [CrossRef]
37. Ghosh, S.; Lee, K.; Moorthy, S. Multiple scale analysis of heterogeneous elastic structures using homogenization theory and Voronoi cell finite element method. *Int. J. Solids Struct.* **1995**, *32*, 27–62. [CrossRef]
38. Belytschko, T.; Xiao, S. Coupling Methods for Continuum Model with Molecular Model. *Int. J. Multiscale Comput. Eng.* **2003**, *1*, 115–126. [CrossRef]
39. Kadowaki, H.; Liu, W. Bridging multi-scale method for localization problems. *Comput. Meth. Appl. Mech. Eng.* **2004**, *193*, 3267–3302. [CrossRef]
40. Wellmann, C.; Wriggers, P. A two-scale model of granular materials. *Comput. Methods Appl. Mech. Eng.* **2012**, 205. [CrossRef]
41. McCune, R.; Armstrong, C.; Robinson, D. Mixed-dimensional coupling in finite models. *Int. J. Numer. Methods Eng.* **2000**, *49*, 725–750. [CrossRef]
42. Ainsworth, M. Essential boundary conditions and multi-point constraints in finite element analysis. *Comput. Methods Appl. Mech. Eng.* **2001**, *190*, 6323–6339. [CrossRef]
43. Yu, Y.; Chan, T.; Sun, Z.; Li, Z. Mixed-Dimensional Consistent Coupling by Multi-Point Constraint Equations for Efficient Multi-Scale Modeling. *Adv. Struct. Eng.* **2012**, *15*, 837–854. [CrossRef]
44. Ren, X.; Li, J. Multi-scale based fracture and damage analysis of steel fiber reinforced concrete. *Eng. Fail. Anal.* **2013**, *35*, 253–261. [CrossRef]
45. Lu, Z.H.; Zhao, Y.G. Empirical stress-strain model for unconfined high-strength concrete under uniaxial compression. *J. Mater. Civ. Eng.* **2010**, *22*, 1181–1186. [CrossRef]
46. Birtel, V.; Mark, P. Parameterized finite element modelling of RC beam shear failure. In Proceedings of the ABAQUS Users' Conference, Cambridge, UK, 23–25 May 2006; pp. 95–107.
47. Hassan, A.M.T.; Jones, S.W.; Mahmud, G.H. Experimental test methods to determine the uniaxial tensile and compressive behaviour of ultra high performance fibre reinforced concrete (UHPFRC). *Constr. Build. Mater.* **2012**, *37*, 874–882. [CrossRef]

Review

The Role of Supplementary Cementitious Materials (SCMs) in Ultra High Performance Concrete (UHPC): A Review

Sungwoo Park [1,†], Siyu Wu [1,†], Zhichao Liu [2,3] and Sukhoon Pyo [1,*]

1 Department of Urban and Environmental Engineering, Ulsan National Institute of Science and Technology (UNIST), Ulsan 44919, Korea; paksungwoo@unist.ac.kr (S.P.); wsy86023@unist.ac.kr (S.W.)
2 State Key Laboratory of Silicate Materials for Architectures, Wuhan University of Technology, Wuhan 430070, China; liuzc9@whut.edu.cn
3 School of Materials Science and Engineering, Wuhan University of Technology, Wuhan 430070, China
* Correspondence: shpyo@unist.ac.kr
† These authors contributed equally to the present study.

Abstract: Although ultra high-performance concrete (UHPC) has great performance in strength and durability, it has a disadvantage in the environmental aspect; it contains a large amount of cement that is responsible for a high amount of CO_2 emissions from UHPC. Supplementary cementitious materials (SCMs), industrial by-products or naturally occurring materials can help relieve the environmental burden by reducing the amount of cement in UHPC. This paper reviews the effect of SCMs on the properties of UHPC in the aspects of material properties and environmental impacts. It was found that various kinds of SCMs have been used in UHPC in the literature and they can be classified as slag, fly ash, limestone powder, metakaolin, and others. The effects of each SCM are discussed mainly on the early age compressive strength, the late age compressive strength, the workability, and the shrinkage of UHPC. It can be concluded that various forms of SCMs were successfully applied to UHPC possessing the material requirement of UHPC such as compressive strength. Finally, the analysis on the environmental impact of the UHPC mix designs with the SCMs is provided using embodied CO_2 generated during the material production.

Keywords: ultra high-performance concrete (UHPC); supplementary cementitious materials (SCMs); sustainability; compressive strength; flowability; shrinkage

Citation: Park, S.; Wu, S.; Liu, Z.; Pyo, S. The Role of Supplementary Cementitious Materials (SCMs) in Ultra High Performance Concrete (UHPC): A Review. *Materials* **2021**, *14*, 1472. https://doi.org/10.3390/ma14061472

Academic Editor: Angelo Marcello Tarantino

Received: 8 February 2021
Accepted: 10 March 2021
Published: 17 March 2021

1. Introduction

Ultra high-performance concrete (UHPC) is one of the leading construction materials with greatly advanced properties compared to conventional concrete. A cementitious mixture of which the compressive strength is over 120 MPa belongs to the UHPC category according to ASTM C1856 [1], while the ACI committee reported that the compressive strength of UHPC should be greater than 150 MPa [2–4]. In addition to the remarkable compressive strength, UHPC, designed based on a particle packing theory, also possesses superior durability compared to conventional concrete with the help of dense microstructure [5–15]. For example, it has strong resistance to water permeability, chloride penetration, freeze and thaw, and chemical attack because of its great material properties. Comprehensive studies on UHPC have been conducted for the last few decades such as the rheological properties of a fresh paste of UHPC [15–20], the effect of fibers [16,19,21–23], mix design [24–29], and structural applications [30–33].

Perhaps, the disadvantage of UHPC comes from the standard ingredient, specifically, a large portion of Portland cement. The cement industry is well known to generate 8–9% of global CO_2 emissions [34]. Even though the structure member size can be smaller with UHPC than conventional concrete because of its high strength [35], UHPC generally contains cement about three times higher than normal concrete by volume [25,36]. Furthermore, not all cement particles in UHPC are hydrated in an extremely low w/c ratio environment. The hydration degree is reported as only 52–61% with a w/c ratio of 0.23–0.33 [37] and the unhydrated cement makes UHPC not eco-friendly [12,38]. Therefore, the possible approach to reduce cement content in UHPC is replacing part of the cement with supplementary cementitious materials (SCMs) such as slag, fly ash (FA), limestone powder (LP), metakaolin (MK), and other SCMs. Most SCMs are industrial by-products or naturally occurring materials: slags are by-products in the production of iron, steel, lithium carbonate, phosphor, or copper; FA is a by-product of a coal power plant; and SF is a by-product from the production of elemental silicon or alloys with silicon. In addition, limestone is a natural occurring material, and MK is obtained from the calcination of kaolinite.

This paper reviewed the effect of various SCMs on UHPC performances. SCMs can be categorized into slag, FA, LP, MK, and other SCMs. The effects of SCMs on compressive strength, flowability, porosity, shrinkage, and environmental impacts of UHPC are discussed. The paper also introduces the method of how to evaluate the environmental effect of each SCM. The objective of the paper is to summarize the current state of SCM application and its performances in UHPC.

2. Summary of SCMs Reviewed

The SCMs reviewed in this paper are summarized in Table 1. The table shows the types of SCMs that have been studied in the literature and their effects on the UHPC's performances. The performances of UHPC investigated are early (at 3 days or earlier) and late compressive strength (later than 3 days), flowability, porosity, and shrinkage. The effects of SCMs were compared to the references in each study. For example, high "Early compressive strength" means that usage of SCMs exhibits higher early compressive strength than a reference mix design. The numbers in the column of "SMC No." are linked to the numbers in the column of "List of SCMs" and the information of SCMs or a combination of SCMs can be found.

When a combination of SCMs is used, the effects of the main SCM are reported. Many studies investigated the usage of combinations of SCMs and it is inefficient to regard each combination as a single different SCM. Therefore, the authors picked one representative SCM in a combination and investigated its effect. The representative SCM in a combination is bolded in the table and others are referred to as a reference. For example, (8) in the "List of SCMs" column in Table 1 used three SCMs: **GGBS (25.5%)** + SF + BA. The compressive strength with the different dosages of GGBS was compared and discussed in the slag section (see Section 3.1). The UHPC with GGBS exhibits lower early age compressive strength compared to the reference with the combination of SF and BA only. In the case of italic texts for the SCMs, the reference is different from the mix designs with SCMs and is not able to be written in the Table 1 because of space limitations. For example, (22) in the "List of SCMs" column in Table 1 compared the effect of *"FFA (20%) + MK (3.8%)"* with the UHPC only with SF. In this case, it is recommended to find more detailed information on the reference mix design through the papers cited. Lastly, the replacement ratio is the weight of the SCM to total solid binder weight, which resulted in the highest compressive strength. For example, (8) in the "List of SCMs" column in Table 1 GGBS is used along with BA and SF. The replacement ratio of 25.5% is obtained from dividing GBBS weight by the sum of cement, GBBS, BA, and SF.

Various combinations of SCMs in UHPC mix designs have been studied by many research groups. Most UHPC studies used SF as a main ingredient and parts of the cement were replaced with other SCMs in the literature. Although there are various combinations, most SCMs are placed in the ordinary categories: slag, FA, LP, and MK. Some other SCMs such as RHA, NP, NMC, DCP, CKD, GGP, BP and FGP are also investigated. The effects of the SCMs on the UHPC performances of early and late compressive strength, flowability, shrinkage, and the environmental impact are discussed comprehensively in the following sections.

Table 1. Summary of the effects of supplementary cementitious materials (SCMs) on the UHPC performances.

Performance		SCM No.	List of SCMs
Early compressive strength ($\leq$3 days)	High	[Slag] 7, 13; [FA] 23, 25; [MK] 32; [O] 35	**[Slag]** (1) **GGBS (60%)** + SF [39] (2) **GGBS (25.5%)** + SF + BA [40] (3) **GGBS (30%)** + SF [41] (4) **GGBS (38.5%)** + SF [42] (5) *GGBS (30%)* [43] (6) **GGBS (20–40%)** + SF [44] (7) **GGBS (23.6%)** + SF [45] (8) **FGGBS (38.5%)** + SF [42] (9) **FGGBS (8.4%)** + SF + BA [40] (10) **SSP (16.9%)** + SF [46] (11) **SSP (15%)** + LP + SF [12] (12) **PS (6.9–34.2%)** + SF + FA [47] (13) **PS (35%)** + SF [48] (14) **PSS (4%)** + SF [49] (15) **LTS (10%)** + SF [50] (16) **CS (16%)** + SF [51] **Fly ash [FA]** (17) **FA (11.8%)** + SF [49] (18) **FA (12.8%)** + SF [40] (19) **FA (15%)** + SF [41] (20) **FA (30%)** + SF [52] (21) **FA (20%)** + SF [53] (22) **FA (38.5%)** + SF [42] (23) **FA (7.4%)** + GGBS + SF [54] (24) *FFA (20%) + MK (3.8%)* [55] (25) **FFA (34.1%)** + SF [56] **Limestone powder [LP]** (26-1) **LP (37.3%)** + SF [57] (26-2) **LP (57.2–78.1%)** + SF [57] (27) **LP (32%)** + SF [58] (28) **LP (4%)** + SF [49] (29) **LP (14%)** + SF + FA [59] (30) **NC (3.2%)** + SF [60] **Metakaolin [MK]** (31) **MK (20%)** + SF [61,62] (32) *MK (16.7%)* [63] (33) **NMK (1%)** + MK [64] (34) **MK (6.9%)** + SF [65] **Others [O]** (35) **RHA (10%)** + SF [66] (36) **NP (11.8%)** + SF [49] (37) **NP (24%)** + SF [67] (38) **NMC (1–9%)** + MK [68] (39) **DCP ($\leq$9%)** + SF + LP [69] (40) **CKD (4%)** + SF [49] (41) **GGP (11.5%)** + SF [70] (42) **FGP (6–15%)** + SF [71] (43) **BP (14%)** + SF + FA [59]
	Low	[Slag] 1, 2, 3, 9, 11, 12, 15; [FA] 18, 19, 21; [LP] 29; [MK] 33; [O] 37, 38, 39, 43	
Late compressive strength (>3 days)	High	[Slag] 1, 2, 5, 6, 7, 9, 12, 14, 15, 16; [FA] 20, 24; [LP] 26-1, 27, 30; [MK] 32; [O] 35, 41, 42	
	Low	[Slag] 3, 4, 8, 10, 11; [FA] 17, 18, 19, 21, 22; [LP] 26-2, 28, 29; [MK] 31, 34, 32; [O] 36, 37, 38, 39, 40, 43	
Flowability	High	[Slag] 2, 4, 6, 7, 8, 9, 10, 12; [FA] 22, 24, 25; [LP] 26-1, 26-2, 28, 29; [O] 41, 42, 43	
	Low	[Slag] 11; [FA] 17, 18; [MK] 33; [O] 36, 37, 38, 39, 40	
Shrinkage	Low	[Slag] 11, 12; [FA] 23, 24; [LP] 26-2, 29; [MK] 32; [O] 43	
	High	[Slag] 6	

3. Effect of SCMs on Material Properties of UHPC

3.1. Slag

Slag is the by-product in pig iron, steel, lithium carbonate, phosphor, or copper plants and their primary oxide components are CaO, SiO_2, and Al_2O_3, although the proportions are not the same among different sources [72]. The effects of slag on UHPC compressive strength, flowability, and shrinkage are summarized in Tables 2–4, respectively. They also include information such as the water to binder (w/b) ratio, the curing method, specimen size, the superplasticizer to binder (SP/b) ratio, the aggregate to binder (Agg/b) ratio, the mixture type, and other solid ingredients. The same types of tables are used for other SCMs in the following subsections.

Table 2. The effect of slag on the compressive strength of UHPC.

SCMs	Compressive Strength (MPa @ Age (% to the Ref.))	w/b Ratio	Curing Method	Specimen Size (mm)	Other Solid Ingredients	Ref.
GGBS (60%) + SF	127 @ 7 (−7.1%) 162 @ 28 (5.2%) 181 @ 90 (8.4%)	0.20	Water	50 cube	Cement (CEM I 52.5 N), Sand	[39]
GGBS (25.5%) + SF + BA	25 @ 1 (−39.0%) 77 @ 3 (−18.1%) 145 @ 28 (0%) 157 @ 91 (0.06%)	0.15	Water	50 cube	Cement (CEM I), Sand, Steel fiber (1 vol.%), Silica powder	[40]
GGBS (30%) + SF	98 @ 3 (−5.7%) 144 @ 28 (−4.0%)	0.16	Water	40 × 40 × 80	Cement (CEM I 42.5 R), Sand, Steel fiber (2 vol.%)	[41]
GGBS (38.5%) + SF	139.4 @ 28 (−16.1%)	0.20	Water and air	100 cube	Cement (CEM I 42.5), Sand, Steel fiber (2 vol.%)	[42]
GGBS (30%)	123 @ 28 (10%) 130 @ 91 (−5.4%)	0.18	Water	40 × 40 × 160	Cement (CEM I 52.5 R), Sand	[43]
GGBS (20–40%) +SF	110–120 @ 28 (0–9%)	0.18	Water	50 cube	Cement (CEM I), Sand	[44]
GGBS (23.6%) + SF	110 @ 3 (0.0%) 125 @ 7 (3.3%)	0.14	Air	40 × 40 × 160	Cement (CEM I 52.5 N), Sand	[45]
FGGBS (38.5%) + SF	163.5 @ 28 (−1.5%)	0.20	Water and air	100 cube	Cement (CEM I 42.5 R), Sand, Steel fiber (2 vol.%)	[42]
FGGBS (8.4%) + SF + BA	13 @ 1 (−68.3%) 101 @ 3 (7.4%) 151 @ 28 (4.1%) 165 @ 91 (5.8%)	0.15	Water	50 cube	Cement (CEM I), Sand, Steel fiber (1 vol.%), Silica powder	[40]
SSP (16.9%) + SF	140 @ 28 (−10.3%)	0.13	Heat, water	100 cube	Cement (CEM I 42.5), Sand, Coarse agg., Steel fiber (1.6 vol.%)	[46]
SSP (15%) + LP + SF	68 @ 1 (−8.7%) 142 @ 28 (−6.4%)	0.16	Water	100 cube	Cement (CEM I 42.5), Sand, Quartz powder, Steel fiber (2 vol.%)	[12]
PS (27.4%) + SF + FA	60 @ 3 (−27.7%) 127.5 @ 28 (2.8%)	0.17	Air	40 × 40 × 160	Cement (CEM I), Sand	[47]
PS (35%) + SF	156.8 @ 3 (3.7%)	0.14	Heat	40 × 40 × 160	Cement (CEM I 52.5), Sand	[48]
PSS (4%) + SF	161 @ 28 (0.0%)	0.15	Water		Cement (CEM I), Sand, Steel fiber (2 vol.%)	[49]
LTS (10%) + SF	98 @ 3 (−4.8%) 146 @ 28 (2.8%) 156 @ 90 (6.8%)	0.18	Water	40 cube	Cement (CEM I 52.5), Sand	[50]
CS (16%) + SF	167 @ 90 (3.1%)	0.15	Water	40 × 40 × 160	Cement (CEM I 52.5 N), Sand	[51]

Table 3. Effect of slag on the flowability of UHPC.

SCMs	Flowability (mm (% to the Ref.))			w/b Ratio	SP/b Ratio	Agg/b Ratio	Type	Ref.
	Slump Flow	Flow Table	Mini Slump					
FGGBS (8.4%) + SF + BA	675 (11.6%)			0.15	0.75%	0.70	Mortar + Steel fiber (1 vol.%)	[40]
GGBS (25.5%) + SF + BA	630 (4.1%)			0.15	0.49%	0.70	Mortar + Steel fiber (1 vol.%)	[40]
SSP (15%) + LP + SF	605 (−0.1%)			0.16	1.80%	1.00	Mortar + Steel fiber (2 vol.%)	[12]
PS (34.2%) + SF + FA		306 (17.2%)		0.17	3.47%	0.90	Mortar	[47]
FGGBS (38.5%) + SF	310 (10.7%)			0.20	3.50%	1.44	Mortar + Steel fiber (2 vol.%)	[42]
GGBS (23.6%) + SF			300 (0.0%)	0.14	0.90%	1.00	Mortar	[45]
GGBS (38.5%) + SF	285 (1.8%)			0.20	3.50%	1.44	Mortar + Steel fiber (2 vol.%)	[42]
GGBS (20–40%) + SF			256 (34.7%)	0.18	2.40%	1.22	Mortar	[44]
SSP (16.9%) + SF		130 (26.8%)		0.13	5.42%	1.25	Mortar	[46]

Table 4. Effect of slag on the shrinkage of UHPC.

SCMs	Shrinkage			w/b Ratio	Binder Weight Ratio			Ref.
	Auto	Dry	Total		Cement	Slag	SF	
SSP (15%) + LP + SF			Low	0.16	0.55	0.35	0.10	[12]
PS (34.2%) + SF + FA	Low			0.17	0.34	0.53	0.13	[47]
GGBS (40%) + SF	High			0.18	0.40	0.40	0.20	[44]

Slag tends to decrease the compressive strength of UHPC at an early age because of its low reactivity. Slag has hydraulic properties and reacts with water [73] and the hydration product of slag is calcium silicate hydrate (CSH) [43,74]. Slag is chemically activated by calcium hydroxide ($Ca(OH)_2$) and gypsum in cement, but its reaction speed is slow, whereas SF reacts with $Ca(OH)_2$ first in UHPC because of its fineness [44,75]. Researchers have also reported that slag decreases the heat of hydration [39,43]. As a result, slag tends to decrease the compressive strength of UHPC at 3 days or earlier. Most of the research revealed that the compressive strengths at 3 days of the UHPCs with slag decrease by around 4.8–18.1% from the reference specimens. Pyo and Kim revealed that the addition of GGBS decreases the compressive strength at 1 day and 3 days by 39% and 18%, respectively [40]. They claimed that GGBS slow down the hydration process and setting time, which causes low early strength. Li et al. reported that SSP improves the workability even though it degrades the mechanical properties and durability of UHPC [12]. The 15% replacement of cement with SSP results in 8.7% lower compressive strength at 1 day than that of the reference specimen. As the SSP has a low hydration activity as well as a retarding effect on the cement hydration, the activity of SSP is lower than OPC and it slowed down the early age hydration. He et al. replaced SF with the LTS that is a by-product in the process of lithium carbonate [50]. The 10% of the LTS to the total binder results in a 4.8% lower compressive strength at 3 days compared to the control mix with SF only. The addition of LTS decreases the early age compressive strength because the pozzolanic reaction of LTS is slower than SF.

Slag could enhance the late age compressive strength of UHPC; the secondary pozzolanic reaction between slag and $Ca(OH)_2$ in the pore solution produces additional CSH, which increases the packing density of the UHPC [44]. Liu et al. found that compressive strength increases up to 9% when GGBS content increased to 40% of the binder because the secondary pozzolanic reaction of GGBS is accompanied by consumption of $Ca(OH)_2$ and the densification of the hardened paste [44]. Abdulkareem et al. reported that the use of GGBS can accelerate the hydration reaction of cement and also improve packing density because its fineness is between those of cement and SF [45]. When GGBS content increased to 23.6% of binder, the compressive strength increased by 3.3% at 7 days because of the improved packing density and higher cement hydration due to the addition of GGBS. Gupta used GGBS with 35% calcium oxide content that contributes to CSH formation, and thus, improves strength development [39]. The 60% replacement of cement with GGBS resulted in a 5.2% increase in the UHPC's compressive strength at 28 days. Yu et al. reported a 10% increase in 28-day compressive strength by replacing 30% of the cement with GGBS [43]. He et al. successfully replaced 10% of cement with LTS and the compressive strength increases by 2.8% and 6.8% at 28 days and 90 days, respectively [50]. The fineness of LTS is between those of cement and SF, which can improve the packing density of UHPC and the pozzolanic reaction of LTS also contributes to the late age compressive strength development. Peng et al. used PS that has a similar glass structure to GGBS [48]. When PS content increased from 30% to 35%, the compressive strength increased by 3.7% because the addition of PS increased both the degree and speed of the pozzolanic reaction, which results in more hydration products in the paste and low porosity. The PS used by Yang et al. has relatively lower reactivity than ordinary slag because of the lower Al_2O_3 content [47]. P_2O_5 in the PS can also retard hydration at an early age. The PS of 27.4% of the binder was found to increase the compressive strength at 28 days by 6.3%. The PS can yield small pores in microstructure at an early age and fill those by the hydration product from the long-term pozzolanic reaction. Edwin et al. proved that CS can be used as an SCM in UHPC increasing the amount of CSH [51]. CS is a by-product of the copper metal smelting process. When CS content increases up to 16% of the binder, the compressive strength of UHPC increases by 3.1% at 90 days. Ahmadm et al. studied the effect of using industrial waste materials like PSS to replace parts of SF [49]. When replacing 20% of SF, the compressive strength of UHPC can still reach 161 MPa at 28 days, which is the same as that of the reference specimen with SF only and flow diameter slightly decreased, but it was still within the acceptable range.

The particle size of slag can be a critical factor in the compressive strength of UHPC. Randl et al. studied the effect of GGBS and FGGBS on the UHPC compressive strength [42]. The GGBS and FGGBS have the Blaine values of 4790 and 5620 cm^2/g, respectively. When 38.5% of cement is replaced with FGGBS, it decreases the compressive strength by 1.5%, whereas GGBS decreases the strength by 16.1% because FGGBS results in a higher packing density than GGBS. The study concluded that the packing density of UPHC is an important factor even more than the hydraulic reactivity of slag. Pyo and Kim compared slags with different particle sizes [40]. The median particle size of FGGBS and GGBS are 2.69 μm and 14 μm, respectively. They reported that the usage of FGGBS increases the compressive strength increased by 7.4% at 3 days, while GGBS decreases the compressive strength with the same dosage of FGGBS. From the hydration heat measurement, they concluded that FGGBS plays a significant role in the early hydration process, which results in a higher compressive strength than the reference specimen.

Some slags may decrease the porosity of UHPC. Liu and Guo studied the effect of SSP on the compressive strength of UHPC [46]. The particle size distribution of SSP is similar to that of cement. It was found that the compressive strength decreased rapidly when the SSP content is high. With the 16.9% replacement ratio of cement with SSP, the compressive strength decreased by 10.3% compared to the reference specimen that contains SF only because SSP increases the proportion of the pores larger than 50 mm by 33%.

The effect of slag on the UHPC shrinkage possibly depends on the type of slag. Table 4 summarizes the effect of slag on the UHPC shrinkage. It has been proved that the addition of GGBS increases the shrinkage of conventional concrete because slag increases the self-desiccation by consuming pore solution (calcium hydroxide) in a small capillary pore structure [76–78]. However, different effects of different types of slags on the UPHC shrinkage have been observed in some studies in the literature. Li et al. found that the total shrinkage of UHPC incorporating SSP is lower than the UHPC without SSP [12]. As the amount of SSP increases, hydration of cementitious materials decreases at early ages, water consumption is reduced, and, thus, the self-desiccation of UHPC becomes weaker. Yang et al. indicated that as the cement replacement ratio with PS increases, the hydration can be slowed down and the cement dilution effect can improve the UHPC volume stability [47]. In other words, the high volume of PS can reduce the autogenous shrinkage of UHPC. On the other hand, Liu et al. found that the addition of GGBS can increase autogenous shrinkage [44]. They insisted that the secondary pozzolanic reaction of GGBS increases the consumption of calcium hydroxide, which increases water consumption. The higher water consumption results in more self-desiccation. They concluded that the secondary pozzolanic reaction of GGBS increases the autogenous shrinkage due to the refined pore structure and the increased depletion of water.

Slag increases the UHPC flowability because of its lower water absorption compared to cement having a slippery surface [79]. Table 3 summarizes the effect of slag on the UHPC flowability. Pyo and Kim reported that the addition of FGGBS and GGBS increases the UHPC flowability by 11.6% and 4.1%, respectively, compared to the reference specimen with SF only [40]. Yang et al. reported that the use of PS can significantly improve the flowability of UHPC. The flowability of UHPC can increase by 17.2% when the cement replacement ratio with PS increased up to 34.2% because it reduces the water absorption [47].

Furthermore, the addition of PS provides the cement dilution effect, which increases the water to cement ratio of UHPC with PS indirectly. Abdulkareem et al. reported that the workability is improved by the addition of GGBS [45]. In the paper, with the increase in GGBS, the dosage of superplasticizer should be reduced to achieve the same level of slump flow. Liu et al. used GGBS to improve the flowability of UHPC. With up to 30% replacement of cement with GGBS, the slump flow can increase by 6.3% because of the smooth surface and lower water absorption of the slag compared to cement [44]. Liu and Guo proved that the addition of SSP can improve the flowability of UHPC. The 16.9% replacement ratio of cement increases the flow diameter by 26.8% because the activity of SSP is lower than that of cement and the water requirement of SSP is also less than that of cement [46]. However, Li et al. found that the 15% replacement of cement with the different SSP source slightly decreases the flowability of UHPC because of the higher specific area than that of cement, which caused higher water demand [12]. Therefore, it should be pointed out that a higher specific area of different sources of slag could decrease the flowability of UHPC. Randl et al. found that the specific area of GGBS and FGGBS can improve the flowability of UHPC [42]. The slump flow of UHPC increased by 1.8% with 38.5% GGBS replacement and the slump flow of UHPC increased by 10.7% with 38.5% FGGBS replacement.

Slag is possibly more helpful than FA to improve the UHPC's compressive strength. Wu et al. reported that when the cement replacement ratio is same, slag exhibits a higher compressive strength of UHPC than FA [41]. The 30% replacement ratio of cement with GGBS results in 5.7% and 4.0% lower at 3 days and 28 days, respectively, compared to the reference specimen containing SF only, while the 30% replacement with FA results in 13.5% and 8.0% lower at 3 days and 28 days, respectively.

Although slag is apt to decrease the early compressive strength of UHPC, many studies have demonstrated that it successfully improves the late compressive strength. The additional CSH produced by the secondary pozzolanic reaction between slag and $Ca(OH)_2$ increases the density of the matrix and the compressive strength, which happens at the late age because of the slow hydration of slag. The disadvantage of the low early compressive strength is assumed to be overcome by adopting heat treatment as it increases the pozzolanic reactivity. The particle size of slag is also an important factor to increase the compressive strength of UHPC; slag of a finer particle size exhibits higher compressive strength. However, the extra grinding work increases the material cost and, therefore, finding reactive slag material seems a more efficient option. Slag decreases the water demand of UHPC because of its lower water absorption compared to cement. It is another critical factor to increase the compressive strength of UHPC because a lower w/c ratio can increase the compressive strength.

3.2. *Fly Ash (FA)*

FA is a by-product of power plants and is collected during the process of coal combustion. The chemical composition and particle size of FA are different from plant to plant, but it is generally a fine spherical powder, which increases the workability of conventional cementitious material. As a pozzolanic material, it is known that FA increases the late age strength of conventional cementitious material. The usage of FA can reduce CO_2 emissions [55,66] and decrease the production cost and energy of concrete [41,49,53,55]. In recent years, many researchers have focused on developing new UHPC mixtures with locally available FA because substituting cement and/or SF with FA can reduce environmental impacts.

The effect of FA on UHPC compressive strength is summarized in Table 5. The compressive strength data reported show around 95 MPa at 3 days, 110–185 MPa at 28 days, and 152–202 MPa at 91 days with a 10–20% replacement of binder materials. It has been shown that the UHPC with FA exhibits a lower compressive strength than those of reference specimens. Ahmad et al. replaced a part of SF with FA, and found that using FA to substitute the SF to up to 11.8% of the binder slightly decreases the compressive strength by 1.9% compared to the reference specimen at 28 days [49]. Although FA degrades the compressive strength of UHPC, the value is higher than the minimum requirement of 150 MPa and the usage of FA can reduce the cost of UHPC. It has been proved that FA can improve many characteristics of high strength mortar. However, Pyo et al. found that replacing 12.8% cement with FA decreased the compressive strength by 48.9% and 6.1% at 1 day and 3 days, respectively, because of the high crystallinity of FA [40]. Wu et al. also investigated the effect of FA as an SCM for concrete, and concluded that the FA has negative effects on the compressive strength of UHPC [41]. The 15% replacement ratio of cement with FA results in 13.5% and 8% lower than those of reference specimen at 3 days and 28 days, respectively. Alsalman et al. adopted that FA can be used as an SCM for UHPC to reduce the cost of UHPC [53]. It was found that adding FA up to 15% of the binder significantly decreases the compressive strength by 33.7% at 1 day because the addition of FA delayed the strength development at early ages. The compressive strength of UHPC becomes similar to that of the reference specimen after the normal curing of 7 days or longer. Randl et al. found that the 38.5% replacement ratio of cement with FA decreases the compressive strength by 24.9% at 28 days compared to the reference specimen, even though the packing density is higher [42]. Therefore, it is inferred that the slow pozzolanic reaction of FA degrades the compressive strength of UPHC. It should be noted, however, that there are occasional studies reporting different trends. Šeps et al. replaced the cement of 30% with FA and it results in 19% higher compressive strength at 28 days than that of the reference specimen containing SF only [52].

Table 5. The effect of FA on the compressive strength of UHPC.

SCMs	Compressive Strength (MPa @ Age (% to the Ref.))	w/b Ratio	Curing Method	Specimen Size (mm)	Other Solid Ingredients	Ref.
FA (11.8%) + SF	158 @ 28 (−1.9%)	0.15	Water	50 cube	Cement (CEM I), Sand, Steel fiber (2 vol.%)	[49]
FA (12.8%) + SF	24 @ 1 (−48.9%) 92 @ 3 (−6.1%) 152 @ 28 (−1.3%) 164 @ 91 (0%)	0.15	Water	50 cube	Cement (CEM I), Sand, Steel fiber (1 vol.%)	[40]
FA (15%) + SF	90 @ 3 (−13.5%) 138 @ 28 (−8%)	0.20	Water	40 × 40 × 80	Cement (CEM I 42.5), Sand	[41]
FA (30%) + SF	125 @ 28 (19%)	0.26	Air	100 cube	Cement (CEM I), Sand, Coarse agg.	[52]
FA (20%) + SF	53.1 @ 1 (−33.7%) 101.5 @ 7 (−1.26%) 114.5 @ 28 (−0.7%) 131.7 @ 56 (2.1%) 152.1 @ 90 (−1.9%)	0.16	Water	50 cube	Cement (CEM I), Sand, Steel fiber (3 vol.%)	[53]
FA (38.5%) + SF	124.7 @ 28 (−24.9%)	0.20	Water and air	100 cube	Cement (CEM I 42.5 R), Sand, Steel fiber (2 vol.%)	[42]
FA (7.4%) + GGBS + SF	281 @ 1 (4.1%)	0.15	Autoclave	50 cube	Cement (CEM I 42.5), Sand	[54]
FFA (20%) + MK (3.8%)	150 @ 28 (26%)	0.20	Water	50 cube	Cement (CEM III)	[55]
FFA (34.1%) + SF	160.3 @ 3 (6.8%)	0.16	Water and steam	50 cube	Cement (CEM I 42.5 R), Sand, Steel fiber (1 vol.%)	[56]

Some of the FFA can improve the compressive strength of UHPC. Ferdosian and Camões introduced the method of how to optimize the UHPC mix design that satisfies the requirements of the compressive strength and the flowability using FFA of which the mean particle size is 4.48 μm [56]. They suggested the eco-efficient mix design that releases the lowest CO_2 and the cost-efficient mix design that maximizes the amount of FFA and sand as well as minimizes the amount of SF. The eco-efficient mix design results in the compressive strength being 6.8% higher than the reference samples using the FFA of 34.1% in the binder.

The ternary use of SCMs including FA also can be a feasible solution to reduce the amount of cement and SF in UHPC. Li found that the ternary use of FA, MK, and cement can provide better compressive strength of UHPC than the binary use of SF and cement [55]. The 20% and 3.8% of cement were replaced with FA and MK, respectively, and the compressive strength increased by 26% at 28 days. Yazıcı et al. found that the ternary SF-FA-GGBS binder system is effective for reducing SF and water demand without sacrificing compressive strength [54]. In the binder system, the 10% replacement ratio of cement with FA increased the compressive strength by 4.1% up to 281 MPa at 1 day under the autoclave curing condition. However, a fundamental understanding of the ternary use of SCMs improving mechanical properties of UHPC is still not clear and requires further research.

Table 6 summarizes the effect of FA on the UHPC flowability, which is controversial among studies. Some researchers reported that FA can improve the workability of UHPC. Li found that the ternary use of FA, MK, and cement can significantly increase the flowability of UHPC by 47% compared to the binary use of cement and SF [55]. Randl et al. found that the addition of FA can increase the flowability of UHPC [42]. The 38.5% replacement ratio of cement with FA can increase the flow diameter of fresh UHPC by 3.6% compared to the reference specimen with SF only. Ferdosian and Camões focused on developing a mixing design method to minimize CO_2 content and material cost with acceptable compressive strength and workability [56]. The 34.1% FA in the cementitious binder satisfied the low limit of flowability of 190 mm. However, degradation of workability by using FA in UHPC was reported. Pyo and Kim found that the 15.7% replacement of silica powder with FA decreases the slump flow by 6.6% compared to the reference specimen [40]. Ahmad et al. studied that the effect of FA replaces the SF in UHPC. It was found that when the use of FA as a replacement of SF and its content is increased to 11.8% of binder, the flowability of UHPC slightly decreases by 8.7% than the reference specimen with SF only but the lower flowability is still acceptable [49].

Table 6. The effect of FA on the flowability of UHPC.

SCMs	Flowability (mm (% to the Ref.))			w/b Ratio	SP/b Ratio	Agg/b Ratio	Type	Ref.
	Slump Flow	Flow Table	Mini Slump					
FA (12.8%) + SF	565 (−6.6%)			0.15	0.75%	0.71	Mortar + Steel fiber (1 vol.%)	[40]
FA (38.5%) + SF	290 (3.6%)			0.20	3.50%	1.44	Mortar + Steel fiber (2 vol.%)	[42]
FFA (20%) + MK (3.8%)		258 (47%)		0.20	1.00%	-	Paste	[55]
FA (11.8%) + SF			210 (−8.7%)	0.15	3.57%	0.90	Mortar + Steel fiber (2 vol.%)	[49]
FFA (34.1%) + SF		190 (0.0%)		0.16	2.50%	1.07	Mortar + Steel fiber (1 vol.%)	[56]

Table 7 summarizes the effect of FA on UHPC shrinkage. It has been shown that the FA can reduce the shrinkage of UHPC. Li et al. found that the ternary use of FA, MK, and cement can reduce the drying shrinkage of UHPC compared to the reference specimen with SF only because the ternary use can reduce water demand [55]. Yazıcı et al. replaced cement with FA and GGBS to reduce the cement amount in UHPC [54]. It was found that when the content of GGBS in the binder is constant, the 10% replacement ratio of cement with FA results in lower shrinkage than the reference specimen with SF only because of the lower amount of cement in UHPC.

The advantage of the usage of FA in UHPC cannot be observed in the compressive strength; most studies using FA reported degradation of the compressive strength of UHPC. The effect of FA on workability is arguable, and this might come from different characteristics of FAs from different sources. Therefore, the purpose of the usage of FA can be limited in reducing material cost or CO_2 emission as discussed in the studies. Perhaps FA can increase the durability of UHPC; however, further studies are required to demonstrate it.

Table 7. The effect of FA on the shrinkage of UHPC.

SCMs	Shrinkage			w/b Ratio	Binder Weight Ratio			Ref.
	Auto	Dry	Total		Cement	FA	SF	
FFA (20%) + MK(3.8%)		Low		0.20	0.77	0.23	-	[55]
FA (8%) + GGBS + SF			Low	0.15	0.64	0.16	0.20	[54]

3.3. Limestone Powder (LP)

The effect of LP in conventional cement and concrete is well known; the use of LP in concrete has various advantages. It can reduce the material cost and CO_2 emission because of having an abundant reservoir. LP has the nucleation effect in early hydration reaction that accelerates the cement hydration [58,80]. It can also physically fill the void and increase the packing density of the system [80]. As a consequence, LP increases the compressive strength of concrete at an early age. However, it may reduce the compressive strength at a late age because it does not have a pozzolanic reaction with cement unlike other SCMs such as slag, FA, and MK, and it requires higher water demand. In this section, the effect of LP in UHPC is reviewed

The effect of FA on UHPC compressive strength is summarized in Table 8. Three different mechanisms of how LP affects the compressive strength of UHPC were observed. First, LP enables the reduction in the amount of superplasticizer to maintain the same flowability. Huang et al. studied the effect of LP on the hydration of UHPC with different cement replacement ratios [58]. The retardation effect caused by the superplasticizer decreases as LP enables the reduction in the amount of superplasticizer by 62.8%, and, as a result, the early compressive strength is not degraded. It was also found that the 32% replacement ratio of cement with LP results in 10.7% and 16.1% higher compressive strength at 28 days and 56 days, respectively.

Table 8. The effect of LP on the compressive strength of UHPC.

SCMs	Compressive Strength (MPa @ Age (% to the Ref.))	w/b Ratio	Curing Method	Specimen Size (mm)	Other Solid Ingredients	Ref.
LP (37.3%) + SF	159.5 @ 28 (4.3%)	0.20	Water	50 cube	Cement (CEM I 52.5 R), Sand	[57]
LP (32%) + SF	165 @ 28 (10.7%) 180 @ 56 (16.1%)	0.13	Sealed	40 × 40 × 160	Cement (CEM I 52.5 N), Sand	[58]
LP (4%) + SF	152 @ 28 (−5.6%)	0.15	Water	50 cube	Cement (CEM I), Sand, Steel fiber (2 vol.%)	[49]
LP (14%) + SF + FA	100 @ 7 (−4.2%) 120 @ 28 (−1.1%) 140 @ 56 (−4%)	0.16	Water	40 × 40 × 160	Cement (CEM I), Sand	[59]
NC (3.2%) + SF	120 @ 7 (9%) 155 @ 28 (15%)	0.16	Water	40 × 40 × 160	Cement (CEM I 42.5), Sand, Steel fiber (2 vol.%)	[60]

Second, LP has a pozzolanic reaction with SF. Li et al. adopted that replacing cement with LP and the optimum content of 37.3% replacement ratio increases the compressive strength of UHPC by 4.3% at 28 days than that of the control mix with SF only [57]. As UHPC with LP has a higher pozzolanic reaction with SF, which contributes to the CSH formation at late ages, strength development can be improved at late ages.

Third, the fine particle-sized LP can accelerate cement hydration. Wu et al. reported the addition of nanoparticles such as NC can improve the mechanical properties and durability of UHPC [60]. It is found that the 3.2% replacement ratio of cement with NC increases the compressive strength of UHPC by 1.1% and 5% at 7 days and 28 days, respectively, because the use of NC accelerates the hydration of cement and makes the microstructure denser due to smaller particle size of NC compared to cement.

However, in some cases, LP can degrade the compressive strength of UHPC. Yang et al. investigated the effect of LP on the hardened properties of the UHPC that contains FA and SF of 24% and 12%, respectively [59]. The 14% replacement ratio of cement with LP decreases the compressive strength of UHPC by 4.2% at 7 days than that of the reference specimen with SF and FA. Although the compressive strengths of UHPC at 28 days and 56 days increase compared to day 7 compressive strength, these two strengths are still 1.1% and 4% lower than the reference specimen, respectively. Even though the LP has the nucleation effect increasing the cement hydration speed, it also dilutes the cement hydration resulting in lower heat of cement hydration. As the amount of LP increases in the low cement binder, the dilution effect becomes more dominant. Ahmad et al. studied the effect of the use of locally available industrial waste material such as LP as a partial substitution of SF [49]. The use of LP decreases the compressive strength of UHPC by 5.65% when the content of LP increases to 4% of the binder compared to the reference specimen with SF only.

LP can significantly improve the workability of UHPC as shown in Table 9. Li et al. insisted that LP can be regarded as a mineral plasticizer that improves the flowability of the UHPC [57]. The 37.3% replacement ratio of cement with LP results in 45.1% higher flowability than that of the reference specimen that contains SF only. The plasticization effect of LP increases the workability of UHPC because of the repulsion between OH^- group localized on the Ca^{2+} surface and its lower water absorption [57]. Yang et al. found that the use of LP as a partial substitution of cement can enhance the flowability of UHPC [59]. The 14% replacement ratio of cement with LP increases the flow diameter by 65.5% than the reference specimen with SF only. This can be attributed to the higher w/c ratio as a part of cement is replaced with LP. Ahmad et al. also found that the use of LP increases the flowability of UHPC by 10.9% when the content of LP is 4% of the binder compared to the reference specimen with SF only [49].

Table 9. The effect of LP on the flowability of UHPC.

SCMs	Flowability (mm (% to the Ref.))			w/b Ratio	SP/B Ratio	Agg/b Ratio	Type	Ref.
	Slump Flow	Flow Table	Mini Slump					
LP (4%) + SF			255 (10.9%)	0.15	3.57%	0.90	Mortar + Steel fiber (2 vol.%)	[49]
LP (14%) + SF + FA		240 (65.5%)		0.16	2.20%	0.85	Mortar	[59]
LP (37.3%) + SF			450 (45.1%)	0.20	1.30%	0.78	Mortar	[57]

LP can lower the shrinkage of UHPC by reducing the amount of cement in UHPC as shown in Table 10. Li et al. found that a 57.2% replacement ratio of cement with LP can improve the total shrinkage of UHPC compared to that of the reference specimen with SF only [57]. The study insisted that the lower amount of cement in UHPC replaced with LP slows down the hydration and reduces the hydration products, and, thus, results in the lower autogenous shrinkage. It should be pointed out, however, that the high content of LP up to 78.1% of the binder provides more free water, and, thus, drying shrinkage increases. In consequence, the total shrinkage decreases because the reduction in autogenous shrinkage is greater than the increase in drying shrinkage. Yang et al. also reported that replacing 14% cement with LP reduces the autogenous and dry shrinkage compared to the reference specimen with SF only [59].

Table 10. The effect of LP on the shrinkage of UHPC.

SCMs	Shrinkage			w/b Ratio	Binder Weight Ratio			Ref.
	Auto	Dry	Total		Cement	LP	SF	
LP (57.2) + SF			Low	0.20	0.39	0.57	0.04	[57]
LP (14%) + SF + FA	Low	Low	Low	0.16	0.49	0.39	0.12	[59]

Although three different mechanisms of how LP increases the compressive strength of UHPC have been proposed, the actual performance of LP in UHPC is debatable. From the literature, it was confirmed that LP increases the workability of UPHC. Therefore, the mechanism of LP to improve the compressive strength of UHPC by reducing water content seems appropriate. The finer LP enhances the compressive strength of UHPC by accelerating the cement hydration. Some studies insisted that the addition of LP decreases the amount of cement in UHPC, which degrades the compressive strength of UHPC. However, their dosages are lower than the other studies showing higher compressive strength with LP, and, therefore, other unknown factors of LP were assumed to degrade the compressive strength.

3.4. Metakaolin (MK)

MK obtained by calcining kaolin has the main chemical composition of alumina and silica, and, therefore, MK is also a pozzolanic material. Studies have reported that MK increases the durability of concrete: low permeability, high resistance against frost, and chemical attack [81–83].

The effect of MK on the UHPC compressive strength is summarized in Table 11. The use of MK only seems to increase the early age compressive strength of UHPC but decreases the late age compressive strength of UHPC. Li et al. found that replacing cement with MK can improve early age compressive strength but the late age compressive strength is decreased compared to UHPC with SF only [63]. It was found that the 16.7% replacement ratio of cement with MK results in 47% higher 1-day compressive strength than the reference specimen with SF only because the use of MK improves the cement hydration at an early age. However, it decreases the 28-day compressive strength by 11.8% compared to the reference sample, of which impact is less significant compared to that of 1-day compressive strength. Tafraoui et al. found a replacement of SF of 20% with MK decreasing the 28-day compressive strength of UHPC by 26.1% with steam curing, and by 5.8% with water curing, respectively [61,62]. The more significant loss of 26.1% compared to 5.8%, even with the same dosage of MK, is because of the usage of the crushed quartz that can lower the compactness by a looseness of granular stacking.

Table 11. The effect of MK on the compressive strength of UHPC.

SCMs	Compressive Strength (MPa @ Age (% to the Ref.))	w/b Ratio	Curing Method	Specimen Size (mm)	Other Solid Ingredients	Ref.
MK (20%) + SF	119, 178, 183 @ 28 (−26.1%, 8.7%, −13.7%) (23 °C, 90 °C, 150 °C)	0.22	Water at 23 °C; and steam at 90 and 150 °C	40 × 40 × 160	Cement (CEM I 42.5), Sand	[61]
MK (20%) + SF	146 @ 28 (−5.8%)	0.22	Water	40 × 40 × 160	Cement (CEM I 52.5 N), Sand	[62]
MK (16.7%)	106 @ 3 (47.0%) 134 @ 28 (−11.8%)	0.20	Water	50 cube	Cement (CEM III), Sand	[63]
NMK (1%) + MK	120 @ 3 (−0.8%) 146 @ 7 (−1.3%) 178 @ 28 (7.9%)	0.20	Heat	100 cube	Cement (CEM I), Sand, Coarse agg.	[64]
MK (6.9%) + SF	163.8 @ 28 (9.3%)	0.25	Sealed	50 cube	GGBS, SF, Potassium, Sand (Alkali-activated material)	[65]

NMK may overcome the degradation of the UPHC compressive strength caused by MK. Muhd Norhasri et al. indicated that the inclusion of NMK in UHPC can achieve a similar compressive strength at early ages compared to the UHPC with MK only [64]. NMK inclusion of 1% in UHPC can increase the compressive strength of UHPC at 28 days by 7.9% than that of the reference specimen with MK only because nano-MK provides a moderate ultra-filling effect in densifying the UHPC. The disadvantage of NMK is that it decreases the workability of UPHC; 1% NMK in UHPC decreases the slump flow by 2.4% because of the higher surface of NMK than that of MK (See Table 12).

Table 12. The effect of MK on the flowability of UHPC.

SCMs	Flowability (mm (% to the Ref.))			w/b Ratio	SP/b Ratio	Agg/b Ratio	Type	Ref.
	Slump Flow	Flow Table	Mini Slump					
NMK (1%) + MK	162 (−2.4%)			0.20	2.00%	1.00	Mortar	[64]

The effect of MK on the shrinkage of UHPC can be different concerning the type of shrinkage measured as shown in Table 13. Li and Rangaraju studied the effect of MK on the shrinkage of UHPC [63]. The addition of MK of 16.7% increases the autogenous shrinkage by 0.16%, but it decreases the drying shrinkage by 0.1%. However, no clear explanation of the different effects of MK on the different types of shrinkage is proposed.

Table 13. The effect of MK on the shrinkage of UHPC.

SCMs	Shrinkage			w/b Ratio	Binder Weight Ratio			Ref.
	Auto	Dry	Total		Cement	MK	SF	
MK (16.7%)	High	Low		0.20	0.83	0.17	-	[63]

MK can be incorporated in alkali-activated material (AAM). Wetzel and Middendorf introduced the UHPC made by AAM. Slag, MK, and SF were mixed with hydroxide solution and glass water [65]. The specimens were cured at 60 °C and exhibit a compressive strength at 28 days over 150 MPa. The alkalinity of AAM is higher than ordinary Portland cement; the pH of AAM is usually over 14, whereas that of ordinary Portland cement is 12.6–13.5. Due to the highly alkaline environment of AAM, SF even increases the workability of UPHC and MK reduces much less than the case of ordinary Portland cement. As a result, AAM concrete shows good workability. MK creates the geopolymer network of Si-O-T (Si, Al) in AAM which increases the chemical attack resistance of UHPC.

MK tends to decrease the compressive strength of UHPC. It decreases the workability of UHPC and its beneficial effect on the shrinkage is not clear. Based on the fact that MK is not naturally stored but needs to be calcined, it also is difficult to find the merits of MK in material cost and CO_2 emission compared to slag, FA, or LP. Therefore, the usage of MK in UHPC seems not suitable. However, another possible application was found; the geopolymer or alkali-activated concrete resulted in a compressive strength of over 150 MPa. As geopolymer is well known for its lower CO_2 emission compared to OPC, developing geopolymer UHPC with MK can be an interesting research subject.

3.5. Other SCMs

Studies adopting other SCMs that do not belong to the SCM categories of slag, FA, LP, and MK to reduce the amount of cement and SF in UPHC are summarized in this subsection. Here are the summaries of the studies reviewed in this paper. Tables 14–16 summarize the effects of other SCMs on the compressive strength, the flowability, and the shrinkage of UHPC, respectively.

Table 14. The effect of other SCMs on the compressive strength of UHPC.

SCMs	Compressive Strength (MPa @ Age (% to the Ref.))	w/b Ratio	Curing Method	Specimen Size (mm)	Other Solid Ingredients	Ref.
RHA (10%) + SF	135 @ 3 (10.6%) 155 @ 7 (5.3%) 185 @ 28 (8.8%) 205 @ 91 (4.1%)	0.18	Moisture	40 cube	Cement (CEM I 52.5 N), Sand	[66]
NP (11.8%) + S	152 @ 28 (−4.3%)	0.15	Water	50 cube	Cement (CEM I), Sand, Steel fiber (2 vol.%)	[49]
NP (24%) + SF	110 @ 7 (−11.4%) 124.5 @ 14 (−6.3%) 130.6 @ 28 (−8.7%) 151 @ 90 (−6.6%)	0.15	Water	100 cube	Cement (CEM I), Sand, Steel fiber (2 vol.%)	[67]
NMC (1–9%) + MK	100 @ 3 (−16.7%) 130 @ 7 (−13.3%) 160 @ 28 (−3.0%) 179 @ 90 (6.5%)	0.20	Heat	100 cube	Cement (CEM I), Sand, Coarse agg.	[68]
DCP ($\leq$ 9%) + SF + LP	45 @ 3 (−0.8%) 65 @ 7 (−0.3%) 100 @ 28 (−0.6%)	0.18	Water	40 × 40 × 160	Cement (CEM I 52.5), Sand	[69]
CKD (4%) + SF	154 @ 28 (−5.6%)	0.15	Water	50 cube	Cement (CEM I), Sand, Steel fiber (2 vol.%)	[49]
GGP (11.5%) +SF	188 @ 28 (15.4%)	0.18	Autoclave	40 × 40 × 160	Cement (CEM I 42.5 R), Sand, Steel fiber (2 vol.%)	[70]
FGP (6%) +SF	125 @ 7 (7.1%) 175 @ 28 (5.0%) 183 @ 56 (4.8%) 196 @ 91 (7.7%)	0.19	Sealed	50 cube	Cement (CEM HS), Sand, Quartz powder	[71]
BP (14%) + SF + FA	90 @ 7 (−16.7%) 120 @ 28 (−1.1%) 130 @ 56 (−10.9%)	0.16	Water	40 × 40 × 160	Cement (CEM I), Sand	[59]

Table 15. The effect of other SCMs on the flowability of UHPC.

SCMs	Flowability (mm (% to the Ref.))			w/b Ratio	SP/b Ratio	Agg/b Ratio	Type	Ref.
	Slump Flow	Flow Table	Mini Slump					
DCP (9%) + SF + LP		255 (−18.9%)		0.18	3.00%	0.90	Mortar	[69]
BP (14%) + SF + FA		230 (58.6%)		0.16	2.20%	0.85	Mortar	[59]
FGP (6%) +SF			225 (18.4%)	0.19	1.25%	1.13	Mortar	[71]
CKD (4%) + SF			220 (−4.3%)	0.15	3.57%	0.90	Mortar + Steel fiber (2 vol.%)	[49]
GGP (11.5%) + SF		200 (4.2%)		0.18	1.90%	1.18	Mortar + Steel fiber (2 vol.%)	[70]
NP (11.8%) + SF			195 (−15.2%)	0.15	3.57%	0.90	Mortar + Steel fiber (2 vol.%)	[49]
NP (24%) + SF		184 (−12.4%)		0.15	3.57%	0.89	Mortar + Steel fiber (2 vol.%)	[67]
NMC (1%) + MK	155 (−6.6%)			0.20	1.00%	1.53	Concrete	[68]

Table 16. The effect of other SCMs on the shrinkage of UHPC.

SCMs	Shrinkage			w/b Ratio	Binder Weight Ratio			Ref.
	Auto	Dry	Total		Cement	SCMs	SF	
BP (14%) + SF + FA	Low	Low		0.16	0.49	0.39	0.12	[59]

3.5.1. Rice Husk Ash (RHA)

RHA obtained by burning rice husk has a very high specific surface area, higher than 250 m^2/g. The small particle size and the amorphous structure of RHA make it a "highly active pozzolan". Van Tuan et al. indicated that cement hydration can be accelerated by the addition of RHA of which mean particle size is 5.6 μm, and it can reduce porosity and improve the compressive strength of concretes [66]. The 10% replacement ratio of cement with RHA can increase compressive strength by 10.6% and 8.8% at 3 days and 28 days, respectively. It was also found when the grinding time increases to produce the fine RHA, the pore structure of RHA is gradually collapsed resulting in the lower porosity of RHA. This collapse of RHA can improve the compressive strength of UHPC. It was also found that SF and RHA has a synergic effect on the compressive strength of UHPC; the SF contributes to the early age compressive strength, while RHA to the late age compressive strength.

3.5.2. Natural Pozzolan (NP)

NP obtained from volcanic rocks is a raw material that shows pozzolanic properties so it can lower both costs and CO_2 emission of concrete. The content of NP up to 11.8% of binder in UHPC decreases the compressive strength at 28 days by 4.3% compared to the reference specimen with SF only [67]. It was found that the replacement of 24% cement with NP decreases the compressive strength at all ages compared to the reference specimen without NP, but the compressive strength of UHPC is still over 150 MPa at 90 days. NP also decreases the flowability by 15.2% because of its higher specific surface area (6666 cm^2/g) than that of Portland cement (3700 cm^2/g)

3.5.3. Nano-Metaclay (NMC)

Norhasri et al. adopted the NMC made from nanoclay which undergoes calcination for 3 h [68]. The particle size of NMC is very small, as much as 20 nm, which increases the water demand and retards cement hydration. The replacement of the cement of 1% with NMC decreases the compressive strength by 16.7%, 13.3%, and 3.0% compared to the reference specimen with MK only at 3 days, 7 days, and 28 days, respectively. However, the use of NMC increases the 90-day compressive strength by 6.5% because it can fill pores and yield a pozzolanic reaction at late ages. The higher surface area of NMC than that of cement and MK led to decreases in workability; the 1% inclusion of NMC in the UHPC paste reduces the slump flow of UHPC by 6.6% compared to the reference specimen.

3.5.4. Dehydrated Cementitious Powder (DCP)

DCP can be obtained from recycled construction waste cementitious materials by heating up to 1000 °C. High temperature dehydrates hydrated products such as ettringite, CSH gel and $Ca(OH)_2$. The dehydrated hydration product will rebuild new hydration products, which are similar to the initial hydration products. Since high temperature is essential to produce DCP from construction wastes, it may not reduce the CO_2 emission; however, it can resolve an issue with a large amount of construction wastes. Qian et al. found that the replacement of cement up to 9% with DCP almost has no significant effect on the compressive strength of UHPC compared to the reference specimen with SF only [69]. However, DCP decreases the flowability by 18.9% because of the higher water demand caused by its larger specific surface area. It was also observed that the internal unstable

CSH structure and the rehydration of CaO and other substances in DCP consume more water after heating treatment.

3.5.5. Cement Kiln Dust (CKD)

Ahmad et al. studied the effect of CKD on the compressive strength of UHPC as a partial substitution of SF [49]. CKD is the fine-grained, solid, and strong alkaline waste removed from cement kiln exhaust gas by air pollution control devices in a cement plant. The content of CKD of up to 4% of the binder can obtain the compressive strength of over 150 MPa at 28 days but still 5.6% lower than that of reference specimen with SF. The addition of CKD decreased the mini-slump value by 4.3% compared to the reference mix. This is attributed to the high CaO content in CKD up to 49.3%. It is reported that the high CaO content in CKD increases the water demand.

3.5.6. Ground Granite Powder (GGP)

GGP can be obtained from stone processing plants. Since GGP is an industrial waste, it can reduce the cost of UHPC by replacing parts of the SF and cement. The pore structure of the cement matrix is improved mainly because GGP is finer than cement, and it can help fill the pores in the hardened cement matrix. It was found that the replacement of 11.5% cement with GGP increases the compressive strength by 15.4% at 28 days than the reference specimen without GGP [70]. Since GGP works as a filler and does not have a pozzolanic reaction in UHPC, the GGP over an optimum amount yielded lower strength. GGP increases the workability of UHPC because GGP lowers the viscosity of the UHPC mortar as it does not react with cementitious material, resulting in increased the flowability by 4.2% compared to the reference specimen.

3.5.7. Basalt Stone Powder (BP)

Yang et al. exploited a BP to reduce the cement amount in UHPC. BP is a type of stone powder obtained from aggregate and its main particle size is around 10–50 μm [59]. The replacement of 14% cement with BP results in a compressive strength of 16.7%, 1.1%, and 10.9% lower at 7, 28, and 56 days, respectively, compared to the reference specimen with SF and FA only because BP has no chemical effect in cement hydration and only plays a role as filler in UHPC. BP can improve the flowability resulting in a 58.3% higher flowability than that of the reference specimen. This can be attributed to the dilution effect of the added BP and its lower water absorption. That BP decreases the shrinkage of UHPC was also found. The BP dosage of 14% to the total binder resulted in lower autogenous and drying shrinkage because the usage of BP reduces the amount of cement in UHPC, and BP can make the microstructure denser so that the surface water evaporates slowly compared to the reference specimen with SF and FA only.

3.5.8. Fine Glass Powder (FGP)

Soliman and Tagnit-Hamou found that the use of FGP as a partial substitution of SF can improve both compressive strength and workability [71]. When replacing SF with FGP up to 6% of the total binder, the 28-day compressive strength of UHPC increases by 5.0% than the reference specimen with SF only. This is attributed to the pozzolanic reaction from SF and FGP. The use of FGP also increases the workability and when the content of FGP up to 6% of the binder the slump flow increases by 18.4% compared to the reference specimen because FGP can decrease the water demand of UHPC with the lower surface area of FGP than that of SF.

RHA and FGP are suitable to improve the compressive strength of UHPC at all ages. The high pozzolanic reactivity of materials resulted in a beneficial effect on the compressive strength. GCP also enhances the compressive strength; however, it works as a filler having no chemical reactions with cement. Other SCMs introduced in this paper degrade the compressive strength, and, therefore, the purpose of their application can be considered as reducing environmental impacts.

4. Environmental Evaluation

The purposes of the usage of SCMs are mainly to reduce material costs and to reduce environmental burdens. Summarizing the comparison of material costs is impractical because the industrial circumstances are different between regions. Therefore, this paper provides a summary of the environmental impact data reviewed in this study.

This paper adopted the embodied carbon dioxide (e-CO_2) and energy consumption (e-Energy) data of the raw materials provided by the previous studies [84–86], as shown in Table 17. The embodied CO_2 and the embodied energy are based on the carbon footprint per unit (kg/kg) of each material and the quantity of non-renewable energy per unit (MJ/kg) of each raw material, respectively. The embodied CO_2 of the UHPC per the unit weight of 1 kg is calculated as the sum of the values obtained by multiplying the carbon footprint values in Table 17 and the mass ratios of each raw material in Table 18; the calculation method of the e-Energy of UHPC is similar to that of embodied CO_2 except using the e-Energy values in Table 17. However, since information is limited in the SCMs of cement, SF, FA, GGBS, MK, and LP, the other SCMs reviewed in Section 3.5 could not be analyzed. The fibers were not taken into account for the calculation because the fiber may dilute the effect of SCMs as not all of the studies applied fibers. Superplasticizers were also not included in the calculation because their dosage in UHPC is relatively very low compared to other ingredients. It is noted that different names of the SCMs are classified to a specific type of SCMs; for example, GGBS, SSP, FGGBS, LTS, SSP, etc. are considered to have the same e-CO_2 and e-Energy values as GGBS in Table 17. Additionally, the influence of the curing method on e-CO_2 and e-Energy was ignored. Therefore, the data provided in this study have potential errors.

Table 17. The embodied carbon dioxide and energy consumption of the raw materials [84–86].

Items	e-CO_2 (kg/kg)	e-Energy (MJ/kg)
Cement [85]	0.8300	4.7270
Water [85]	0.0003	0.0060
River sand [85]	0.0010	0.0220
Crushed stone [85]	0.0070	0.1130
Slag [85]	0.0190	1.5880
Fly ash [85]	0.0090	0.8330
Limestone powder [85]	0.0170	0.3500
Metakaolin [85]	0.4000	3.4800
Silica fume [84]	0.0140	0.1000
Sodium silicate [86]	1.5140	18.3000

Table 18. The summary of the e-CO_2 and the e-Energy of the UHPC reviewed in this study.

Category	Binder Mix Design	Water (wt.%)	Binder (wt.%)						Aggregate (wt.%)		e-CO_2 (kg/kg)	e-Energy (MJ/m^3)	Ref.
			Cement	Slag	FA	LP	MK	SF	Fine	Coarse			
Slag	LTS (10%) + SF	9	38	5	0	0	0	5	43	0	0.321	1.906	[50]
	PSS (4%) + MS	7	39	2	0	0	0	8	44	0	0.324	1.885	[49]
	GGBS (30%)	7	26	11	0	0	0	0	56	0	0.218	1.415	[43]
	SSP (16.9%) + SF	5	28	7	0	0	0	7	39	13	0.239	1.487	[46]
	FGGBS (8.4%) + SF + BA	8	37	5	9	0	0	5	37	0	0.307	1.890	[40]
	CS (16%) + SF	7	32	8	0	0	0	10	43	0	0.265	1.636	[51]
	GGBS (23.6%) + SF	7	28	11	0	0	0	8	47	0	0.238	1.525	[45]
	GGBS (20%) + SF	7	25	8	0	0	0	8	51	0	0.211	1.335	[44]
	SSP (15%) + LP + SF	7	25	7	0	9	0	5	46	0	0.215	1.362	[12]
	PS (35%) + SF	7	23	16	0	0	0	7	47	0	0.199	1.382	[48]
	GGBS (38.5%) + SF	8	18	15	0	0	0	5	55	0	0.152	1.091	[42]
	FGGBS (38.5%) + SF	8	18	15	0	0	0	5	55	0	0.152	1.091	[42]
	GGBS (30%) + SF	7	21	14	0	0	0	12	46	0	0.178	1.228	[41]
	GGBS (25.5%) + SF + BA	8	21	15	9	0	0	9	38	0	0.183	1.341	[40]
	PS (27.4%) + SF + FA	8	22	11	9	0	0	6	44	0	0.190	1.325	[47]
	GGBS (60%) + SF	11	16	32	0	0	0	5	37	0	0.138	1.262	[39]

Table 18. *Cont.*

Category	Binder Mix Design	Water (wt.%)	Binder (wt.%)						Aggregate (wt.%)		e-CO_2 (kg/kg)	e-Energy (MJ/m^3)	Ref.
			Cement	Slag	FA	LP	MK	SF	Fine	Coarse			
FA	FA (38.5%) + SF	8	18	0	15	0	0	5	55	0	0.150	0.981	[42]
	BA (15.7%) + SF	8	37	0	9	0	0	9	38	0	0.306	1.821	[40]
	FA (12.8%) + SF	8	37	0	7	0	0	9	38	0	0.313	1.847	[40]
	FFA (34.1%) + SF	7	27	0	15	0	0	2	48	0	0.228	1.425	[56]
	FA (20%) + SF	8	30	0	8	0	0	2	52	0	0.250	1.496	[53]
	FA (20%) + MK (3.8%)	17	64	0	17	0	3	0	0	0	0.541	3.252	[55]
	FA (30%) + SF	8	23	0	10	0	0	0	30	29	0.193	1.203	[52]
	FA (11.8%) + SF	7	39	0	6	0	0	4	44	0	0.328	1.923	[49]
	FA (7.4%) + GGBS + SF	7	30	4	4	0	0	9	47	0	0.250	1.520	[54]
	FA (15%) + SF	8	28	0	7	0	0	12	46	0	0.233	1.391	[41]
LP	LP (32%) + SF	9	39	0	0	22	0	9	21	0	0.326	1.917	[58]
	NC (3.2%) + SF	8	38	0	0	2	0	10	42	0	0.318	1.823	[60]
	LP (37.3%) + SF	10	29	0	0	19	0	3	39	0	0.241	1.427	[57]
	LP (14%) + SF + FA	8	24	0	12	7	0	6	42	0	0.206	1.297	[59]
	LP (4%) + SF	7	39	0	0	2	0	8	44	0	0.327	1.874	[49]
MK	NMK (1%) + MK	7	33	0	0	0	4	0	20	36	0.290	1.730	[64]
	MK (20%) + SF	9	33	0	0	0	8	0	49	0	0.309	1.870	[61]
	MK (16.7%)	8	34	0	0	0	7	0	51	0	0.310	1.856	[63]
	MK (6.9%) + SF (1)	10	0	21	0	0	2	2	50	0	0.240	3.162	[65]

The values of water, solid binder, and aggregate are the mass ratio. (1) The specimen is an alkali-activated material and its mix design was deduced based on the mixing ratio described in the paper. It was assumed that the mass ratio of sodium silicate used is approximately 0.15, and that the mass ratio of fine aggregate is 0.50.

The relationship between e-CO_2 and e-Energy is almost linear as shown in Figure 1 indicating that the energy used to produce the material also generates CO_2 proportionally. Therefore, the e-CO_2 data are used to investigate the environmental impact of the UHPC mix designs reviewed in this study. Figure 2 shows the summary of the e-CO_2 and the 28-day compressive strength of the UHPCs. The bar graph corresponds to the 28-day compressive strength of the left *Y*-axis and the line plot to the e-CO_2 data of the right *Y*-axis, and the hatched ones mean that the specimen was thermally treated. The data are divided by the type of SCMs used and, then, sorted by the 28-day compressive strength in descending order. More detailed information on the e-CO_2 and e-Energy can also be found in Table 18. It can be concluded that the 28-day compressive strength of UHPC is not always correlated to the e-CO_2 data. This implies the possibilities of optimizing the UHPC mix design for higher compressive strength as well as the lower e-CO_2 of UHPC. However, the investigation on the e-CO_2 and e-Energy of various types of SCMs should be preceded for the accurate analysis. Slag seems to have a lower environmental impact compared to other SCMs because of its higher dosage. Therefore, the applicable dosage of raw SCM material is also an important factor contributing to the decrease in the e-CO_2 of UHPC. It is believed that the summary data can show a comprehensive understanding of which SCMs are more efficient to reduce the environmental impact and to have higher compressive strength.

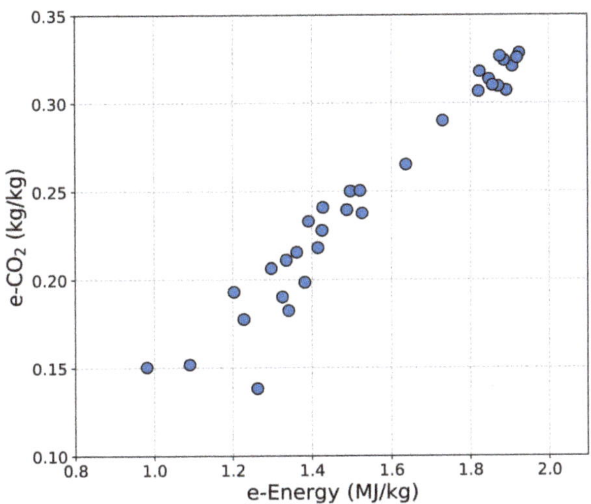

Figure 1. The linear relationship between e-CO_2 and e-Energy.

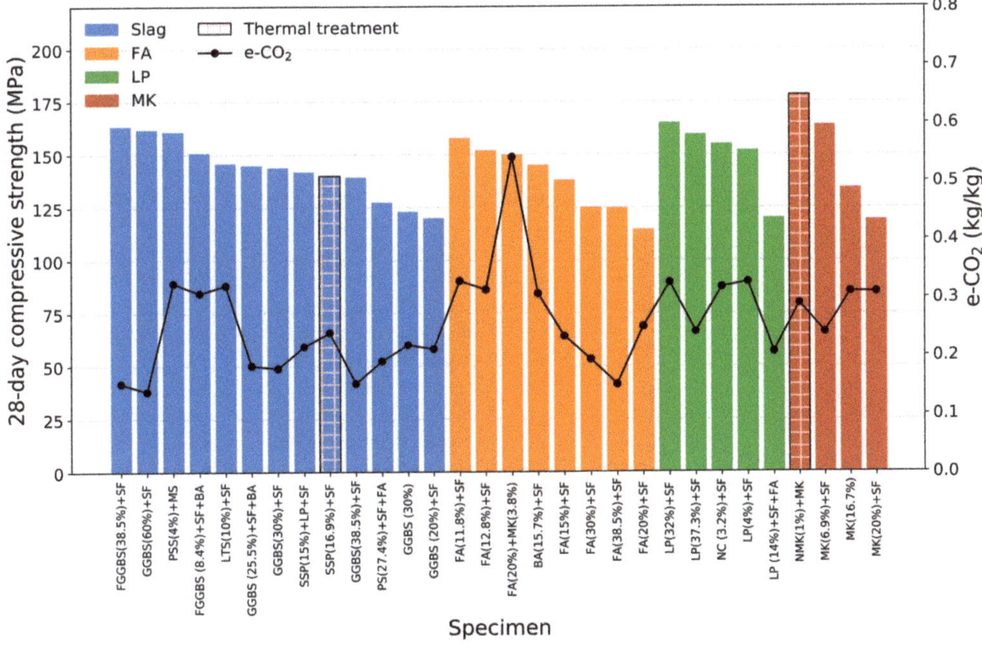

Figure 2. The 28-days compressive strength and the e-CO_2 of the UHPC reviewed in this paper.

5. Conclusions

This paper reviewed the effect of SCMs on the properties of UHPC. The various types of SCMs such as slag, FA, LP, MK, and others were successfully applied to UHPC, satisfying material requirements such as compressive strength. Based on the discussions, their effects are summarized as follows:

(1) The main purposes of the usage of SCMs are to decrease the material cost and the environmental impact caused during material production by a partial replacement of cement or silica fume. Since most SCMs are industrial by-products from plants or naturally occurring resources, the usage of SCMs corresponds well to this purpose; it was confirmed that the e-CO_2 of UHPC is lower when the dosage of an SCM is higher.

(2) Slag tends to decrease the compressive strength of UHPC at an early age because of the slow hydration of slag, but it increases the late age compressive strength through the pozzolanic reaction between slag and $Ca(OH)_2$ that increases the packing density of the UHPC. The finer particle size of slag exhibits higher compressive strength. Slag also increases the workability of UHPC because of its lower water absorption compared to cement.

(3) FA degrades the compressive strength of UHPC; however, some of the FFA can enhance compressive strength. The ternary use of SCMs including FA can be another feasible option to reduce the amount of cement in UHPC. The effect of FA on the workability of UPHC is different among studies. It is also proved that FA is effective to reduce the shrinkage of UHPC.

(4) LP enhances the compressive strength of UHPC with the three mechanisms: (i) LP decreases the water demand of UHPC, that is, it increases the workability of UHPC, (ii) LP has a pozzolanic reaction with SF, which increases the late age compressive strength, and (iii) LP can accelerate the cement hydration. However, some cases that LP degrades the compressive strength of UHPC were observed. LP can decrease the shrinkage of UHPC by reducing the amount of cement in UHPC.

(5) MK seems to increase the early age compressive strength of UHPC, but decreases the late age compressive strength. It was confirmed that the MK of the finer particle size can overcome the degradation of the early age compressive strength. It was reported that MK decreases the autogenous shrinkage while it increases the drying shrinkage. Another application of MK was found; the alkali-activated material synthesized using slag, MK, and sodium silicate solution results in the proper compressive strength over 150 MPa.

(6) Other SCMs are also introduced. RHA has a synergic effect on the compressive strength of UHPC resulting in the higher compressive strength at both early and late age compared to the reference specimen only with SF. NP decreases the compressive strength of UHPC at all ages; however, it results in the compressive strength of UHPC over 150 MPa at 90 days. NMC increases the late age compressive strength of UHPC because it yields a pozzolanic reaction at late ages. DCP and CDK degrade the compressive strength of UHPC because they increase the water demand. GCP is a good source of SCM; it improves both the compressive strength at 28 days and the flowability of UHPC. GCP does not chemically react in UHPC but works as a filler. BP was confirmed to decrease the compressive strength of UHPC, but it increases the workability. Partial substitution of SF with FGP can improve both the compressive strength because of its pozzolanic reaction and advance the workability of UHPC because of the lower surface area compared to SF.

Although this paper examined the effect of extensively various SCMs on UHPC properties, the properties themselves are limited in compressive strength, flowability, shrinkage, and environmental impact. Due to the small number of studies found in the literature, other important properties of UHPC such as tensile strength, modulus of elasticity, and fracture energy were not able to be summarized in this paper. Therefore, further studies should be proceeded for a review of the effect of SCMs on those properties.

Author Contributions: Conceptualization, S.P. (Sungwoo Park), S.W., S.P. (Sukhoon Pyo); methodology, S.P. (Sungwoo Park), S.W.; software, S.W.; writing—original draft preparation, S.P. (Sungwoo Park), S.W.; writing—review and editing, Z.L., S.P. (Sukhoon Pyo); visualization, S.P. (Sungwoo Park); supervision, S.P. (Sukhoon Pyo); project administration, S.P. (Sukhoon Pyo); funding acquisition, S.P. (Sukhoon Pyo). All authors have read and agreed to the published version of the manuscript.

Funding: This work was supported by the National Research Foundation of Korea (NRF) grant funded by the Korean government (MSIT) (No. 2021R1C1C1008671). The opinions expressed in this paper are those of the authors and do not necessarily reflect the views of the sponsors.

Institutional Review Board Statement: Not applicable.

Informed Consent Statement: Not applicable.

Data Availability Statement: The data presented in this study are available on request from the corresponding author.

Conflicts of Interest: The authors declare no conflict of interest.

Abbreviations

BA	Bottom Ash
BP	Basalt stone Powder
CKD	Cement Kiln Dust
CS	Copper Slag
DCP	Dehydrated Cementitious Powder
FA	Fly Ash
FFA	Fine Fly Ash
FGGBS	Fine Ground Granulated Blast-furnace Slag
FGP	Fine Glass Powder
GGBS	Ground Granulated Blast-furnace Slag
GGP	Ground Granite Powder
LP	Limestone Powder
LTS	Lithium Slag
MK	Metakaolin
NC	Nano Calcium carbonate
NMC	Nano Metaclay
NMK	Nano Metakaolin
NP	Natural Pozzolan
OPC	Ordinary Portland Cement
PS	Phosphorous Slag
PSS	Pulverized Steel Slag
RHA	Rice Husk Ash
SF	Silica Fume
SSP	Steel Slag Powder

References

1. ASTM C1856/M-17. *Standars Practice for Fabricating on Testing Specimens of Uhpc*; ASTM: West Conshohocken, PA, USA, 2017; pp. 1–8. [CrossRef]
2. Schmidt, M.; Fehling, E. Ultra-high-performance concrete: Research, development and application in Europe. *ACI Spec. Publ.* **2005**, *228*, 51–78.
3. Meng, W.; Khayat, K.H.; Bao, Y. Flexural behaviors of fiber-reinforced polymer fabric reinforced ultra-high-performance concrete panels. *Cem. Concr. Compos.* **2018**, *93*, 43–53. [CrossRef]
4. Meng, W.; Khayat, K.H. Improving flexural performance of ultra-high-performance concrete by rheology control of suspending mortar. *Compos. Part B Eng.* **2017**, *117*, 26–34. [CrossRef]
5. Piérard, J.; Dooms, B.; Cauberg, N. Evaluation of durability parameters of UHPC using accelerated lab tests. In Proceedings of the 3rd International Symposium on UHPC and Nanotechnology for High Performance Construction Materials, Kassel, Germany, 7–9 March 2012; pp. 371–376.
6. Graybeal, B.; Tanesi, J. Durability of an ultrahigh-performance concrete. *J. Mater. Civ. Eng.* **2007**, *19*, 848–854. [CrossRef]
7. Pyo, S.; Koh, T.; Tafesse, M.; Kim, H.K. Chloride-induced corrosion of steel fiber near the surface of ultra-high performance concrete and its effect on flexural behavior with various thickness. *Constr. Build. Mater.* **2019**, *224*, 206–213. [CrossRef]
8. Piérard, J.; Dooms, B.; Cauberg, N. Durability evaluation of different types of UHPC. In Proceedings of the RILEM-fib-AFGC International Symposium on Ultra-High Performance Fiber-Reinforced Concrete, Marseille, France, 1–3 October 2013; pp. 275–284.
9. Abbas, S.; Soliman, A.M.; Nehdi, M.L. Exploring mechanical and durability properties of ultra-high performance concrete incorporating various steel fiber lengths and dosages. *Constr. Build. Mater.* **2015**, *75*, 429–441. [CrossRef]

10. Pyo, S.; Abate, S.Y.; Kim, H.K. Abrasion resistance of ultra high performance concrete incorporating coarser aggregate. *Constr. Build. Mater.* **2018**, *165*, 11–16. [CrossRef]
11. Liu, J.; Song, S.; Wang, L. Durability and micro-structure of reactive powder concrete. *J. Wuhan Univ. Technol. Sci. Ed.* **2009**, *24*, 506–509. [CrossRef]
12. Li, S.; Cheng, S.; Mo, L.; Deng, M. Effects of steel slag powder and expansive agent on the properties of ultra-high performance concrete (UHPC): Based on a case study. *Materials* **2020**, *13*, 683. [CrossRef]
13. Alkaysi, M.; El-Tawil, S.; Liu, Z.; Hansen, W. Effects of silica powder and cement type on durability of ultra high performance concrete (UHPC). *Cem. Concr. Compos.* **2016**, *66*, 47–56. [CrossRef]
14. Pyo, S.; Tafesse, M.; Kim, H.; Kim, H.K. Effect of chloride content on mechanical properties of ultra high performance concrete. *Cem. Concr. Compos.* **2017**, *84*, 175–187. [CrossRef]
15. Dils, J.; Boel, V.; De Schutter, G. Influence of cement type and mixing pressure on air content, rheology and mechanical properties of UHPC. *Constr. Build. Mater.* **2013**, *41*, 455–463. [CrossRef]
16. Lowke, D.; Stengel, T.; Schießl, P.; Gehlen, C. Control of rheology, strength and fibre bond of UHPC with additions-effect of packing density and addition type. In *Ultra-High Performance Concrete and Nanotechnology in Construction, Proceedings of the Hipermat 2012, 3rd International Symposium on UHPC and Nanotechnology for High Performance Construction Materials, Kassel, Germany, 7–9 March 2012*; Kassel University Press GmbH: Kassel, Germany, 2012; pp. 215–224.
17. Khayat, K.H.; Meng, W.; Vallurupalli, K.; Teng, L. Rheological properties of ultra-high-performance concrete—An overview. *Cem. Concr. Res.* **2019**, *124*, 105828. [CrossRef]
18. Choi, M.S.; Lee, J.S.; Ryu, K.S.; Koh, K.-T.; Kwon, S.H. Estimation of rheological properties of UHPC using mini slump test. *Constr. Build. Mater.* **2016**, *106*, 632–639. [CrossRef]
19. Pyo, S.; Wille, K.; El-Tawil, S.; Naaman, A.E. Strain rate dependent properties of ultra high performance fiber reinforced concrete (UHP-FRC) under tension. *Cem. Concr. Compos.* **2015**, *56*, 15–24. [CrossRef]
20. Wang, R.; Gao, X.; Huang, H.; Han, G. Influence of rheological properties of cement mortar on steel fiber distribution in UHPC. *Constr. Build. Mater.* **2017**, *144*, 65–73. [CrossRef]
21. Wille, K.; El-Tawil, S.; Naaman, A.E. Properties of strain hardening ultra high performance fiber reinforced concrete (UHP-FRC) under direct tensile loading. *Cem. Concr. Compos.* **2014**, *48*, 53–66. [CrossRef]
22. Ghafari, E.; Costa, H.; Júlio, E. RSM-based model to predict the performance of self-compacting UHPC reinforced with hybrid steel micro-fibers. *Constr. Build. Mater.* **2014**, *66*, 375–383. [CrossRef]
23. Wu, Z.; Shi, C.; He, W.; Wang, D. Static and dynamic compressive properties of ultra-high performance concrete (UHPC) with hybrid steel fiber reinforcements. *Cem. Concr. Compos.* **2017**, *79*, 148–157. [CrossRef]
24. Richard, P.; Cheyrezy, M. Reactive Powder Concretes With High Ductility and 200–800 Mpa Compressive Strength. In *Concrete Technology: Past, Present, and Future, Proceedings of the V. Mohan Malhotra Symposium ACI SP-144, San Francisco, CA, USA, 21–23 March 1994*; American Concrete Institute: Detroit, MI, USA, 1994.
25. Richard, P.; Cheyrezy, M. Composition of reactive powder concretes. *Cem. Concr. Res.* **1995**, *25*, 1501–1511. [CrossRef]
26. Kim, H.; Koh, T.; Pyo, S. Enhancing flowability and sustainability of ultra high performance concrete incorporating high replacement levels of industrial slags. *Constr. Build. Mater.* **2016**, *123*, 153–160. [CrossRef]
27. Pyo, S.; Kim, H.K.; Lee, B.Y. Effects of coarser fine aggregate on tensile properties of ultra high performance concrete. *Cem. Concr. Compos.* **2017**, *84*, 28–35. [CrossRef]
28. Wille, K.; Boisvert-Cotulio, C. Material efficiency in the design of ultra-high performance concrete. *Constr. Build. Mater.* **2015**, *86*, 33–43. [CrossRef]
29. Pyo, S.; Tafesse, M.; Kim, B.J.; Kim, H.K. Effects of quartz-based mine tailings on characteristics and leaching behavior of ultra-high performance concrete. *Constr. Build. Mater.* **2018**, *166*, 110–117. [CrossRef]
30. Bajaber, M.A.; Hakeem, I.Y. UHPC evolution, development, and utilization in construction: A review. *J. Mater. Res. Technol.* **2021**, *10*, 1058–1074. [CrossRef]
31. Xue, J.; Briseghella, B.; Huang, F.; Nuti, C.; Tabatabai, H.; Chen, B. Review of ultra-high performance concrete and its application in bridge engineering. *Constr. Build. Mater.* **2020**, *260*, 119844. [CrossRef]
32. Bae, Y.; Pyo, S. Ultra high performance concrete (UHPC) sleeper: Structural design and performance. *Eng. Struct.* **2020**, *210*, 110374. [CrossRef]
33. Bae, Y.; Pyo, S. Effect of steel fiber content on structural and electrical properties of ultra high performance concrete (UHPC) sleepers. *Eng. Struct.* **2020**, *222*, 111131. [CrossRef]
34. Worrell, E.; Price, L.; Martin, N.; Hendriks, C.; Meida, L.O. Carbon dioxide emissions from the global cement industry. *Annu. Rev. Energy Environ.* **2001**, *26*, 303–329. [CrossRef]
35. Odler, I.; Rößler, M. Investigations on the relationship between porosity, structure and strength of hydrated Portland cement pastes. II. Effect of pore structure and of degree of hydration. *Cem. Concr. Res.* **1985**, *15*, 401–410. [CrossRef]
36. Rossi, P. Influence of fibre geometry and matrix maturity on the mechanical performance of ultra high-performance cement-based composites. *Cem. Concr. Compos.* **2013**, *37*, 246–248. [CrossRef]
37. Yu, R.; Spiesz, P.; Brouwers, H.J.H. Mix design and properties assessment of ultra-high performance fibre reinforced concrete (UHPFRC). *Cem. Concr. Res.* **2014**, *56*, 29–39. [CrossRef]

38. Korpa, A.; Kowald, T.; Trettin, R. Phase development in normal and ultra high performance cementitious systems by quantitative X-ray analysis and thermoanalytical methods. *Cem. Concr. Res.* **2009**, *39*, 69–76. [CrossRef]
39. Gupta, S. Development of ultra-high performance concrete incorporating blend of slag and silica fume as cement replacement'. *Int. J. Civ. Struct. Eng. Res.* **2014**, *2*, 35–51.
40. Pyo, S.; Kim, H.K. Fresh and hardened properties of ultra-high performance concrete incorporating coal bottom ash and slag powder. *Constr. Build. Mater.* **2017**, *131*, 459–466. [CrossRef]
41. Wu, Z.; Shi, C.; He, W. Comparative study on flexural properties of ultra-high performance concrete with supplementary cementitious materials under different curing regimes. *Constr. Build. Mater.* **2017**, *136*, 307–313. [CrossRef]
42. Randl, N.; Steiner, T.; Ofner, S.; Baumgartner, E.; Mészöly, T. Development of UHPC mixtures from an ecological point of view. *Constr. Build. Mater.* **2014**, *67*, 373–378. [CrossRef]
43. Yu, R.; Spiesz, P.; Brouwers, H.J.H. Development of an eco-friendly Ultra-High Performance Concrete (UHPC) with efficient cement and mineral admixtures uses. *Cem. Concr. Compos.* **2015**, *55*, 383–394. [CrossRef]
44. Liu, Z.; El-Tawil, S.; Hansen, W.; Wang, F. Effect of slag cement on the properties of ultra-high performance concrete. *Constr. Build. Mater.* **2018**, *190*, 830–837. [CrossRef]
45. Abdulkareem, O.M.; Ben Fraj, A.; Bouasker, M.; Khelidj, A. Mixture design and early age investigations of more sustainable UHPC. *Constr. Build. Mater.* **2018**, *163*, 235–246. [CrossRef]
46. Liu, J.; Guo, R. Applications of Steel Slag Powder and Steel Slag Aggregate in Ultra-High Performance Concrete. *Adv. Civ. Eng.* **2018**, *2018*, 1–8. [CrossRef]
47. Yang, R.; Yu, R.; Shui, Z.; Gao, X.; Xiao, X.; Zhang, X.; Wang, Y.; He, Y. Low carbon design of an Ultra-High Performance Concrete (UHPC) incorporating phosphorous slag. *J. Clean. Prod.* **2019**, *240*, 118157. [CrossRef]
48. Peng, Y.Z.; Huang, J.; Ke, J. Preparation of Ultra-High Performance Concrete Using Phosphorous Slag Powder. *Appl. Mech. Mater.* **2013**, *357–360*, 588–591. [CrossRef]
49. Ahmad, S.; Hakeem, I.; Maslehuddin, M. Development of UHPC Mixtures Utilizing Natural and Industrial Waste Materials as Partial Replacements of Silica Fume and Sand. *Sci. World J.* **2014**, *2014*, 713531. [CrossRef] [PubMed]
50. He, Z.; Du, S.; Chen, D. Microstructure of ultra high performance concrete containing lithium slag. *J. Hazard. Mater.* **2018**, *353*, 35–43. [CrossRef]
51. Edwin, R.S.; De Schepper, M.; Gruyaert, E.; De Belie, N. Effect of secondary copper slag as cementitious material in ultra-high performance mortar. *Constr. Build. Mater.* **2016**, *119*, 31–44. [CrossRef]
52. Šeps, K.; Broukalová, I.; Chylík, R. Cement Substitutions in UHPC and their Influence on Principal Mechanical-Physical Properties. *IOP Conf. Ser. Mater. Sci. Eng.* **2019**, *522*, 012009. [CrossRef]
53. Alsalman, A.; Dang, C.N.; Micah Hale, W. Development of ultra-high performance concrete with locally available materials. *Constr. Build. Mater.* **2017**, *133*, 135–145. [CrossRef]
54. Yazıcı, H.; Yiğiter, H.; Karabulut, A.Ş.; Baradan, B. Utilization of fly ash and ground granulated blast furnace slag as an alternative silica source in reactive powder concrete. *Fuel* **2008**, *87*, 2401–2407. [CrossRef]
55. Li, Z. Drying shrinkage prediction of paste containing meta-kaolin and ultrafine fly ash for developing ultra-high performance concrete. *Mater. Today Commun.* **2016**, *6*, 74–80. [CrossRef]
56. Ferdosian, I.; Camões, A. Eco-efficient ultra-high performance concrete development by means of response surface methodology. *Cem. Concr. Compos.* **2017**, *84*, 146–156. [CrossRef]
57. Li, P.P.; Brouwers, H.J.H.; Chen, W.; Yu, Q. Optimization and characterization of high-volume limestone powder in sustainable ultra-high performance concrete. *Constr. Build. Mater.* **2020**, *242*, 118112. [CrossRef]
58. Huang, W.; Kazemi-Kamyab, H.; Sun, W.; Scrivener, K. Effect of cement substitution by limestone on the hydration and microstructural development of ultra-high performance concrete (UHPC). *Cem. Concr. Compos.* **2017**, *77*, 86–101. [CrossRef]
59. Yang, R.; Yu, R.; Shui, Z.; Gao, X.; Han, J.; Lin, G.; Qian, D.; Liu, Z.; He, Y. Environmental and economical friendly ultra-high performance-concrete incorporating appropriate quarry-stone powders. *J. Clean. Prod.* **2020**, *260*, 121112. [CrossRef]
60. Wu, Z.; Shi, C.; Khayat, K.H.; Xie, L. Effect of SCM and nano-particles on static and dynamic mechanical properties of UHPC. *Constr. Build. Mater.* **2018**, *182*, 118–125. [CrossRef]
61. Tafraoui, A.; Escadeillas, G.; Lebaili, S.; Vidal, T. Metakaolin in the formulation of UHPC. *Constr. Build. Mater.* **2009**, *23*, 669–674. [CrossRef]
62. Tafraoui, A.; Escadeillas, G.; Vidal, T. Durability of the Ultra High Performances Concrete containing metakaolin. *Constr. Build. Mater.* **2016**, *112*, 980–987. [CrossRef]
63. Li, Z.; Rangaraju, P.R. Development of UHPC using a ternary blend of ultra. In *International Interactive Symposium on Ultra-High Performance Concrete;-Fine Class F Fly Ash, Meta-Kaolin and Portland Cement*; Iowa State University Digital Press: Ames, IA, USA, 2016.
64. Muhd Norhasri, M.S.; Hamidah, M.S.; Mohd Fadzil, A.; Megawati, O. Inclusion of nano metakaolin as additive in ultra high performance concrete (UHPC). *Constr. Build. Mater.* **2016**, *127*, 167–175. [CrossRef]
65. Wetzel, A.; Middendorf, B. Influence of silica fume on properties of fresh and hardened ultra-high performance concrete based on alkali-activated slag. *Cem. Concr. Compos.* **2019**, *100*, 53–59. [CrossRef]
66. Van Tuan, N.; Ye, G.; Van Breugel, K.; Fraaij, A.L.A.; Bui, D.D. The study of using rice husk ash to produce ultra high performance concrete. *Constr. Build. Mater.* **2011**, *25*, 2030–2035. [CrossRef]

67. Ahmad, S.; Mohaisen, K.O.; Adekunle, S.K.; Al-Dulaijan, S.U.; Maslehuddin, M. Influence of admixing natural pozzolan as partial replacement of cement and microsilica in UHPC mixtures. *Constr. Build. Mater.* **2019**, *198*, 437–444. [CrossRef]
68. Norhasri, M.S.M.; Hamidah, M.S.; Fadzil, A.M. Inclusion of nano metaclayed as additive in ultra high performance concrete (UHPC). *Constr. Build. Mater.* **2019**, *201*, 590–598. [CrossRef]
69. Qian, D.; Yu, R.; Shui, Z.; Sun, Y.; Jiang, C.; Zhou, F.; Ding, M.; Tong, X.; He, Y. A novel development of green ultra-high performance concrete (UHPC) based on appropriate application of recycled cementitious material. *J. Clean. Prod.* **2020**, *261*, 121231. [CrossRef]
70. Zhang, H.; Ji, T.; He, B.; He, L. Performance of ultra-high performance concrete (UHPC) with cement partially replaced by ground granite powder (GGP) under different curing conditions. *Constr. Build. Mater.* **2019**, *213*, 469–482. [CrossRef]
71. Soliman, N.A.; Tagnit-Hamou, A. Partial substitution of silica fume with fine glass powder in UHPC: Filling the micro gap. *Constr. Build. Mater.* **2017**, *139*, 374–383. [CrossRef]
72. Oner, A.; Akyuz, S. An experimental study on optimum usage of GGBS for the compressive strength of concrete. *Cem. Concr. Compos.* **2007**, *29*, 505–514. [CrossRef]
73. Regourd, M.; Thomassin, J.H.; Baillif, P.; Touray, J.C. Blast-furnace slag hydration. Surface analysis. *Cem. Concr. Res.* **1983**, *13*, 549–556. [CrossRef]
74. Manmohan, D.; Mehta, P.K. Influence of pozzolanic, slag, and chemical admixtures on pore size distribution and permeability of hardened cement pastes. *Cem. Concr. Aggregates* **1981**, *3*, 63–67.
75. Thomassin, J.H.; Goni, J.; Baillif, P.; Touray, J.C.; Jaurand, M.C. An XPS study of the dissolution kinetics of chrysotile in 0.1 N oxalic acid at different temperatures. *Phys. Chem. Miner.* **1977**, *1*, 385–398. [CrossRef]
76. Liu, Z.; Hansen, W. Aggregate and slag cement effects on autogenous shrinkage in cementitious materials. *Constr. Build. Mater.* **2016**, *121*, 429–436. [CrossRef]
77. Chern, J.-C.; Chan, Y.-W. Deformations of concretes made with blast-furnace slag cement and ordinary portland cement. *Mater. J.* **1989**, *86*, 372–382.
78. Darquennes, A.; Staquet, S.; Delplancke-Ogletree, M.-P.; Espion, B. Effect of autogenous deformation on the cracking risk of slag cement concretes. *Cem. Concr. Compos.* **2011**, *33*, 368–379. [CrossRef]
79. Virgalitte, S.J.; Luther, M.D.; Rose, J.H.; Mather, B.; Bell, L.W.; Ehmke, B.A.; Klieger, P.; Roy, D.M.; Call, B.M.; Hooton, R.D.; et al. *Ground Granulated Blast-Furnace Slag as a Cementitious Constituent in Concrete*; American Concrete Institute ACI Report 233R-95; American Concrete Institute: Farmington Hills, MI, USA, 1995.
80. Bonavetti, V.; Donza, H.; Menéndez, G.; Cabrera, O.; Irassar, E.F. Limestone filler cement in low w/c concrete: A rational use of energy. *Cem. Concr. Res.* **2003**, *33*, 865–871. [CrossRef]
81. Bakera, A.T.; Alexander, M.G. Use of metakaolin as supplementary cementitious material in concrete, with focus on durability properties. *RILEM Tech. Lett.* **2019**, *4*, 89–102. [CrossRef]
82. Vejmelková, E.; Pavlíková, M.; Keppert, M.; Keršner, Z.; Rovnaníková, P.; Ondráček, M.; Sedlmajer, M.; Černý, R. High performance concrete with Czech metakaolin: Experimental analysis of strength, toughness and durability characteristics. *Constr. Build. Mater.* **2010**, *24*, 1404–1411. [CrossRef]
83. Ramezanianpour, A.A.; Bahrami Jovein, H. Influence of metakaolin as supplementary cementing material on strength and durability of concretes. *Constr. Build. Mater.* **2012**, *30*, 470–479. [CrossRef]
84. Kuruşcu, A.O.; Girgin, Z.C. Efficiency of Structural Materials in Sustainable Design. *J. Civ. Eng. Archit.* **2014**, 8.
85. Long, G.; Gao, Y.; Xie, Y. Designing more sustainable and greener self-compacting concrete. *Constr. Build. Mater.* **2015**, *84*, 301–306. [CrossRef]
86. Turner, L.K.; Collins, F.G. Carbon dioxide equivalent (CO2-e) emissions: A comparison between geopolymer and OPC cement concrete. *Constr. Build. Mater.* **2013**, *43*, 125–130. [CrossRef]

Article

A Numerical Study on Structural Performance of Railway Sleepers Using Ultra High-Performance Concrete (UHPC)

Moochul Shin [1], Younghoon Bae [2,*] and Sukhoon Pyo [3,*]

[1] Department of Civil and Environmental Engineering, Western New England University, Springfield, MA 01119, USA; moochul.shin@wne.edu
[2] Korea Railroad Research Institute, 176 Chuldobangmulgwan-ro, Uiwang-si 16105, Korea
[3] Department of Urban and Environmental Engineering, Ulsan National Institute of Science and Technology (UNIST), Ulsan 44919, Korea
* Correspondence: yhbae@krri.re.kr (Y.B.); shpyo@unist.ac.kr (S.P.)

Abstract: This numerical study investigates the structural performance of railway sleepers made of ultra high-performance concrete (UHPC). First, numerical concrete sleepers are developed, and the tensile stress-strain relationship obtained from the direct tension test on the UHPC coupons is used for the tensile constitutive model after applying a fiber orientation reduction factor. The numerical sleeper models are validated with the experimental data in terms of the force and crack-width relationship. Second, using the developed models, a parametric study is performed to investigate the performance of the UHPC sleepers while considering various design/mechanical/geometrical parameters: steel fiber contents, size of the cross-section, and diameter and strength of prestressing (PS) tendons. The simulation results indicate that the size of the cross-section has the most impacts on the performance, while the effect of yielding strengths of PS tendons is minimal among all the parameters. Engineers need to pay attention to efficiency and an economical factor when using a larger cross-section, since sleepers with larger cross-sections can be an over-designed sleeper. This study suggests an economical design factor for engineers to evaluate what combination of parameters would be economical designs.

Keywords: ultra high-performance concrete (UHPC); railway sleeper; static bending test; numerical simulation; structural performance

Citation: Shin, M.; Bae, Y.; Pyo, S. A Numerical Study on Structural Performance of Railway Sleepers Using Ultra High-Performance Concrete (UHPC). *Materials* **2021**, *14*, 2979. https://doi.org/10.3390/ma14112979

Academic Editor: Bahman Ghiassi

Received: 4 May 2021
Accepted: 28 May 2021
Published: 31 May 2021

1. Introduction

In a railway track structure system, sleepers (or ties) perform critical functions by transferring and distributing train loadings from rail to ballast or concrete slab. The critical components undergo repeated train loading and impact loading; however, the exact load transfer mechanism within the sleeper is still unclear due to uneven ballast support conditions and irregular surface conditions of rail and wheels. Due to inaccurately identified loads and support conditions, various parts of sleepers can have damages such as center-binding crack, and flexural and/or shear cracks at the rail-seat section. Nowadays, the railway industry has paid more attention to how to improve the service life of sleepers not only because of increasing axle loads, speed, and traffic volume, but also because of increasing maintenance costs including expensive sleeper replacing costs [1].

Concrete has been widely used for manufacturing sleepers in the world [2], and various attempts have been carried out to complement the brittle nature of the material. Cracks in concrete sleepers have been widely investigated and identified that it is mainly attributed to the material brittleness, particularly under dynamic loadings [1,3]. Although the tensile crack development in concrete is inevitable, it is revealed that the crack propagation can be effectively controlled by using various types of discontinuous reinforcements such as steel fibers [4]. Similarly, various efforts have also been made for concrete sleeper applications to enhance material ductility using fibers [5–8]. For example, Ramezanianpour et al. [5]

used polypropylene fiber to enhance the tensile and flexural strength, and the durability of concrete by reducing chloride diffusion, water penetration, and sorptivity. Shin et al. [6] concluded that the use of 0.75% of steel fibers results in enhanced static and impact flexural capacity and toughness. Yang et al. [7] revealed that the concrete sleepers reinforced with steel fibers showed increased flexural and fatigue capacity at the rail-seat section compared with conventional concrete sleepers with conventional stirrups, since the concrete sleepers with steel fibers can mitigate crack propagation and prevent brittle shear failure.

For an efficient massive concrete sleeper production process in a precast concrete facility, high strength concretes are generally used in order to promote early demolding and applying prestressing forces. Therefore, many national and international standards require specified minimum compressive strength of concrete for sleeper applications, e.g., C45/55 MPa in European standards [9], 50 MPa in Australia [10], and C50/60 MPa in the International Union of Railways recommendation [11]. Ultra high-performance concrete (UHPC) is one of the most advanced cement-based materials showing a compressive strength at 28 days higher than 150 MPa [12], which possesses strong potentials to extend the service lives of structures with various engineering merits including high ductility [13], durability [14], abrasion resistance [15], and impact resistance [16]. Recently, the authors revealed that the adoption of UHPC in railway sleepers resulted in stable structural behavior and outstanding crack resistance capability even after initial cracks developed [17].

Wide-width concrete sleepers are one of the special types of sleepers, which can significantly reduce the burden of ballast and track substructures due to a larger contact area. The large contact areas of the wide sleepers enable to reduce vibration values, extends maintenance intervals, and the life of the track system [18,19]. With these advantages, the range of applications of the wide concrete sleepers is getting higher from general lines to highly loaded areas such as transitional zones between earthwork, bridges, and tunnels.

Recently, there have been great efforts to study the numerical models of reinforced concrete structures by considering the nonlinearity of concrete behavior, bond strength, the stochastic natures of concrete, etc. Sucharda et al. [20] presented the nonlinear behavior of reinforced concrete beams without shear reinforcement using a stochastic model. In their concrete model, they incorporated the uncertainties in the concrete properties and studied the sensitivity to input parameters including fracture energy, G_f. Instead of performing a direct tension test on concrete, they conducted splitting, three-point bending, and four-point bending tests. In their study, they reported that the ratio of the maximum to minimum loads is not necessarily corresponding to the limit of the input parameters. Valikhani et al. [21] studied numerical modeling of bonding of regular concrete and UHPC since UHPC can be used for the repair of concrete structures. In order to model the interface between concrete and UHPC, they used a zero-thickness volume element with post-failure tension-separation laws. They demonstrated the importance of the interface between two different materials. Shin and Yu [22] presented a numerical study on the splitting performance of prestressed concrete prisms by incorporating bond-slip behavior of prestressed concrete using a cohesive element. They used a user-defined material model to describe the bond-slip behavior at the interface.

In this research, a numerical model of wide-type UHPC sleepers with respect to different amounts of fiber contents are developed and compared to the experimental tests. A direct tension test is performed and used for obtaining nonlinear properties of UHPC in tension. Using the developed models, a parametric study is performed to investigate the structural performance of the sleepers with respect to the content of steel fibers, the diameter of the prestressing tendons, and the yielding strength of PS tendons. The commercial finite element program ABAQUS is used in this study [23].

2. Mix Design and Fabrication of UHPC Sleeper

Wide-width concrete sleepers were manufactured using UHPC mixtures with three levels of fiber contents, mainly 0.5%, 1.0%, and 1.5%. The detailed mix proportion of the UHPC mixtures and the mixing procedures can be found in Bae and Pyo [8]. The compressive strength was evaluated using a 50 mm cubic specimen and averaged from at least three specimens. The compressive strengths of 0.5%, 1.0%, and 1.5%-UHPC specimens at 28 days were 149, 160, and 159 MPa, respectively. In addition, tensile strength results were adopted from Pyo et al. [24], where the tensile behavior of the similar UHPC mixture without ground granulated blast furnace slag (GGBFS) was characterized by following the JSCE recommendation [25]. The thickness of the tested tensile specimens was 30 mm-thick according to the recommendation. Figure 1 shows the averaged stress-strain relationships of UHPC with three levels of fiber contents under the direct tension test. For numerical constitutive models, the experimentally obtained constitutive relationship data were used to calibrate the uniaxial tensile behavior of three different concrete models. The solid lines represent the experimental data [24] and the dashed lines the numerical models.

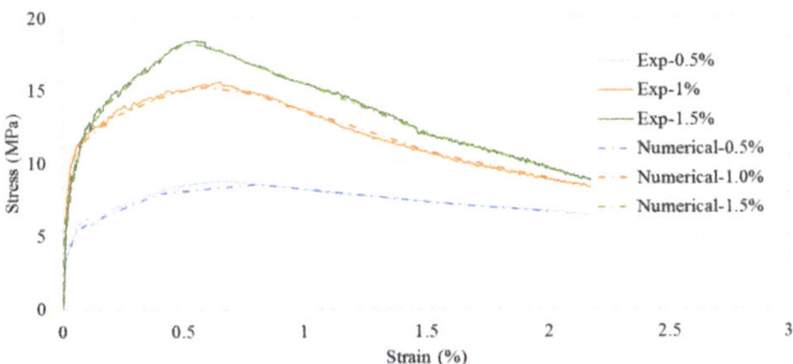

Figure 1. Experimentally obtained averaged stress-strain relationships under the direct tensile test on the UHPC with various fiber volume contents and the corresponding numerical constitutive models.

Figure 2 shows the detailed layout of the fabricated UHPC sleepers in the previous research [8,17], in which four PS tendons with diameters of 9.2 mm were used. Six sleepers with the 1.0% fiber volume case and three sleepers with the 0.5% and 1.5% fiber volume cases each were fabricated and tested. The fresh UHPC mixture was cast in the mold with external vibration, similar to the conventional sleeper production protocol. The casted UHPC sleepers were demolded after 24 h of curing and the sleepers were air-cured for an additional 24 h. Then, the prestress forces were introduced with the post-tensioning method. It is important to note that the prestress force was introduced without a post-tension duct and a thin layer of coating was applied to the surface of the PS tendons.

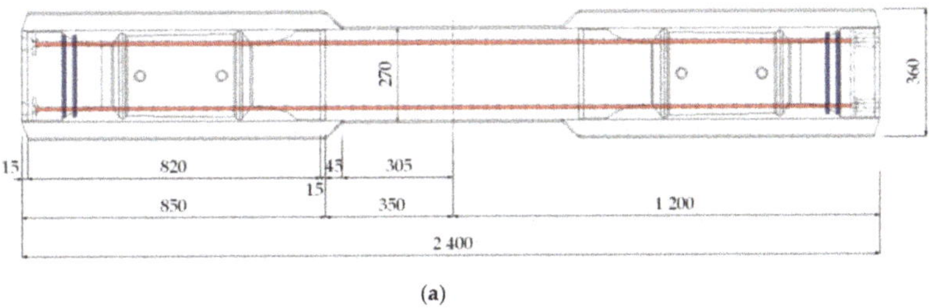

(a)

Figure 2. *Cont.*

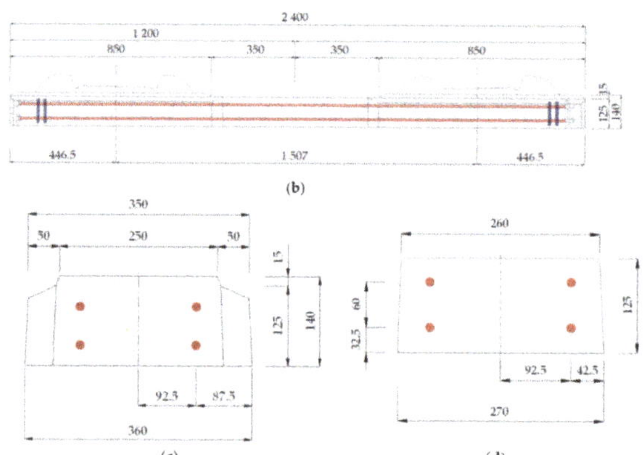

Figure 2. Geometrical dimension of L-150 series sleeper (unit: mm): (**a**) top view; (**b**) front view; (**c**) rail-seat section; (**d**) center section.

3. Finite Element Modeling

The brittle cracking model available in ABAQUS [23] is adopted to describe the brittle failure nature of the concrete. Since the direct tensile stress and strain relationships were available for three different levels of the steel fiber contents, the direct stress after cracking and direct cracking strain data were computed and used to define concrete cracking behaviors. Figure 1 shows the comparisons between the numerical and experimental stress-strain relationships for the cracking models. The inelastic tensile strain is computed by Equation (1).

$$\epsilon_t^{in} = \epsilon_t^{current} - \frac{\sigma_t^{current}}{E_o}, \quad (1)$$

where ϵ_t^{in} is the inelastic strain (direct cracking strain) in tension, $\epsilon_t^{current}$ is the total strain, $\sigma_t^{current}$ is the current stress level (direct stress after cracking), and E_o is the initial elastic modulus of concrete [23]. For the simplicity and the elastic nature of the UHPC (with the compressive strength of 150 MPa), the compression region of concrete is modeled as a linear elastic model. Table 1 summarizes the important mechanical properties of concrete and prestresssing tendon in the models. For the post-cracking behavior of the UHPC, the direct stress onset of cracking was found to be 3.17 MPa, 6.52 MPa, and 5.58 MPa for the UHPC with steel fiber content 0.5%, 1.0%, and 1.5%, respectively, from the direct tension test. It is important to note that the direct stress onset of cracking of the UHPC with 1.0% steel fiber content is slightly higher than that of the UPHC with 1.5% steel fiber content. However, the ultimate strength of the UHPC with 1.5% steel fiber content is the highest (see Figure 1).

In this study, a 2D model with plane stress elements (four-node plane stress element) was adopted to describe the concrete body of a sleeper. The width of the sleeper is separately defined at the various regions; the width of 360 mm was assigned to the rail-seat area, and the width of 270 mm at the center section. The PS tendons were modeled as 1D truss model (two-node linear truss element) with a specific area (132.95 mm^2 = 2 × 66.48 mm^2) at the specific heights. The prestressing tendons were fully embedded into the concrete body. Figure 3 shows the 2D numerical model developed in ABAQUS (ABAQUS 6.14, Dassault Systèmes Simulia Corp, Providence, RI, USA). The total number of the elements and the nodes were 1840 and 1985, respectively. Then, 69,000 N (1038 MPa) of the prestressing force was assigned to each tendon. A pin and roller boundary conditions were assigned at 197 mm and 697 mm nodes from the free end. A point load was applied at 447 mm from the free end on the top surface at the rail-seat section, similar to the experimental test. An explicit dynamics analysis was performed for a quasi-static process [23].

Table 1. Summary of the material properties.

Concrete	Young's modulus	51.0 GPa
	Compressive strength	150 MPa
	Poisson's ratio	0.2
	Tensile strength (steel fiber 0.5%)	8.82 MPa
	Tensile strength (steel fiber 1.0%)	15.6 MPa
	Tensile strength (steel fiber 1.5%)	18.4 MPa
	Direct stress onset of cracking (steel fiber 0.5%)	3.17 MPa
	Direct stress onset of cracking (steel fiber 0.5%)	6.52 MPa
	Direct stress onset of cracking (steel fiber 0.5%)	5.58 MPa
Steel tendon	Young's modulus	200 GPa
	Yielding strength	1275 MPa
	Poisson's ratio	0.3

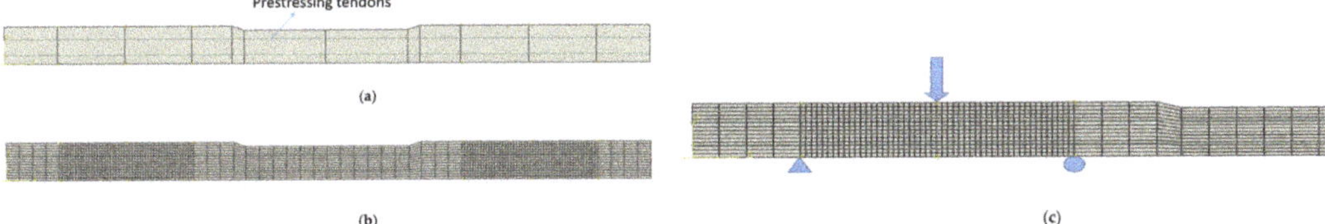

Figure 3. The 2D numerical sleeper model (**a**), its mesh (**b**), and the boundary conditions (**c**).

4. Comparisons with Experimental Data

4.1. Summary of the Testing at the Rail-Seat Section

A quasi-static three-point bending test according to European standards [9], was conducted on three UHPC sleepers with 0.5% steel fiber contents, six sleepers with 1.0% steel fiber contents, and three sleepers with 1.5% steel fiber contents. The centerline of the actuator is placed at 447 mm away from the free end of the sleeper, and the supports were placed 500 mm away from each other. Figure 4 shows a testing setup of the static three-point bending test. The reference test load, Fr_o of 126.8 kN, was computed [17]. The force and crack-width relationship of each sleeper was obtained and compared to each other. Overall, the higher the steel fiber contents are, the higher load capacities become. The 1.5% UHPC sleepers showed the highest failure forces and were able to mitigate the crack propagation. In Section 4.2.2, the experimental force and crack-width relationships together with numerical results are presented. The detail of the experimental tests and results can be found in the previous study [8].

Figure 4. Static bending test setup at the rail-seat section.

4.2. *Validation of the Numerical Sleeper Models*

4.2.1. Fiber Orientation Reduction Factor

The fiber orientation effect was considered, herein, when calibrating the experimentally obtained tensile constitutive relationships of UHPC depicted in Figure 1. It is well known that the tensile capacities of fiber reinforced concrete including UHPC principally depends on the fiber properties including the distribution and volume fraction [26,27]. It should be pointed out that due to the relatively thin specimen used in the tensile test (30 mm) [24], 19.5 mm long steel fibers tend to be aligned in a two-dimensional manner. On the other hand, the discontinuous steel fibers can be assumed to be three-dimensionally oriented in the 140 mm thick rail-seat section of the sleepers. The fiber orientation factors, α, are known to be $2/\pi$ and 0.5 for two-dimensional and three-dimensional fiber orientations, respectively [27,28]. Therefore, it is logical to adopt 0.785 ($=\frac{0.5}{2/\pi}$) as the fiber orientation reduction factor in this numerical study. Figure 5 shows the adopted stress-strain curves in the numerical analysis after considering 0.785 of the reduction factor, α. After the proportional limit of the obtained stress-strain relationships, the strength is reduced by 21.5% of the original strengths. Therefore, the constitutive relationships with α of 0.785 were used in the numerical simulations.

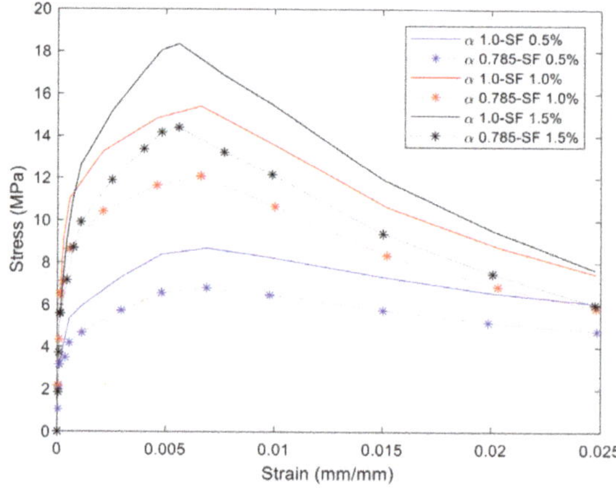

Figure 5. Stress and strain curves after applying the reduction factor of 0.785 for tension.

4.2.2. Validation of the Numerical Models

Three numerical sleeper models were prepared: (1) with 0.5% steel fiber contents, (2) with 1.0% steel fiber contents, and (3) with 1.5% steel fiber contents. A point load is applied at the center of the rail-seat section of the models. The applied load and the crack widths were monitored and compared with the experimental tests. Figures 6–8 show the comparisons of the force and crack-width curves between the numerical simulations and the experimental tests. Overall, the numerical models agree well with the experimental test results. The figures demonstrate that the sleeper models incorporated in this study are capable of capturing the initial stiffness, yielding of the steel tendons, cracking of the concrete, and the capacity of the sleepers due to the different level of steel fiber contents. It is also worthwhile to note that the fiber orientation factor, α, of 0.785 is able to describe the strength change in the UHPC of the sleepers from the coupon tests. In addition, the 1% steel fiber UHPC sleeper tends to overestimate the strength, while 0.5% and 1.5% steel fiber UHPC sleepers underestimate the ultimate strengths as compared to the experimental results.

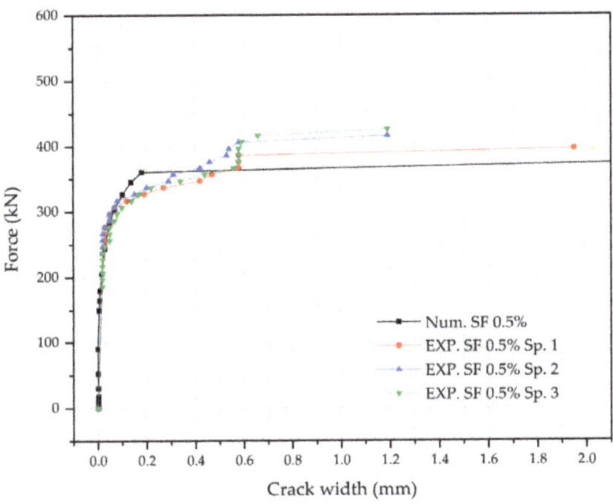

Figure 6. Comparison of the force and crack-width curves with 0.5% steel fiber UHPC at the rail-seat section.

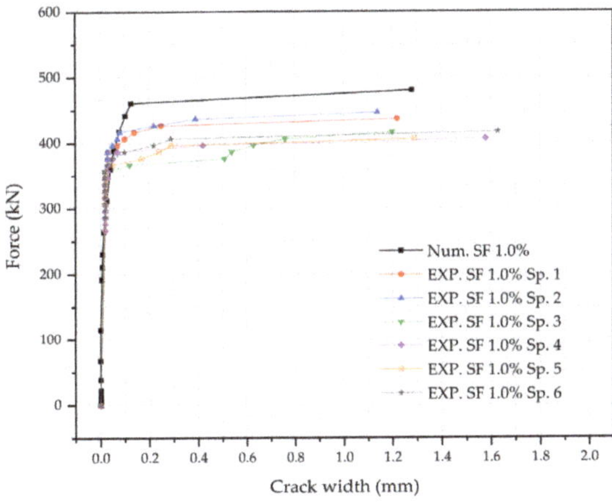

Figure 7. Comparison of the force and crack-width curves with 1.0% steel fiber UHPC at the rail-seat section.

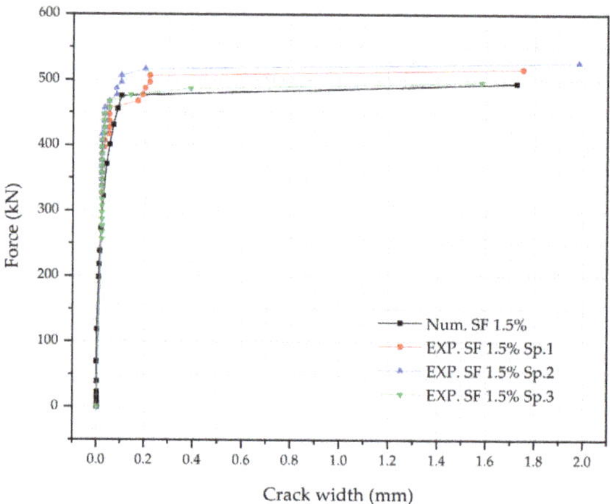

Figure 8. Comparison of the force and crack-width curves with 1.5% steel fiber UHPC at the rail-seat section.

5. Parametric Study

5.1. Design of Input Parameters

A parametric study was conducted with respect to the cross-sectional dimensions of sleepers and different types of steel tendons and steel fiber contents in the UHPC using the developed numerical sleeper models. In this parametric study, the structural performance of the UHPC sleepers were explored in terms of the crack width, the load capacities, the safety factor, and an economical design factor.

The important mechanical and geometrical parameters of the UHPC sleepers considered, herein, are as follows: (1) the cross-sectional dimensions, (2) the diameter of the steel tendons, (3) the yielding strength of the PS tendons, and (4) the steel fiber content of the UHPC. Table 2 summarizes the input values of each parameter. When the height of the cross-section at the rail-seat (h_r) changes, the height of the cross-section (h_c) changes accordingly. In addition, the locations of the steel tendons on top (P_1) and bottom (P_2) have to be adjusted (see Figure 9). Three different heights at the rail-seat section have been explored: 140 mm (L-type), 165 mm (M-type), and 195 mm (H-type). L, M, and H stands for lower, medium, and high height of the cross-sections, respectively. In the railway industry in South Korea, a 9.2 mm diameter tendon with 1080 MPa of the yielding strength has been commonly used. However, there is a growing interest in adopting larger diameter tendons and/or high strength steel such as 11.0 mm and 1275 MPa of the yielding strength when manufacturing prestressed concrete sleepers. Three different levels of steel fiber contents (0.5%, 1.0%, and 1.5%) are also explored. The total number of the simulation cases is 21, and Table 3 summarizes the 21 different simulation cases. Specimen numbers 1~7 were designed to have 0.5% of steel fiber of the UHPC, specimen numbers 8~14, 1.0%, and specimen numbers 15~21, 1.5%, respectively.

Table 2. Summary of the input parameters.

Steel fiber contents (%)	0.5	1.0	1.5
Yielding stress of prestressing tendon (f_y) (MPa)	1 080	1 275	
Diameter of prestressing tendon (φ) (mm)	9.2	11.0	
Cross-sectional parameters	L-Type	M-Type	H-Type
Height of the rail-seat section, h_r (mm) (h_{r1}, mm)	140 (125)	165 (150)	195 (180)
Height of the center section, h_c (mm)	125	150	180
Location of the prestressing tendon, P_1 (mm)	32.5	35	50
Location of the prestressing tendon, P_2 (mm)	60	75	80

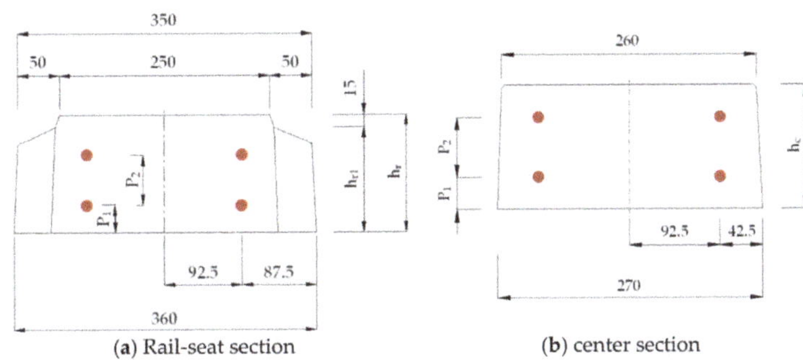

(**a**) Rail-seat section (**b**) center section

Figure 9. The schematics of the UHPC sleeper sections.

Table 3. Summary of the numerical concrete sleeper models and their nomenclatures.

Sp. No.	Steel Fiber 0.5%	Sp. No.	Steel Fiber 1.0%	Sp. No.	Steel Fiber 1.5%
No.1	L-φ9.2-f_y1 275	No.8	L-φ9.2-f_y1 275	No.15	L-φ9.2-f_y1 275
No.2	L-φ11.0-f_y1 275	No9	L-φ11.0-f_y1 275	No.16	L-φ11.0-f_y1 275
No.3	L-φ11.0-f_y1 080	No.10	L-φ11.0-f_y1 080	No.17	L-φ11.0-f_y1 080
No.4	M-φ09.2-f_y1 275	No.11	M-φ9.2-f_y1 275	No.18	M-φ9.2-f_y1 275
No.5	M-φ09.2-f_y1 080	No.12	M-φ9.2-f_y1 080	No.19	M-φ9.2-f_y1 080
No.6	H-φ09.2-f_y1 275	No.13	H-φ9.2-f_y1 275	No.20	H-φ9.2-f_y1 275
No.7	H-φ09.2-f_y1 080	No.14	H-φ9.2-f_y1 080	No.21	H-φ9.2-f_y1 080

5.2. Analysis Results

5.2.1. Cross-Sectional Dimensions: L, M and H

In order to discuss the effect of the cross-section sizes (L, M, and H), three simulation results were presented in Table 4 and Figure 10 as examples: (1) L/9.2/1275/1.0%, (2) M/9.2/1275/1.0%, and (3) H/9.2/1275/1.0%. In this discussion, the only variable is the size of the cross-section, when other parameters are kept constant: the diameter of the tendons is 9.2 mm, the yielding strength of the tendon is 1275 MPa, and the steel fiber content is 1.0%. In general, the larger the cross-section is, the greater the loading capacity of the sleepers becomes. In the figure, the simulation result of the sleeper with 140 mm of h_r, 9.2 mm of the diameter, 1275 MPa of f_y (steel), and 1% of the steel fiber content is represented by the black square line (L/9.2/1275/1%). ΔF_1 means the change in the applied load required between the force (Fr_r) when the crack width is about 0.01 mm and the corresponding force ($Fr_{0.05}$) when the crack width reaches about 0.05 mm. Higher

ΔF_1 is observed from the larger section sleepers. This means that the larger cross-section sleepers are capable of delaying crack propagations. In other words, when the cross-section of the sleeper gets larger, the moment of inertia becomes greater, which results in increased flexural rigidity and sustains higher moments without significant damages. After the crack width reached 0.05 mm, the secant and tangent modulus of the force-crack width diagram were gradually reduced. At approximately 0.12 mm crack width, the PS tendons reached the yielding. Soon after the yielding of the prestressing tendons, the sleepers reached the failure (Fr_B) of the rail-seat section due to the significantly reduced flexural rigidity. Similar trends were observed when the steel fibers were 0.5% and 1.5% as well. The force and crack-width graphs of other cases are presented in Section 5.2.3.

Table 4. Summary of the L, M, and H-type sleepers with the following parameters: 9.2 mm diameter, fy of 12,175 MPa and 1% steel fiber.

Simulation Case	Rail-Seat Section Area (mm^2)	Force (kN)		Crack Width (mm)	100Fr_B/Area	ΔF_1 (kN) = ($Fr_{0.05}$ − Fr_r)	ΔF_2 (kN) = (Fr_B − $Fr_{0.05}$)	$Fr_B/2.5Fr_0$
L/9.2/1 275/1.0%	47 200	Fr_r	230.4	0.008 6	1.02	158.4	91.2	1.51
		$Fr_{0.05}$	388.8	0.056 4				
		Fr_B	480.0	1.28				
M/9.2/1 275/1.0%	56 075	Fr_r	300.0	0.009	1.12	206.2	118.8	1.97
		$Fr_{0.05}$	506.2	0.051				
		Fr_B	625.0	2.94				
H/9.2/1 275/1.0%	66 725	Fr_r	400.8	0.009 4	1.25	276.5	158.7	2.63
		$Fr_{0.05}$	676.3	0.052 8				
		Fr_B	835.0	1.988				

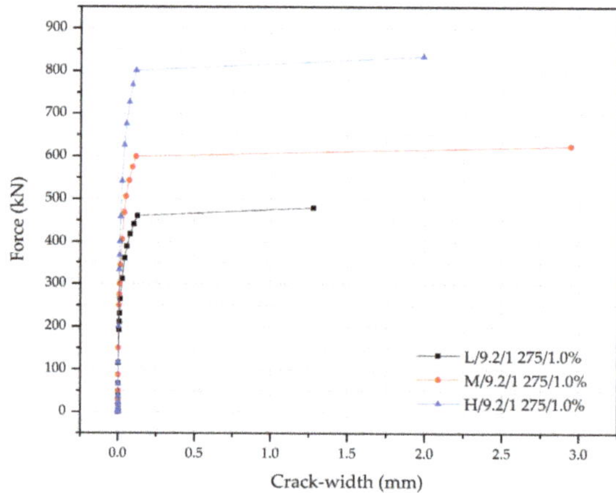

Figure 10. The force-crack width diagram of the L, M, and L-type sleepers with 9.2 mm diameter, 1% steel fiber, and fy of 1275 MPa.

The ratio of the cross-sectional area of the M-type sleeper to the L-type sleeper, and the ratio of the cross-sectional area of the H-type sleeper to the L sleeper are 1.19 and 1.41, respectively, while the ΔF_1 ratios of the M to L sleeper and H to L sleeper were 1.30 and 1.74, respectively. This means that the increased capacity ratios of the sleepers were higher than the increased area ratios. The safety factor of each sleeper can be computed by $\frac{F_{rB}}{2.5F_{ro}}$, where Fr_B and Fr_o is the force at the failure and the design reference force; Fr_o is 126.8 kN and 2.5 is the dynamic factor [17]. L, M, and H's safety index was found to be 1.51, 1.97, and 2.64. Too large a safety index means the sleeper is over-designed. This study suggests an economical design factor, which can be computed by $100Fr_B$/Area. When this index is close to 1, the sleeper is structurally sound and economical. The $100Fr_B$/Area index value of the L, M, and H sleepers were found to be 1.02, 1.12, and 1.25, which indicate that the L-type sleeper is the most economical design.

5.2.2. The Diameter and the Yielding Strength of PS Tendons

Table 5 and Figure 11 shows the simulation results with respect to the diameter and the yielding strength of the PS tendons when the steel fiber content was kept at 1.0%. Two different diameters of the PS tendons are explored: (1) 9.2 mm (smaller diameter), and (2) 11.0 mm (larger diameter). In addition, 1080 MPa and 1275 MPa of the yielding strength, f_y are considered. As examples, five simulations are presented in Table 5 and Figure 11: (1) L/9.2/1275/1.0%, (2) L/11.0/1275/1.0%, (3) L/11.0/1080/1.0%, (4) H/9.2/1275/1.0%, and (5) H/9.2/1080/1.0%. Given that the cross-sections and the steel fiber contents are kept constant, about 20% higher yielding strength PS tendons results in only 4.4% and 9.5% increase in ΔF_2 for H/9.2 types, and L/11.0 types, respectively. This is due to the area of the PS tendons to the area of the cross-sectional area of concrete being relatively low for the H/9.2 type. When using the larger diameter PS tendons, the load capacities of the sleepers increase accordingly. When comparing the results between L/9.2/1275/1.0% and L/11.0/1275/1.0%, the area of the larger diameter PS tendons is 1.43 times to that of the smaller diameter tendons; and the increase in Fr_r, $Fr_{0.05}$, and Fr_B is 20%, 11%, and 20%, respectively. These results indicate that the use of the larger diameter tendons would be more efficient than the use of the higher strength PS tendons in terms of the load increase capacities. In addition, these simulation results give some insights on whether sleeper (or crosstie) engineers would like to use a combination of (1) smaller diameter with higher strength PS tendons or (2) larger diameter with lower strength PS tendons. L/11.0/1080/1.0% case shows higher load capacities and safety factors than those from L/9.2/1275/1.0%. However, when engineers prefer an economical design, L/9.2/1275/1.0% can also be adopted since the safety factor is 1.51 and the $100Fr_B$/Area index is close to 1.0.

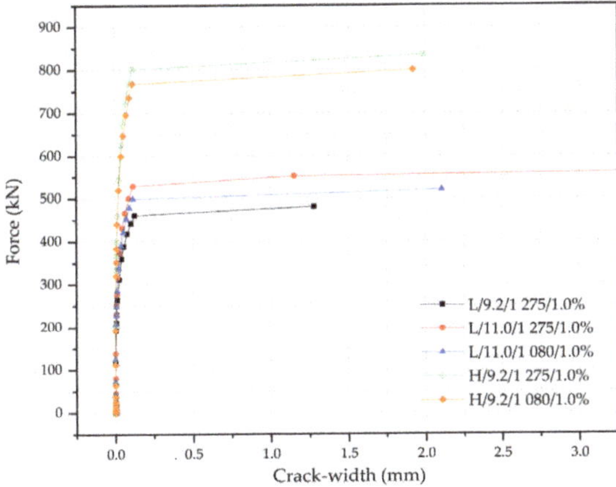

Figure 11. Force and crack width diagrams with respect to the diameter and the yielding strength of the PS tendons (the steel fiber content was kept at a 1.0% constant).

Table 5. Summary of the simulation results with respect to the diameter and the yielding strength of the PS tendons (the steel fiber content was kept at a 1.0% constant).

Analysis Case	Rail-Seat Section Area (mm^2)	Force (kN)		Crack Width (mm)	$100Fr_B$/Area	ΔF_1 (kN) = ($Fr_{0.05}$ − Fr_r)	ΔF_2 (kN) = (Fr_B − $Fr_{0.05}$)	$Fr_B/2.5Fr_0$
L/9.2/1275/1.0%	47 200	Fr_r	230.4	0.0086	1.02	158.4	91.2	1.51
		$Fr_{0.05}$	388.8	0.0564				
		Fr_B	480.0	1.28				
L/11.0/1275/1.0%	47 200	Fr_r	276.0	0.0098	1.22	155.3	142.7	1.81
		$Fr_{0.05}$	431.3	0.0452				
		Fr_B	575.0	7.292				
L/11.0/1080/1.0%	47 200	Fr_r	249.6	0.0115	1.10	171.6	98.8	1.64
		$Fr_{0.05}$	421.2	0.0532				
		Fr_B	520.0	2.104				
H/9.2/1275/1.0%	66 725	Fr_r	400.8	0.0094	1.25	276.5	158.7	2.63
		$Fr_{0.05}$	676.3	0.0528				
		Fr_B	835.0	1.988				
H/9.2/1080/1.0%	66 725	Fr_r	384.0	0.0084	1.12	264.0	152.0	2.52
		$Fr_{0.05}$	648.0	0.0539				
		Fr_B	800.0	1.922				

5.2.3. Steel Fiber Contents

This section presents the simulation results with respect to the steel fiber contents (i.e., 0.5%, 1.0%, and 1.5%). Table 6 and Figure 12 show the summary of the results of the six simulations used as examples: (1) L/9.2/1275/0.5%, (2) L/9.2/1275/1.0%, (3) L/9.2/1275/1.5%, (4) H/9.2/1275/0.5%, (5) H/9.2/1275/1.0%, (6) H/9.2/1275/1.5%. In general, as the steel fiber content increases, the load capacities, the safety factor, and the economic design factor increase. For the smaller cross-section sleepers (L-type cases), the use of 1.0% and 1.5% steel fiber contents results in the significant increase in the performance compared to that of the sleeper with 0.5% steel fiber content. The performance of L/9.2/1275/1.0% and L/9.2/1275/1.5% are similar to each other, and the increase in the load capacities are only 3~6%; furthermore, the safety factor only increases to 1.56 (1.5% of the steel fiber) from 1.51 (1.0% of the steel fiber). The performance of the L-type-0.5% steel fiber sleeper is significantly lower than that of the sleepers with the higher steel fiber contents. As observed, $\frac{F_{rB}}{2.5F_{r0}}$ is only 1.18 and $100Fr_B$/Area is 0.79 for L/9.2/1275/0.5%. For the larger cross-section sleepers (H-type cases), the trends are similar to those from the L-type cases. The steel fiber 1.0% and 1.5% sleepers show good performance while the difference between two cases is not as great as the L-types. The H-type-0.5% steel fiber sleeper shows lower load capacities and safety factors when compared to those of the higher steel fiber content sleepers; $100Fr_B$/Area is 0.94, which is still less than 1.0. When casting a smaller cross-section UHPC sleeper (L-type case), the use of 0.5% steel fiber content is not adequate. In addition, the difference in the performance of the sleepers between 1.0 and 1.5% steel fiber UHPC in terms of the force and crack-width at the rail-seat is not much different. Therefore, 1.0% steel fiber UHPC can be an economical design choice. For the larger cross-section sleepers, all three steel fiber contents would be acceptable. However, instead of H-type 0.5% sleeper, M-type sleepers could be a good alternative since M-type sleepers shows the similar performance while they are more economical. As an example, the performance of M/9.2/1275/1.0% is similar to that of H/9.2/1275/0.5% in term of

Fr_r, $Fr_{0.05}$, Fr_B, the economical design factor, and the safety factor (see Tables 4 and 6). The economical design factor, $100Fr_B$/Area of the M-type was found to be 1.12, while that of the H-type was 0.94.

Table 6. Summary of the simulation results with respect to the 0.5%, 1.0%, and 1.5% steel fiber contents.

Analysis Case	Rail-Seat Section Area (mm^2)	Force (kN)		Crack Width (mm)	$100Fr_B$/Area	ΔF_1 (kN) = ($Fr_{0.05} - Fr_r$)	ΔF_2 (kN) = ($Fr_B - Fr_{0.05}$)	$Fr_B/2.5Fr_0$
L/9.2/1 275/0.5%	47 200	Fr_r	180.0	0.0103	0.79	101.3	93.7	1.18
		$Fr_{0.05}$	281.3	0.0487				
		Fr_B	375.0	2.282				
L/9.2/1 275/1.0%	47 200	Fr_r	230.4	0.0086	1.02	158.4	91.2	1.51
		$Fr_{0.05}$	388.8	0.0564				
		Fr_B	480.0	1.28				
L/9.2/1 275/1.5%	47 200	Fr_r	217.8	0.0108	1.05	183.1	94.1	1.56
		$Fr_{0.05}$	400.9	0.0532				
		Fr_B	495.0	1.724				
H/9.2/1 275/0.5%	66 725	Fr_r	300.0	0.0105	0.94	168.7	156.3	1.97
		$Fr_{0.05}$	468.7	0.0454				
		Fr_B	625.0	3.137				
H/9.2/1 275/1.0%	66 725	Fr_r	400.8	0.0094	1.25	276.5	158.7	2.63
		$Fr_{0.05}$	676.3	0.0528				
		Fr_B	835.0	1.988				
H/9.2/1 275/1.5%	66 725	Fr_r	388.5	0.0109	1.32	273.8	220.7	2.79
		$Fr_{0.05}$	662.3	0.0452				
		Fr_B	883.0	2.224				

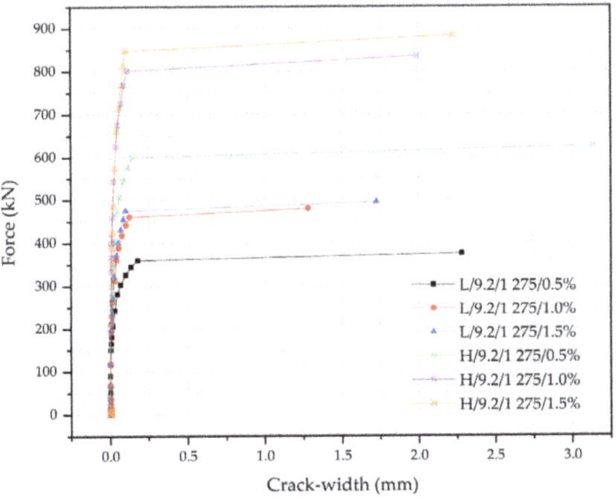

Figure 12. Force and crack-width relationships of the L- and H-type sleepers with respect to the three different steel fiber contents.

Figures 13–15 show the force and crack-width relationship of all 21 numerical simulations. Regardless of the steel fiber contents, the increase in size would have the most significant impacts on the performance of the UHPC sleepers. The improvement can be further enhanced with the combination of the higher steel fiber contents, the larger diameter PS tendons, and the higher strength PS tendons. However, the use of a larger cross-section with 1.5% steel content, 11.0 mm diameter PS tendons of 1275 MPa yielding strength is a clearly over-designed sleeper. This study also indicates that the improvement due to the higher strength PS tendons would be minimum among all the design parameters considered. It is also interesting to note that there are cases that would show similar performance even though the cross-section sizes are different. For example, M/9.2/1080/0.5% and L/11.0/1275/0.5% show similar performance, as well as M/9.2/1275/1.0% and H/9.2/1275/0.5%.

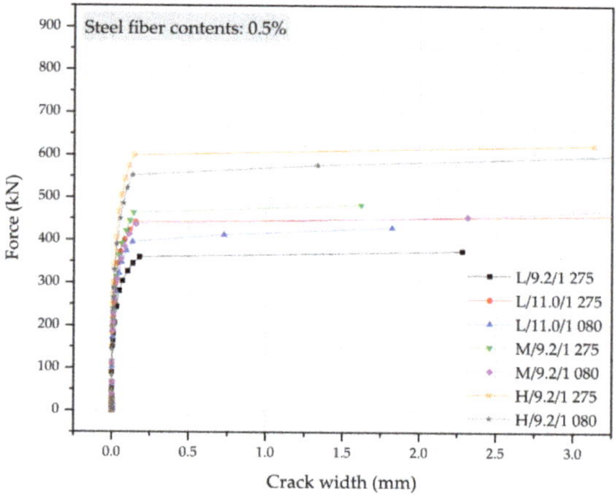

Figure 13. Simulation results of the force and crack-width curves with 0.5% steel fiber UHPC at the rail-seat section.

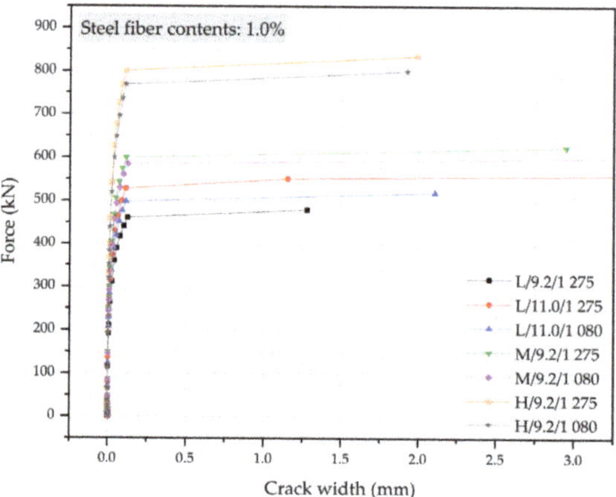

Figure 14. Simulation results of the force and crack-width curves with 1.0% steel fiber UHPC at the rail-seat section.

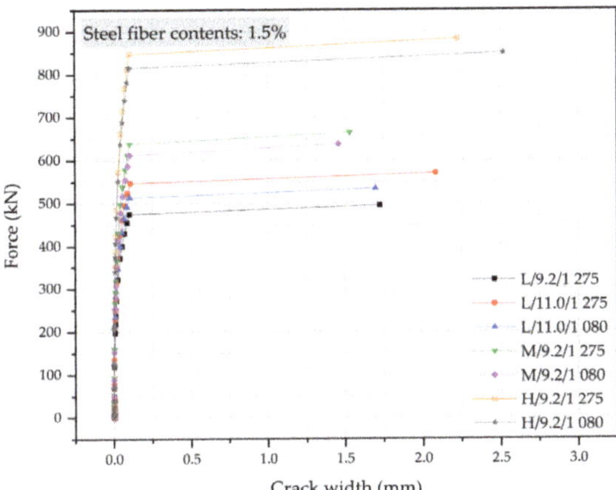

Figure 15. Simulation results of the force and crack-width curves with 1.5% steel fiber UHPC at the rail-seat section.

6. Discussion

In this numerical study, a 2D prestressed concrete sleeper model with a brittle cracking constitutive model was developed and validated with experimental data. In general, the numerical results were compatible with the experimental results in terms of the force and crack-width relationship at the rail-seat section. The obtained tensile stress-strain relationships of UHPC with different steel fiber contents [24] were directly used to define the cracking stress and cracking strain of the brittle cracking model. In this process, an orientation reduction factor of 0.785 was applied to the post-cracking behavior of all three UHPCs. As it has been known to the community, 2D and 3D orientation reduction factors are $2/\pi$, and 0.5. Therefore, a single orientation reduction factor (0.785) was adopted in this study for the simulations. However, the fiber reduction factor could be dependent upon the distribution, orientation, and volume fraction of the fibers in a concrete mix. Therefore, it is challenging to use a just deterministic approach for the factor while the concrete properties including the reduction factor has the stochastic characters. With the 0.785 reduction factor, 1.0% steel fiber sleepers predict the ultimate strength higher than the experimental test. This shows that there is room for improvement. A stochastic approach can be applied to evaluate the range in the performance of the sleepers using UHPC. It should be, however, pointed out that additional experimental research should be conducted to achieve a statistically meaningful dataset to adopt the stochastic approach, especially for the fiber orientation effect. In addition, the usage of a single orientation reduction factor would be beneficial for sleeper engineers and structural engineers to practically design concrete structures with fiber reinforcements.

7. Conclusions

This numerical study focuses on investigating the performance of UHPC sleepers with respect to various design/mechanical/geometrical parameters. The parameters include the steel fiber contents, the size of the cross-section, the diameter and yielding strength of the PS tendons. The key observations and findings of this research can be summarized as follows:

1. The developed numerical 2D-UHPC sleeper model was capable of representing the force and crack-width relationships. Three UHPC direct tension tests with the 0.5%, 1.0%, and 1.5% steel fiber contents were used for the UHPC tensile constitutive models.
2. The fiber orientation factor, α, of 0.785 is used to represent the realistic stress-strain behavior of the UHPC in 3D as opposed to the thin coupon test where the fibers are well aligned in a 2D manner.

3. The numerical analysis results indicate that the bigger the cross-section is, the higher the load capacities and the safety factor become. However, using a too large cross-section can result in uneconomical design sleepers. The economical design factor, $100Fr_B/Area$ is computed to evaluate the economical factor of the UHPC sleeper. When $100Fr_B/Area$ is close to 1.0, the UHPC sleeper is economical.
4. There are growing interests in using a larger diameter tendon and/or a higher strength tendon. This study recommends using a larger diameter tendon with a lower strength for an economical design.
5. A steel fiber content of 0.5% tends to yield to lower strengths UHPC sleepers relative to the 1.0% and 1.5% steel fiber content sleepers.
6. Some M-type sleepers with 1.0% steel fiber UHPC show similar performance to H-type sleepers with 0.5% steel fiber UHPC.

This numerical study was able to provide insights on the effects of the design parameters for developing concrete sleepers using UHPC. Additional research needs to be conducted to investigate the overall behavior of UHPC sleepers, including bending at the center-section of the sleepers and the effect of the variability of concrete properties including an orientation reduction factor.

Author Contributions: Conceptualization, M.S., Y.B. and S.P.; methodology, M.S., Y.B. and S.P.; software, M.S.; validation, M.S. and Y.B.; formal analysis, M.S. and Y.B.; investigation, M.S., Y.B. and S.P.; resources, M.S., Y.B. and S.P.; data curation, M.S. and Y.B.; writing—original draft preparation, M.S.; writing—review and editing, M.S., Y.B. and S.P.; visualization, M.S. and Y.B.; supervision, S.P.; project administration, Y.B. and S.P.; funding acquisition, Y.B. and S.P. All authors have read and agreed to the published version of the manuscript.

Funding: The research described herein was sponsored by a grant from R&D Program of the Korea Railroad Research Institute, Republic of Korea. This work is also supported by the Korea Agency for Infrastructure Technology Advancement (KAIA) grant funded by the Ministry of Land, Infrastructure and Transport (Grant 21CTAP-C152046-03). The opinions expressed in this paper are those of the authors and do not necessarily reflect the views of the sponsors.

Institutional Review Board Statement: Not applicable.

Informed Consent Statement: Not applicable.

Data Availability Statement: The data presented in this study are available on request from the corresponding authors.

Conflicts of Interest: The authors declare no conflict of interest.

References

1. Sadeghi, J.; Kian, A.R.T.; Khabbazi, A.S. Improvement of mechanical properties of railway track concrete sleepers using steel fibers. *J. Mater. Civil. Eng.* **2016**, *28*, 04016131. [CrossRef]
2. Ferdous, W.; Manalo, A. Failures of mainline railway sleepers and suggested remedies–review of current practice. *Eng. Fail. Anal.* **2014**, *44*, 17–35. [CrossRef]
3. Kaewunruen, S.; Remennikov, A. Dynamic properties of railway track and its components: A state-of-the-art review. In *New Research on Acoustics*; Weiss, B.N., Ed.; Hauppauge, Nova Science: New York, NY, USA, 2008; Volume 28, pp. 197–220.
4. Pyo, S.; Alkaysi, M.; El-Tawil, S. Crack propagation speed in ultra high performance concrete (UHPC). *Constr. Build. Mater.* **2016**, *114*, 109–118. [CrossRef]
5. Ramezanianpour, A.A.; Esmaeili, M.; Ghahari, S.A.; Najafi, M.H. Laboratory study on the effect of polypropylene fiber on durability, and physical and mechanical characteristic of concrete for application in sleepers. *Constr. Build. Mater.* **2013**, *44*, 411–418. [CrossRef]
6. Shin, H.O.; Yang, J.M.; Yoon, Y.S.; Mitchell, D. Mix design of concrete for prestressed concrete sleepers using blast furnace slag and steel fibers. *Cem. Concr. Compos.* **2016**, *74*, 39–53. [CrossRef]
7. Yang, J.M.; Shin, H.O.; Yoon, Y.S.; Mitchell, D. Benefits of blast furnace slag and steel fibers on the static and fatigue performance of prestressed concrete sleepers. *Eng. Struct.* **2017**, *134*, 317–333. [CrossRef]
8. Bae, Y.; Pyo, S. Effect of steel fiber content on structural and electrical properties of ultra high performance concrete (UHPC) sleepers. *Eng. Struct.* **2020**, *222*, 111131. [CrossRef]

9. EN 13230-2. *Railway Applications-Track-Concrete Sleepers and Bearers-Part 2: Prestressed Monoblock Sleepers*; European Committee for Standardization (CEN): Brussles, Belgium, 2016.
10. AS-1085.14. *Railway Track Material Part 14: Prestressed Concrete Sleepers*; Standard Australia: Sydney, Australia, 2012.
11. UIC. *713R. Design of Monoblock Concrete Sleepers*; UIC Leaflet, International Union of Railways: Paris, France, 2004.
12. Hashim, D.T.; Hejazi, F.; Lei, V.Y. Simplified Constitutive and Damage Plasticity Models for UHPFRC with Different Types of Fiber. *Int. J. Concr. Struct. Mater.* **2020**, *14*, 1–21. [CrossRef]
13. Pyo, S.; El-Tawil, S.; Naaman, A.E. Direct tensile behavior of ultra high performance fiber reinforced concrete (UHP-FRC) at high strain rates. *Cem. Concr. Res.* **2016**, *88*, 144–156. [CrossRef]
14. Park, S.; Wu, S.; Liu, Z.; Pyo, S. The role of supplementary cementitious materials (SCMs) in ultra high performance concrete (UHPC): A review. *Mater.* **2021**, *14*, 1472. [CrossRef] [PubMed]
15. Pyo, S.; Abate, S.Y.; Kim, H.K. Abrasion resistance of ultra high performance concrete incorporating coarser aggregate. *Constr. Build. Mater.* **2018**, *165*, 11–16. [CrossRef]
16. Pyo, S.; El-Tawil, S. Capturing the strain hardening and softening responses of cementitious composites subjected to impact loading. *Constr. Build. Mater.* **2015**, *81*, 276–283. [CrossRef]
17. Bae, Y.; Pyo, S. Ultra high performance concrete (UHPC) sleeper: Structural design and performances. *Eng. Struct.* **2020**, *210*, 110374. [CrossRef]
18. Auersch, L.; Said, S.; Knothe, E.; Rücker, W. The dynamic behavior of railway tracks with under sleeper pads, finite-element boundary-element model calculations, laboratory tests and field measurements. In Proceedings of the 9th European Conference on Structural Dynamics (EURODYN 2014), Porto, Portugal, 30 June–2 July 2014; pp. 805–812.
19. Chandra, S.; Shukla, D. Sustainability of Railway Tracks. In *Sustainability Issues in Civil Engineering*; Springer: Singapore, 2017; pp. 91–104.
20. Sucharda, O.; Mateckova, P.; Bilek, V. Non-Linear Analysis of an RC Beam Without Shear Reinforcement with a Sensitivity Study of the Material Properties of Concrete. *Slovak J. Civ. Eng.* **2020**, *28*, 33–43. [CrossRef]
21. Valikhani, A.; Jahromi, A.J.; Mantawy, I.M.; Azizinamini, A. Numerical Modeling of Concrete-to-UHPC Bond Strength. *Materials* **2020**, *13*, 1379. [CrossRef] [PubMed]
22. Shin, M.; Yu, H. Numerical Evaluation of Splitting Performance of Prestressed Concrete Prisms With Larger Diameter Prestressing Wires. In Proceedings of the 2019 ASME Joint Rail Conference, Snowbird, UT, USA, 9–12 April 2019. [CrossRef]
23. ABAQUS. *ABAQUS Documentation*; Dassault Systemes: Providence, RI, USA, 2012.
24. Pyo, S.; Kim, H.K.; Lee, B.Y. Effects of coarser fine aggregate on tensile properties of ultra high performance concrete. *Cem. Concr. Compos.* **2017**, *84*, 28–35. [CrossRef]
25. Japan Society of Civil Engineers (JSCE). *Recommendations for Design and Construction of High Performance Fiber Reinforced Cement Composites with Multiple Fine Cracks (HPFRCC)*; Concrete Engineering Series; Springer: Tokyo, Japan, 2008.
26. Naaman, A.E. Engineered steel fibers with optimal properties for reinforcement of cement composites. *J. Adv. Concr. Technol.* **2003**, *1*, 241–252. [CrossRef]
27. Pyo, S.; Wille, K.; El-Tawil, S.; Naaman, A.E. Strain rate dependent properties of ultra high performance fiber reinforced concrete (UHP-FRC) under tension. *Cem. Concr. Compos.* **2015**, *56*, 15–24. [CrossRef]
28. Wille, K.; Kim, D.J.; Naaman, A.E. Strain-hardening UHP-FRC with low fiber contents. *Mater. Struct.* **2011**, *44*, 583–598. [CrossRef]

Article

Influence of Fiber Addition on the Properties of High-Performance Concrete

Szymon Grzesiak [1], Matthias Pahn [1,*], Milan Schultz-Cornelius [1], Stefan Harenberg [2] and Christoph Hahn [3]

1 Department of Civil Engineering, Technical University of Kaiserslautern, 67663 Kaiserslautern, Germany; szymon.grzesiak@bauing.uni-kl.de (S.G.); milan.schultz-cornelius@bauing.uni-kl.de (M.S.-C.)
2 Implenia Schalungsbau GmbH, 67240 Bobenheim-Roxheim, Germany; stefan.harenberg@implenia.com
3 Master Builders Solutions Deutschland GmbH, 68199 Mannheim, Germany; christoph.hahn@mbcc-group.com
* Correspondence: matthias.pahn@bauing.uni-kl.de

Abstract: High performance fiber-reinforced concrete (HPFRC) has been frequently investigated in recent years. Plenty of studies have focused on different materials and types of fibers in combination with the concrete matrix. Experimental tests show that fiber dosage improves the energy absorption capacity of concrete and enhances the robustness of concrete elements. Fiber reinforced concrete has also been illustrated to be a material for developing infrastructure sustainability in RC elements like façade plates, columns, beams, or walls. Due to increasing costs of the produced fiber reinforced concrete and to ensure the serviceability limit state of construction elements, there is a demand to analyze the necessary fiber dosage in the concrete composition. It is expected that the surface and length of used fiber in combination with their dosage influence the structure of fresh and hardened concrete. This work presents an investigation of the mechanical parameters of HPFRC with different polymer fiber dosage. Tests were carried out on a mixture with polypropylene and polyvinyl alcohol fiber with dosages of 15, 25, and 35 kg/m^3 as well as with control concrete without fiber. Differences were observed in the compressive strength and in the modulus of elasticity as well as in the flexural and splitting tensile strength. The flexural tensile strength test was conducted on two different element shapes: square panel and beam samples. These mechanical properties could lead to recommendations for designers of façade elements made of HPFRC.

Keywords: high performance fiber reinforced concrete (HPFRC); polypropylene fiber (PP); polyvinyl alcohol fiber (PVA); compressive strength; residual flexural strength; splitting tensile strength

Citation: Grzesiak, S.; Pahn, M.; Schultz-Cornelius, M.; Harenberg, S.; Hahn, C. Influence of Fiber Addition on the Properties of High-Performance Concrete. *Materials* **2021**, *14*, 3736. https://doi.org/10.3390/ma14133736

Academic Editor: Alessandro P. Fantilli

Received: 31 May 2021
Accepted: 29 June 2021
Published: 3 July 2021

1. Introduction

Since the development of concrete, RC constructions allow for more and more filigree and lightweight elements with the contemporary growth of structure loads [1]. Reduced cross-sections of components are associated with advanced technologies and materials based on higher material properties [2]. With an increase in concrete specifications like compressive strength, the post crack behavior of concrete becomes worse. In concrete compositions, different kinds of fibers are added to avoid brittle fracture behavior and ameliorate the ductility of those materials [3]. Fiber reinforcement concrete (FRC) has already been used successfully in many horizontal and vertical structural as well as non-structural elements [4]. For example, using fiber reinforcement together with traditional steel bar reinforcement decreases crack propagation and displacement of concrete slabs like industrial floors [5]. In buildings and bridges in seismic areas, fiber reinforced concrete improves the behavior of structural parts like columns, beams, or walls [6].

Recently, FRC has also been used for the production of pre-cast elements in which the fibers—in combination with ultra-high-performance concrete—enhance the durability of cracked concrete [7]. As studies show, a decisive role is played by the dense microstructure of concrete pre-cast elements. This can be ensured by low water–cement and water–binder ratios. Mateckova et al. [8] report that high-performance concrete with its dense structure presents higher resistance to chemical penetration in comparison to ordinary concrete. The character of used materials in HPC can improve the acid attack by FRC through better integrity of the binder matrix to fiber inclusion. Ali et al. [9] showed the influence of fibers and silica fume on the mechanical and durability performance of concrete concerning a reduction in the materials' permeability.

The possible benefits of using HPFRC are in sustainable resource management. A good example for the use of FRC are façade panels in building constructions, leading to a considerable reduction in the material volume [10]. Therefore, the major advantage of fiber-reinforced concrete elements is the reduced thickness, thus leading to a reduction in CO_2 footprint. Without steel rebars, façade panels can be just a few centimeters thick [11]. For concrete elements without steel reinforcement, a corrosion protection system like concrete cover can be omitted [12].

Vertical exterior elements of buildings exposed to environmental factors are investigated for structural performance under gravity and wind load. Concrete elements such as façade panels for certain boundary conditions are under flexural load. The wind pressure is distributed as area load, which causes tensile and compressive stresses in the cross-section of the building's façade and results in deformation [13]. Therefore, exterior elements are designed to transfer loads to the main structural system of the building. This is why the flexural tensile strength of fiber reinforced concrete is one important design parameter [14]. Moreover, the impact of panel behavior and damage to the FRC is evaluated as a safety factor.

To understand the behavior of HPC and UHPC as well as concrete in general with fiber addition, scientific measurement methods in experimental research study have recently been published [15]. Typical fibers for HPFRC are made of steel [16], carbon [17], or polymers [18]. Examining the available research literature demonstrates that another material like wool [19], basalt [20], or glass [21] could be successfully added to concrete. The fibers differ according to their origin, mechanical properties, and their corrosion resistance [22]. The analysis of this material focuses particularly on a unique combination of concrete and fiber reinforcement [23]. The damage process and the mechanical properties of HPFRC can be taken into account for different dimensions and different shapes of samples [24]. Results of the experimental tests should implement the anisotropy of fiber orientation in the concrete matrix [25]. The location of deformation and the position of the cracked zone can lead to difficulties during examination [26]. By means of a clip gauge, it is possible to estimate the behavior of specimens in the cracked region. The clip gauge is used to measure the crack mouth opening displacement (CMOD) [27]. The details of the classic experimental setup and the examination with a clip gauge will be presented in the Section 2.

The aim of this work was to investigate the influence of fiber addition on the properties of high-performance concrete. As is known from other publications, the addition of fiber does not always have a positive effect on the mechanical properties. Other studies have also focused on other types of fibers in connection with HPFRC. Therefore, in this study, it was proposed to also consider the fiber type and shape variation of the concrete sample. Furthermore, this paper expands the database with an overview of the mechanical properties in HPFRC with polypropylene and polyvinyl alcohol fiber. It is suspected that the surface and length of used fiber influence the structure of fresh and hardened concrete. Optimization and a better quantity of fiber dosage will allow for a reduction and also a better use of the materials in the concrete mix design. Mechanical parameters of HPFRC enable the economical design of filigree, safe, and lightweight elements.

2. Experimental Program

2.1. Materials

In this study, only fine aggregates and particles were used to improve the homogeneity of high-performance concrete [28]. In the following sections, the properties of the fibers and the concrete mix design are discussed.

2.1.1. Fiber

In this study, the influence of two fiber types on the fresh concrete and the mechanical properties of hardened concrete is investigated: polypropylene fibers (PP) and polyvinyl alcohol fibers (PVA) (see Figure 1). The characteristic of PP- and PVA-fibers are given in Table 1. Compared to steel, both fibers are corrosion-resistant.

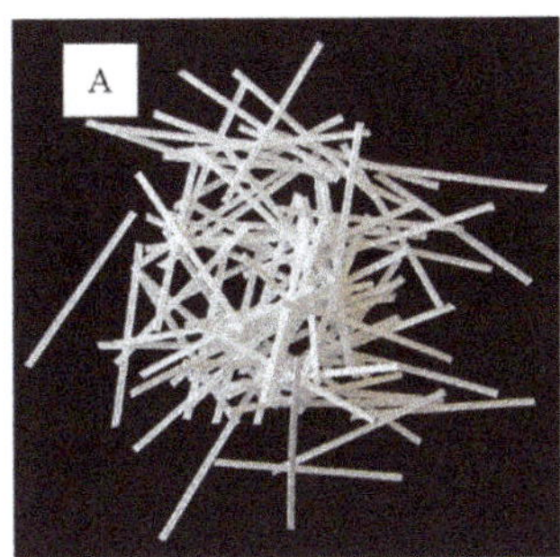

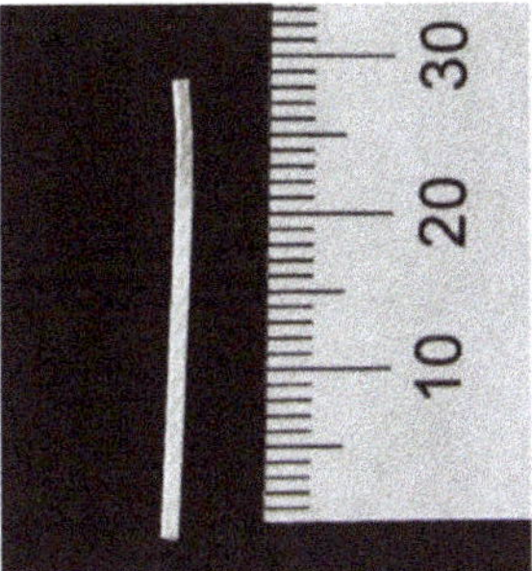

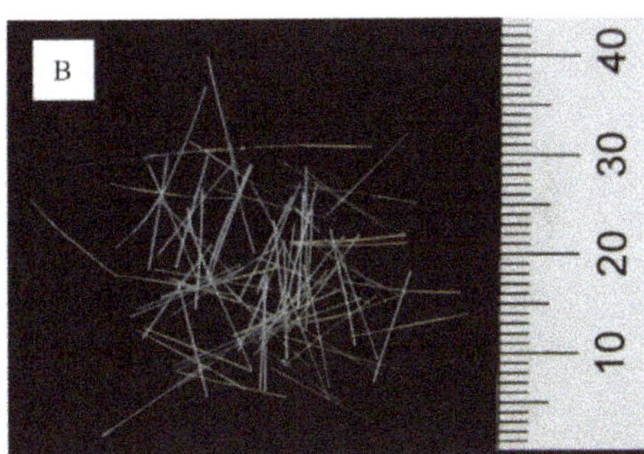

Figure 1. (**A**) Polypropylene fiber MasterFiber 235 SPA (PP). (**B**) Polyvinyl alcohol fiber MasterFiber 401(PVA) used in the present study.

Table 1. Mechanical parameters of the used fiber.

Type of Fiber	Tension Strength [MPa]	E-Modul [MPa]	Diameter [mm]	Length [mm]	Specific Gravity [kg/m³]
MasterFiber 235 SPA (PP)	500	>8.000	0.70	30	910
MasterFiber 401 (PVA)	800	29.000	0.16	12	1.300

2.1.2. Concrete Mix Design

For this experimental research, five different concrete mixtures with the same amount of cement, aggregates, and additives (Table 2) were prepared. They only differed in the fiber type (Table 1) and fiber dosage (Table 3). Mix ID 1 contained 35 kg/m^3 of the MasterFiber 401 (PVA). For comparison with the MasterFiber 235 SPA (PP), Mix ID 2 contained 35 kg/m^3. To investigate the influence of the fiber dosage, Mix IDs 3, 4, and 5 contained 25 kg/m^3, 15 kg/m^3, and 0 kg/m^3 of fibers, respectively. The mixtures were prepared by using a 55-L-capacity horizontal forcing type concrete mixer.

Table 2. Concrete mix design.

Material	Raw Density [kg/m³]	Weight [kg/m³]
Cement (CEM I 42.5 R)	3100	650
Aggregate 0 to 3 mm	2600	990
Silica fume	700	50
Limestone powder	2700	415
Plasticizer MasterGlenium ACE 430	1060	18
Water	1000	210

The appropriated HPFRC was produced with cement CEM I 42.5 R [29], quartz sand 0.1/0.6 [30], limestone powder, and silica fume. The limestone powder and the silica fume provide a dense microstructure of concrete and are used as fillers. The aggregates were dried sand and basalt [31]. Figure 2 presents the grain size curve of the material used in the present study [32]. For better workability, a plasticizer, MasterGlenium ACE 430, was used [33]. The water–cement (w/c) and water–binder (w/b) ratio in the concrete composition was 0.323 and 0.238, respectively.

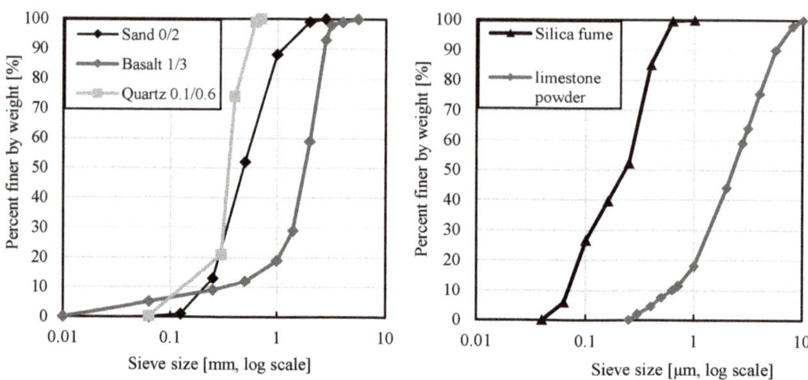

Figure 2. Grading of the fines for the material used in the present study. Sand, basalt, and quartz own studies. Silica fume [34], limestone powder [35].

2.2. Testing Procedure

In this study, tests were carried out on fresh and hardened high-performance concrete with and without fibers. The compressive strength tests on the hardened concrete were performed on 150 mm cubes according to EN 12390-1 [36]. Cylinders of 150 mm in diameter and 300 mm in height were used to test the modulus of elasticity and splitting tensile strength according to EN 12390-3 [37].

Table 3. Overview of tested samples.

Series	Fiber Dosage [kg/m^3]	Number of Specimens
panel specimens (250 × 250 × 35 mm^3) in accordance with EN 12467 [38]	0	3
	15	3
	25	3
	35	3
beam specimens (100 × 100 × 400 mm^3) in accordance with EN 14651 [39]	0	3
	15	3
	25	3
	35	3

A test program based on the different fiber dosage was conducted (Table 3). The flexural tensile strength test was conducted on square panel specimens (250 × 250 × 35 mm^3) in accordance with EN 12467 [38] and beam specimens with a cross-section (100 × 100 × 400 mm^3) in accordance with EN 14651 [39]. For comparison purposes in both tests, we used the classical measurement method with the force–displacement relationship. Simultaneously, for the measurements of beam deformation, we applied the test method of crack mouth opening displacement (CMOD), which has been recently mentioned in the experimental investigations of HPFRC [27,40]. For this purpose, the bottom surface of the beam specimens had a notch with a depth of 17 mm and a width of 5 mm according to the procedure in EN 14651 [39] (see Figure 3). The notch was milled for the determination of strain during the crack initiation in the middle of the span. Near the notch were glued metal plates to attach the clip gauge (see Figure 4). The clip gauge measures the displacement between two points that are on two different edges of the crack [41].

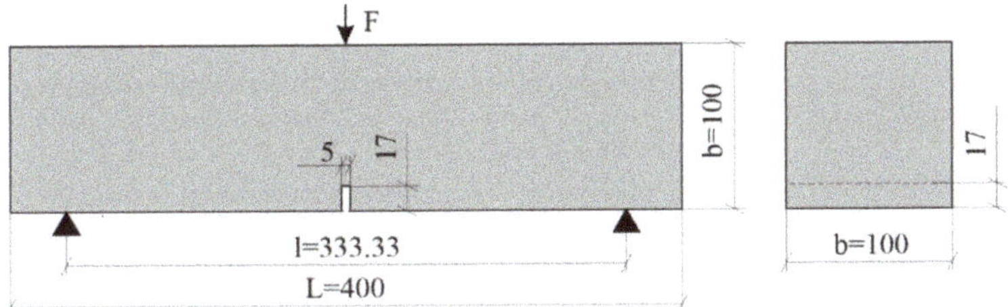

Figure 3. Beam specimen geometry. Dimensions in [mm].

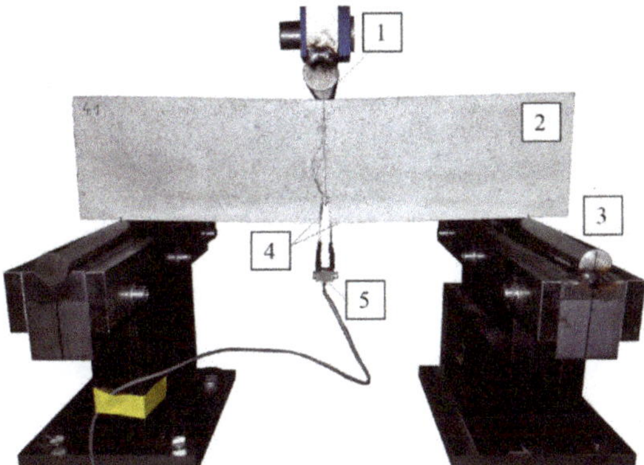

Figure 4. Experimental setup: 1 = point of application of symmetrical load, 2 = HPFRC beam specimen, 3 = supports, 4 = glued on metal plates to attach the clip gauge, 5 = clip gauge.

As shown in a study by Bi et al. [42], the distribution of fiber during concrete flow influences the mechanical properties of HPFRC. Mostly, fibers are randomly distributed but computer tomography shows that fibers accumulate in the upper part of the specimens. It has been observed that the flexural strength of HPFRC is higher with more fibers located in the tension zone. Therefore, during the bending tests, the upper side of the beams is turned backwards.

Three-point bending tests were carried out on plates and beams in order to be able to estimate the flexural strength [38]. The stresses f_L were calculated according to the *fib* Model Code 2010 [43], as follows:

$$f_L = \frac{3 \cdot \mathrm{F} \cdot l}{2 \cdot b \cdot h^2} \tag{1}$$

where F represents the cylinder force in [N]; b and h are the width and height of the specimens, respectively; and l is the distance of the supports. The distance for the plates equaled $l = 0.75 \cdot L$ and for the beams, $l = b/0.3$. Dimensions are provided in [mm].

In their study, Schultz-Cornelius [24] investigated flexural strength for many different thicknesses of façade panels separately. They observed an effect in size for specimens with a depth below 50 mm. The tests showed a linearly increasing bending tensile strength with decreasing thickness. In order to examine the impact of fiber dosage on the bending tensile strength, only panels with a thickness of 35 mm were tested.

The specimens were cured and stored at a temperature of 20 °C and a relative air humidity of 60%. Each test on hardened concrete was performed after 30/31 days. The measurements were processed by Catman AP software. All bending tests were performed based on controlled displacement with the same testing speed and the same measuring rate of clip gauge.

3. Analysis of the Results

In this study, the properties of fresh concrete as well as the mechanical parameters of hardened concrete like compressive strength, modulus of elasticity, bending, and splitting tensile strength were investigated. The test results are discussed in the following sections.

3.1. Properties of Fresh Concrete

The slump test was performed according to EN 12350-5 [44]. The results presented in Table 4 show that both PP and PVA fibers performed well during the slump flow test. The slump flow diameter was measured on the base plate in two directions, from which the average value was calculated. It should be mentioned that the plasticizer MasterGlenium ACE 430 dosage in each concrete mixture was identical. With increasing fiber dosage, the workability of HPFRC increases. Tests show that the slump flow measure increased from 595 mm to 650 mm based on a PP fiber dosage of 15 kg/m^3 and 35 kg/m^3. The phenomenon of higher fiber dosage and a concurrently larger slump flow diameter was also observed in investigations of self-compacting concrete with PP and steel fiber batches [45]. For HPFRC with PVA fibers, a smaller slump flow diameter is recognized than for HPFRC with PP fibers. If a PVA fiber with a smaller diameter is used, the mixtures tend to absorb much more water and hence change the consistency of fresh concrete. This is due to the high specific surface area of PVA fibers. Chen [46] reported that fine fibers were responsible for reducing the mixture workability and suggested a combination of small and medium fibers for the best balance of fresh concrete and hardened material properties. The progress of the slump experiment and a picture of a fresh concrete mixture are shown in Figure 5.

Table 4. Fresh concrete properties of the investigated concrete mixtures.

Mix ID	Type of Fiber	Fiber-Dosage	Air Void	Bulk Density [st. dev]	Slump Flow Diameter [st. dev]
		[kg/m^3]	[%]	[kg/m^3]	[mm]
1	MasterFiber 401 (PVA)	35	3.8	2288 [6]	418 [25]
2	MasterFiber 235 SPA (PP)	35	4.2	2239 [8]	650 [42]
3		25	3.8	2251 [21]	635 [21]
4		15	3.5	2278 [16]	595 [7]
5		0	2.8	2307 [20]	565 [7]

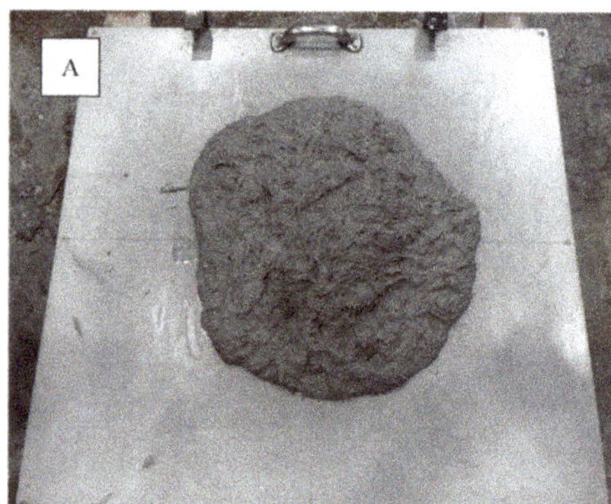

Figure 5. (**A**) Slump experiment for MasterFiber 401 (PVA). (**B**) Fresh concrete with MasterFiber 401 (PVA). (**C**) Fresh concrete with MasterFiber 235 SPA (PP).

Table 4 shows the air depending on the fiber dosage of each mixture. This was determined according to EN 12350-7 [47]. The concrete with less pore volume had a higher bulk density. Furthermore, the air impacts mechanical properties like compressive strength and modulus of elasticity. Hassan [48] stated that concrete with lower porosity was resistant against chloride penetration. Furthermore, it showed a lower fluid and gas permeability, which impacts the frost resistance of HPFRC.

3.2. Compressive Strength

Figure 6 shows the compressive strength of the concrete after 28 days with different fiber dosages. The results showed that the compressive strength decreased with increasing fiber dosage. Lower dosages of fiber in normal concrete, in contrast, behave differently (e.g., under 30 MPa). In general, the mechanical properties of concrete with a lower compressive strength will be improved through the addition of fibers [49,50]. The space between concrete and fibers is filled with air, which already starts to accumulate during the mixing process between the adhesive surfaces. The air voids are linked to the small interaction between the PP and the cement paste. A small air volume reduces the compressive strength of high-performance concrete. Furthermore, the higher fiber dosage reduces the volume of the concrete–matrix dosage and at the same time reduces the compressive strength and modulus of elasticity of the concrete.

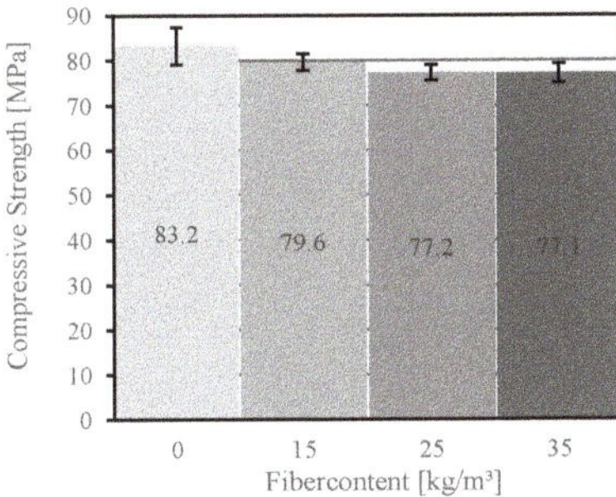

Fiber-dosage	f_c [st. dev]
[kg/m³]	[MPa]
0	83.2 [4.2]
15	79.6 [1.9]
25	77.2 [1.7]
35	77.1 [2.2]

Figure 6. Compressive strength results of concrete with different fiber dosages of MasterFiber 235 SPA (PP).

3.3. Modulus of Elasticity

The modulus of elasticity was determined according to EN 12390-13 [51]. Figure 7 shows the test results for the cylindrical specimens [36]. The modulus of elasticity's value depends on the fiber dosage. An increase in the volume fraction of PP fiber in concrete leads to a decrease of the modulus of elasticity, which is due to the compressive strength of HPFRC. Similar results were obtained in a study on fiber reinforced concrete with a compressive strength up to 100 MPa [52].

3.4. Splitting Tensile Strength

The addition of fibers to the concrete mix has a major impact on the tensile strength of FRC. The test results showed an increase from 4 MPa for concrete without any fibers to 6.9 MPa for concrete with a fiber dosage of 35 kg/m^3 (see Figure 8). Furthermore, the results showed that the difference in the splitting tensile strength between a fiber dosage of 25 kg/m^3 and 35 kg/m^3 was lower compared to a fiber dosage of 15 kg/m^3 and 25 kg/m^3. It can be assumed that this effect is caused by the strength of the fiber–matrix interface. The strength depends on the volume of the concrete surrounding the fibers. With increasing fiber dosage, this volume decreases and hence the pull-out-strength of the fibers in the cracking zone is reduced. It can therefore be concluded that the splitting tensile strength was at most 72.5% higher due to the addition of fibers.

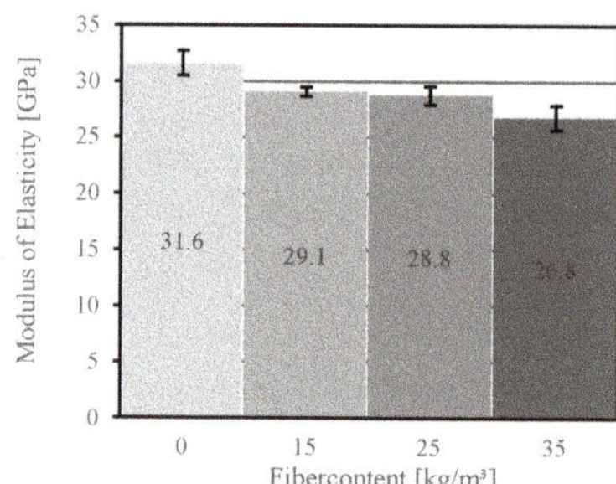

Fiber-dosage	E_c [st. dev]
[kg/m³]	[GPa]
0	31.6 [1.1]
15	29.1 [0.4]
25	28.8 [0.8]
35	26.8 [1.1]

Figure 7. Modulus of elasticity of concrete with different fiber dosages of MasterFiber 235 SPA (PP).

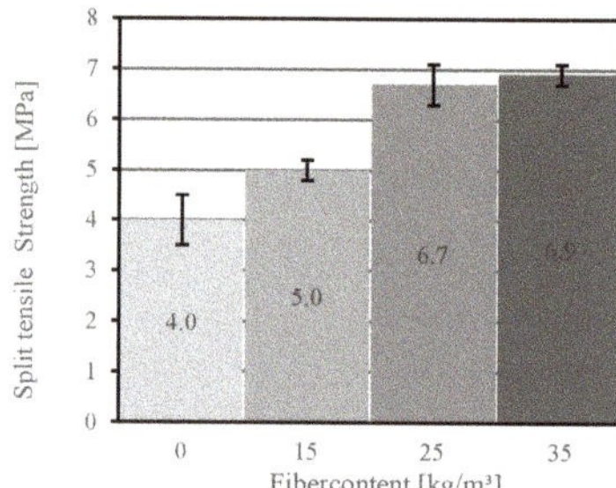

Fiber-dosage	f_{ct} [st. dev]
[kg/m³]	[MPa]
0	4.0 [0.5]
15	5.0 [0.2]
25	6.7 [0.4]
35	6.9 [0.2]

Figure 8. Split tensile strength of concrete with different fiber dosages of MasterFiber 235 SPA (PP).

3.5. Residual Flexural Strength

For façade panels, the most important mechanical property is the bending tensile strength [53]. Fibers in concrete enhance the concrete's mechanical properties. They absorb post-crack energy and improve ductility of the FRC. For thin concrete elements like façade panels, a higher bending tensile strength leads to an improved resistance against area loads caused by wind and impacts.

The experimental setup and the dimensions of the test specimens are discussed in Section 3. The plates according to EN 12467 [38] reached higher bending-tensile strengths up to 10.5 MPa for a fiber dosage of 35 kg/m^3. The test results with average values for plates with PP fiber are presented in Figure 9.

The stress–deflection curves in Figure 9 show a linear behavior at the beginning of the loading. After the first crack, stress deflection curves are non-linear. In HPFRC, the deflection, and concurrently the stress, still increases. Once the maximum value of stress has been reached, they decrease slowly. A similar effect of decreasing stresses after the first crack was described by the Association Francaise de Genie Civil [54] (AFGC). In concrete without fibers, the stress–deflection curves reached 5.61 MPa after the first crack and then collapsed. In contrast, the specimens with fiber dosage had an inclined plateau phase. This behavior is facilitated by a bond length of the PP fibers in the fiber–matrix interface, which is activated as soon as a crack occurs. The longer the bond of the anchored fiber, the greater the pull-out force. This behavior is the so-called bridging effect [55]. According to the first crack [38], it was possible to estimate the bending tensile strength of the concrete with fiber and concrete without fiber. The results with different fiber dosages are presented in Table 5.

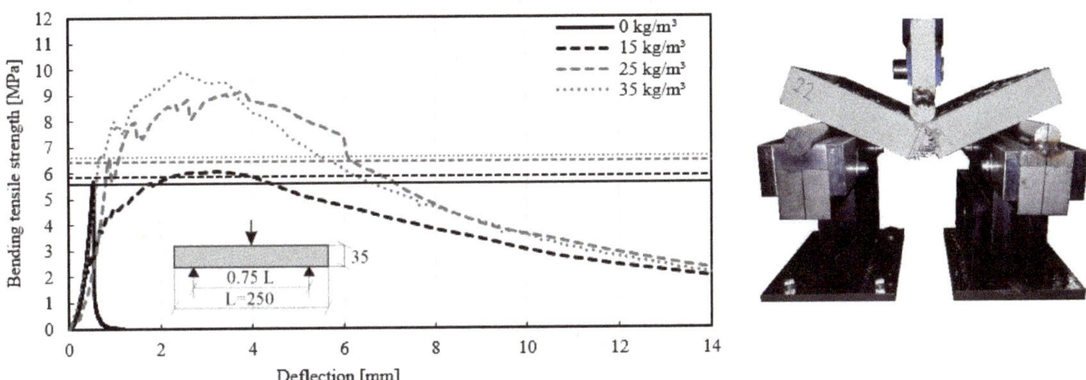

Figure 9. Bending-tensile strengths of concrete with different fiber dosages MasterFiber 235 SPA (PP) in accordance with EN 12467 [38].

Table 5. Bending tensile strength of concrete with different fiber dosages of MasterFiber 235 SPA (PP) in accordance with EN 12467 [38].

Mix ID	Type of Fiber	Fiber-Dosage [kg/m^3]	f_L [st. dev] [MPa]	Percentage Increase [%]
2	MasterFiber 235 SPA (PP)	35	6.62 [0.20]	118
3		25	6.43 [0.16]	115
4		15	5.85 [0.38]	104
5		0	5.61 [0.40]	100

The test results in Table 5 clearly reflect the addition of fiber dosage: an increasing bending tensile strength was observed with higher fiber dosage. The tests revealed an increase of up to 4% for a fiber dosage of 15 kg/m^3, 15% for 2d5 kg/m^3, and 18% for 35 kg/m^3. The experiments showed that by adding fibers, the failure mode changed from brittle to ductile. Similar to the study conducted by Kahanji [56], the variation of the fiber dosage has an enormous influence on the post-crack behavior of HPFRC.

The same results as for the panels were achieved with beams according to EN 14651 [39]. Figure 10 presents the test results with CMOD and Figure 11 shows the results with a deflection in the middle span of the specimen. The stress–strain diagram shows that the strain increased immediately after crack initiation. A comparison of the test results obtained for mixtures ID 2 (35 kg/m^3) and ID 3 (25 kg/m^3) showed similar bending tensile strengths. However, the dosage of fiber quantity for mixtures ID 2 and 3 differed. By comparing these two fiber dosages, an increase in the bending tensile strength of up to 200% was evident. Mixture ID 4 showed increases of up to 84%. The strain of mixture ID 4 measured with the clip gauge reached a plateau. Following this crack initiation, the measured strains increased with higher load capacity. For concrete without fibers (mixture ID 5), the stress–strain curve reached 3.14 MPa, then declined and reached zero.

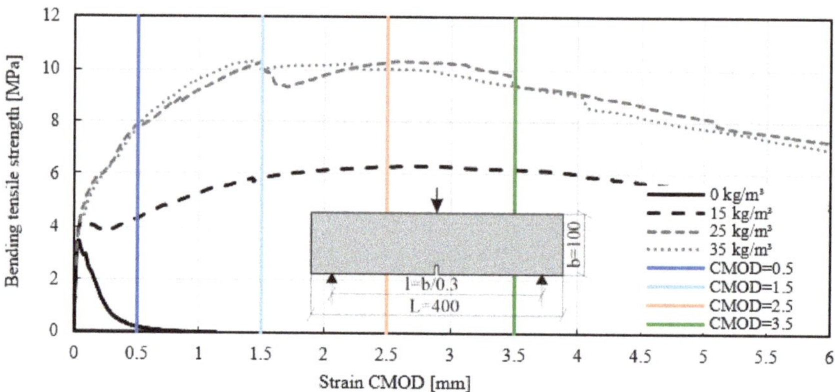

Figure 10. Stress–strain curves of the beam test-series with different fiber dosages MasterFiber 235 SPA (PP). Clip gauge strain in accordance with EN 14651 [39]. The vertical axes are CMOD 0.5 to CMOD 3.5.

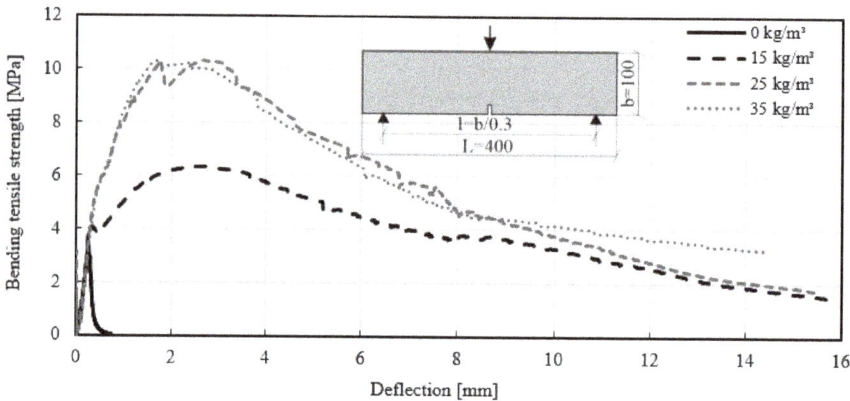

Figure 11. Stress–deflection curves of the beam test-series with different fiber dosages MasterFiber 235 SPA (PP).

The test method for fiber reinforced concrete presented in EN 14651 [39] enables the calculation of flexural tensile strength. The stresses f_L in limit of proportionality (LOP) were calculated according to the equation for the test method in concrete with metallic fiber [39,57]. The required force was determined in the case of crack opening CMOD = 0.05 mm (see Figure 12). The flexural tensile strength (limit of proportionality) was calculated and listed in Table 6 (characteristic values with k_s = 2.336). The correlation of the stress–strain curves showed that with a fiber addition between 15 kg/m^3 and 35 kg/m^3, the results of LOP were similar. A significant difference was found in the post crack behavior due to the addition of fibers to the mix. The samples with Mix ID 5 failed when the maximum flexural stresses were reached. This means that the CMOD with a value higher than 0.5 mm could not be determined.

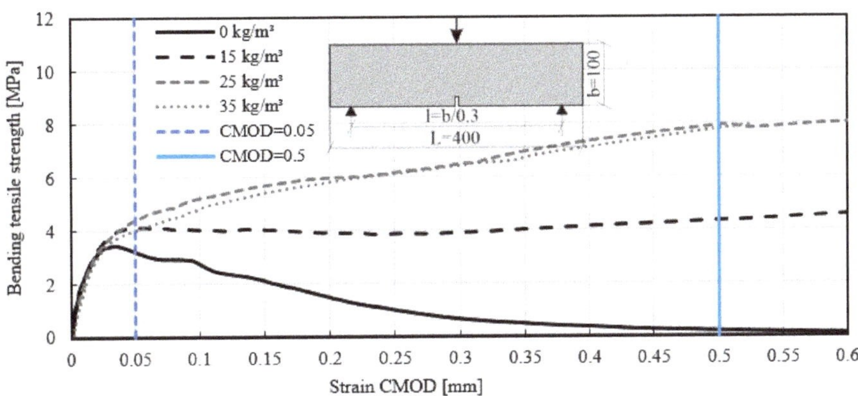

Figure 12. Stress–strain curves of the beam test-series with different fiber dosages MasterFiber 235 SPA (PP). Clip gauge strain in accordance with EN 14651 [39].

Table 6. Flexural tensile force according EN 14651 [39].

Mix ID	Fiber-Dosage [kg/m^3]	Type of Fiber	f_L LOP [MPa]	Percentage Increase [%]	f_L at Prescribed CMODj Values in MPa, [st. dev] 0.5 [mm]	1.5 [mm]	2.5 [mm]	3.5 [mm]
2	35	MasterFiber 235 SPA (PP)	3.89 [0.07]	124	5.66 [0.14]	8.57 [0.07]	8.03 [0.10]	7.34 [0.11]
3	25		4.10 [0.06]	131	6.30 [0.08]	8.04 [0.09]	9.16 [0.04]	9.24 [0.01]
4	15		4.04 [0.03]	129	3.06 [0.15]	4.17 [0.14]	4.78 [0.14]	4.94 [0.10]
5	0		3.14 [0.09]	100	0.11 [0.22]	-	-	-

Based on experimental stress–strain curves, parameter f_L was evaluated at four different CMOD values: 0.5, 1.5, 2.5, and 3.5 mm. The residual bending tensile strength f_L for different fiber dosages was calculated and listed in Table 6 (characteristic values). This study allows for the following conclusions to be drawn in post-crack bending tensile strengths. With a fiber addition of 25 kg/m^3 and 35 kg/m^3, the maximum bending tensile strength varied between 6.30 MPa and 5.66 MPa for CMOD of 0.5 mm and 9.24 MPa–7.34 MPa for CMOD of 3.5 mm. For the fiber addition of 15 kg/m^3, the residual post-cracking strength reached 3.06 MPa for CMOD of 0.5 mm and 4.94 MPa for CMOD of 3.5 mm.

Figure 13 shows the stress–deflection curves of different fiber types. The first curve reflects the long MasterFiber 235 SPA (PP), and the second the short MasterFiber 401 (PVA). The short fiber is also thinner than the long fiber, which affects the mix design with a much higher number of fibers in the concrete. Both concrete mixtures were made of the same raw materials and comprised a fiber dosage of 35 kg/m^3. Although the PP fiber was longer than the PVA fiber, a similar maximum bending tensile strength was achieved with both types of material, regardless of the fiber length. In the study by Yoo [58], the authors reported that the use of a longer fiber led to higher flexural strength than the shorter fibers. In the experimental test, only strains as a function of the length of the fiber were detected. The stress–deflection curves for MasterFiber 235 SPA (PP) showed significantly better results than the stress–deflection curves for MasterFiber 401 (PVA) following a deflection of 6 mm. The stress–deflection curve of fiber type MasterFiber 401 (PVA) revealed a higher increase in deflection compared to fiber type MasterFiber 235 SPA (PP), which may indicate unfavorable adhesion forces between PVA fiber and the matrix. Shorter fibers pull-out of the matrix faster than longer fibers. This is attributed to the bonding forces between the fibers and the concrete matrix. The 30 mm long fibers provided a better friction range than the 12 mm long fibers and also provided a better stress transfer in the matrix. However, the smaller fiber had a higher tensile strength. Different types of fibers reflect the bonding behavior between the fibers and the surrounding concrete [59].

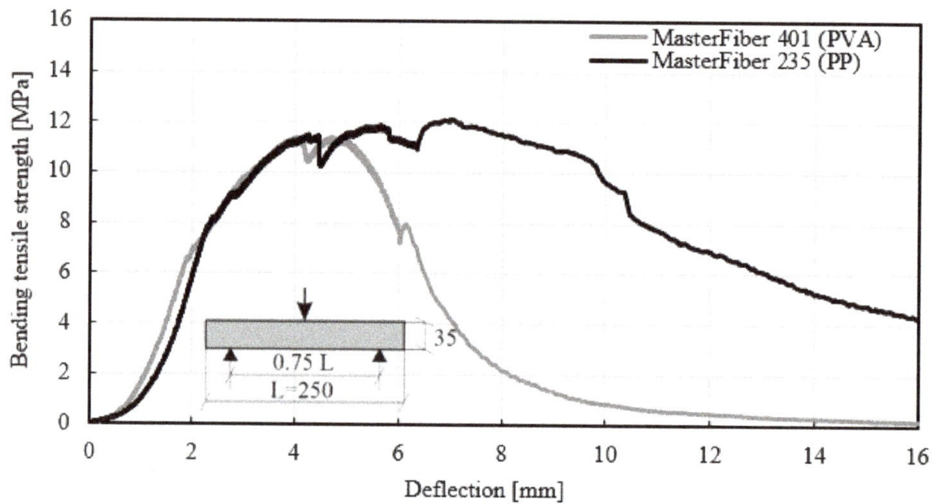

Figure 13. Bending-tensile strength of concrete with different fiber types (MasterFiber 235 SPA and MasterFiber 401) for a fiber dosage of 35 kg/m^3 in accordance with EN 12467 [38].

Figures 14–16 clearly show the cracking patterns of selected specimens in a three-point bending test. The main crack occurred in the middle of the specimen. During the loading tests, the crack width increased and lead to breakage or pull-out of the fiber. Due to the high strain rate of the materials, two halves of a concrete slab held together. There was no brittle failure because PP or PVA fibers under bending tensile load prevented the opening of a crack and finally prevented the sudden destruction of the concrete. The fibers, which were distributed along the axis of the beam, improved the bending tensile strength of the concrete. Some of the fibers in the crack zone reached their tensile strength.

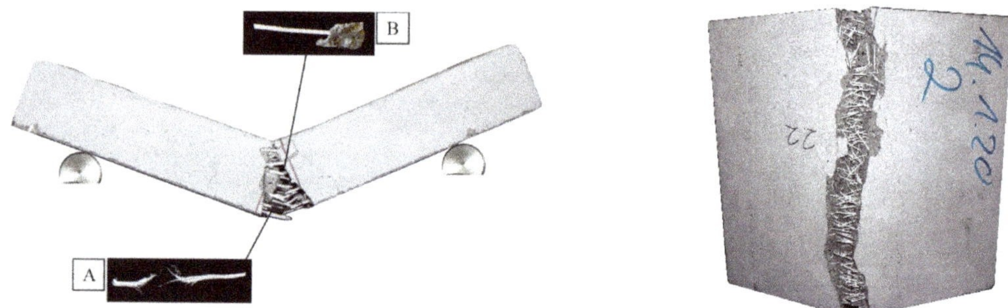

Figure 14. Typical failure modes for specimens [38] with MasterFiber 235 SPA (PP). (**A**) Images of the tensile fracture face of a fiber. (**B**) Images of the pull-out fracture face of a fiber.

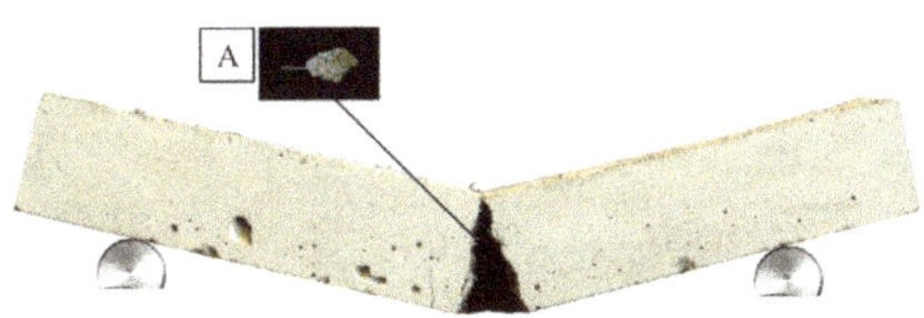

Figure 15. Typical failure modes for specimens [38] with MasterFiber 401 (PVA). (**A**) Images of the pull-out fracture face of a fiber.

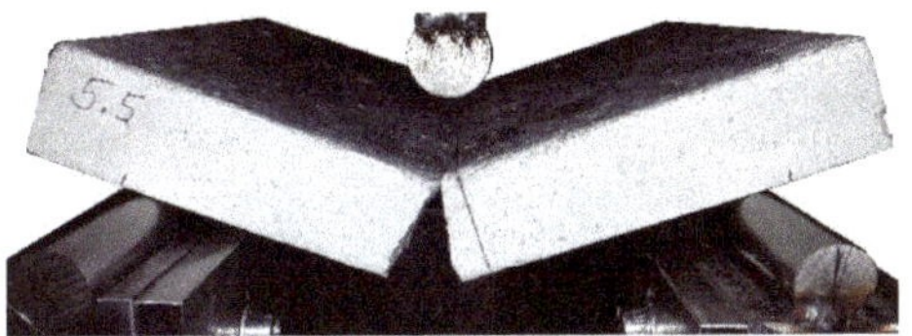

Figure 16. Typical failure modes for specimens [38] without fiber.

4. Discussion

As tests of the fresh and hardened concrete show, HPFRC is strongly affected by the density of used fiber and the presence of the air voids. The bulk density, compressive strength, and modulus of elasticity decreased with fiber addition. This effect can be attributable to fiber dosage in the concrete mix. Liu et al. [60] analyzed the permeability of carbon fiber reinforced concrete and observed the same impact based on the water–cement ratio. The fiber improves the impermeability of concrete only for the w/c ratio of 0.25. The higher ratio reduces the impermeability of hardened concrete. Richardson [61] studied the differences between plain concrete and concrete with fiber additions. Due to the higher air content in the fiber-reinforced mixtures compared to normal concrete, the compressive strengths differed from each other.

Another tendency showed in some cases of the normal strength concretes where the higher dosage increased compressive strength and modulus of elasticity. Kilmartin-Lynch et al. [62] tested a concrete mix with a compressive strength of 50.34 MPa and reached higher values for the dosage of recycled polypropylene fibers. The increase of higher fiber dosage can be noted because the fibers became more densely spaced, which therefore increased compressive strength. Sekhar Das et al. [63] showed that the compressive strength of 36.9 MPa initially increased with fiber content up to 0.5% and then decreased with further use of fibers. This content corresponds to a fiber dosage of 4.55 kg/m^3, which was not tested in the presented study. Therefore, the conclusion is that the correlation between compressive strength and modulus of elasticity depends on fiber dosage. For further studies, a lower fiber content should be tested.

In the experiment, the fiber dosage improved the flexural properties of concrete. The flexural strength increased the maximal 31% for a fiber dosage of 25 kg/m^3 in comparison to the plain concrete. Rostami et al. [64] reported that the highest flexural strength was 94% relative to the control sample. Such a large difference may be due to the length of the polypropylene fiber. In their experiment, Rostami et al. used longer PP fibers of 48 mm. Similar observations were made by Zhou et al. [65] who reported that PP improved the flexural strength of concrete. Unfortunately, the tests were carried out on another type of fiber with a maximal length of 18 mm.

Not only are the values of the flexural strength reflected in the studies on FRC, but the failure modes for specimens in flexural tests show similarities with other research. Abbas et al. [66] tested tunnel lining segments using UHPC with fiber dosage where the crack developed during the progression of load in the middle part of the sample.

5. Conclusions

The objective of this work was to investigate how fiber dosage affects the mechanical parameters of high-performance fiber reinforced concrete. This study allows for the following conclusions to be drawn in the area of material properties:

- The percentage of air voids in the concrete corresponds to the compressive strength and the modulus of elasticity of the concrete. A significant difference was found in the compressive strength of the concrete due to the addition of fibers to the mix. The fiber addition of 15 kg/m^3 in the concrete composition reduced the compressive strength from 83.2 MPa to 79.6 MPa. The higher fiber dosage showed a similar trend. Furthermore, it reduced compressive strength and the modulus of elasticity of the concrete.
- PP and PVA fibers have proven to be effective in increasing the splitting tensile strength of concrete, which allows better utilization of material capacities and has an impact on the production costs of FRC members. The comparison showed that the dosage of fibers increased from 4.0 MPa to 5.0 MPa (for 15 kg/m^3), 6.7 MPa (25 kg/m^3), and 6.9 MPa (35 kg/m^3).
- The analysis of the bending tensile tests revealed differences between MasterFiber 401 (PVA) and MasterFiber 235 SPA (PP) in the post-crack phase. MasterFiber 235 SPA is intended to be used because of its higher ductility.
- The bridging effect, which improves the safety of the concrete components, was identified in the bending tensile test. The bending tensile strength of concrete with added fibers increased by up to 18% compared to materials without fibers. Some of the fibers reached their tensile strength and were no longer involved in the transfer of the load. The pull-out effect of the fiber changed the brittle fracture behavior of concrete into the ductility behavior of these materials.
- In the present study, the highest PP fiber dosage examined in the concrete composition amounted to 35 kg/m^3. However, the addition of more than 25 kg/m^3 of fibers to the concrete mix had less influence on the bending tensile strength of the concrete. This concrete mix had an overcritical fiber dosage and was characterized by tensile strain-hardening behavior. A comparison of the stress–deflection curves with the addition of 25 kg/m^3 and 35 kg/m^3 of fibers revealed that the cracking behavior of concrete for these two fiber contents did not differ significantly.
- Further study of HPFRC comprises more mechanical experiments. New attempts will be focused on anchorage techniques for façade plates in building construction. A higher load capacity for the steel anchor system with a higher fiber dosage is expected.

Author Contributions: Conceptualization, S.G., M.P., M.S.-C. and S.H.; methodology, S.G., M.P., M.S.-C. and S.H.; investigation, S.G. and S.H.; writing—original draft preparation, S.G. and S.H.; writing—review and editing, S.G., M.P., M.S.-C., S.H. and C.H.; supervision, S.G., M.P., M.S.-C., S.H. and C.H.; project administration, M.P., M.S.-C. and C.H.; funding acquisition, M.P., M.S.-C. and C.H. All authors have read and agreed to the published version of the manuscript.

Funding: This research was funded by the Master Builders Solutions Deutschland GmbH.

Institutional Review Board Statement: Not applicable.

Informed Consent Statement: Not applicable.

Data Availability Statement: Not applicable.

Acknowledgments: This research work was supported by Master Builders Solutions Deutschland GmbH. The authors would like to thank the staff of the Civil Engineering Department of Technical University in Kaiserslautern for the support extended during the experimental works carried out in the laboratory.

Conflicts of Interest: The authors declare no conflict of interest.

References

1. Hegger, J.; Zell, M.; Horstmann, M. Textile Reinforced Concrete–Realization in applications. In *Tailor Made Concrete Structures—Walraven & Stoelhorst, Proceedings of the International FIB symposium Amsterdam, The Netherlands, 19–22 May 2008*; Taylor & Francis Group: London, UK, 2008; pp. 357–362.
2. Schultz-Cornelius, M.; Pahn, M. Development of an innovative experiment set-up for filigree (U)HPC-facades. In Proceedings of the IV German-Polish PhD Symposium Kaiserslautern, Annweiler am Trifels, Germany, 2–5 July 2014; pp. 145–151.
3. Bund, B.; Breit, W. Prediction and verification of the distribution of fibers in fine grain systems. In Proceedings of the IV German-Polish PhD Symposium Kaiserslautern, Annweiler am Trifels, Germany, 2–5 July 2014; pp. 1–8.
4. Pinkerton, L.; Stecher, J.; Novak, J. Twisted Steel Micro Reinforcement. *Concr. Int.* **2013**, *35*, 1–14.
5. Juhasz, P.K.; Schaul, P. Design of Industrial Floors—TR34 and Finite Element Analysis (Part 2). *J. Civ. Eng. Archit.* **2019**, *13*, 512–522. [CrossRef]
6. Boita, I.-E.; Dan, D.; Stoian, V. Seismic Behaviour of Composite Steel Fibre Reinforced Concrete Shear Walls. *IOP Conf. Series: Mater. Sci. Eng.* **2017**, *245*, 22006. [CrossRef]
7. Lachance, F.; Charron, J.-P.; Massicotte, B. Development of Precast Bridge Slabs in High-Performance Fiber-Reinforced Concrete and Ultra-High-Performance Fiber-Reinforced Concrete. *ACI Struct. J.* **2016**, *113*. [CrossRef]
8. Mateckova, P.; Bilek, V.; Sucharda, O. Comparative Study of High-Performance Concrete Characteristics and Loading Test of Pretensioned Experimental Beams. *Crystals* **2021**, *11*, 427. [CrossRef]
9. Ali, B.; Raza, S.S.; Hussain, I.; Iqbal, M. Influence of different fibers on mechanical and durability performance of concrete with silica fume. *Struct. Concr.* **2021**, *22*, 318–333. [CrossRef]
10. Yoo, D.-Y.; Yoon, Y.-S. A Review on Structural Behavior, Design, and Application of Ultra-High-Performance Fiber-Reinforced Concrete. *Int. J. Concr. Struct. Mater.* **2016**, *10*, 125–142. [CrossRef]
11. Heinzle, G.; Freytag, B.; Linder, J. Rissbildung von biegebeanspruchten Bauteilen aus Ultrahochfestem Faserbeton. *Beton Und Stahlbetonbau* **2009**, *104*, 570–580. [CrossRef]
12. Marcos-Meson, V.; Michel, A.; Solgaard, A.; Fischer, G.; Edvardsen, C.; Lund, S. Corrosion resistance of steel fiber reinforced concrete–a literature review. In *Performance-Based Approaches for Concrete Structures: Proceedings*; Fib Symposium: Cape Town, South Africa, 2016.
13. Herr, C.M.; Lombardi, D.; Galobardes, I. Parametric Design of Sculptural Fibre Reinforced Concrete Facade Components. In Proceedings of the 23rd International Conference of the Association for Computer-Aided Architectural Design Research in Asia (CAADRIA), Beijing, China, 17–19 May 2018; 2018; 2, pp. 319–328.
14. Abdel-Mooty, M.; Shaaban, S. Nonlinear dynamic response of RC building fasade panels to impact loads. *Struct. Under Shock. Impact XII* **2012**, *126*, 281–291. [CrossRef]
15. Baby, F.; Graybeal, B.; Marchand, P.; Toutlemonde, F. Identification of UHPFRC tensile behaviour: Methodology based on bending tests. In Proceedings of the Symposium on Ultra-High Performance Fiber-Reinforced Concrete, UHPFRC, Marseille, France, 1–3 October 2013.
16. Banyhussan, Q.; Yildirim, G.; Anil, Ö.; Erdem, R.; Ashour, A.; Sahmaran, M. Impact resistance of deflection-hardening fiberreinforced concretes with different mixture parameters. *Struct. Concr.* **2019**, 1–29. [CrossRef]
17. Foglar, M.; Hajek, R.; Fladr, J.; Pachman, J.; Stoller, J. Full-scale experimental testing of the blast resistance of HPFRC and UHPFRC bridge decks. *Constr. Build. Mater.* **2017**, *145*, 588–601. [CrossRef]
18. Ehrenbring, H.Z.; Quinino, U.C.D.M.; Oliveira, L.F.S.; Tutikian, B.F. Experimental method for investigating the impact of the addition of polymer fibers on drying shrinkage and cracking of concretes. *Struct. Concr.* **2019**, *20*, 1064–1075. [CrossRef]
19. Onuaguluchi, O.; Banthia, N. Plant-based natural fiber reinforced cement composites: A review. *Cem. Concr. Compos.* **2016**. [CrossRef]
20. Jalasutram, S.; Sahoo, D.R.; Matsagar, V. Experimental investigation of the mechanical properties of basalt fiber-reinforced concrete. *Struct. Concr.* **2017**. [CrossRef]
21. Shakor, P.; Pimplikar, S. Glass Fiber Reinforced Concrete Use in Construction. *Int. J. Technol. Eng. Syst.* **2011**, *2*.
22. Granju, J.; Balouch, S. Corrosion of steel fiber reinforced concrete from the cracks. *Cem. Concr. Res.* **2005**, *35*, 572–577. [CrossRef]
23. Hassan, A.; Jones, S.; Mahmud, G. Experimental test methods to determine the uniaxial tensile and compressivebehaviour of ultra high performance fiber reinforced concrete (UHPFRC). *Constr. Build. Mater.* **2012**, *37*, 874–882. [CrossRef]
24. Schultz-Cornelius, M.; Pahn, M. Influence of the Size Effect on the Flexural Tensile Strength of Filigree UHPC Components. In Proceedings of the Young Researchers Symposium, Kaiserslautern, German, 14–15 April 2016.
25. Duque, L.F.M.; Graybeal, B. Fiber orientation distribution and tensile mechanical response in UHPFRC. *Mater. Struct.* **2016**, *50*. [CrossRef]
26. Graybeal, B.; Baby, F.; Marchand, P.; Toutlemonde, F. Direct and Flexural Tension Test Methods for Determination of the Tensile Stress-Strain Response of UHPC. In *Ultra-High Performance Concrete and Nanotechnology in Construction, Proceedings of the Hipermat 2012 3rd International Symposium on UHPC and Nanotechnology for High Performance Construction Materials, Kassel, Germany, 7–9 March 2012*; Kassel University Press GmbH: Kassel, Germany, 2008; pp. 395–402.

27. Prudencio, L.; Austin, S.; Jones, P.; Armelin, H.; Robins, P. Prediction of steel fiber reinforced concrete under flexure from an inferred fiber pull-out response. *Mater. Struct.* **2006**, *39*, 601–610. [CrossRef]
28. Sakr, M.A.; Sleemah, A.A.; Khalifa, T.M.; Mansour, W.N. Shear strengthening of reinforced concrete beams using prefabricated ultra-high performance fiber reinforced concrete plates: Experimental and numerical investigation. *Struct. Concr.* **2019**, *20*, 1137–1153. [CrossRef]
29. *EN 197-1: Cement–Part 1: Composition, Specifications and Conformity Criteria for Common Cements*; Beuth Verlag GmbH: Berlin, Germany, 2011.
30. EN 13139. *Aggregates for Mortar*; Beuth Verlag GmbH: Berlin, Germany, 2002.
31. EN 12620. *Aggregates for Concrete*; Beuth Verlag GmbH: Berlin, Germany, 2008.
32. EN 933-1. Tests for Geometrical Properties of Aggregates—Part 1: Determination of Particle Size Distribution—Sieving Method. 2012. Available online: https://standards.iteh.ai/catalog/standards/cen/100b983f-85a4-4a80-934c-e93c584dbdb4/en-933-1-2012 (accessed on 7 June 2021).
33. Sofi, A.; Sinha, R.; Bhattacharya, S. Mechanical Properties of Concrete Containing Polypropylene Fiber and Silica Fume. *Int. J. Civ. Eng. Technol.* **2017**.
34. Ludwig, H.-M.; Linß, E. *Report: Grading of the Fines for Microsilica*; FIB—F. A. Finger-Institut für Baustoffkunde: Weimar, Germany, 2012.
35. SH Minerals GmbH. *Report: Grading of the Fines for Limestone Powder: Product: Sh_compact I*; SH Minerals GmbH: Heidenheim, Germany, 2020.
36. *EN 12390-1: Testing Hardened Concrete–Part 1: Shape, Dimensions and Other Requirements for Specimens and Moulds*; Beuth Verlag GmbH: Berlin, Germany, 2019.
37. *EN 12390-3: Testing Hardened Concrete—Part 3: Compressive Strength of Test Specimens*; Beuth Verlag GmbH: Berlin, Germany, 2019.
38. *EN 12467:2012+A1:2016, Fiber-Cement Flat Sheets–Product Specification and Test Methods*; Beuth Verlag GmbH: Berlin, Germany, 2016.
39. *EN 14651:2005+A1:2007, Test Method for Metallic Fiber Concrete-Measuring the Flexural Tensile Strength (Limit or Proportionality (LOP), Residual)*; Beuth Verlag GmbH: Berlin, Germany, 2007.
40. Pająk, M.; Ponikiewski, T. Flexural behavior of self-compacting concrete reinforced with differenttypes of steel fibers. *Constr. Build. Mater.* **2013**, *47*, 397–408. [CrossRef]
41. Harenberg, S.; Pahn, M.; Malárics-Pfaff, V.; Dehn, F.; Caggiano, A.; Schicchi, D.; Yang, S.; Koenders, E. Digital image correlation strain measurement of ultra-highperformance concrete-prisms under static and cyclic bendingtensile stress. *Struct. Concr.* **2019**, *20*, 1220–1230. [CrossRef]
42. Bi, J.; Zhao, Y.; Guan, J.; Huo, L.; Qiao, H.; Yuan, L. Three-dimentional modeling of the distribution and orientation of steel fibers during the flow of self-compacting concrete. *Struct. Concr.* **2019**, *20*, 1722–1733. [CrossRef]
43. *Fib Model Code for Concrete Structures*; Ernst & Sohn: Berlin, Germany, 2010.
44. *EN 12350-5: Testing Fresh Concrete–Part 5: Flow Table Test*; Beuth Verlag GmbH: Berlin, Germany, 2009.
45. Aslani, F.; Sun, J.; Bromley, D.; Ma, G. Fiber-reinforced lightweight self-compacting concrete incorporating scoria aggregates at elvated temperatures. *Struct. Concr.* **2019**, *20*, 1022–1035. [CrossRef]
46. Chen, Y.; Matalkah, F.; Weerasiri, R.; Balachandra, A.; Soroushian, P. Despersion of Fibers in Ultra-High-Performance Concrete. *Concr. Int.* **2017**, *39*, 45–50.
47. *EN 12350-7: Testing Fresh Concrete-Part 7: Air Dosage Pressure Methods*; Beuth Verlag GmbH: Berlin, Germany, 2009.
48. Hassan, M.; Abo Sabah, S.; Bunnori, N.; Megat Johari, M. Fluid transport properties of normal concrete overlay composite. *Structural Concrete* **2019**.
49. Aslani, F.; Hou, L.; Nejadi, S.; Sun, J.; Abbasi, S. Experimental analysis of fiber-reinforcement recycled aggregate self-compacting concrete using waste recycled concrete aggregates. *Struct. Concr.* **2019**, *20*, 1771–1780. [CrossRef]
50. Meesala, C.R. Influence of different types of fibreon the properties of recycled aggregate concrete. *Struct. Concr.* **2019**, *20*, 1656–1669. [CrossRef]
51. *EN 12390-13: Testing Hardened Concrete-Part 13: Determination of Secant Modulus of Elasticity in Compression*; Beuth Verlag GmbH: Berlin, Germany, 2019.
52. Zhang, R.; Jin, L.; Tian, Y.; Dou, G.; Du, X. Static and dynamic mechanical properties of eco-friendly polyvinyl alcohol fiber-reinforced ultra-high-strength concrete. *Struct. Concr.* **2019**, *20*, 1051–1063. [CrossRef]
53. Hegger, J.; Will, N.; Voss, S. Textile Reinforced Concrete Facades. Concrete Structures: The Challenge of Creativity fib Symposium. *Struct. Concr.* **2004**.
54. Association Francaise de Genie Civil. *Bétons Fibrés á Ultra-Hautes Performances*; AFGC: Paris, France, 2013.
55. Karatas, M.; Dener, M.; Benli, A.; Mohabbi, M. High temperature effect on the mechanical behaviour of steel fiber reinforced self-compacting concrete containing ground pumice powder. *Struct. Concr.* **2019**, *20*, 1734–1749. [CrossRef]
56. Kahanji, C.; Ali, F.; Nadjai, A. Structural performance of ultra-high-performance fiber-reinforced concrete beams. *Struct. Concr.* **2017**, *18*, 249–258. [CrossRef]
57. *DAfStb Guideline 2010, Deutscher Ausschuss für Stahlbeton e. V.–DafStb*; Beuth Verlag GmbH: Berlin, Germany, 2010.
58. Yoo, D.-Y.; Banthia, N.; Yoon, Y.-S. Impact Resistance of Reinforced Ultra-High-Performance Concrete Beams with Different Steel Fibers. *ACI Struct. J.* **2017**, *114*. [CrossRef]

59. Lee, D.H.; Han, S.-J.; Kim, K.S.; LaFave, J.M. Shear capacity of steel fiber-reinforced concrete beams. *Struct. Concr.* **2017**, *18*, 278–291. [CrossRef]
60. Liu, R.; Xiao, H.; Liu, M.; Li, Y.; Geng, J. Effect of carbon fiber on properties of concrete withdifferent W/C and its air-entraining models. *Struct. Concr.* **2021**, *22*. [CrossRef]
61. Richardson, A.E. Compressive strength of concrete with polypropylene fibre additions. *Struct. Surv.* **2006**, *24*, 138–153. [CrossRef]
62. Kilmartin-Lynch, S.; Saberian, M.; Li, J.; Roychand, R.; Zhang, G. Preliminary evaluation of the feasibility of using polypropylene fibres from COVID-19 single-use face masks to improve the mechanical properties of concrete. *J. Clean. Prod.* **2021**, *296*, 126460. [CrossRef] [PubMed]
63. Das, C.S.; Dey, T.; Dandapat, R.; Mukharjee, B.B.; Kumar, J. Performance evaluation of polypropylene fibre reinforced recycled aggregate concrete. *Constr. Build. Mater.* **2018**, *189*, 649–659. [CrossRef]
64. Rostami, R.; Zarrebini, M.; Abdellahi, S.B.; Mostofinejad, D.; Abtahi, S.M. Investigation of flexural performance of concrete reinforced with indented and fibrillated macro polypropylene fibers based on numerical and experimental comparison. *Struct. Concr.* **2021**, *22*, 250–263. [CrossRef]
65. Zhou, X.; Zeng, Y.; Chen, P.; Jiao, Z.; Zheng, W. Mechanical properties of basalt and polypropylene fibre-reinforced alkali-activated slag concrete. *Constr. Build. Mater.* **2021**, *269*, 121284. [CrossRef]
66. Abbas, S.; Nehdi, M.L. Mechanical Behavior of Ultrahigh-Performance Concrete Tunnel Lining Segments. *Materials* **2021**, *14*, 2378. [CrossRef]

MDPI
St. Alban-Anlage 66
4052 Basel
Switzerland
Tel. +41 61 683 77 34
Fax +41 61 302 89 18
www.mdpi.com

Materials Editorial Office
E-mail: materials@mdpi.com
www.mdpi.com/journal/materials

www.ingramcontent.com/pod-product-compliance
Lightning Source LLC
LaVergne TN
LVHW071612170726
843515LV00009B/2375